Impressum:

hpk media – eine Marke der
HPK Products H.-Peter Kerzel, 50767 Köln
E-Mail: hpkproducts@gmx.de
Internet: www.hpkproducts.xilando.de

Einzelplatz Lizenz
Software: Erlaubt die Installation und den Betrieb auf einer Domain bzw. Subdomain. eBook, Hörbuch: Erlaubt die private Nutzung und Installation auf einem Rechner. Die Dokumente dürfen für den privaten Gebrauch auch ausgedruckt werden. Buch: Ausschließlich für den privaten Gebrauch dürfen Kopien angefertigt werden.

1. Auflage / Copyright © 2019

Alle Rechte vorbehalten. Vervielfältigungen, auch auszugsweise, nur mit schriftlicher Genehmigung.
Die Inhalte dieses Buches wurden mit größter Sorgfalt erstellt. Trotzdem können weder Autor noch Verlag für eventuelle Verluste oder Nachteile, die durch die Anwendung des Inhaltes entstehen, haftbar gemacht werden.

Wichtige Hinweise zur Verwendung der Patentinformationen:

Bitte beachten Sie bei der Verwendung der in diesem Buch angebotenen Informationen folgende Hinweise:

Patentschriften fallen nach deutschem Recht unter "amtliche Dokumente", sind daher gemeinfrei. Sie können die in den Patentschriften enthaltenen Informationen zu privaten Zwecken nutzen, die Patentgegenstände dürfen zur privaten Verwendung nachgebaut werden.

Jegliche gewerbliche Nutzung der Patentinformationen ist strengstens untersagt. Bei Zuwiderhandlung drohen juristische Konsequenzen und hohe Schadenersatzforderungen seitens der Patentinhaber.

Sollten uns Verstöße gegen diese Bedingungen bekannt werden, informieren wir unverzüglich die betreffenden Rechteinhaber.

Alle hier dargestellte Patente sind recherchiert in den Datenbanken des deutschen, europäischen und amerikanischen Patentamts. Detaillierte Quellenhinweise finden Sie in jeder einzelnen Schrift.

50 berühmte historische Erfindungen der Welt

Patentschriftensammlung

hpk media

Dec. 3, 1968 S. OLDBERG ET AL 3,414,292

INFLATABLE SAFETY DEVICE

Filed July 1, 1967 3 Sheets-Sheet 1

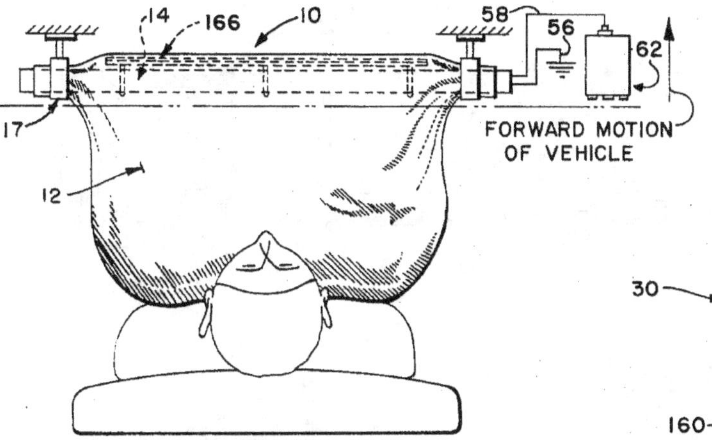

FORWARD MOTION OF VEHICLE

FIG. 1

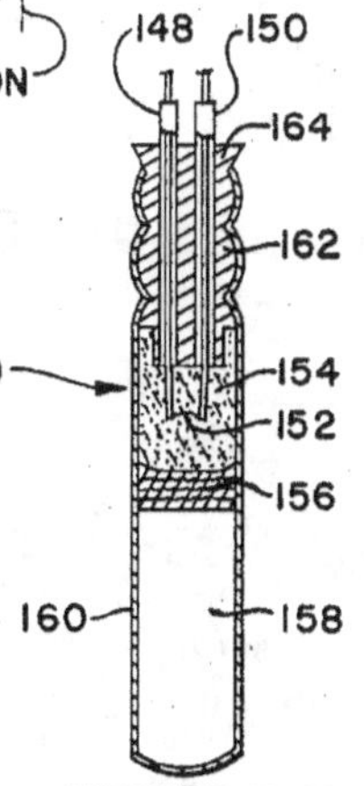

FIG. 13

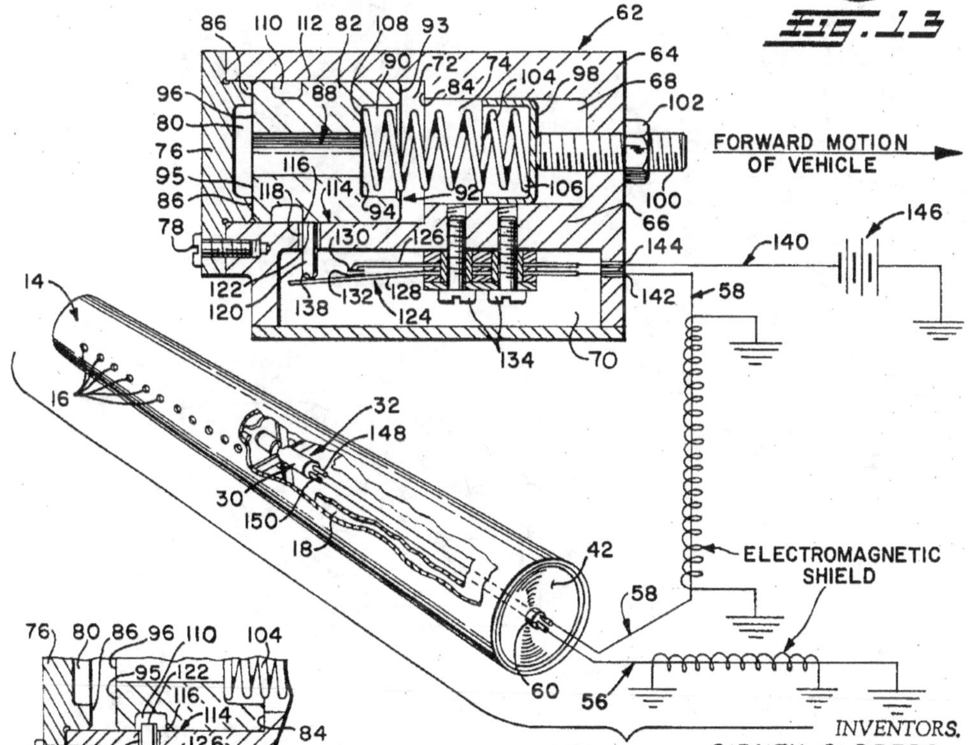

FORWARD MOTION OF VEHICLE

ELECTROMAGNETIC SHIELD

FIG. 2

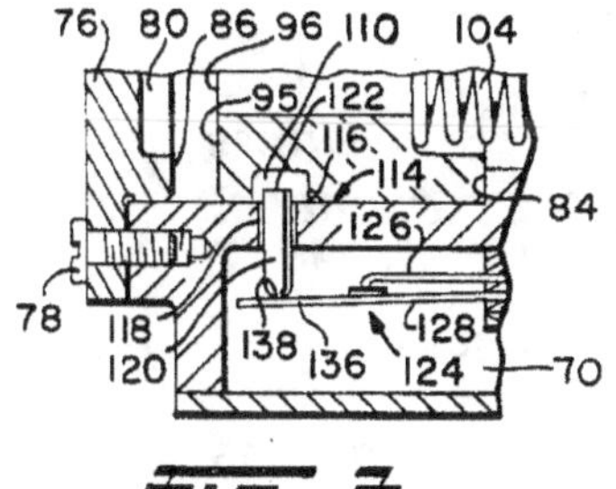

FIG. 3

INVENTORS.
SIDNEY OLDBERG
WILLIAM R. CAREY

BY Teagno & Toddy

ATTORNEYS

Dec. 3, 1968 S. OLDBERG ETAL 3,414,292

INFLATABLE SAFETY DEVICE

Filed July 1, 1967 3 Sheets-Sheet 2

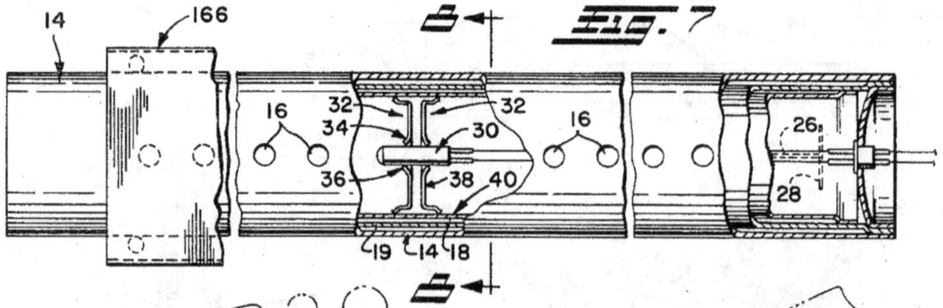

Fig. 7

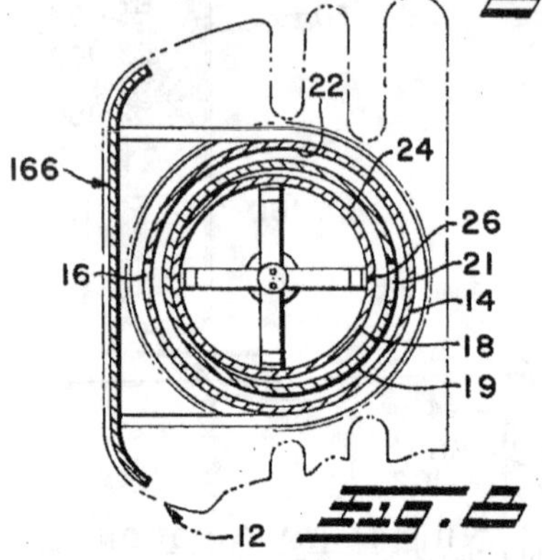

Fig. 8

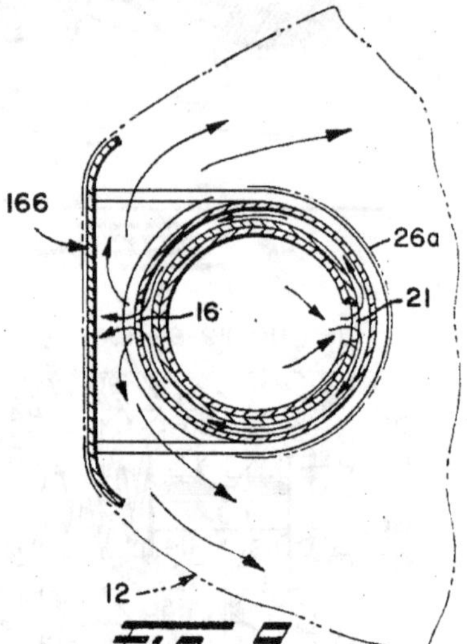

Fig. 9

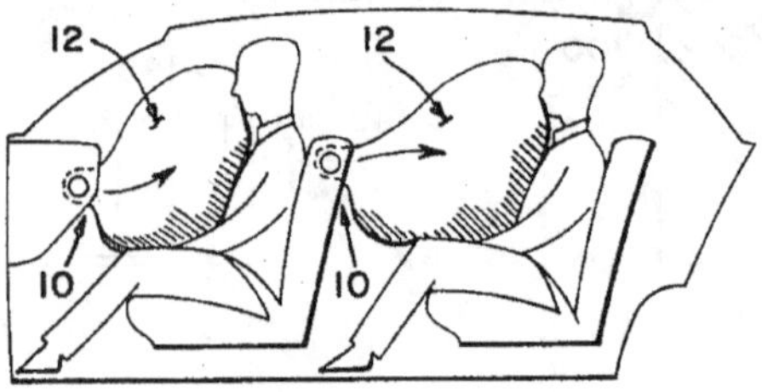

Fig. 10

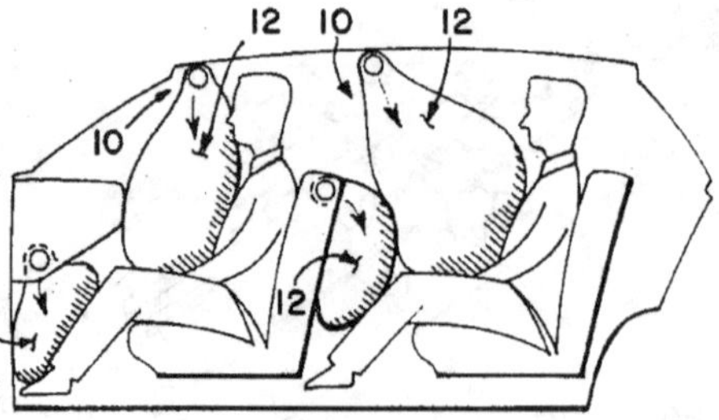

Fig. 11

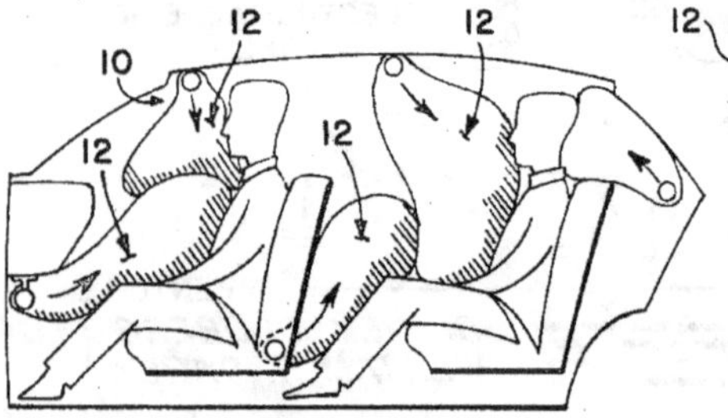

Fig. 12

INVENTORS
SIDNEY OLDBERG
WILLIAM R. CAREY

BY Teague & Toddy

ATTORNEYS

Dec. 3, 1968 S. OLDBERG ET AL 3,414,292

INFLATABLE SAFETY DEVICE

Filed July 1, 1967 3 Sheets-Sheet 3

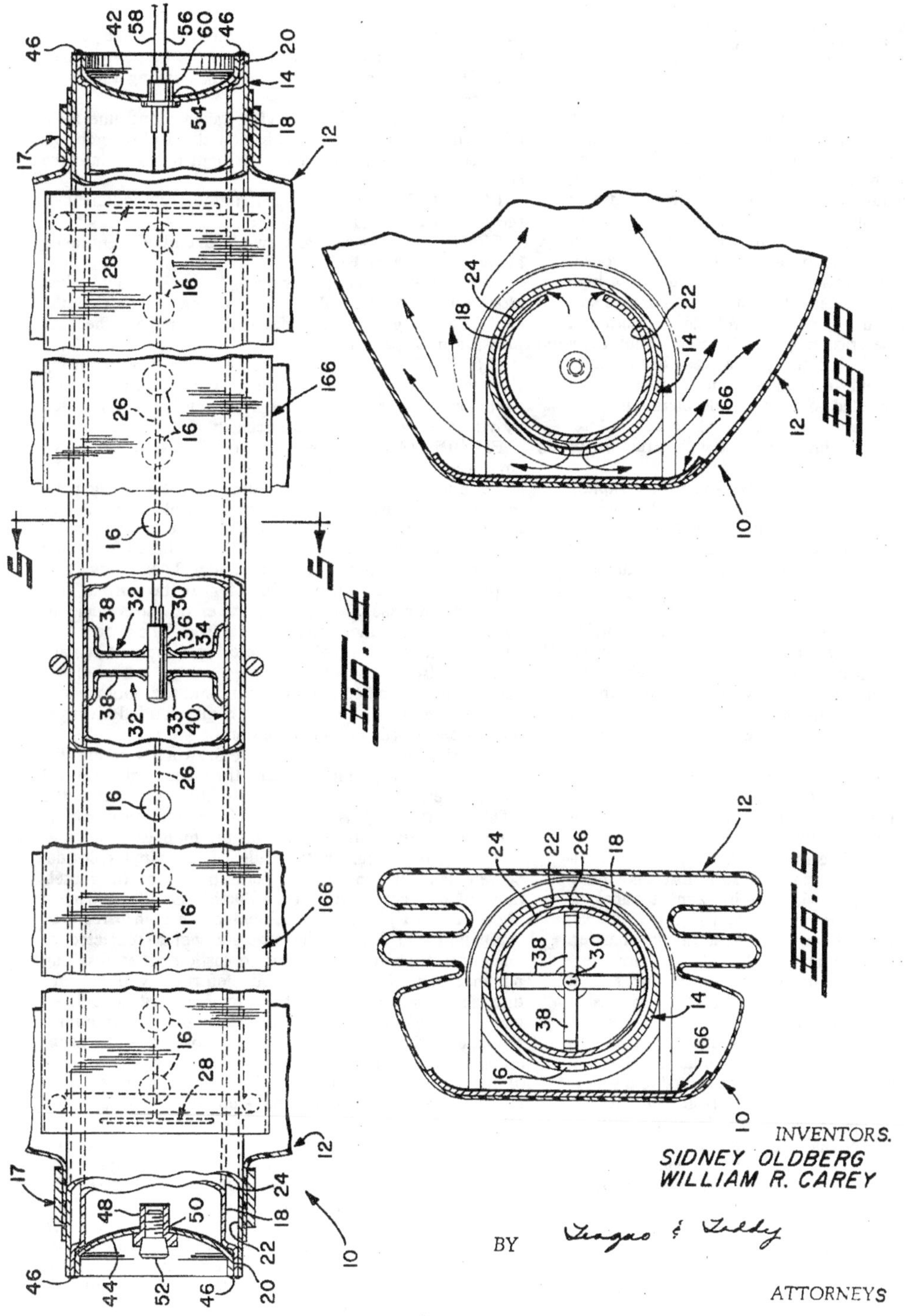

INVENTORS.
SIDNEY OLDBERG
WILLIAM R. CAREY

BY Teagno & Toddy

ATTORNEYS

United States Patent Office

3,414,292

Patented Dec. 3, 1968

3,414,292

INFLATABLE SAFETY DEVICE

Sidney Oldberg, Birmingham, and William R. Carey, Farmington, Mich., assignors to Eaton Yale & Towne Inc., Cleveland, Ohio, a corporation of Ohio

Filed July 1, 1966, Ser. No. 562,289

29 Claims. (Cl. 280—150)

This invention relates to an inflatable safety crash bag that provides protection for vehicle passengers and is especially suited for aircraft, automobiles, speedboats, and other vehicles which travel on land, sea, or in the air, wherein the occupant thereof is protected from rapid or violent deceleration of the vehicle in which he is riding.

The object of this invention is to provide a rapidly inflatable confinement or crash bag that serves as an occupant restraint and that is automatically inflated by a high pressure gas which is released by an inertia responsive mechanism upon sensing a rapid vehicle deceleration, causing the inflatable crash bag to be inflated within a few milliseconds. This rapid inflation of the crash bag member will effectively restrain forward movement of a vehicle passenger when subjected to the rapid deceleration of an accidental stop, whereby the human passenger is prevented from coming in abrupt contact with structure of the vehicle.

Another object of this invention is to provide a rapidly inflatable ductile shield which is interposed between a structural member of a vehicle and the body of the passenger, whereby the forward momentum of the passenger is substantially absorbed or dampened by the inflatable shield.

A further object is the provision of an elongated rupturable reservoir or container for pressurized fluid having an explosive charge positioned therein; and with an inflatable bag being attached to the container so that rupture of the container along a stress riser therein, in response to firing of the explosive charge, will communicate the interior of the inflatable bag with the container through the stress riser.

In order for an inflatable crash container to substantially reduce injury to a vehicle passenger subjected to a sudden impact or rapid deceleration as the vehicle is traveling 30 m.p.h., the inflatable crash container must be fully inflated within a time range of from 40 to 60 milliseconds. Tests have shown it takes 130 milliseconds for a person's body to collide with the instrument panel or windshield of a conventional vehicle when the vehicle, traveling 30 m.p.h., collides with a stationary barrier. These tests have shown a passenger's body will move forward 4 to 6 inches relative to the passenger compartment in the time allowed (40 to 60 milliseconds) for expansion of the inflatable crash container. Thus, it is apparent the inflatable crash container must be fully inflated at the end of 60 milliseconds after sensing of a sudden deceleration by an inertia mechanism operatively connected to a source of pressurized fluid which inflates the crash container.

Other objects and purposes of this invention will be apparent to persons acquainted with vehicle passenger safety apparatus of this general type upon reading the following specification in conjunction with accompanying drawings.

In the drawings:

FIGURE 1 is a plan view of a vehicle compartment showing the inflatable crash bag of the present invention and the relationship between a vehicle occupant and the inflated bag.

FIGURE 2 is a diagrammatic view showing a detailed cross-section of the inertia responsive device electrically connected to an explosive charge disposed in a high pressure fluid container.

FIGURE 3 is a fragmentary sectional view of the inertia responsive switch mechanism in a closed or energizing position.

FIGURE 4 is an elevation view of an embodiment of the present safety device showing the arrangement of the baffle means with respect to the fluid discharge openings.

FIGURE 5 is a sectional view taken along line **5—5** of FIGURE 4.

FIGURE 6 is a sectional view taken along line **5—5** of FIGURE 4, however, in this view the explosive charge has been detonated to release the high pressure into the crash bag.

FIGURE 7 is an elevation of another embodiment of the present invention.

FIGURE 8 is a sectional view of the embodiment taken along line **8—8** of FIGURE 7.

FIGURE 9 is a sectional view of the embodiment taken along line **8—8** of FIGURE 7, however, in this view the explosive charge has been detonated to release the high pressure into the crash bag.

FIGURES 10, 11 and 12 are elevation views showing the various positions a crash bag may be placed in a vehicle to protect a passenger from impact and rebound caused by violent deceleration of an accidental stop.

FIGURE 13 is a cross-section of the explosive charge employed in the present invention.

Certain terminology will be used in the following description for convenience in reference only and will not be limiting. The words "upwardly," "downwardly," "rightwardly" and "leftwardly" will designate directions in the drawings to which reference is made. The words "forwardly" and "rearwardly" will refer to the normal forward and reverse direction of travel of the vehicle to which the safety mechanism is attached. The words "inwardly" and "outwardly" will refer to directions toward and away from, respectively, the geometric center of the device and designated parts thereof. Said terminology will include the words specifically mentioned, derivatives thereof and words of similar import.

Referring in greater detail to the drawings wherein like numerals indicate similar parts throughout the several views, reference number **10** of FIGURE 1 discloses the inflatable crash bag assembly which includes an inflatable confinement or thermosetting or thermoplastic polymer crash bag or container **12** made of polyvinyl chloride, polyethylene or some other thermoplastic or thermosetting polymer composition or a ductile material having a high degree of plasticity which exhibits a high percent of elongation and low tensile strength which subjected to loading in tension. Such a thermoplastic or thermosetting polymer bag will be irreversibly deformed without rupture when a vehicle occupant is thrown against the inflated bag as a result of a rapid deceleration and thus minimize rebound of the occupant. It is well known physical properties such as hardness, modulus, flexibility, and rigidity of the thermoplastic or thermosetting polymer composition can be controlled by the amount of plasticizer or other compounding ingredients utilized therewith.

The inflatable crash bag **12** may completely enclose or only partially surround a diffuser member or first outer container **14** which has a plurality of fluid discharge openings **16**, as shown in FIGURE 2, so constructed and arranged to direct fluid flow therethrough into the interior of bag **12**. A plurality of bands or clamps **17** may releasably secure the crash bag **12** about outer container **14**.

A fluid reservoir or an inner container **18**, as shown in FIGURES 4, 5 and 6, has flange portions **20** which are deformed or swaged radially outwardly for frictionally contacting the interior wall surface **22** of outer container **14**, thus the outer surface **24** of container **18** is spaced from the inner surface **22** of container **14**, thereby creating a first fluid flow path therebetween. A reduced cross-section **26** for producing a stress riser forms an integral part of inner container **18**. A stress riser being defined as a structure having a notch or any abrupt change in cross-section, whereby the maximum stress will occur at this change in cross-section and this maximum stress will be greater than the stress calculated by elementary formulas based upon simplified assumptions as to the stress distribution.

The stress riser **26** extends substantially along the entire longitudinal length of container **18** and is generally parallel to the geometric axis thereof. The manner of producing a stress riser **26** may be accomplished by numerous well-known and conventional means; for example, forming a groove on the surface of container **18**, welding two ends of a metal plate which form container **18** together utilizing a weld-metal having a lower tensile strength than the parent metal or by brazing, etc. The longitudinally extending stress riser **26** intersects a plurality of similar stress risers **28** which extend generally transverse or normal to stress riser **26** and these stress risers **28** are usually located adjacent opposite ends of container **18**. However, the stress risers **28** are not necessary to the operation of the present invention.

An explosive means in the form of an explosive charge **30** is generally positioned centrally within container **18** by a plurality of spider support elements **32** with each element **32** having a centrally disposed hub **34** including a bore **36** for slidably receiving and frictionally engaging the outer surface of explosive charge **30**. A plurality of tab elements **38** extend radially outwardly from hub **33** and frictionally contact the inner surface **40** of container **18** to centrally position the explosive charge **30**. It is readily apparent other means of positioning the explosive charge may be utilized without departing from the scope of the present invention.

A plurality of dished heads **42** and **44** as shown in FIGURE 4 are inserted in opposite ends of container **14** and are seam-welded at **46** to provide a fluid tight seal between heads **42** and **44** and inner container **18**. An internally threaded coupling **48** is welded in opening **50** of dished head **44** or otherwise secured thereto to form a fluid tight connection therewith. The removable threaded male plug **52** and coupling **48** provides means whereby container **18** may be charged with a high pressure relatively inert fluid; for example, nitrogen, air, carbon dioxide, or some relatively inert gas mixture which is compatible to human beings. The opposite dished head **42** contains a feed thru connector **54** having electrical wires **56** and **58** extending therethrough and connecting the explosive charge **30** to a source of electrical energy; in this case, the battery of the vehicle. A fluid seal **60** contains a plurality of holes for receiving and supporting wires **56** and **58** in addition to providing a fluid seal between the wires and dished head **42**.

An inertia responsive device **62**, as shown in FIGURE 2, has a housing **64** divided by wall **66** into chambers **68** and **70**. Chamber **68** has an enlarged portion **72** of greater cross-sectional area than the remaining portion **74** of chamber **68**. The enlarged chamber portion **72** is located adjacent removable cover **76** which is removably secured to housing **64** by a plurality of threaded studs or screws **78**. The removable cover **76** closes one end of chamber **68** and has a recess **80** in axial alignment with and having the same cross-sectional area and configuration as chamber portion **68**. An inertia mass or weight **82** is reciprocally disposed within the enlarged portion **72** for movement in one or another direction. A plurality of stop or abutment surfaces **84** and **86** formed by the surface areas which interconnect the differential cross-sectional areas created by chambers **68**, **74** and **80**, limit the longitudinal movement of the inertia mass element **82**. The inertia mass **82** has an elongated body with a bore **88** extending generally therethrough and having its geometric axis aligned with chamber portions **68**, **70** and **80**. An enlarged recess **90** located on one end **92**, of inertia mass **82** has a greater cross-sectional area than bore **88** and possesses a relatively flat bottom wall **94** which intersects bore **88**. A portion of the remaining surface area **93** of one end **92** abuts surface **84** whenever the mass **82** moves its complete longitudinal extend in said one direction and a portion of surface area **95** on the opposite end **96** of mass **82** abuts surface **86** when the mass **82** moves its complete longitudinal extent in another direction.

A spring retainer **98** is disposed in chamber **68** and is adjustably positioned therein by set screw **100** which threadedly engages housing **64** and is secured in position by a lock nut **102**. This adjustment feature is neither essential or required in the present device; inasmuch as a spring **104** could be designed to have the proper deflection rate and could completely fill chamber portion **68**. The spring or biasing member **104** has one end **106** bearing against spring retainer **98** while the opposite end **108** bears against bottom wall **94** to bias surface area **95** of intertia mass **82** into abuting engagement with surface **86**.

An annular recess **110**, as shown in FIGURES 2 and 3, located adjacent opposite end **96** interrupts the outer surface **112** of inertia mass **82** and a cam surface **114** located on outer surface **112**, extends from the inner edge **116** of the recess **110** toward one end **92**. The dividing wall portion **66** has a hole **118** containing a cam follower **120** having an end surface **122** extending therethrough for contacting cam surface **114**. An electrical switch **124** including two resilient leaf members **126** and **128** have at least two opposed contact surfaces **130** and **132** connected respectively thereto. A plurality of bolts **134** threadedly connect switch **124** to wall portion **66** thereby providing an anchor for members **126** and **128**. A plurality of insulating and spacing elements are provided to electrically insulate members **126** and **128** from each other respectively, and the wall portion **66** and bolts **134**. The leaf member **128** has a portion **136** extending beyond contact **132** for contacting the other end surface **138** of cam follower **120**. Since cam follower **120** is slightly longer than would be required for a normal contacting or switch closing relationship between opposed contacts **130** and **132**, the electrical switch **124** is in the normally opened position when surface **122** of cam follower **120** is engaging cam surface **114**.

Electrical wires 58 and 140 extend through respective openings **142** and **144** in a side wall of chamber **70**, in sealed relationship therewith. One end of wire **140** is connected to a source of electrical energy; for example, the battery **146** of a vehicle, with the opposite end being connected to electrical switch leaf member **126** and wire **58** is connected to electrical switch leaf member **128**, with the opposite end thereof being connected to leg wire **148** of explosive charge **30**.

The explosive charge shown in FIGURE 13 includes two leg wires **148** and **150** interconnected by a bridge wire **152** which acts as an electrical resistance to the flow of electric current therethrough, to produce sufficient heat in response to a predetermined current, to ignite the ignition mix **154** which subsequently fires the primer charge **156**, thereby exploding the base charge **158**. A metal jacket **160** partially encloses the above mentioned components to form explosive charge **30** and a rubber plug **162** acts as a closure for open end **164** of the metal jacket **160** and additionally supports and insulates leg wires **148** and **150** from metal jacket **160**. The wire **56** electrically interconnects electrical explosive leg wire **150** to ground thereby completing the electric circuit. Both electric wires **56** and **58** are electrically shielded from any external magnetic field created by an outside source; for example, a radio transmitter. The purpose of this electromagnetic shield is to prevent premature or impromptu detonation of the explosive charge **30**.

FIGURES 4–6 disclose another embodiment of the present invention. In the inflatable bag assembly **10** of

FIGURES 4–6 a baffle **166** is fixedly disposed by conventional means oposite fluid discharge outlets **16**. The purpose of baffle **166** is to (1) redirect and divide the fluid flow being discharged through openings **16**, and (2) to absorb, attenuate, alleviate, or dampen the sonic shock waves, or a concentration of forces if any, created by firing the explosive charge **30** and thereby prevent rupture of inflatable bag **12**. The inflatable crash bag assembly **10** may be mounted by any number of conventional ways not shown on the structure of a vehicle. However, it has been found advantageous to provide an elongate recess in the dash or instrument panel of a vehicle, for mounting assembly **10** therein. The FIGURES 10, 11 and 12 in the drawing disclose various positions the assembly **10** may be mounted within a vehicle passenger compartment; for example, a recess may be provided in the front or rear seat in which the crash bag assembly may be positioned.

FIGURES 7, 8 and 9 disclose still another embodiment of the present invention showing a limiting member **19** interjacent spaced apart walls **22** and **24** for limiting outward deformation of inner container **18**. The purpose of the limiting member **19** is to insure an adequate predetermined fluid flow path is maintained between outer wall **22** and inner wall **24**. As illustrated in FIG. 9, the intermediate member **19** restricts movement of at least one portion **26*a*** of the inner container **18** upon activation of the explosive. As a result, the configuration of the opening formed in the inner container **18** is controlled. The limiting member **19** as shown in FIGURE 7 is preferably an intermediate container positioned between inner container **18** and outer container **14**. However, a plurality of spaced rings, bands, or other similar structure may be employed with equal effectiveness. In the embodiment shown in FIGURE 7, the intermediate container or limiting member **19** has a plurality of aligned apertures **21** disposed opposite stress riser **26** and angularly positioned approximately 180° relative to the openings **16** in outer container **14**. This angular relationship of apertures **21** relative to openings **16** requires the fluid that is released from inner container **18** to reverse its direction of flow before being discharged through openings **16** into the interior of crash bag **12**.

In the operation of this safety device, the inertia device **62** is mounted on the vehicle in advance of the passenger compartment. It is obvious when the vehicle encounters rapid or violent deceleration the potential energy retained by the inertia weight **82**, in response to the linear speed of the vehicle, will be converted into kinetic energy and weight **82** will overcome the resistance of spring **104** and move forwardly until a portion of surface **93** contacts stop **84**. With the weight **82** in its forward position, the end surface **122** of cam follower **120** completely clears corner **116** of cam surface **114**. Accordingly, resilient leaf portion **136** biases cam follower **120** upwardly into annular recess **110** as shown in FIGURE 3, and simultaneously brings contacts **130** and **132** together to close the normally open switch **124**, thereby energizing the electric circuit connecting battery **146** to explosive charge **30**. The purpose of annular recess **110** and cam follower **126** is to prevent instantaneous rebound or rearward movement of weight **82** upon impact with stop **84**. Therefore, in the present invention the rearward movement of weight **82** subsequent to the forward movement thereof is interrupted by the interlocking of cam follower **120** within recess **110**. The purpose of which is to assure sufficient electrical current can be built up in the electric circuit to insure positive foring of explosive charge **30**. The cam surface **114** may be beveled (not shown) or the surface **122** may be beveled or both surfaces **114** and **122** may be beveled (not shown). By utilizing the beveled surfaces noted above, the inertia weight **82**, in its rearward movement would be temporarily delayed or restrained due to frictional surface contact between surfaces **114** and **122**. Thus, the switch **124** would be restrained in a closed position for a sufficient length of time to insure positive firing of the explosive charge **30**. The detonation of explosive charge **30** produces sonic waves with an accompanying instantaneous increase in pressure within inner container **18** for opening inner container **18** along longitudinally extending stress riser **26** and simultaneously outwardly deforming the inner container **18** toward outer container **14**. Thus, pressurized fluid flows from the inner container **18** along the flow path created by spaced walls **22** and **24** and is subsequently discharged through aperture **16** into the interior of inflatable bag **12**. In the event baffle **166** is employed, the fluid as it leaves aperture **16** impinges there against and is redirected to at least another flow path which is substantially transverse to the flow path emanating from aperture **16**.

From the above description it should be readily apparent that applicant's have provided a new and improved inflatable safety crash bag for use by vehicle occupants. It should be also apparent that certain modifications, changes and adaptations may be made in the structure of the disclosed, and it is hereby intended to cover all such modifications, adaptations, and constructions which fall within the scope of the appended claims.

Having described our invention, we claim:

1. Safety apparatus for protecting an occupant of a vehicle, said safety apparatus comprising an expandable occupant restraint having a contracted condition, a reservoir for containing a supply of fluid, means operable to provide for fluid flow from said reservoir to said occupant restraint to dispose said occupant restraint adjacent an occupant of a vehicle during a collision for limiting movement of the occupant during such collision, said occupant restraint comprising confinement means formed of a material which is substantially irreversibly expandable to increase the volume thereof in response to movement of the occupant relative to the confinement means to absorb at least a part of the kinetic energy of the occupant created by such movement and thereby minimize rebound of the occupant therefrom.

2. Safety apparatus as defined in claim **1** wherein said material of which said confinement means is formed comprises a ductile material having a high degree of plasticity and a high percent of elongation when subjected to loading in tension.

3. Safety apparatus as defined in claim **1** further including a diffuser member associated with said reservoir for directing the flow of fluid to said confinement means to effect expansion thereof.

4. Safety apparatus as defined in claim **1** wherein said means operable to provide for fluid flow from said reservoir to said occupant restraint comprises explosive means for effecting the formation of an opening in said reservoir.

5. Safety apparatus for protecting an occupant of a vehicle, said safety apparatus comprising a confinement having a collapsed inoperative condition and an expanded operative condition, said confinement when in said expanded operative condition being effective to restrain movement of an occupant of a vehicle during a collision, a reservoir for containing a supply of fluid for expanding said confinement, explosive means operable to effect the formation of an opening in said reservoir by effecting movement of at least one portion of said reservoir to provide for flow of fluid therefrom, and means for controlling the configuration of the opening formed in said reservoir by restricting movement of said portion of said resrevoir upon detonation of said explosive means.

6. Safety apparatus as defined in claim **5** wherein said means for controlling the configuration of said opening formed in said reservoir comprises a member associated with said reservoir and positioned adjacent thereto.

7. Safety apparatus as defined in claim **5** wherein said means for controlling the configuration of said opening comprises a member enclosing at least that portion of said reservoir in which said opening is formed.

8. Safety apparatus as defined in claim **1** further including a diffuser member defining a chamber in which at least a part of said reservoir is located and said means for controlling the configuration of the opening formed in said reservoir comprises a member interposed between said diffuser member and said reservoir.

9. Safety apparatus as defined in claim **8** wherein said diffuser member is disposed to change the direction of the fluid flow from said reservoir and further including a baffle member located in the path of fluid flow from said diffuser member and for redirecting said fluid flow to said confinement.

10. Safety apparatus as defined in claim **8** wherein said diffuser member has a longitudinal axis and a plurality of outlet openings spaced along the longitudinal axis thereof.

11. Safety apparatus for protecting an occupant of a vehicle, said safety apparatus comprising a confinement having a collapsed inoperative condition and an expanded operative condition, said confinement when in said expanded operative condition being effective to restrain movement of an occupant of a vehicle during a collision, first and second baffle members, a reservoir for containing a supply of fluid for expanding said confinement, means for effecting the formation of an opening in said reservoir to provide for fluid flow therefrom in a first general direction toward said first baffle member, said first baffle member having a surface portion in the path of fluid flow from said reservoir for redirecting at least a substantial portion of said fluid flow in another direction toward said second baffle member, said second baffle member having a surface portion for changing the direction of said fluid flow to substantially said first general direction and to said confinement with said first and said second baffle members being effective to absorb a portion of the kinetic energy of said flowing fluid, and said reservoir being located intermediate said surface portions of said first and said second baffle members and being at least in part in the path of fluid flow in said another direction toward said second baffle member.

12. Safety apparatus as defined in claim **11** wherein said first baffle member comprises a diffuser member defining a chamber in which said reservoir is at least in part located.

13. Safety apparatus as defined in claim **11** wherein said means for effecting the formation of an opening in said reservoir comprises explosive means operable to effect movement of at least one portion of said reservoir to provide for flow of fluid therefrom and further including means for controlling the configuration of the opening formed in said reservoir by restricting movement of said portion of said reserovir upon detonation of said explosive means.

14. Safety apparatus as defined in claim **13** wherein said means for controlling the configuration of the opening comprises a member interposed between said diffuser member and said reservoir.

15. Safety apparatus for protecting an occupant of a vehicle, said safety apparatus comprising a confinement having a collapsed inoperative condition and an expanded operative condition, said confinement when in said expanded operative condition being effective to restrain movement of an occupant of a vehicle during a collision, structure operable to supply fluid for expanding said confinement, means operable to effect the formation of an opening in said structure to provide for flow of fluid from said structure, and a member disposed in the path of flow of fluid from said structure for controlling the formation of the opening therein.

16. Safety apparatus as defined in claim **15** further including a diffuser member located in the path of fluid flow.

17. Safety apparatus as defined in claim **66** wherein said diffuser member defines a chamber in which at least a part of said structure is located.

18. Safety apparatus as defined in claim **66** wherein said diffuser member and said structure are located in said confinement.

19. Safety apparatus as defined in claim **66** wherein at least a part of said diffuser member has a plurality of spaced apart openings therein.

20. Safety apparatus as defined in claim **19** wherein said part of said diffuser member is located in said confinement.

21. Safety apparatus for protecting an occupant of a vehicle, said safety apparatus comprising a confinement having a collapsed inoperative condition and an expanded operative condition, said confinement when in said expanded operative condition being effective to restrain movement of an occupant of a vehicle during a collision, a reservoir for containing a supply of fluid for expanding said confinement, explosive means for effecting the formation of an opening in said reservoir, a diffuser member for diffusing and directing said fluid flow to said confinement, and a member disposed intermediate said diffuser member and said reservoir for controlling the formation of the opening in said reservoir.

22. Safety apparatus as defined in claim **21** wherein said explosive means effects the formation of an opening in said reservoir by effecting movement by at least one portion of said reservoir and said member is operable to control the configuration of the opening formed in said reservoir by restricting movement of said portion of said reservoir.

23. Safety appaartus as defined in claim **21** wherein said diffuser member defines a chamber in which at least a part of said reservoir and a part of said member are located.

24. Safety apparatus for protecting an occupant of a vehicle, said safety apparatus comprising a confinement having a collapsed inoperative condition and an expanded operative condition, said confinement when in said expanded operative condition being effective to restrain movement of an occupant of the vehicle during a collision, structure operable to supply fluid for expanding said confinement, explosive means for effecting the formation of an opening in said structure, and a member operatively associated with said structure and disposed intermediate said structure and said confinement for controlling the formation of the opening in said structure.

25. Safety apparatus for protecting an occupant of a vehicle, said safety apparatus comprising a confinement having a collapsed inoperative condition and an expanded operative condition, said confinement when in said expanded operative condition being effective to restrain movement of an occupant of the vehicle during a collision, said confinement including means for dissipating the kinetic energy of the occupant to minimize rebound of the occupant therefrom, structure operable to supply fluid for expanding said confinement, explosive means operable to provide for flow of fluid from said structure, a diffuser member disposed in the path of flow of fluid from said structure and effective to control said flow to said confinement, and a member interposed between said confinement and said structure for controlling the formation of an opening in said structure by said explosive means.

26. Safety apparatus as defined in claim **25** wherein said diffuser member has a plurality of spaced apart openings located within said confinement.

27. Safety apparatus as defined in claim **26** wherein said diffuser member has a surface for directing at least a portion of the flow of fluid in a direction opposite the direction in which the fluid flows from said structure and further including a baffle member having a surface for changing the direction of flow of fluid to substantially the general direction in which the fluid flows from said structure.

28. Safety apparatus as defined in claim **26** wherein said confinement is made of a material which is substantially irreversibly expandable to increase the volume thereof in response to movement of the occupant relative to the confinement.

29. Safety apparatus for protecting an occupant of a vehicle, said safety apparatus comprising a confinement having a collapsed inoperative condition and an expanded operative condition, said confinement when in said expanded operative condition being effective to restrain movement of an occupant of the vehicle during a collision, a fluid reservoir for containing a supply of fluid for expanding said confinement, means operable to provide for flow of fluid from said reservoir, a diffuser member disposed in the path of flow of fluid from said reservoir and effective to control said flow, said diffuser member defining a chamber in which said reservoir is at least in part located and into which fluid flows from said reservoir, and said diffuser member including a plurality of spaced apart openings therein disposed within said confinement for providing for flow of fluid from said chamber through said diffuser member and to said confinement, said means operable to provide for flow of fluid from said reservoir comprising explosive means operable to effect the formation of an opening in said reservoir and further including a member interposed between said diffuser member and said reservoir for controlling the formation of said opening by said explosive means.

References Cited

UNITED STATES PATENTS

3,243,822	4/1966	Lipkin	2—2
2,834,606	5/1958	Bertrand	280—150
2,850,291	9/1958	Ziccardi	280—150
2,957,415	10/1960	Lazari	102—28
3,197,234	7/1965	Bertrand.	
3,224,924	12/1965	Von Ardenne et al.	161—68

BENJAMIN HERSH, *Primary Examiner.*

E. SIEGEL, *Assistant Examiner.*

Feb. 15, 1966 J. W. RYAN 3,234,689

DOLL CONSTRUCTION FOR NATURAL MOVEMENTS AND POSITIONS

Filed June 8, 1962 3 Sheets-Sheet 1

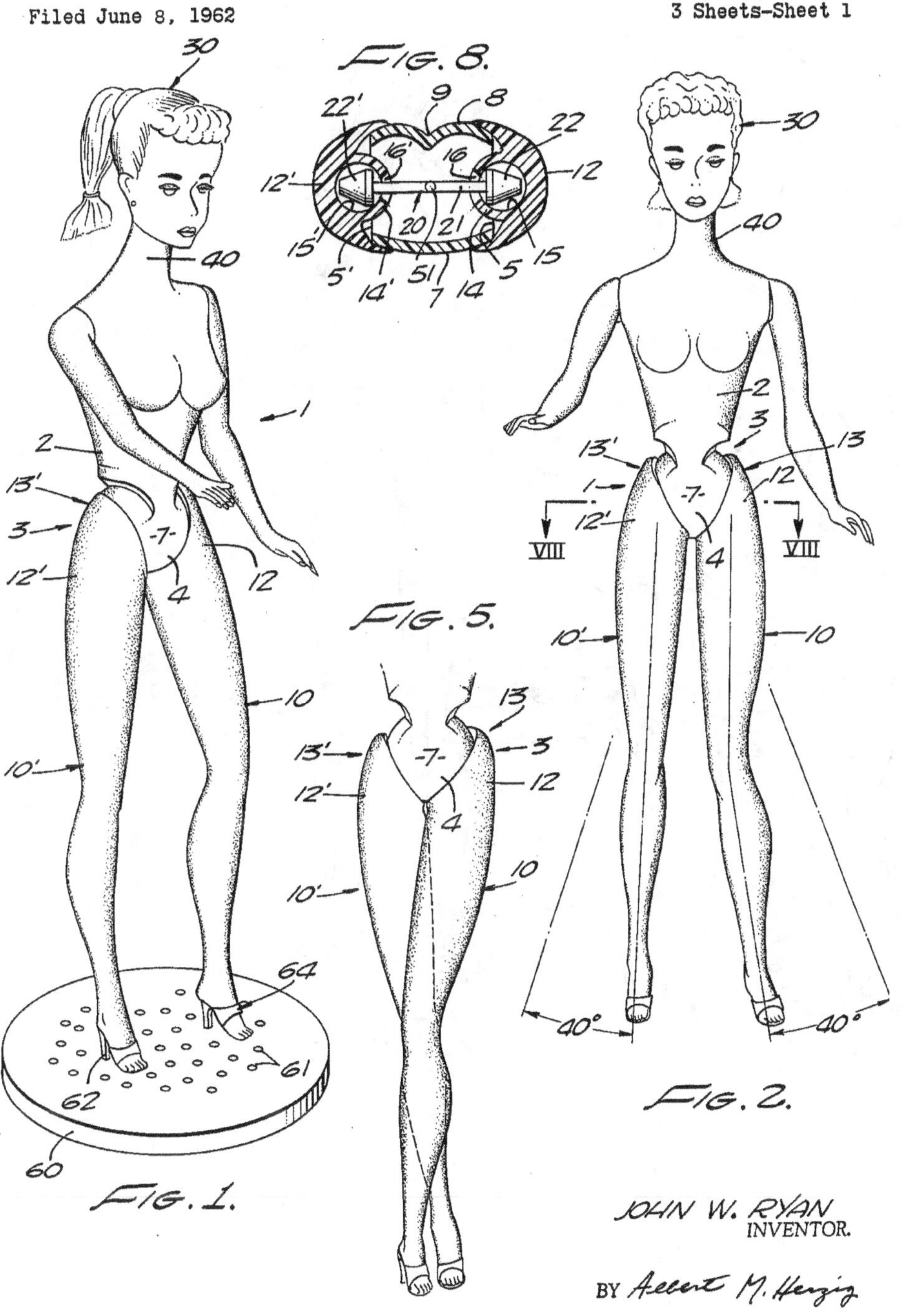

JOHN W. RYAN
INVENTOR.

BY Albert M. Herzig

ATTORNEY

Feb. 15, 1966 J. W. RYAN 3,234,689

DOLL CONSTRUCTION FOR NATURAL MOVEMENTS AND POSITIONS

Filed June 8, 1962 3 Sheets-Sheet 2

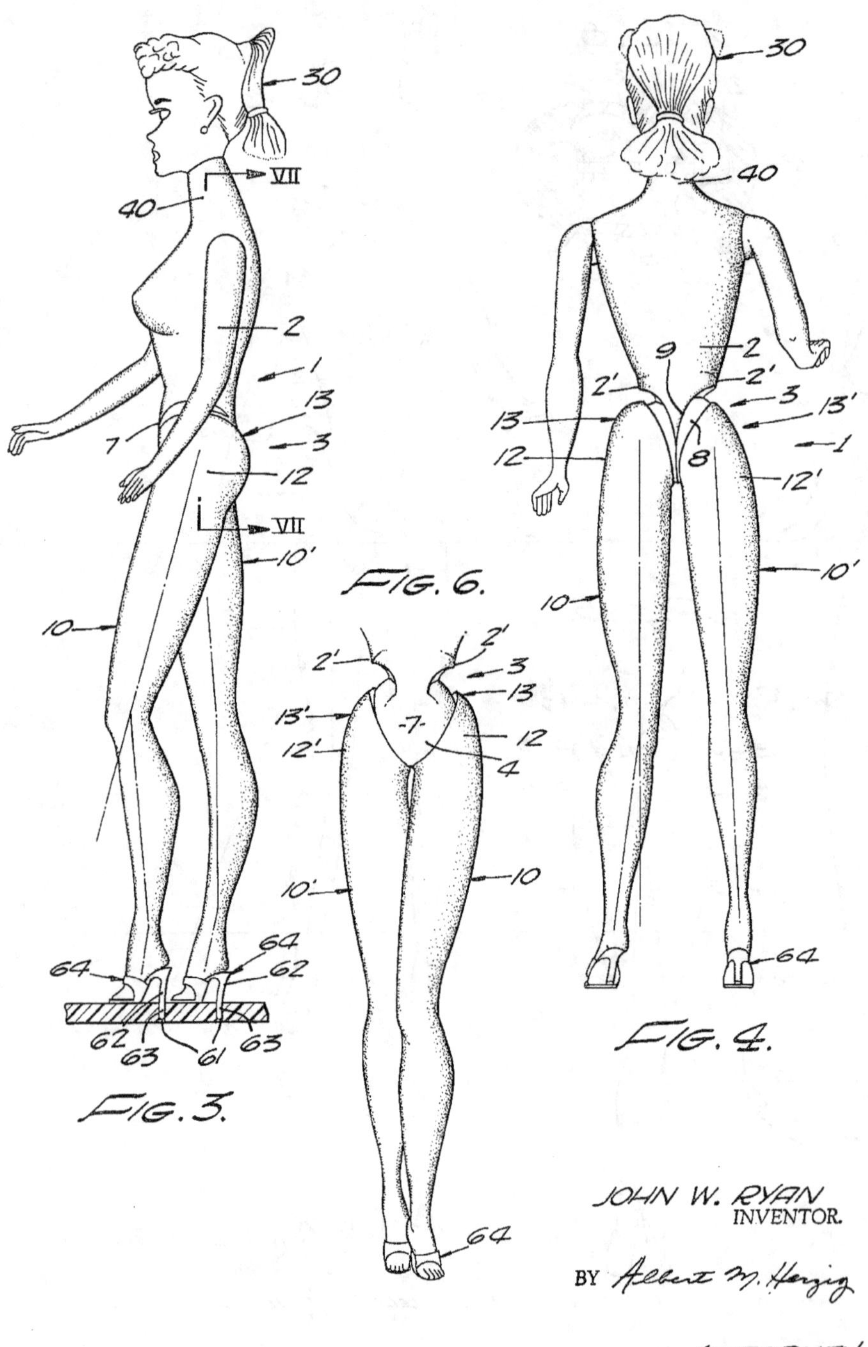

JOHN W. RYAN
INVENTOR.

BY Albert M. Herzig

ATTORNEY

Feb. 15, 1966 J. W. RYAN 3,234,689

DOLL CONSTRUCTION FOR NATURAL MOVEMENTS AND POSITIONS

Filed June 8, 1962 3 Sheets-Sheet 3

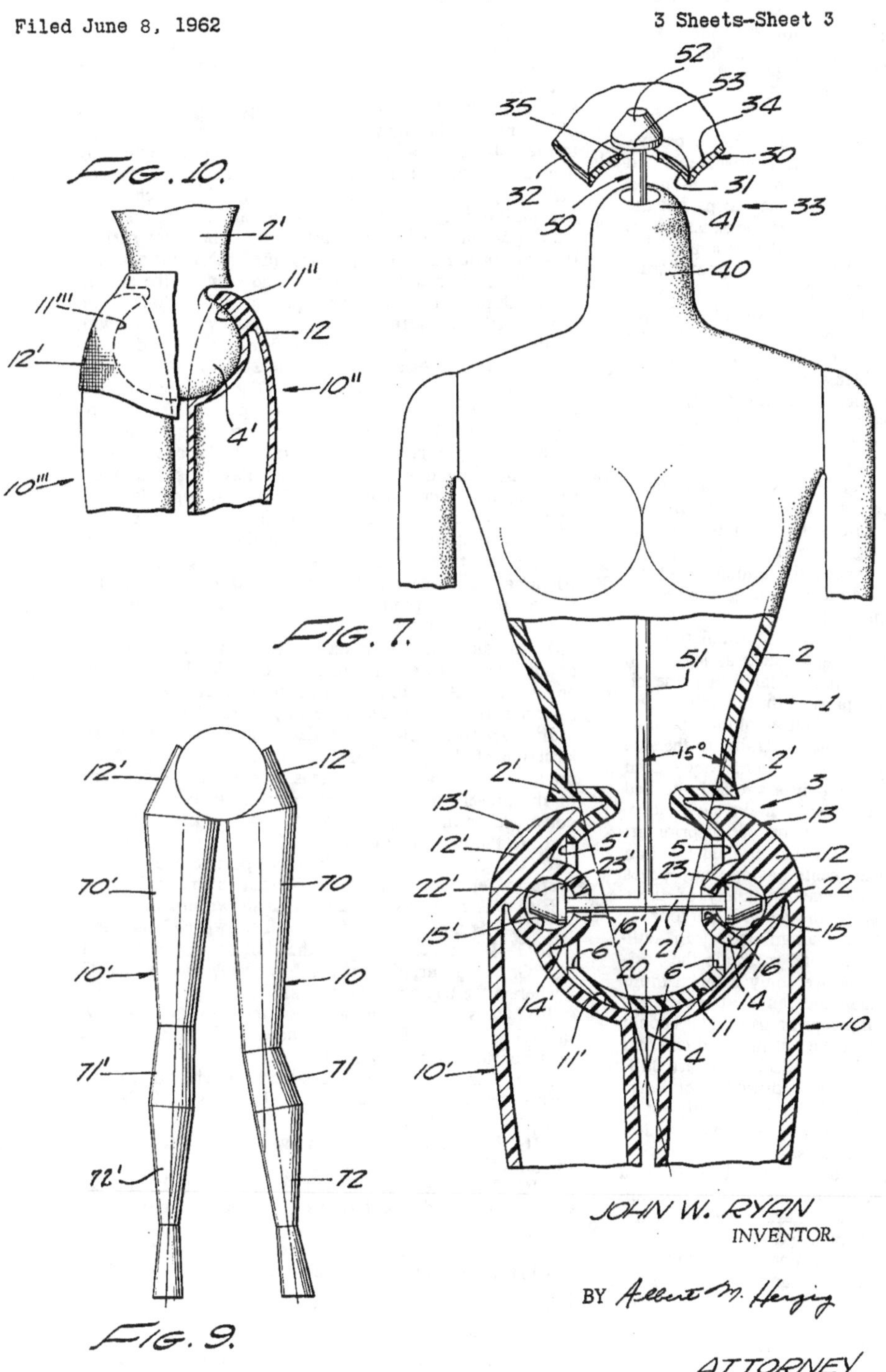

United States Patent Office

3,234,689

Patented Feb. 15, 1966

3,234,689

DOLL CONSTRUCTION FOR NATURAL MOVEMENTS AND POSITIONS

John W. Ryan, Bel Air, Calif., assignor to Mattel, Inc., Hawthorne, Calif., a corporation of California

Filed June 8, 1962, Ser. No. 201,059
14 Claims. (Cl. 46—161)

The present invention refers to a novel doll construction adapted to attain numerous natural movements and positions. More particularly, the present invention relates to doll construction which is adapted to permit the doll trunk appandages, i.e. the doll's arms, legs, and head, to reproduce closely the movement and positions of the human body. The term "trunk" is used to designate herein the central portion or torso of the body.

As is generally known, the construction of dolls for children's toys is a well developed art with many varieties of dolls, both male and female, having been produced. One of the main problems involved in production of dolls has been the achievement of a realistic reproduction of humans, particularly the movements and positions of the human appendages. It has been found particularly difficult to reproduce the various movements and positions of human appendages simply and inexpnesively, since such appendages are normally free to move in many different directions and to obtain a wide range of positions. For example, the human leg may rotate parallel to a plane bisecting the human trunk up to an angle of at least about 40 degrees to said bisecting plane and about its longitudinal axis.

A common solution to the problem of simulating the movement of the human appendages, such as legs, is to rotate the leg about an axis perpendicular to a plane bisecting the human trunk and passing through the lower portion of the trunk. However, such a solution permits the rotation of the leg only in a plane parallel to the side of the body. Many other forms of joints have been utilized, but they normally have only a similarly limited degree of movement and consequently make such movement look very unnatural. Also, many of the prior art doll joints, in order to obtain even such limited freedom of movement required joint elements having sharp edges and other irregularities. Consequently, many of these joints bore little resemblance to the smooth contour of the corresponding joints of the human body. In addition, such prior art in all joints could maintain the positions of the human appendages over only a much more limited range of positions than the range of movement of the appendages. For example, prior art doll legs could be rotated sideways to a large angle but required external force to maintain such positions. Also, prior art dolls a straight-out position. Such leg positions combined normally utilized symmetrical pairs of legs rigidly set in with the limited movement of the hip joint made the doll assume very stiff and unnatural poses.

Consequently, an object of the present invention is a doll construction which is adapted to permit the doll appendages to reproduce closely the natural movements and positions of the human body appendages.

Another object of the present invention is a doll joint having a smooth contour closely resembling the corresponding portion of the human body over a wide range of positions of the appendage.

Still another object of the present invention is a doll construction which is simple and inexpensive to manufacture and yet result in a rugged, long-lasting construction.

Still another object of the present invention is a doll with legs adapted to permit the doll to assume numerous natural positions.

Other objects and advantages of the present invention will be readily apparent from the following description and drawings which illustrate exemplary embodiments of the present invention.

In general, the present invention involves a doll construction adapted to permit the doll appendages, particularly its legs, to reproduce closely the movements and positions of the human body appendages. The doll of the present invention involves an appendage attached to the doll trunk having a curved recess therein extending over substantially the whole inner surface of the portion of the appendage adjacent to the doll trunk. Such recess has substantially the same radius of curvature as has the trunk portion to which it is attached. The outer surface of said portion of the appendage preferably has substantially the same radius of curvature adjacent to the adjoining trunk portion as said trunk portion. The doll's trunk has a portion adjacent said appendage having a boss with a curved surface matingly received in said appendage recess and the recesses cover about 20% to 40% of the total surface area of said boss. The trunk portion is adapted to form with the appendage portion a joint having a substantially smooth contour closely resembling the corresponding region of the human body over a wide range of appendage positions and having substantially uniform frictional forces maintaining the appendage position over the entire range of appendage positions. The present invention also includes fastening means for maintaining said trunk boss and appendage recess in mating frictional relationship. The joint and fastening means are adapted to permit the appendage to independently rotate parallel to a plane bisecting the doll trunk, angularly to said bisecting plane up to an angle of at least about 40 degrees and about the longitudinal axis of said appendage. Also, the doll arms may be constructed and fastened to the doll the same way as the legs and head.

In order to facilitate an understanding of the doll construction of the present invention, reference will now be made to the appended drawings of specific embodiments of the present invention. Such drawings should not be construed as limiting the invention which is properly set forth in the appended claims.

In the drawings, FIG.1 is a perspective view of the preferred embodiment of the present invention.

FIG. 2 is a front view of the doll shown in FIG. 1.

FIG. 3 is a side view of the doll shown in FIG. 1.

FIG. 4 is a rear view of the doll shown in FIG. 1.

FIG. 5 is a variation of the front view shown in FIG. 2 wherein the legs are shown crossed.

FIG. 6 is another variation of the front view shown in FIG. 2 wherein the legs are shown with the feet placed on a substantially straight line.

FIG. 7 is an enlarged cross-sectional view of the hip portion of FIG. 3 taken along the lines VII—VII of FIG. 3.

FIG. 8 is a cross-sectional view of FIG. 2 taken along the lines VIII—VIII of FIG. 2.

FIG. 9 is a geometrical representation of the hip and leg portions of the preferred embodiment of the present invention.

FIG. 10 is a partial cross-sectional view of the hip portion of a doll showing another embodiment of the present invention.

In FIGS. 1–8, the doll **1** has legs **10** and **10′** respectively attached to its trunk **2**. Legs **10** and **10′** have substantially spherical recesses **11** and **11′** respectively therein extending over substantially the whole inner surface of the upper portions of the legs **10** and **10′** adjacent to the doll trunk **2**. The pelvic portion **3** of trunk **2** adjacent legs **10** and **10′** has a single generally spheroidal boss **4** matingly received in the leg recesses **11** and **11′** with said recesses covering about 40% to 80% of the spheroidal surface defined by said boss **4**. The resulting edges of recesses **11** and **11'** are flexible and the edges of said recesses **11** and **11'** lie in planes at an angle in the range of about 5 degrees to 25 degrees to a fore-and-aft vertical plane bisecting the doll trunk when the axis of the doll trunk and the upper portions of the legs are oriented perpendicular to a flat surface on which the doll is standing (see FIG. 7, for example, where the angle is approximately 15°). More particularly, boss **4** has a horizontal cardioidal cross section as shown in FIG. 8 although it may also be described as generally elliptical. The pelvic portion **3** is adapted to form with the upper portions **12** and **12'** of legs **10** and **10'** joints **13** and **13'** respectively having substantially smooth contours closely resembling the corresponding hip joints of the human body over the usual range of leg positions and having substantially uniform frictional forces maintaining the leg position over the entire range of such leg positions. The recesses **11** and **11'** have substantially the same radius of curvature as the lateral portions of boss **4** and the outer surfaces of upper portions **12** and **12'** have substantially the same radius of curvature adjacent to the adjoining portions of boss **4** as said boss portions.

Recesses **11** and **11'** of legs **10** and **10'** have knobs **14** and **14'** respectively projecting inwardly from their central portions. Boss **4** has truncated portions **5** and **5'** with substantially circular holes **6** and **6'** respectively therethrough for receiving the recess knobs **14** and **14'** respectively. The holes **6** and **6'** and knobs **14** and **14'** are adapted to limit the rotation of legs **10** and **10'** angularly to a plane bisecting the doll trunk and to limit the rotation of the legs about their longitudinal axis to ranges natural to the human body. Thus the legs **10** and **10'** as shown in FIG. 2 may be rotated angularly to the plane bisecting the doll trunk to an angle of about 40 degrees. Similarly, the legs **10** and **10'** may be rotated about their longitudinal axis, either rearwardly or frontwardly, into abutting relationship.

Fastening means **20** for maintaining the pelvic boss **4** and the leg recesses **11** and **11'** in mating frictional relationship include an elastic bar **21** having frusto-conical end portions **22** and **22'** respectively with the bases of said end portions **23** and **23'** respectively being attached to the bar **21**. The end portions **22** and **22'** are received in substantially spherical cavities **15** and **15'** in the knobs **14** and **14'** respectively and the entrances **16** and **16'** of cavities **15** and **15'** respectively have frusto-conical shape slightly smaller than the conical end portions **22** and **22'**. Consequently, when the conical end portions **22** and **22'** are inserted in said cavities, they are compressed and then expanded to become engaged within the cavities. Such engagement permits rotation of the legs **10** and **10'** into numerous natural positions and forms sufficient frictional contact between the adjoining trunk and leg surfaces so that such positions are maintained without the aid of external force. The hip joints **13** and **13'** and the fastening means **20** are adapted to permit legs **10** and **10'** to independently rotate laterally of the doll trunk up to an angle of at least about 40 degrees and also about their own longitudinal axes into mutual abutting relationship.

The front section **7** of pelvic portion **3** has a radius of curvature substantially larger than the remaining portions of the pelvic boss so that the doll as a whole has only a slightly curved stomach which does not protrude beyond the front of the legs. The rear section **8** of pelvic portion **3** has a central vertical crevice **9** with rounded edges so that the crevice is adapted to receive the upper leg portions when the legs are rotated rearwardly about their longitudinal axis. The contour of front section **7** and rear section **8** forms a pelvic boss having a horizontal generally cardioidal cross-section as shown in FIG. 8.

The side waist portions **2'** of the doll trunk **2** flare outwardly and downwardly in the direction of the pelvic portions and are then undercut inwardly to permit the free rotation of legs **10** and **10'**. The undercut of the side waist portions **2'** is generally horizontal but preferably downward to fit the displacement of the upper portions of the legs **10** and **10'** when they are swung outwardly at an angle to a fore-and-aft vertical plane bisecting the doll trunk. The legs **10** and **10'** are rigid; however, preferably one of the legs, e.g. leg **10** is bent at an angle in the range of about 10 degrees to 30 degrees; for example, about 20 degrees in a plane parallel to a plane bisecting the doll trunk when the upper leg portion is maintained parallel to said bisecting plane. Preferably, the lower leg portion of said bent leg, i.e. the portion below the knee, is also bent at an angle in the range of about 1 degree to 15 degrees; for example, about 10 degrees in a plane angularly to a plane bisecting the doll body when the upper leg portion is maintained parallel to said bisecting plane. The combination of at least one bent leg and the hip joints of the present invention are adapted to permit the doll to assume numerous natural positions, such as being able to cross her legs (FIG. 5) or to assume the standard poses of fashion models, e.g. FIGS. 1–4 or placing the feet on a substantially straight line, one in front of the other (FIG. 6).

It has been found that the ratio of the length of the doll trunk to the diameter of the pelvic boss should fall within a definite range so that the doll may be able to closely reproduce the movement and positions of the human legs while retaining natural human proportions and contours. This range has been found to be about 2.4 to 4.0. Specifically, when the doll trunk is about 3–3½ inches in length, the pelvic boss should range about ⅞–1¼ inches in diameter, and preferably the diameter is about 1–1⅛ inches.

The head **30** attached to the doll trunk **2** has a spherical recess **31** in the portion **32** attached to the neck portion **40** of doll trunk **2**. The neck portion **40** of trunk **2** adjacent to the head **30** has a spherical top **41** matingly received in the head recess **31**. The neck portion **40** is adapted to form with the adjacent portion **32** of the head **30**, a joint **33** having a smooth contour closely resembling the corresponding neck joint of the human body. The neck top **41** and the head recess **31** are maintained in mating relationship by means of an elastic fastening means **50**. Fastening means **50** is adapted to permit the head **30** to rotate independently in a plane parallel to a plane bisecting the doll trunk, angularly to said bisecting plane up to an angle of at least about 40 degrees and about the longitudinal axis of the head **30**. Fastening means **50** includes an elastic bar **51** which may be attached to bar **21** or to means connecting the arms (not shown). Bar **51** has a conical end portion **52** with a base **53** attached to bar **51**. Conical end portion **52** is received in cavity **34** in the portion **32** of the head **30** matingly fitted to the adjoining neck portion **41**. The entrance **35** of the cavity **34** has a frusto-conical shape slightly smaller than the conical end portion **52** of bar **51**. Consequently, when the conical end portion **52** is inserted into cavity **34** it is compressed and then expanded to become engaged therewith while permitting the rotation of the head and maintaining of the head in any desired position.

In addition, doll **1** may stand on a platform **60** having a plurality of holes **61** therein. Conveniently the doll **1** is selectively engaged in holes **61** by means of elongations **63** of its shoe heels **62**. Alternatively, the shoes **64** may have additional downwardly directed pegs (not shown) thereon for rotatably mounting the doll **1** on platform **60**. By so mounting the doll of the present invention on a platform the various doll leg positions may be accurately arranged in any desired manner and the doll will remain standing in such position. Also many variations of each position may be obtained merely by rotating the doll trunk which causes the legs to adjust to the trunk position while the feet remain in fixed positions. In other words, one or more pegs between the heel or sole and platform may be provided.

In FIG. 9, a geometrical representation of the legs of the preferred embodiment of the present invention is shown to illustrate their preferred proportions in relation to the hips and the result of bending the right leg. The thigh portions **70** and **70'** of legs **10** and **10'** respectively are frusto-conical in shape with their bases adjacent upper leg portions **12** and **12'** respectively. The ratio of the diameter of the base to the diameters of the apexes of the thigh portions **70** and **70'** is in the range of about 1.6 to 2.7, while the ratio of their heights to the diameters of their bases is in the range of about 2.2 to 3.3. The knee portions **71** and **71'** of legs **10** and **10'** respectively are frusto-conical in shape with their apexes at the knee adjacent the thigh portions **70** and **70'** respectively. The ratio of the diameter of the bases to the diameters of the apexes of the knee portions **71** and **71'** is in the range of about 0.8 to 1.9, while the ratio of their heights to the diameters of their bases is in the range of about 1.2 to 2.3.

Finally, the calf portions **72** and **72'** of legs **10** and **10'** respectively are frusto-conical in shape with their bases adjacent the bases of the knee portions **71** and **71'** respectively. The ratio of the diameters of the bases to the diameters of the apexes of the calf portions **72** and **72'** is in the range of about 2.1 to 3.2, while the ratio of their heights to the diameters of their bases in the range of about 2.0 to 3.1

In FIG. 10 is illustrated the doll construction of the present invention wherein the boss **4'** is substantially elliptical and the fastening means is an elastic garment **80** worm by the doll. The elastic garment **80** is stretched to cover the upper leg portions **12** and **12'** adjacent to the doll trunk **2'** and the adjoining portion of the doll trunk, i.e. boss **4'**.

Also, the recesses **11''** and **11'''** of the legs **10''** and **10'''** respectively are simply matingly received on the pelvic boss in such fashion that the legs can be rotated forward and backward and side to side or any combination of such movements. In addition, each of the legs may be rotated about its longitudinal axis while the hip joint maintains a substantially smooth contour closely resembling the hip joint of the human body. Similarly, the frictional forces are maintained substantially uniform for each leg position over the entire range of leg positions.

There are many features in the present invention which clearly show the significant advance the present invention represents over the prior art. Consequently, only a few of the more outstanding features will be pointed out to illustrate the unexpected and unusual results attained by the present invention.

One of the features of the present invention is the joint between the leg and the pelvic portion of the doll body. Such hip joint is adapted to permit the leg to reproduce closely the movement of the human leg while retaining a substantially smooth contour closely resembling the human hip joint. By utilizing a single, substantially spherical boss on the doll trunk fitted into recesses in the upper portions of the legs, the hip movement is closely reproduced and maintained over a wide range of leg positions. In addition, by utilizing a pelvic boss having a horizontal cardioidal cross-section, both the normal contour of the body and the normal movement of the hip joint are more closely obtained. By the use of such hip joint the legs may be moved quite freely to the front and to the back into abutting relationship, but their movement is more restricted from side to side and about their longitudinal axis. It should be noted that such action fits relatively close to the freedom of movement of the normal hip joint.

Another feature of the present invention is that the hip joint of the present invention in combination with the bent form of at least one of the legs permits the legs to obtain a wide range of positions, such as crossing the legs, putting the feet one in front of the other on a substantially straight line, and many other positions. In addition, the hip configuration permits the joint to maintain substantially the same frictional engagement over the whole range of its leg positions. For example, one of the legs may be bent upwardly to the side to an angle of, say, 30 degrees, and such position is maintained without the aid of external force.

Still another feature of the present invention, as illustrated in the drawings, is the proportion and location of the various parts of the doll of the present invention. Thus, for example, the proportions of the leg contours and the pelvic boss are preferably adapted to facilitate the movement and positioning of the legs. Similarly, the straight line connecting the pelvic boss to the horizontal axis between the doll shoulders is preferably located behind the doll trunk axis to facilitate the upright positioning of the doll trunk while the leg positions are varied over a wide range.

Still another feature of the present invention is the unusual fastening means which may be combined with the joint of the present invention. For example, by utilizing the fastening means illustrated in FIGS. 7 and 8, freedom of movement of the appendages is obtained very simply while maintaining uniform frictional engagement with the adjoining trunk portion. Also the fastening means shown in FIG. 10 permits simple exchange of legs on the doll, e.g. legs have different degrees of bending.

It will be understood that the foregoing description and drawings are only illustrative of the present invention and it is not intended that the invention be limited thereto. Many other specific embodiments of the present invention will be obvious to one skilled in the art in view of this disclosure. All substitutions, alterations and modifications of the present invention which come within the scope of the following claims or to which the present invention is readily susceptible without departing from the spirit of the scope of this disclosure are considered part of the present invention.

I claim:

1. In a doll construction: a body having a trunk; the lower extremity of said trunk comprising a single bulbous boss with substantially the entire outer surface thereof defining a smooth and continuously convex surface of generally spheroid shape; a pair of legs, each having a concave recess extending over substantially the entire inner surface of the upper portion thereof, each recess receiving a lateral portion of said bulbous boss and being substantially complementary in shape to the surface thereof with said recesses each covering about 20% to 40% of the external surface area of said boss; the juncture between the outer surface of the upper portion of each leg and the boundary edge of its recess defining a relatively thin edge and said outer surface being configured to substantially the same curvature as the adjacent exposed portions of said boss and adjacent portions of said body to define therewith a surface closely simulating the surface of the pelvic region of a human body; and fastening means holding said boss and recesses in frictional mating relation while permitting said legs to swing in any direction independently of each other on said boss and laterally outwardly from a fore-and-aft vertical plane bisecting said doll trunk up to an angle of at least about 40 degrees from said plane.

2. A doll construction as defined in claim **1** wherein the boundary edge of each recess lies in a plane at an angle in the range of about 5 degrees to 25 degrees to said fore-and-aft plane bisecting the doll trunk when the axis of said doll trunk and the upper portion of the legs adjacent to the doll trunk are oriented perpendicularly to a flat surface on which the doll is standing.

3. A doll construction as defined in claim **1** wherein said boss is substantially circular in vertical section in a plane extending laterally of said trunk.

4. A doll construction as defined in claim **1** wherein said boss is formed with its laterally opposed sides truncated.

5. A doll construction as defined in claim 1 wherein said boss is generally elliptical in horizontal section.

6. A doll construction as defined in claim 1 wherein said fastening means is a removable elastic garment worn by the doll and stretched to embrace said upper portions of said legs and said bulbous boss.

7. A doll construction as defined in claim 1 wherein said boss has an opening extending laterally therethrough the lateral extremities of which are substantially circular, a knob projecting from the central portion of each leg recess and into said bore, said bore and knob functioning to limit the said swinging of the legs outwardly of said plane and about the longitudinal axes of said legs to ranges natural to the human body.

8. A doll construction as defined in claim 1 wherein the ratio of the length of the doll trunk to the diameter of said bulbous boss is in the range of about 2.4 to 4.0.

9. A doll construction as defined in claim 1 wherein the front surface of said bulbous boss has a substantially larger radius of curvature than the remainder thereof.

10. A doll construction as defined in claim 1 wherein the rear surface of said bulbous boss has a central vertical crevice with rounded edges adapted to receive the rear edges of said upper leg portions when the legs are rotated rearwardly about their longitudinal axes.

11. A doll construction as defined in claim 1 wherein said bulbous boss has a horizontal cardioidal cross section.

12. A doll construction as defined in claim 1 wherein the side waist portion of the doll trunk adjacent said bulbous boss is undercut inwardly to accommodate the upper portions of the legs when said legs are swung outwardly of said plane.

13. A doll construction as defined in claim 1 wherein said legs are rigid, with the lower portion of one of said legs being bent, at the knee thereof, at an angle in the range of about 10 degrees to 30 degrees relative to the upper portion thereof.

14. A doll construction as defined in claim 13 wherein the lower portion of the bent leg is further bent laterally outwardly at an angle in the range of about 1 degree to 15 degrees to said upper leg portion.

References Cited by the Examiner

UNITED STATES PATENTS

130,068	7/1872	Parent	46—161
1,855,992	4/1932	Schaeffer	46—173
2,021,115	11/1935	Jackson	46—173 X
2,215,500	9/1940	Greneker	46—161 X
2,804,721	9/1957	Cohn	46—173
3,009,284	11/1961	Ryan	46—32

FOREIGN PATENTS

369,103 2/1923 Germany.

RICHARD C. PINKHAM, *Primary Examiner.*

DELBERT B. LOWE, *Examiner.*

B. F. Palmer,

Artificial Leg.

No. 4,834. Patented Nov. 4, 1846

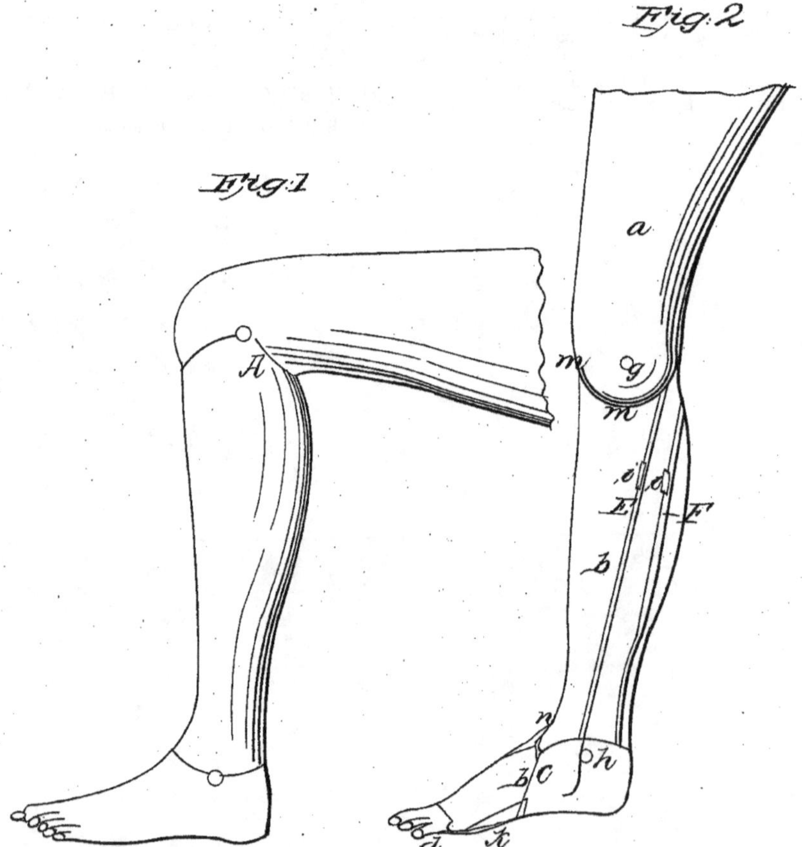

AM. PHOTO-LITHO. CO. N.Y. (OSBORNE'S PROCESS.)

UNITED STATES PATENT OFFICE.

BENJ. F. PALMER, OF MEREDITH, NEW HAMPSHIRE.

ARTIFICIAL LEG.

Specification of Letters Patent No. 4,834, dated November 4, 1846.

To all whom it may concern:

Be it known that I, Benjamin Franklin Palmer, of Meredith, in the county of Belknap and State of New Hampshire, have invented a new and useful Improvement in Artificial Legs and Feet; and I do declare that the following is a full, clear, and exact description thereof, reference being had to the accompanying drawing, which forms a part of this specification, in which—

Figure 1, is a perspective view. Fig. 2, is a vertical section on a line parallel with the foot.

The nature of my invention consists in new and improved articulations of the knee, ankle, and toes, in constructing the various parts in such new manner as to leave no opening in the exterior of the leg about the joints, and in supplying and arranging tendons and springs in such manner as to give more elasticity, strength, durability and freedom of motion to the limb than are to be found in artificial legs heretofore known and used.

The construction of my leg is as follows. *a*, *b*, *c*, and *d*, are made of any suitable material, *a*, being that part of the leg above the knee, *b* that part between the knee and ankle, *c*, the foot, and *d*, the toes. *a*, is hollow, and so constructed as to receive within it the stump of the natural limb (in case of amputation above the knee) which may rest in any desirable manner according to circumstances, *b*, is also hollow, and receives within it the natural limb in case of amputation below the knee. The lower end of *a*, is a hemisphere, (and may be varied in order to give a correct likeness of any limb at the knee) which rolls within the shell of *b*, upon the pivot, *g*, which passes through it equidistant from the points *m*, *m*. That portion of *b*, about the joint is made very thin in the anterior part, and cut away in the rear, to give sufficient motion to the knee, as shown by the line A, in Fig. 1.

z, is a smoothly polished metallic bolt which passes through an aperture in *a*, and through two metallic plates that are fastened, one upon each side of *b*. This bolt does not move within the metallic plates when the leg is in use, but may be taken out at any time, by means of a nut upon one end, or otherwise. Two other metallic plates are fastened to the lower part of *b*, (one upon each side) which pass down a little below its lower extremity, upon the sides of the foot, *c*. The bolt, *h*, passes through these metallic plates and the foot, in a proper place for the ankle. This bolt moves within the foot, but not within the metallic plates when the leg is in operation. That part of *c*, in the rear of the bolt, *h*, is so constructed that its upper extremity moves within the shell of *b*, (which is very thin at this place, and of any desirable shape) so closely as to leave no external aperture when the joint assumes all its various positions. Upon the instep is fitted a casing of green hide, which passes up far enough to conceal all cavity about the anterior part of the ankle, as shown at the point *n*. These articulations are free from that degree of friction attendant upon those heretofore in use, are not liable to contract and expand, do not rattle when moved quickly, are of much greater strength and durability, and together with the joint give a more perfect exterior to the limb. *c*, and *d*, are united by a socket joint, a semicircular groove in *d*, receiving a corresponding convexity on the front end of *c* in which it is secured by a wire passing through both and united at the ends or in any convenient manner.

E, and F, are tendons; both entirely in the interior of the foot and leg. E, is secured in a groove, or by means of apertures in the rear of *a*, near the lower end, and passing through *b*, is secured in two apertures in *c*, just forward of the pivot *h;* or it may be attached to the lower part of *b*, in any suitable manner. Its function is to stop the motion of the knee at a proper time, when the foot is moving forward. It also tends to obviate the too rapid downward motion of the anterior part of the foot at the touch of the heel, if attached to the foot forward of the bolt *h;* and it should be thus attached in all cases where there is a sufficient length of the natural limb remaining to govern the knee readily, as it gives better motion and more elasticity to the entire limb. This tendon is used only in case of an artificial knee. F, is attached to *b*, in any suitable manner, as far from the foot as circumstances will permit, and to the foot near the extremity of the heel. These tendons may be of gut, and fastened within the thimbles *i*, *i*. The function of this tendon is to act in conjunction with E, in giving elasticity to the limb, and to keep the front lower extremity of *b*, from coming in contact with *c* at the recess *j*,—k is a metallic spring concealed within the covering of the foot, one end of which is fastened to the bottom of *d*, just forward of the toe joint. A metallic wire or gut cord, *l*, is attached to the other end of this spring, which passes through the foot and is fastened in any suitable way to the anterior part of *b*. This spring answers a fourfold purpose. It acts conjunctively with the tendon E, in obviating the rapid and unnatural downward motion of the "ball" of the foot, when the heel touches the floor, it keeps the ankle in its proper position, it stops the motion of the ankle when it has moved sufficiently far, by coming in contact with the foot, and regulates the motion of the toes, moving them like natural ones at every step. The whole exterior of the limb may be covered with leather, which should be varnished. The limb may be attached in any of the usual forms.

Having thus fully described my improved artificial leg and foot what I claim therein as new and desire to secure by Letters Patent, is—

1. The long tendon E, the spring *k*, and the cord *l*, respectively combining and acting upon the parts *a*, *b*, *c*, and *d*, substantially in the manner and for the purpose herein set forth.

2. I also claim the improved manner of forming the knee joint uniting the parts *a* and *b*, to each other, by means of the hemisphere at the lower end of *a*, the partial concave beveled to a thin edge on the front side of the upper end of *b*, and the pivot *z*, combined and operating substantially in the manner herein set forth; for the purpose of obviating noise or friction in working the joint, and producing a perfect contour thereof.

3. I also claim the improved manner of forming the ankle joint uniting the parts *b*, and *c*, to each other,—the rear side of the lower end of *b*, being beveled to a thin edge passing over and inclosing the heel portion of that part of *c*, in the rear of the joint pivot *h*, and the front upper part of *c*, at *n*, being brought to a thin edge and overlapping the lower end of the front side of *b*, substantially as herein set forth,—thus forming a pliable joint that will work without noise, and preserve its contour in all positions.

BENJAMIN FRANKLIN PALMER.

Witnesses:
RICHARD CLEMENT,
BENGE. BORDMAN.

2 Sheets—Sheet 1.

L. L. LANGSTROTH.

Bee Hive.

No. 9,300.

Patented Oct. 5, 1852.

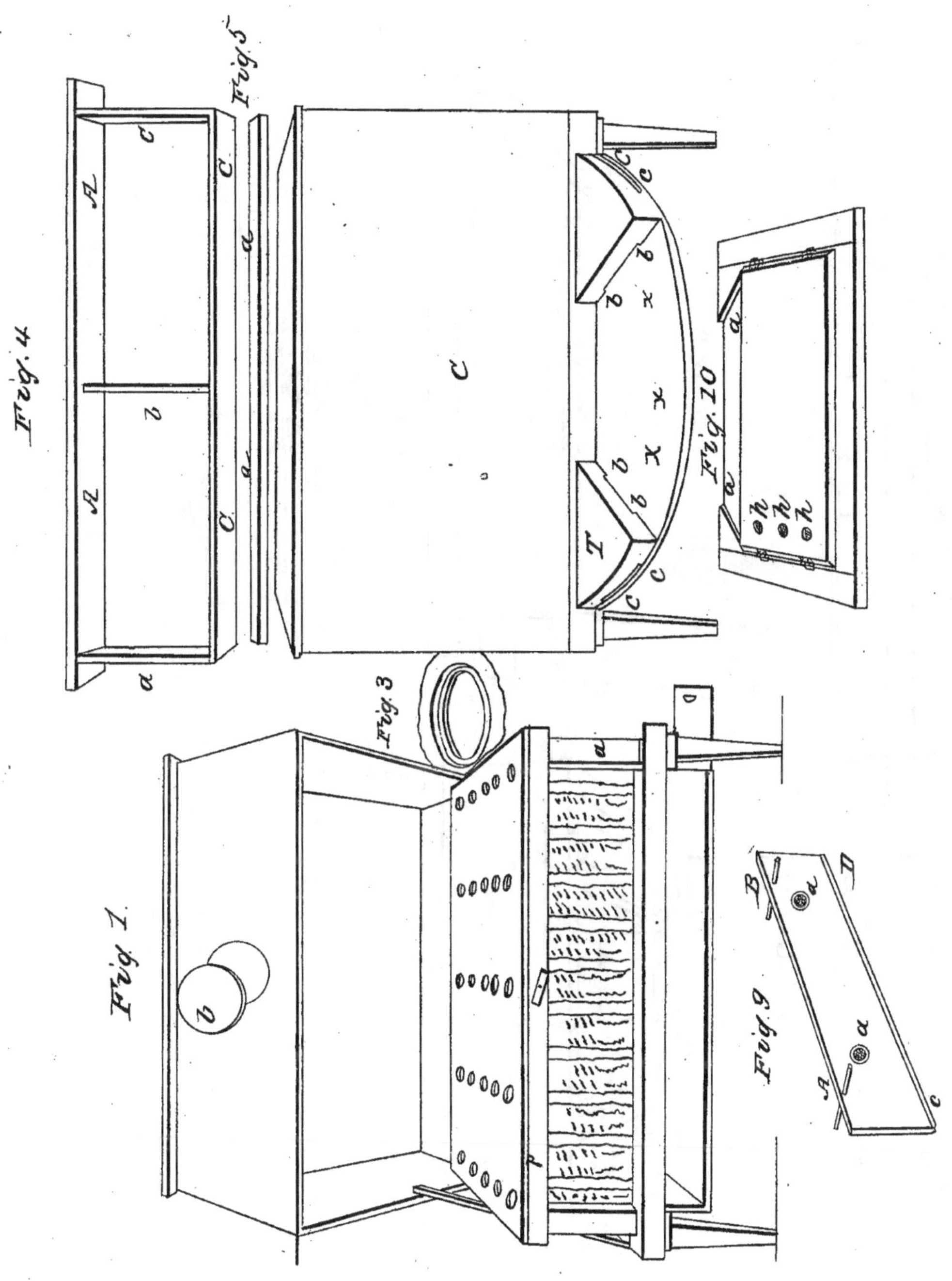

N. PETERS, Photo-Lithographer, Washington, D. C.

2 Sheets—Sheet 2.

L. L. LANGSTROTH.

Bee Hive.

No. 9,300. Patented Oct. 5, 1852.

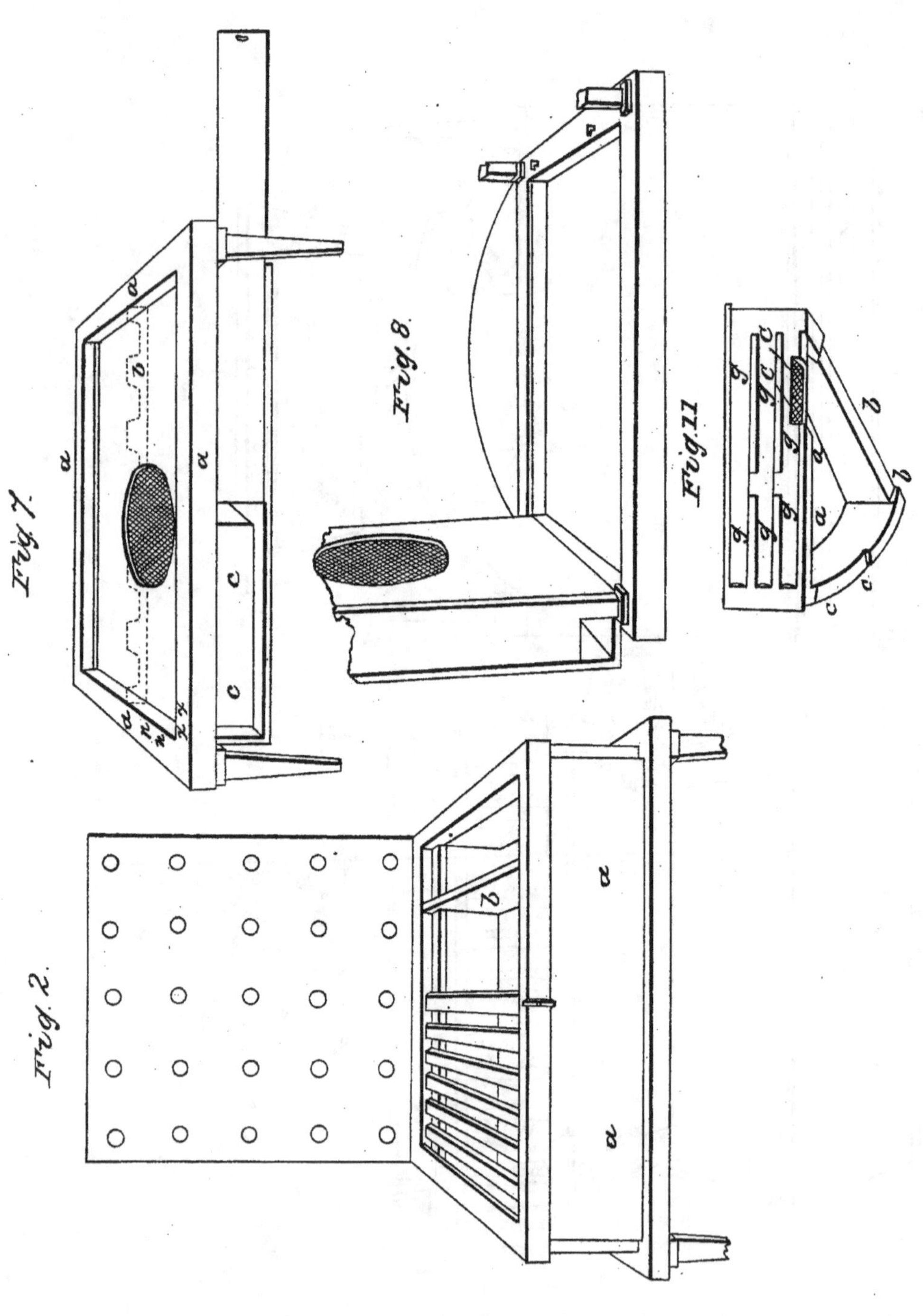

N. PETERS. Photo-Lithographer, Washington, D. C.

UNITED STATES PATENT OFFICE.

LORENZO L. LANGSTROTH, OF PHILADELPHIA, PENNSYLVANIA.

BEEHIVE.

Specification forming part of Letters Patent No. 9,300, dated October 5, 1852; Reissued May 26, 1863, No. 1,484.

To all whom it may concern:

Be it known that I, Lorenzo L. Langstroth, of Philadelphia, in the county of Philadelphia, State of Pennsylvania, have invented a new and Improved Mode of Constructing Beehives; and I do hereby declare that the following is a full and exact description thereof, reference being had to the accompanying drawings, and to the letters of reference marked thereon.

The nature of my invention consists: 1st, in affording the bees a more thorough protection against the bee-moth. 2d, in affording them a more effectual protection against extremes of heat and cold, sudden changes of temperature, and dampness in their hives. 3rd, in enabling the apiarian when desirable rapidly to multiply his colonies. 5th, in enabling him to obtain his surplus honey in the most convenient, beautiful and salable forms. 5th, in enabling him to perform all necessary operations without injuring a single bee. 6th, in enabling the most timid to remove the surplus honey without danger to themselves.

Movable frames.—To be able to remove the combs from the hive without mutilating them or seriously disturbing the bees, will secure the following advantages in the management of an apiary:—(1st.) The combs may at any time be readily examined for any purpose and thoroughly cleansed from the larvæ of the bee-moth. (2nd.) When the brood combs become too old, they may be renewed. (3rd.) Feeble colonies may be strengthened, by transferring to them from stronger colonies, combs containing honey and maturing brood. (4th.) Spare queens in embryo may be secured, either for creating artificial swarms or for supplying colonies which have lost their queen. (5th.) The queen of a hive may be easily caught, for any purpose. (6th.) Spare comb and honey may be easily removed. (7th.) Combs unsuitable for rearing workers may be removed and not allowed as in ordinary hives permanently to interfere with the prosperity of the colony. (8th.) New colonies may be multiplied to an extraordinary extent, the apiarian being made independent of the uncertainties and perplexities of natural swarming. In short to be able easily to remove the combs, and to transfer them from one hive to another, is indispensable to the scientific and most profitable management of bees.

Apiarians have aimed to effect these ends by placing movable bars or slats in the top of the hive, to which the combs may be attached. There are two difficulties in the practical working of these bars which have hitherto prevented their general use:

(1st.) The bees often attach the same comb partly on one bar and partly on another thus preventing its easy removal. (2nd.) They are compelled to attach the combs to the sides of the hive and these attachments must be severed before the bars can be removed. This often seriously annoys the bees and wastes the time both of the apiarian and the bees who must with much labor refasten all such combs in whatever hives they are put. To obviate these disadvantages I have invented a movable frame or compound bar—see, drawing Figure 4. A, is about one inch wide and one-quarter of an inch thick—C, C—*a*, *b* and *c*, are about one-quarter of an inch thick and about one inch wide. These should be about three eighths of an inch space between *a* and *c*, and the sides and C C and the bottom board of the hive this will prevent the bees from attaching the frame to the sides or bottom board of the hive, hindering its easy removal, and will allow them to pass freely between the sides and the bottom board, and the frame so as to afford no lurking place for moths or worms.—A series of such frames are placed as in Fig. 2—about half an inch apart on rabbets made in the front and back of the hive or in any other convenient way.

Fig. 5, *a*, *a*—is a small strip placed over the frames to hold them down, and to prevent the bees from attaching the ends of the frames to the blocks between which they rest, thus destroying the play which is necessary to their easy removal. This strip is just as wide as the rabbets on which the frames rest—(about half an inch)—and as thick as the depth of the rabbets below the sides of the hive and the top bars of the movable frames—the object of which depth is hereafter described. The construction of these frames is such as to induce the bees to attach each comb to a single frame. If I have clean worker comb I attach a small piece by dipping it into melted wax to A, and

A. If I have none I draw a line of melted wax lengthwise across the center of the inside surface of A A, a and c—and on both sides of b. The bees commencing their combs at f—f— will build them in a true direction toward b, in order to attach them to it—and as they cannot fasten them to the sides of the hive, they will carry them also in a true direction toward a, and c. The bees in attaching their combs to the common bar will frequently make slight deviations from the line of the bar, as they are able to attach the combs anywhere to the sides of the hive. These deviations soon lead to the building of combs partly on one bar and partly on another. As the combs are never attached to the sides or bottom of the box or hive which contains my movable frames. they may at all times be easily removed. If the operation is carefully performed the bees seldom resent it but seem to be intent only on filling themselves with honey with which to return to their hive. If they exhibit any signs of anger a few whiffs of smoke from burning rags or paper or a slight sprinkling with sugar water will at once quiet and subdue them.

As there is a stratum of air always interposed between the combs and the sides and buttom board of the hive the bees are much more effectually guarded against extremes of heat and cold and the pernicious effects of condensed moisture than they can be in hives of the usual construction. These frames may be easily put together with small brads. It is obvious that the mode of constructing and inserting them in the hive may be varied, while the same important ends may be secured.

The rabbets on which the frames rest may be made about three quarters of an inch wide so as to give them a proper support and yet leave sufficient play to loosen them when they are to be removed from the hive and to regulate at will the distances between the combs.

Shallow chamber or covered way.—The arrangement for taking the spare honey of a hive should possess at least the following requisites: (1st) It should allow the bees, together with the heat and odor of the hive, to pass from all parts of the main hive, with the fewest impediments, to the receptacles for storing surplus honey. (2nd) It should be as well adapted to secure the storing of surplus honey in small, as in large receptacles. (3rd) It should enable the apiarian to remove the spare honey without injury to the bees, or danger to himself. (4th) It should admit of enlarging or contracting the space for storing surplus honey, so that the hive may be a swarming or non-swarming hive at the pleasure of its owner. I have invented a shallow chamber or covered way, which secures these and other advantages.

The bees, before entering the spare honey receptacles, are made to enter a chamber about half an inch deep, or just shallow enough to prevent them from occupying it with comb, they pass from this chamber, by suitable apertures, into the honey receptacles. This shallow chamber may be connected in various ways with the main hive, but to answer all its purposes to the best advantage, it ought to be over the top of a hive, furnished with bars. Fig. 2, shows the way in which I usually construct this shallow chamber. The rabbets which receive the movable frames or bars, are deep enough to form this chamber between the frames and the cover A. This cover is elevated in Fig. 2, and is shown in its proper position for holding the spare honey receptacles, in Fig. 1. It should be about half an inch thick, and clamped to prevent warping. It is fastened to its place by screws or hooks. Fig. 3, shows one of the apertures admitting the bees from the chamber into the honey receptacles. It is made by boring to a slight depth into the cover with a circular bit, and then through, with a smaller one so as to form a sunken ledge for a piece of zinc cut out with a punch, or anything else suitable for a cover. Any number of these or any other apertures, of any size, may be made in this cover, without, as in other hives, interfering with the arrangement of the combs below, or being liable to be more or less obstructed by the bees, and they made by opened or shut each one independently.

The bees will be able to pass, with the greatest ease, from between all the ranges of comb, into this chamber, just as though it were the upper part of their main hive, and from the chamber to any of the honey receptacles, instead of losing much valuable time, in the height of the gathering season, by crowding through a few inconvenient passages, as in the common hive.

When surplus honey is taken from the bees in large quantities at once, they are often so discouraged, that they refuse, for a long time, to enter the receptacles again, whereas if a few small receptacles are removed from time to time, they are stimulated to increased activity. The shallow chamber while it greatly facilitates the storing of spare honey in large boxes, is particularly adapted to the securing of it in small receptacles, in the most beautiful, convenient and salable forms. If desired, it may all be stored in quart, pint or even half pint tumblers, from which it may be removed, without being disfigured, in ornamental forms to grace the table.

There are other and more important reasons why the common arrangement for taking spare honey, cannot be applied successfully to small receptacles: (1st.) The building of comb requires a great amount of animal heat to be maintained by the bees, and hence they always work best when thep can economize their heat to the best advantage. But this they cannot advantageously do in tumblers or small receptacles communicating with the main hive through apertures made in its solid top or sides, because such apertures do not freely admit the heat from the main colony, and the bees are too few in number to keep up the requisite temperature. The shallow chamber, like the part of a warm room nearest to the ceiling is itself, in the working season, always full of the warmest air of the hive, and keeps the small receptacles filled with the same.

(2nd) Bees are always anxious to work in large numbers, so that they can easily intercommunicate with each other. The common arrangement for inducing them to work in small receptacles, is opposed to this instinct. But the shallow chamber affords a place of repose for multitudes of bees engaged in elaborating the wax to be used in the small receptacles above. As soon as a bee has a pellicle of wax formed in its wax pouch, it ascends into a receptacle to assist in comb-building. A constant succession of bees are thus ascending and descending, and they work in the small receptacles with scarcely any more isolation, and with almost as much rapidity, as though they were engaged in filling their main hive.

If a sufficient number of receptacles, or sufficient space in a few, is given to the bees—they will ordinarily be prevented from swarming. The honey may be removed in these receptacles, without any injury to the bees, or danger to their owner. If a slide of tin or zinc is pushed between the receptacle and the opening into the shallow chamber, and allowed to remain for five minutes or longer, the bees perceiving that they are separated from their companions, will gorge themselves with honey to be carried back to the hive. Then, if they are allowed to fly, they seek only to escape to their home, a bee when filled with honey, never, under any circumstances, acting on the offensive. This shallow chamber answers other important purposes. It prevents the bees from fastening with their pupolis the cover of the chamber, to the frames or bars of the hive—and thus enables it to be easily removed, to give access to the combs. It gives a space of air between the combs and the top of the hive, to guard the bees against extremes of weather, and sudden changes of temperature. It enables us to keep the feeder, in cool weather, filled with the heat of the hive—and it furnishes a warm and accessible place for feeding destitute colonies, in winter, with sugar or barley candy—which has been found to be the best and cheapest winter food for them.

By leaving one of the apertures in the cover of the chamber open during the cold weather, the moisture from the hive will escape, instead of condensing as it does upon the solid top of the common hive, to the great injury of the bees. Having already described the movable frames, it will be seen that the combs in my hive are kept some distance from the sides, top and bottom of the box or hive which contains them, and are thus, to a very extraordinary degree protected against moisture as well as sudden changes of temperature.

When small receptacles are used for storing the surplus honey, they are not near so apt to be occupied by the queen for breeding—as she generally prefers to keep her brood in large masses of comb.

In tumblers or round glasses just small enough to prevent the building of more than one comb—from the cylindrical shape of this comb—the cells will be of such various depths that they will from this cause be less likely to be occupied with brood.

Divider.—When a small swarm is put into a hive of the usual size, being unable to concentrate their animal heat, they cannot work to the best advantage. If they are placed in a small hive incapable of enlargement, the next season if not sooner, it is found to be too small. To obviate these difficulties, I have invented a divider to be inserted from the top of the hive between any of the ranges of comb, and to extend down to the bottom board of the hive. By its use, the hive may be adjusted to the size of the swarm, and yet may at any time be restored to its usual dimensions. This divider is also of great importance in rearing young queens, as a colony may be divided by it, and the part without a queen be compelled to rear young ones.

It is very serviceable in compelling destitute swarms which have not filled their hives with comb, to store up the food fed to them in the fall, instead of consuming it to build new comb. It will also enable them the better to concentrate their animal heat in winter, by shutting them off from the empty part of the hive. Fig. 9 shows the shape of this divider, A and B are the pins which rest upon the frames or bars. The bottom C D is chamfered to an edge, to prevent the crushing of the bees when the bottom board is opened and shut. *a*, *a*, are small holes covered with perforated zinc which may be screwed so as to open or shut these holes. By means of these ventilators new colonies put into different parts of the hive and separated by the divider may be made to have the same scent so as to be readily united—and colonies which have been divided into two or more families, so as to induce the part or parts without a queen to rear young ones, having the same odor may at any time be united so as to form again one vigorous stock.

Late in the fall several feeble colonies may be put into different divisions in the hive, and in a few days allowed to unite so as to form one vigorous stock—or they may be preserved as independent families, and the hive thus made up of several colonies may be used as a nursery for rearing queens for artificial swarming. Separate entrances can easily be given one on each of the four sides of the hive—so that four colonies may be made to tenant the hive. This divider may be made of thin wood or of zinc or glass or any other suitable material. Fig. 2[b] shows a divider with shoulders resting on the rabbets which support the bars when movable frames are not used or when no strips are used to confine the frames.

Platform.—The mode of supporting the hive should possess, at least, the following requisites: (1st) It should facilitate the easy handling of the hive for safe transportation or any other purpose. (2nd) It should allow it to be easily reëstablished in any new place without delay and expense in changing fixtures. (3rd) It should permit the bottom board of the hive to be easily opened and shut without injury to the bees. (4th) It should admit of the safe and speedy hiving of new swarms. (5th) It should afford no crevices for lurking places to the bee moth or its larvæ. I have invented a platform or mode of supporting my hives which possesses these and other desirable requisites. As this platform supports the hive and all its fixtures I shall describe in connection with it such parts of the hive as have not been previously, or are not to be hereafter, described.

Fig. 7, *a*, *a*, *a*, *a*, is a frame about one inch and a half thick, mortised or otherwise firmly put together. Its inside length and breadth are exactly the same with those of the box, Fig. 2, containing the combs. This box must be fastened down to the frame or platform, so that there may be no crevices between them, and so that the platform adds its own depth to the depth of the box. The box may be made of wood, but I very much prefer, for several important reasons, to have it of glass, set in a frame, Fig. 1. The posts *a* of this frame are about inch an a half square (their height depending upon the capacity desired to be given on the hive). They are mortised or strongly fastened into the platform, on its corners, so as to be just even with the inside edges of this platform *x*, *x*, *y*, *y*, Fig. 7. The top parts of this frame, are halved or rabbeted at the ends, so as to be firmly fastened down on the tops of the posts, by nails or screws. They are inch and a half wide, and thick enough to give a firm support to the frames after the rabbets have been made in front and rear for the movable frames and shallow chamber. The clear space between these posts, and between the under sides of the rails and the platform, should exceed by about one eighth of an inch the length and breadth of the panes of glass, so that there may be no difficulty from slight variations, of fitting them in. These panes of glass are then pushed in to the inside edges of the platform, and are prevented from going in any farther by small tins or brads. The bottom edge of the glass rests upon the platform. Thin strips of wood, about three quarters of an inch wide, are pushed in against the glasses, on the tops and sides, so as to close all apertures, while they form a support against which a second set of glasses, of the same size with the first, abut. These are fastened with small pieces of tin. A small amount of paint put upon the edges of the thin strips will keep the spaces between the glasses air tight. This arrangement enables us with ease to remove and replace broken glass in hives occupied by bees.

The extra cost of glass over wood is not great, and is much more than compensated by its advantages:

(1st.) The rational curiosity of the apiarian is gratified, and he is able at all times to ascertain the condition of his colonies.

(2nd.) A very great degree of protection is given to the bees against extremes and sudden changes of temperature. The space of perfectly confined air between the double glass, is one of the very best non-conductors of heat, and thus obviates the objection commonly urged against the use of glass. I have used double glass, for many years, and have found the bees to be better protected in such hives than in any others, unless where the wood is very thick, or doubled, and then the hives are both clumsy and expensive. Small covers of wood may be placed against the posts (see *a*, *a*, Fig. 2) and fastened by a button, leaving about half an inch space between them and the glass. This space, for greater protection, may be filled with tow, waste cotton. or any good non-conductor of heat.

(3rd) Another very important advantage of the double glass sides is the great protection which they furnish to the bees against the injurious, and often fatal effects of condensed and frozen moisture. A strong colony of bees is seldom if ever killed by cold alone—even in the thinnest hive. Their hives, however, are often filled with frost, and the bees being prevented from passing to the frosty combs containing their stores, perish, even in hives abundantly supplied with honey. Where the whole colony is not destroyed—the dampness, from the condensed moisture and thawing frost in common hives, often causes the combs to mold, destroys many bees and injures seriously the health of the whole colony. By means of the movable frames the comb in my hives are prevented from being attached to the sides, as they are by the common arrangement. If, then, the moisture can be prevented from condensing any where over the bees, where it may fall upon them or their combs—and if wherever it condenses it may be easily discharged from the hives—it cannot seriously annoy them.

The arrangement used in my hive, or one on substantially the same principles, is I believe the only one by which the bees can be thoroughly protected at all seasons, from the injurious effects of dampness in their hives—which in climates of cold winters is an enemy scarcely less formidable than the bee moth itself: (1st) My plan by giving them uncommon protection against extremes and sudden changes of temperature—and by affording a vent above the hive for the escape of moisture, diminishes as far as possible the condensation of moisture in the hive. (2nd) By making the sides of the hive of glass, they are always the coldest parts of the hive, and the moisture will be condensed upon them, in preference to any other part of the interior—just as it is upon the walls rather than the floor or ceiling of a cold room. But as the bees are kept away from the sides of the hive—this moisture cannot annoy them, or wet their combs. It cannot penetrate glass as it does wooden sides—to cause a protracted dampness—but runs down their smooth surfaces, and may thus be easily discharged from the hive.

I have thus succeeded in making the worst property of glass, its rapid conduction of heat—one of its best, for the purposes of a bee hive—and in successfully employing a material which unites so many desirable requisites.

The inside dimensions of the box containing the combs, as I usually construct it, are as follows: Lenth, eighteen inches and one eighth. This will give room for twelve movable frames. Breadth, twelve inches and one eighth. The extra eighth of an inch is to enable me to have sufficient play for slight variations in the glass. Depth below the rabbets nine or ten inches—including in this the depth of the platform. Whatever shape is used the size of the movable frames should always be uniform so that they may fit in every hive in the same apiary.

There must be room, on the upper surface of the platform *a, a, a, a,* Fig. 7, not only for the box which has just been described, but for an outside cover of this box, made of about three quarter inch stuff—between which and the box, there ought to be about an inch of clear space. Fig. 6 shows this cover in its proper position on the hive. It ought to be in the clear, not less than seven inches higher than the board which covers the shallow chamber, and while the top of it projects slightly on all sides, it should slope from the front to the rear, to carry off the rain, and at the same time discharge it away from the alighting board.

In Fig. 1 a small opening with its cover is shown on the back of the cover to the hive, which may be used when it is desired to have a current of air through the hive, to carry off moisture or for any other purpose. It gives access also to a small button which fastens down the cover. A lock may be used where it is judged necessary.

Fig. 10 shows the under side of the platform; *a, a,* is a sunken entrance for the bees, about three eighths of an inch deep, which is covered by a projecting alighting board, extending about eight inches in front of the hive. It is screwed to the front under surface of the platform, Fig. 8, and is shown in its proper position, Fig. 6. It should be well clamped to prevent warping

The bottom board to the hive may be a box, as is represented in Fig. 8, where it is dropped, or Fig. 7 where it is closed. It has gauze wire on its upper surface next to the bees, Fig. 7, and two sides (only one of which is seen in the figure) to regulate the admission of air. *c, c* is a block which may be pushed into the box, just far enough to allow air to pass between it and the slide, so that when the slide is shut far enough partially to hide the block, the bees may have air given to them, without light. This box bottom board will give the bees great protection against extremes and sudden changes of weather. By filling it in winter with sawdust, or some other good non-conductor of heat, still greater protection may be given.

Fig. 10 shows a solid bottom board made of inch and a quarter stuff, and well clamped to prevent warping. The front edge of the bottom board and the back edge of the alighting board are chamfered off to an edge, so that there may be no harbor for moths or worms. *h, h, h,* are holes made like Fig. 3, the ledge is sunk deep enough to hold gauze wire ventilators, and a cover like that of Fig. 3.

Ventilation beyond what is effected by the bees themselves, is often of much importance in the hands of the skilful apiarian, to the ignorant and careless it is worse than useless. Such are advised never to open the ventilators *h, h, h,* unless when their bees are shut up, for any reason, in the hive.

The bottom board (whether a box or solid) should exceed or fall short of by about one-eighth of an inch, the inside dimensions of the platform and is hinged so as to project all around about one-sixteenth of an inch over the under side of the platform or to recede the same distance from it. Fig. 10 shows the arrangement when it recedes, Fig. 8 when it projects. In Fig. 8 buttons are shown which keep it short, in Fig. 10 hinges which are fastened by a movable pin. As the bottom board is clamped on the ends where the fastenings are put it will not shrink so as to disarrange them and if there is any shrinkage from front to rear this may be easily remedied, by very slightly shifting the hinges or the alighting board. Fig. 8 shows the place for attaching the legs. These legs are about eight or ten inches long. They firmly support the whole superstructure, and enable the apiarian to move it about with ease and safety, and to reëstablish it in any place that he chooses, without the necessity of additional fixtures. By making the back legs about one and a half inches longer than the front ones (the backward slant of the outside cover must be greater than this) the whole hive will slant forward, and the bees will have all the advantages of hives whose bottom boards are inclined planes.

The arrangement which my platform enables me to give to the bottom board and entrance secures several important advantages:

(1st.) The outside cover of the hive may be raised without disturbing the bees that are entering or leaving the hive, and the very act of elevating it for inspecting the hive shows a protecting shield between the timid apiarian and the bees. Fig. 1 shows a rod which rests in a notch on the platform, and which supports the cover in its elevated position.

(2nd.) The entrance being a covered one, and inclining forward effectually prevents the wet from beating into the hive.

(3rd.) The bottom board, whether projecting or receding, may be easily opened and shut without hurting the bees and without affording the lurking places usual in common hives for moths or worms, while at the sime time, it is not below the level of the entrance so as to compel the bees at great loss of time, to drag their dead or anything that they carry out of the hive "up hill."

(4th.) When the bottom board projects, the moisture will run out between the chamfered edges of the bottom board and the alighting board. When the bottom board recedes the moisture will be discharged at once, from the cold sides of the hive on which it has condensed, without wetting the bottom board. The small amount of air all around admitted by this opening, will tend to keep the hive dry. Hives which are shut up too closely in cold weather are generally so damp as to cause the comb to mold.

(5th.) The platform arrangement in combination with the bottom board and entrance, enables me to give uncommon protection to the hive against the moth and worm while all the other advantages enumerated are preserved. There are no crevices large enough for moths or worms to harbor between the bottom board and the platform, whether the bottom board is a projecting or receding one and as the platform itself stands upon legs, there are no places between it and its supports, where moths or worms can secrete themselves.

During the whole of the breeding season, the bottom board should be lowered from time to time (this, from its construction, may be done without injury or serious annoyance to the bees,) and cleansed of all the particles of wax, bee-bread, etc., which fall upon it, and which are the favorite places for the eggs of the bee-moth and the nurture of her young until they have attained sufficient size to ascend into the combs. The timid apiarian by using the receding bottom-board, may dispense with lowering it. Let him pass his knife in the space all around the bottom-board and keep it clear of pupolis, and as there will be no crevice between the platform and the bottom-board which is inaccessible to the bees, the moths and worms will have no harbor. In the winter season, the bottom board should be dropped occasionally, and cleared of dead bees, or else when the bees carry out their dead—if there is snow on the ground—falling with them as they usually do, before they detach themselves from them, they are chilled and perish, and thus large numbers are lost at that season when the preservation of every bee is important. As the bees are quiet in cold weather, this operation may be performed by the most timid.

(6th.) When a new swarm is to be hived, the bottom board may be dropped so as to form an inclined plane up which the bees will readily ascend into the hive, so as to be easily and quickly hived, without any risk that any of them will be crushed or that their queen will be killed.

(7th) The platform arrangement gives a cheap support to my hives. As each hive has its own cover, and will stand firmly upon its legs, there is no necessity for expensive bee-houses and other cumbersome fixtures to shelter and support the hives. Bees, if properly protected, flourish best in the open air, and the bee houses and other fixtures in general use are favorite places of concealment for moths and worms.

The bottom board and platform should be made of thoroughly seasoned wood, and should be well painted with the Ohio Mineral Paint or some other paint possessing its valuable properties and containing no white lead, so offensive to bees. As the mineral paint thoroughly protects the wood against moisture, and soon hardens to a species of slate or stone, defying the ovi-depositor of the bee moth, I advise that all cracks in the interior of the hive and all corners or places which are not perfectly air tight and which the bees are wont to fill with pupolis (a favorite nidus for the eggs of the moth,) should be filled with this same paint. The outside cover of the hive and all the wood exposed to the weather should be thoroughly painted with the same material.

Combined moth and worm trap.—A trap to answer its professed object should possess at least the following requisites: (1st.) It should be near the only entrance by which the moth can get into the hive—and should during the moth season be filled with the heat and smell of the hive—or else the moth will avoid the trap. (2nd.) It should be made inaccessible to the bees, or else the moth will not so readily enter it. (3rd.) It should not allow the moth to pass directly from the trap into the hive, or else she will often enter the hive, instead of stopping to lay her eggs in the trap. (4th.) It should be so made and placed as to entrap the full grown worm when it leaves the hive to find a seucre place for spinning its cocoon, while at the same time it furnishes an inviting place for the moth to deposit her eggs. (5th.) It should be simple, requiring but little time for its management. I have invented a trap containing these and other desirable requisites.

In Fig. 6 the trap is seen in its proper position on the alighting board. Two are used for a hive—and they are made of wood or any other suitable material. Under the traps T, T, are cavities for holding a small piece of old comb—the refuse from the bottom boards or anything else in which the moth will deposit her eggs. In Fig. 11 —*a*, *a*—this cavity is shown. *b*, *b*—*c*, *c*, are entrances for the moth about one eighth of an inch deep, too small to admit a bee. —*d*, *d*— is a piece of gauze wire admitting the heat and odor of the hive from the covered entrance to the trap, and yet excluding the moth from the hive. From the entrances *b*, *b*, and *c*, *c* Fig. 6 the heat and scent of the hive are constantly proceeding to attract the moth. These entrances resemble the crevices between the bottom board and hive, by means of which the moth so frequently gains admission to the common hive. If the moth alights anywhere upon the alighting board *x*, *x*, *x*—Fig. 6— her instinct leads her at once to glide to the openings *b*, *b*, instead of passing on to the entrance of the hive over a surface, which, from the shape of the trap, is constantly narrowing, so as to be more and more strongly guarded by the bees. If the bees are not numerous, or the moth is very troublesome, the traps may be moved closer together, so as to contract the entrance and guard the hive more effectually. The full grown worm or larva of the bee moth, finds no suitable place in a well made hive, occupied by a flourishing colony, for spinning its cocoon, to undergo its transformation to a moth. In leaving the hive to find such a place it will, in accordance with its instincts, make for any crevices admitting of a hiding place, just as it crawls between the edges of ordinary hives and their bottom boards. Grooves may be made in the trap— Fig. 11—*g*, *g*—or on the front edge of the platform against which the trap abuts—or anywhere else, protected by the trap from the bees, and yet made easily accessible to the worms, into which the worms will readily enter to spin their cocoons.

The traps should be inspected about once a week during the moth season—and the worms and eggs destroyed. These traps serve other useful ends—by moving them closer together, the entrance to the hive may be contracted to guard against robber bees or entirely closed, and yet air will be admitted under the traps to the hive.

The peculiar shape of the trap, at all times, directs the bees to the entrance—and thus saves much time lost in searching for it—while it enables the apiarian to enlarge or contract the entrance to the hive, without perplexing the bees or wasting their time. Although the bee returns to her home with unerring precision from her flights abroad, yet unless the entrances are very judiciously arranged, they lose much valuable time in searcing for them after they have alighted. This is more especially the case, when the entrances are altered in any way.

These traps may be constructed in a great variety of ways and yet be made to answer substantially all their ends, provided their triangular shape is preserved. Simple triangular blocks will answer every purpose as a guide to the alighting bees and a protection against robbing bees.

What I claim as my invention, and desire to secure by Letters Patent is.

1. The use of a shallow chamber substantially as described, in combination with a perforated cover for enlarging or diminishing at will the size and number of the spare honey receptacles.

2. The use of the movable frames A, A, Fig. 4—or their equivalents—substantially as described—also their use in combination with the shallow chamber with or without my arrangement for spare honey receptacles.

3. A divider substantially as described in

combination with a movable cover allowing the divider to be inserted from above between the ranges of comb.

4. The use of the double glass sides in a single frame substantially as and for the purposes set forth.

5. The construction of the trap for excluding moths and catching worms, so arranged as to increase or diminish at will the size of the entrance for bees, substantially in the manner and for the purposes set forth.

L. L. LANGSTROTH.

Witnesses:
E. D. Saunders,
W. J. P. White.

[First Printed 1912.]

Jan. 30, 1968 J. C. VITTONE 3,366,411

VEHICLE FOR OFF-THE-ROAD SERVICE

Original Filed Jan. 8, 1964 4 Sheets-Sheet 1

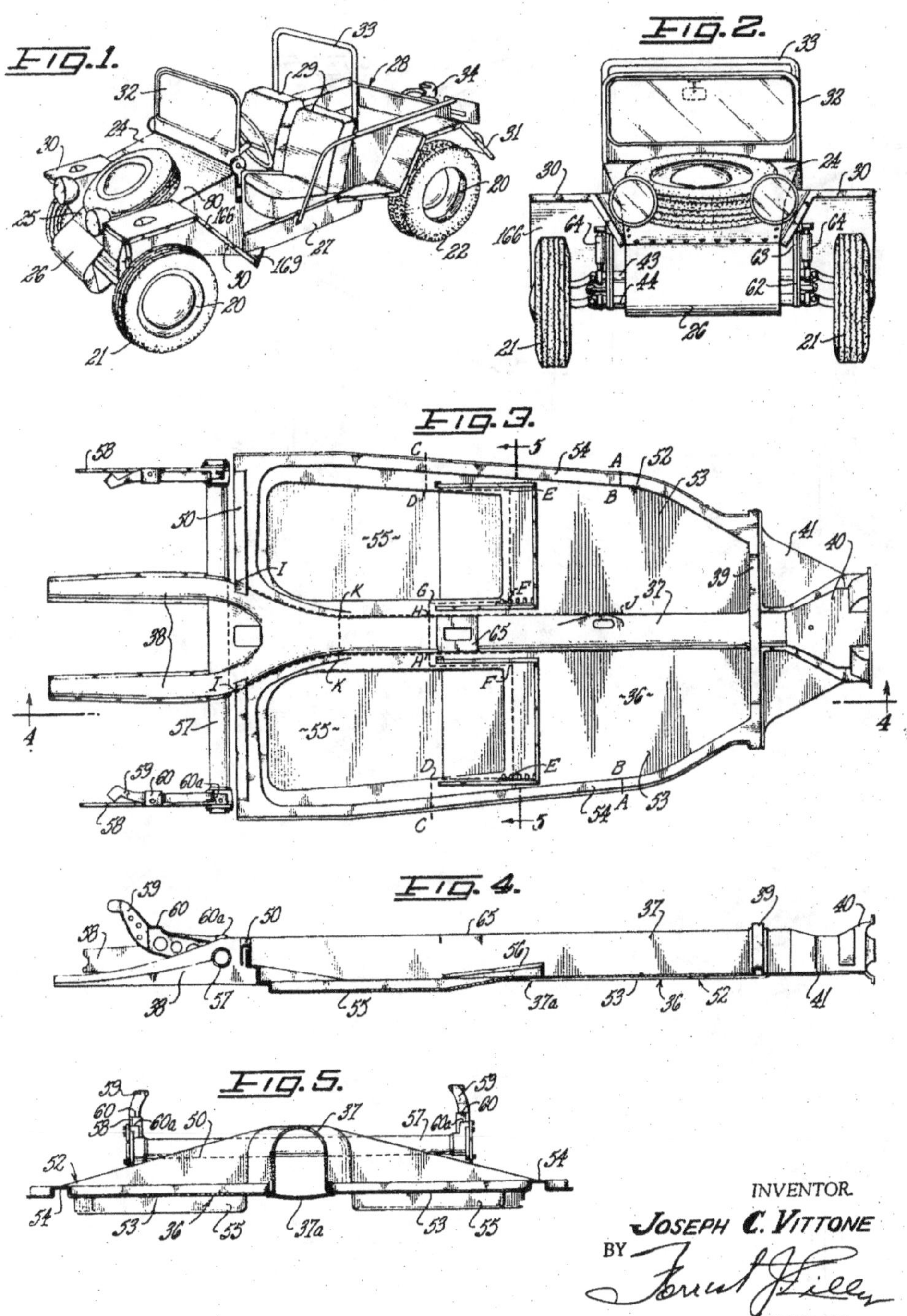

VEHICLE FOR OFF-THE-ROAD SERVICE

Original Filed Jan. 8, 1964

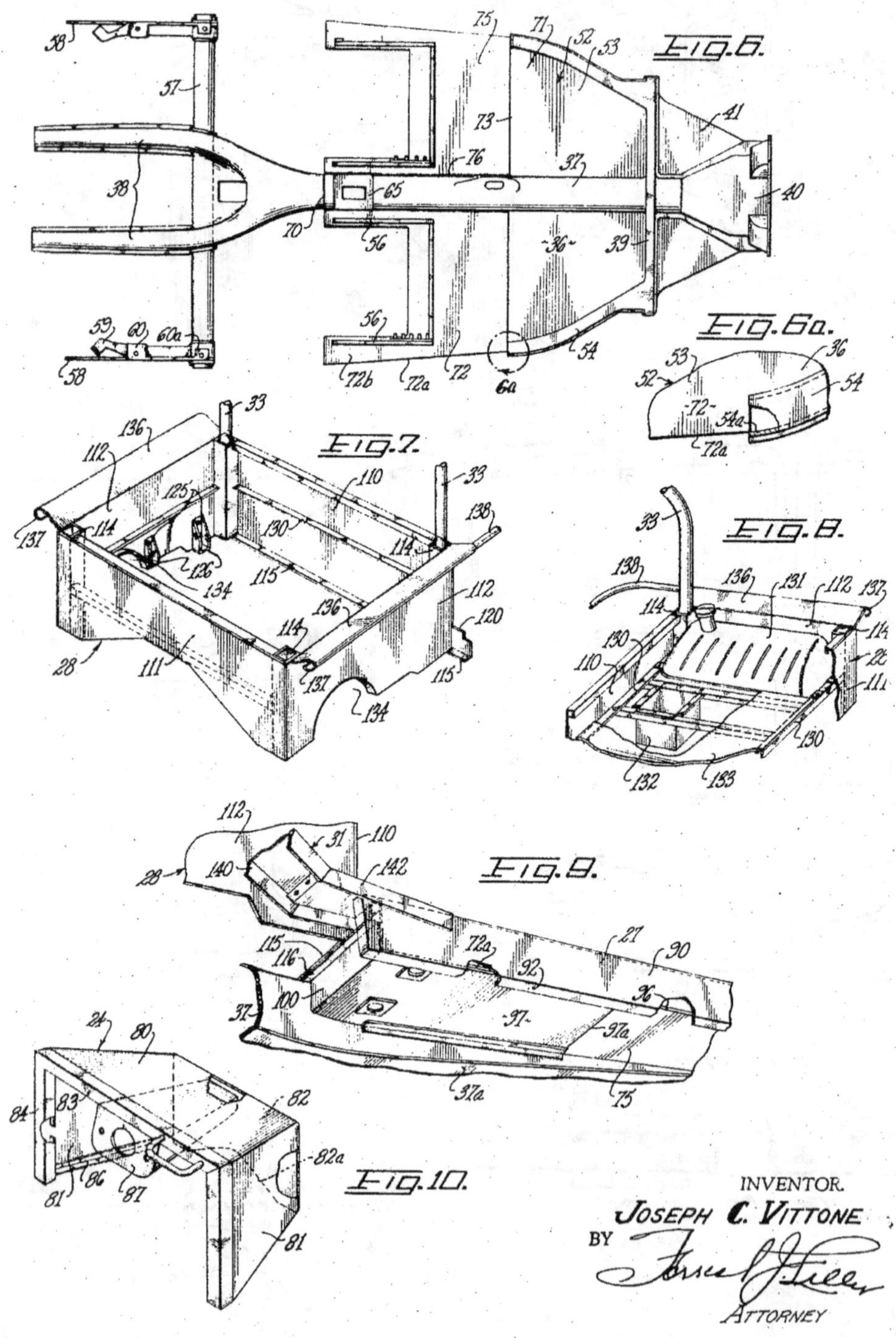

INVENTOR.
JOSEPH C. VITTONE
BY
ATTORNEY

Jan. 30, 1968 J. C. VITTONE 3,366,411

VEHICLE FOR OFF-THE-ROAD SERVICE

Original Filed Jan. 8, 1964 4 Sheets-Sheet 3

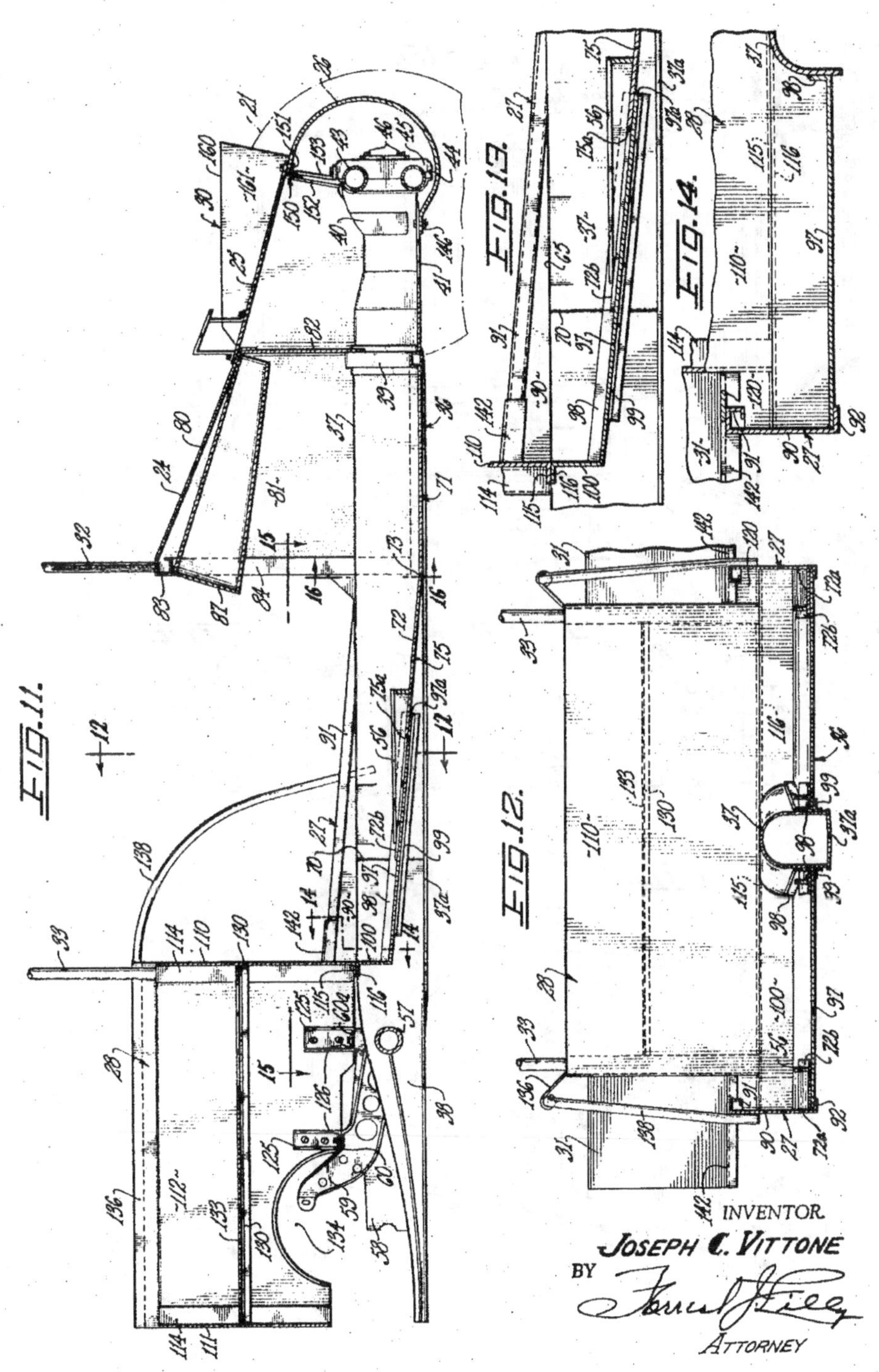

Jan. 30, 1968 J. C. VITTONE **3,366,411**

VEHICLE FOR OFF-THE-ROAD SERVICE

Original Filed Jan. 8, 1964 4 Sheets-Sheet 4

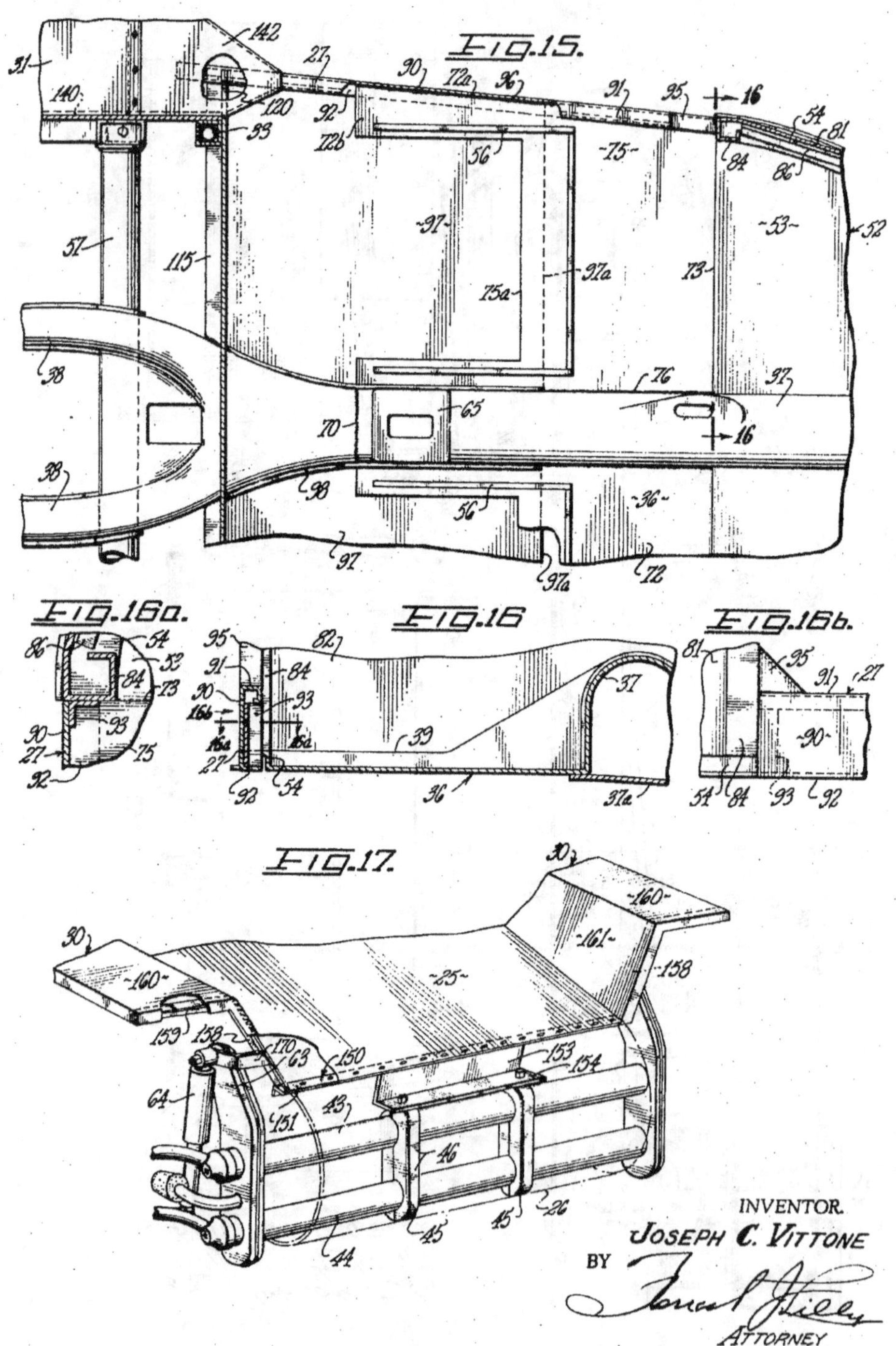

United States Patent Office 3,366,411
Patented Jan. 30, 1968

3,366,411

VEHICLE FOR OFF-THE-ROAD SERVICE

Joseph C. Vittone, P.O. Box 169,
Riverside, Calif. 92502

Original application Jan. 8, 1964, Ser. No. 336,463. Divided and this application Mar. 17, 1966, Ser. No. 554,234
9 Claims. (Cl. 296—28)

This application is a division of my application entitled, Method of Reconstructing a Frame, Ser. No. 336,463, filed Jan. 8, 1964, now abandoned.

This invention relates generally to motor vehicles, more particularly of the off-the-road type, intended for use over rough and rugged roadless terrain, up steep hills, over rocks, dry washes, through gullies, or on bad or severely rutted roads.

For such uses, special vehicles have been constructed in the past, and have usually been of the four-wheel-drive type. Vehicles for this type of service are generally fairly expensive.

A general object of the present invention is the provision of a light and inexpensive two-wheel-drive vehicle for off-the-road service of the kind mentioned.

The vehicle contemplated by the invention is one which may be driven on or off the highway, and when off the highway, is one which can be driven through rough and hilly back country, either for the sport of negotiating difficult terrain in a power vehicle, or to penetrate back country areas otherwise accessible only by long and hard hiking or by horseback, or possibly by special motorcycle.

The vehicle contemplated by the invention is of a very light and compact two-passenger type, with provision for carrying camping, hunting or fishing gear, but otherwise characterized by small size, low cost, and unusual simplicity. The vehicle contemplated by the invention is further characterized by high and even spectacular performance over rugged terrain in the type of service described.

Objects of the invention are the provision, by a simply executed procedure, and at low cost, of a vehicle meeting the conditions mentioned in the foregoing, and characterized further by a short wheel base, exceptional road clearance, rugged construction capable of withstanding the type of service contemplated, and by certain protective features against hazards such as boulders to be driven over, or sharp rises or changes in the inclination of the terrain which might be struck by a vehicle traveling thereover.

The basic concept of the invention is the construction of a vehicle of the type mentioned from a wrecked or used sedan of a certain well-known European make, the Volkswagen, currently easily available at low cost, and the invention in one aspect comprises a reconstruction process by which such a vehicle can be easily, and comparatively inexpensively, radically modified to achieve an off-the-road vehicle having the desired features heretofore mentioned. In another aspect, it is a broad object to provide a unique off-the-road car, of simple and unique construction, having these desirable features.

For driving over rough or rocky terrain, it is, of course, desirable that large road clearance be provided. It is also desirable that the vehicle have a comparatively short wheel base. Thus, a vehicle of large road clearance but long wheel base, traversing a sharp crest at the top of a rise, for example, can easily strike its underside at a point between the wheels, whereas a short wheel base vehicle of the same clearance would completely clear the same crest.

My invention includes the discovery that the characteristic standard platform frame of the Volkswagen is, fortuitously, suited to certain unobvious modifications which

I have found to be possible, easily made, and entirely practicable, and which convert the original vehicle into one altered to meet the basic requirements of an off-the-road vehicle. In practicing the invention, the entire original body of the original vehicle is removed from the chassis and discarded, the chassis being retained. The unique, characteristic platform frame of the Volkswagen is preserved, but shortened by cutting and removing a twelve inch section of its so-called backbone or tunnel, and a large portion of the original frame, and pan or floor, is cut away and discarded. This results in shortening the wheel base of the chassis by a foot, reducing it to 82½ inches, which I have found to be both possible and ideal. The pan or floor of the frame is originally partly at and partly below the bottom level of the frame tunnel, and in rebuilding the frame, this pan is rebuilt and repositioned so that the front portion thereof remains at the bottom level of the frame tunnel, but approximately the rearward two-thirds thereof rise on an angle to a higher and higher elevation on the frame tunnel. The progressively increasing road clearance in the rearward direction as so provided is of material importance, it being a matter of experience that the great majority of collisions between the pan and high spots in the terrain driven over occur within the rearward area of the pan. The frame tunnel, of course, remains below the road clearance level afforded by the elevated pan. However, the informed and qualified driver, seeing a high spot or boulder ahead, can generally drive so that the tunnel, which is only a relatively few inches in width, will pass to one side or the other of the obstruction.

The frame being thus shortened and partly cut away, as mentioned, components for completing a novel frame are fabricated and welded to the remainder of the origanal frame, and in the course of this procedure, the pan or floor is elevated, as already described.

A feature of the new vehicle frame is a curved heavy-gauge steel plate at the front which is utilized as a bumper. Unlike ordinary bumpers, however, this bumper is contained between and within the outside circles defined by the front tires. Terminating short of the forwardmost portions of the tires, the tires will necessarily strike a large obstruction before the bumper does. However, the hood and the front fenders also terminate well rearwardly of the forwardmost portions of the tires, so that in many cases, the tires function as bumpers. The bumper plate serves to protect the frontal area of the vehicle between the tires against collision. In traveling up and down over very rugged territory, it is easily possible for the bumper plate, if positioned out in front of the tires, to collide with a sharp rise or mound of earth, or a boulder. But by positioning the bumper plate back between the tires, in a somewhat withdrawn position, the tires, one or the other, or both, will generally encounter the rise or mound before the bumper plate, and will thus elevate the vehicle over the rise, rather than permitting such a collision. This unique location for the bumper is an important feature of the invention.

The invention will better understood from the following detailed description of an illustrative embodiment of the invention, and showing the construction thereof from the chassis of an original sedan. Reference for this purpose is had to the accompanying drawings, in which:

FIG. 1 is a perspective view of a completed embodiment of an off-the-road vehicle in accordance with the invention;

FIG. 2 is a front elevational view thereof;

FIG. 3 is a plan view of the so-called platform frame of a standard make of vehicle from which the frame of the vehicle of the invention is constructed, and showing thereon the location of certain cuts to be made in the practice of the present invention;

FIG. 4 is a section taken on line 4—4 of FIG. 3;

FIG. 5 is a transverse section taken on line 5—5 of FIG. 3;

FIG. 6 is a plan view of the frame of FIG. 3 after the performance of certain modifications in accordance with the invention;

FIG. 6a is an enlarged detail of a portion of the frame of FIG. 6 as shown within the dashed circle 6a of FIG. 6;

FIG. 7 is a perspective view of the box of the vehicle of the invention, shown detached from the vehicle;

FIG. 8 is a perspective view of a portion of the box of the invention, shown with the addition of a gas tank, battery box and floor;

FIG. 9 is a fragmentary perspective view looking upward from underneath the modified frame of the invention from a point approximately opposite the right rear wheel;

FIG. 10 is a perspective view of a cowl in accordance with the invention;

FIG. 11 is a longitudinal section through the modified vehicle of the invention, taken in a plane represented by line 11—11 on FIG. 12;

FIG. 12 is a transverse section taken on line 12—12 of FIG. 11;

FIG. 13 is an enlarged fragmentary view taken from FIG. 11;

FIG. 14 is a section taken on broken line 14—14 of FIG. 11;

FIG. 15 is a view taken in accordance with section line 15—15 of FIG. 11;

FIG. 16 is a section taken on line 16—16 of FIG. 15;

FIG. 16a is a fragmentary detail taken on section line 16a—16a of FIG. 16;

FIG. 16b is a fragmentary detail taken in the direction of the arrow 16b in FIG. 16; and

FIG. 17 is a perspective view of a portion of the front end of the vehicle of the invention, the bumper plate being removed and shown in phantom lines.

Reference is first directed primarily to FIGS. 1 and 2, showing an illustrative vehicle in accordance with the invention, made by conversion of a chassis, i.e., the frame, running gear, front and rear suspension, steering and braking mechanism, etc., from a Volkswagen sedan. The original engine may also be used, though it is also possible to use a more powerful engine from certain other rear drive vehicles such as a Porsche or Corvair. The original body has been discarded, and the frame extensively modified. New body parts are welded to and function as structural parts of the new or modified frame. As shown in FIG. 1, the vehicle has wheels 20, and these are preferably equipped with large front and rear "snow" tires 21 and 22, respectively, for improved traction. The vehicle has a forwardly sloping cowl 24, a correspondingly sloping flat hood or hood plate 25 (see also FIG. 11), curved heavy-gauge sheet metal bumper plate or nose 26, low side rails 27 which are structurally integrated into the frame, sturdy sheet metal box 28 which is also structurally integrated into the frame, and constitutes a part thereof, seats 29, front and rear fenders 30 and 31, respectively, folding windshield 32 mounted on cowl 24, and roll bar 33. At the rear, mounted in the usual way on the frame fork, is the engine, fragmentarily appearing at 34, in FIG. 1.

FIGS. 3–5 are views of a standard, unmodified platform frame 36, such as is recoverable from a used Volkswagen sedan. This frame is modified in various ways, as presently to be described, to furnish the unique frame of the present vehicle. The platform frame 36 as shown in FIGS. 3–5 has been in use for many years and is very familiar to those acquainted with the art. It must, however, be briefly described. In general terms, this frame, as originally manufactured, comprises esentially a tubular sheet metal "tunnel" or "backbone" 37, somewhat in the form, in cross section, of an inverted U (FIG. 5), the bottom of which is closed and completed by a metal plate 37a welded in place. The rearward end of the tunnel is forked, the furcations 38 receiving the vehicle transmission and final drive unit, not illustrated, between them. As well known, and not necessary to illustrate herein, the transmission and final drive unit are mounted to the fork of the frame, and the engine is bolted to the transmission and final drive unit case. These latter relationships remain undisturbed in the new vehicle.

Near the front end of the frame, the tunnel 37 merges with a transversely extending so-called "front cross member" 39, projecting laterally outwardly from opposite sides of the tunnel. Joined to this front cross member and extending forwardly therefrom is the "frame head" 40, which includes a floor level reinforcement plate 41 joined to cross member 39. Frame head 40 is adapted to engage the usual upper and lower front torsion bar tubes 43 and 44 of the front suspension, and to be rigidly clamped thereto by clamp bars 45 and screws 46 (FIGS. 11 and 17). Just rearwardly of the beginning point of the fork in the tunnel, there is joined to the tunnel furcations 38 so-called "frame end plate" 50, extending transversely of the tunnel and laterally outward on opposite sides thereof. A floor, "pan," or platform 52 is provided, comprised of two sheet steel floor plates 53 welded into position between the front cross member 39 and the frame end plate 50, along the bottom edges of the tunnel 37, one on each side of the latter. The longitudinal edges of the platform 52 are provided with raised box sections 54 for stiffness, and the original vehicle body (not shown) rests on and is bolted to these box sections 54. The rearward portions of floor plates or pans 53 are shown as formed with the usual depressions 55 for gain of foot room for the rear seat passengers, dropping to a level below the lower level of the tunnel, and thereby reducing road clearance by a small but seriously important extent for the purpose of the car of the present invention. The depressions 55 are surrounded at the front by front seat rail assemblies 56 mounted on the floor pans 53 in the positions shown.

Immediately to the rear of frame end plate 50, the frame bifurcations 38 support a rear cross tube 57, which is in effect a part of the frame, and which contains rear torsion bars, not shown, but understood to be connected by flat spring plates 58 to the rear axle, also not shown. The extremities of cross tube 57 brackets 59, which provide certain pads or steps 60 and 60a to which the original body is bolted and also act as upper supports for the rear shock absorbers, not shown. As seen best in FIG. 17, the front torsion bar tubes 43 and 44 are connected at the ends by fixtures 62 which have upward extensions or brackets 63 to which the upper ends of the front shock absorbers 64 are connected. These brackets 63 are utilized in the invention as later described.

The frame and suspension components so far mentioned are well known in the art and no further illustration or description will be required.

To practice the invention, a used or partially wrecked Volkswagen sedan is provided, with the frame, such as illustrated in FIGS. 3–5, front and rear suspension, transmission, running gear, and other components in serviceable condition. The sedan body is removed from the chassis and discarded, and all components disassembled from the platform frame, which then appears as in FIG. 3. This frame is then cut, slit or severed along the dashed construction lines in FIG. 3, resulting in removal of the portions of the pan or floor 52 enclosed within the dashed lines, as well as a twelve inch long section of the tunnel 37. Thus, according to a preferred and illustrative procedure, in accordance with the invention, transverse cuts are made through the box sections 54 of the frame on each side, as from A to B, in a plane located substantially 18 inches to the rear of front cross member 39. Transverse cuts C–D are also made in floor pan 53 on each side, through the box sections 54 and inwardly to D, just aft of the seat rails 56, and approximately two-thirds of the way from front cross member **39** to frame end plate **50**. At D, which is a point just inside the outer seat rail, and just outside the floor depression **55**, the cut turns forward to E, just ahead of the beginning point of the floor depression. The cut is then made in a transverse direction to F, close to the inner seat rail, just inside the inner line of the floor depression. The cut then extends rearward to G, in line with C–D. It is then extended transversely to H, immediately adjacent tunnel **37**, and then extended rearwardly, along the tunnel, and along the front portion of the tunnel fork to I, just rearwardly of the frame end plate **50**. The frame end plate **50** is cut entirely off the tunnel, and discarded, along with the severed portion of the floor pan from C to I.

Slits are cut in the remaining portions of the pan from points H forwardly along the tunnel to points J, transversely aligned with the cuts A–B. Transverse cuts are made across and entirely through tunnel **37**, the forward cut in a vertical plane through points H and H, just rearward of handbrake lever mounting **65**, and the rearward cut along line K—K, in a vertical plane, twelve inches to the rear of the front cut. A twelve inch section of the tunnel is thus removed, and is discarded.

The two cut ends of the tunnel are then joined and welded to one another, as indicated at **70** in FIG. 6. After smoothing and painting, this weld joint is no longer evident. As well known, there are certain conduits, cables and control rods in the tunnel, and these are all shortened by twelve inches to correspond with the shortened tunnel.

The portion of box section **54** on each outer edge of the now remaining portion **71** of the original pan, between cuts A–B and C–D, is then hammered out flat, and into the same plane as the then remaining flat floor portion **72** of the pan, thus widening this floor portion **72** rearwardly of cut A–B. This widened floor portion **72** is then trimmed or cut longitudinally so as to form a new side edge **72*a***, extending rearwardly along a line which is a prolongation of the line of the inside surface of the outer wall **54*a*** of the remaining portions of box section **54** (FIGS. 6 and 6*a*). It will be seen that the edge **72*a*** is in the position formerly occupied by the inside surface of the outer wall of the now cut-away portion of box section **54**.

The now remaining portion **71** of the original floor pan is bent upwardly on a transverse line **73** which extends between the cuts or slits A–B. The portion **75** of the pan rearwardly of line **73** thus extends rearwardly at an inclination of a few degrees, such as seen in FIGS. 11 and 13. This portion **75** of the pan is supported from below as later to be described, and may be welded to the tunnel, as indicated at **76** in FIG. 6.

The modified frame thus appears at this stage as seen in FIG. 6.

The aforementioned cowl **24** is then mounted on the frame. This cowl, fabricated of 18 gauge sheet steel, comprises a downwardly and forwardly sloping top **80**, vertical, forwardly converging sides **81**, and a vertical front wall **82**. The adjoining edges of these members are formed with welded lap joints to afford a strong structure. The edges of the top and side walls of the cowl facing the driver's position are formed with reinforcing box section formations **83** and **84**, respectively. The cowl is shaped and dimensioned to fit down on top of the frame box section **54** and front cross member **39** (FIGS. 6, 11 and 15). The lower horizontal edge portions of sides or side walls **81** rest on the frame members **54**, and also have welded thereto angle members **86** which engage the frame members **54**, and the side walls **81** and the angle members **86** are welded to the frame members **54**. Front cowl wall **82** is shaped as at **82*a*** (FIG. 10) to fit cross member **39**, and engages and may, if desired, be welded to the latter. The cowl is here shown to mount an instrument housing and panel structure **87**, and also mounts the aforementioned windshield **32**.

Special frame side rails **27** are fabricated of 18 gauge sheet steel, and are welded to the cowl and to the rearward ends of the remaining portions of the frame box section members **54** which are now to serve as relatively stiff front end boundary frame members leading back to the added frame side rails **27**. These side rails **27** are preferably channel members approximately six inches in depth, and 38¼ inches long, and comprise vertical side walls **90**, with inwardly turned box sections **91** along the upper edges thereof, and inwardly turned flanges **92** along the lower edges thereof. They are mounted in a slightly inclined position, rising rearwardly, at the same angle as heretofore given for the sloping pan section **75**, i.e., a few degrees. The front ends of these side rails abut the rearward ends of the remaining portions of frame box sections **54** and also the lower portions of the vertical box sections **84** on the side walls of the cowl **24** (FIGS. 1, 15, 16, 16*a* and 16*b*). An angle member **93** is welded inside the forward end of each side rail **27**, so that one of its flanges abuts the cowl box section **84** (FIGS. 16 and 16*a*). The end edges of the side rail are welded to the abutting edges of frame section **54** and cowl box section **84**, and the flange abutting box section **84** is also welded thereto. For additional strengthening, a triangular gusset **95** is placed in the corner between the top of each side rail **27** and the abutting cowl section **84**, and is welded to these members.

The supporting arrangement for the rearward ends of the rails **27** will be considered presently. Attention is directed at this point to FIGS. 12 and 13, showing that the outer edge portions of the rearwardly projecting floor portions or members **72*b***, defined on the outside by the edges **72*a***, overlie and are supported by the lower flanges **92** of the side rails. The edge portions of members **72*b*** are welded to the side rails, as at **96** (FIG. 15).

A new rearward floor plan **97**, fabricated from 18 gauge sheet steel, is inserted under the rearward portion **72*b*** of upwardly inclined pan portion **75** on each side of the frame tunnel, its forward edge **97*a*** overlapping the edge **75*a*** of pan portion **75**, as clearly shown in FIG. 15. The outer edge portions of the pan members **97** engage over and are welded to the bottom flanges **92** of side rails **27** (FIG. 12). The new rearward floor pan members **97** are thus upwardly inclined at the same angle as the side rails and the pan members **75**. The inner edges of pan members **97** are provided with flanges **98** which engage and are welded to the tunnel. FIGS. 11 and 13 show that these inner edges of pan members **97** engage the tunnel along an upwardly inclined line. It will be appreciated that the upwardly inclined side rails and pan arrangement provide a gradually increased road clearance in the rearward direction on both sides of the tunnel. Preferably, for additional support of the pan members **97**, angle members **99** are welded to the underside of the inner edge portions thereof and to the tunnel, as seen best in FIG. 9. The pan members **97** are thus welded to the frame tunnel **37** and to the side rails **27**, tying these members rigidly together. The structure is then further stiffened and stabilized by means of rearward wall members **100** bent upwardly from the pan members **97**, these members **100** being cut to fit against the rails **27** and the tunnel **37**, and being welded thereto. A stiffened structure is thus afforded between the rearward end portions of the rails. These rearward end portions of the rails are then sturdily mounted to the furcations **38** of the "backbone" or tunnel portion of the shortened platform frame, utilizing for this purpose certain of the "skin" structure of the aforementioned "box" **28**, and also the previously mentioned body support pads or steps **60** and **60*a***. These arrangements will next be described.

Box **28** is fabricated, as are all the frame components, from 18 gauge sheet steel. It comprises a front wall **110**, coplanar with rearward pan walls **100**, together with a rearward wall **111** and side walls **112**. These box walls meet and are welded to one another edge-to-edge, and square, box-braces **114** are welded into the corners of the box. The horizontal lower edge of front wall **110** has a rearwardly turned flange **115**, which engages and is welded to rearwardly turned flanges **116** on the upper edges of rearward vertical pan walls **100**. The front wall of the box is thus structurally integrated to the pan walls **100**, and through the latter to the rails **27**. Front box wall **110** also has, projecting laterally outward from its two lower corners, a pair of structural bracing tabs **120** which engage under, and are welded to, the box section portions **91** of side rails **27** (FIG. 14). The flange **115** along the bottom edge of box front wall **110** will be seen to continue along the tabs **120**, which are thus strengthened.

The side walls **112** of the box have welded to the inner surfaces thereof a pair of mounting brackets **125** which include vertically braced bottom walls **126** which are aligned with and engage and are welded to the aforementioned body steps **60** and **60*a*** on the bracket arms **59**, it being recalled that the latter are fixed on the ends of cross tube **57** fixed to the frame furcations **38**. This cross tube **57** is strong, and sturdily supported from the furcations **38**, having been initially designed to assume the support of the rearward end of the original sedan body. It will be seen that the side rails **27** securely tie the cowl structure and the front part of the original platform frame to the box structure mounted on the rearward frame bracket arms **59** which originally supported the sedan body, and that the weight on the floor and side rails is transferred at the rear through portions of the front wall **110** and side walls **112** of the box to these frame bracket arms **59**. The side rails comprise, in effect, low sides for the vehicle, and act also as front-to-rear structural frame components paralleling the frame tunnel, and integrated to the latter through the structures heretofore explained. It is a novel feature of the invention that the rearward end portions of rails **27** are structurally integrated, by welding, to the vertical rearward pan walls **100**, which are in turn welded to front box wall **110**, and are also directly integrated, by welding to lateral front box wall tabs **120**, and thus to front box wall **110**. the pan walls **100** and front box wall **110** thus assume stressing from the side rails **27** and pan and the loading thereon. This stressing of box wall **110** continues through the side box walls **112** to the mounting brackets **125** which rest on and are welded to the frame supported steps **60** and **60*a***. The stress loading at the rear is thus carried from the pan and the side rails to the frame members **59** through the pan and box walls **100**, **110** and **112**, in a strong "stressed-skin" or monocoque fashion.

Box **28** is shown to be completed by flanges **130** supporting a gas tank **131**, a battery box **132**, and a floor **133**. Box side walls **112** are formed, concentrically with the rear axles, with arcuate cutouts **134** to afford necessary clearance and access, and the rearward box wall **111** is shaped to afford necessary clearance as indicated. The upper portions of side walls **112** preferably have outwardly bent portions **136**, rolled over at their upper edges as at **137** to receive the rearward end portions of hand rails **138**, the opposite ends of which are welded to side rails **27** (FIG. 1). The two front box braces **114** of the box **28** receive and support the lower end portions of roll bar **33**.

The aforementioned rear fenders **31** are flanged, as at **140** in FIGS. 9 and 15, and are securely fastened through these flanges, as by bolting, to the box sides **112**. A downwardly flanged sheet metal step **142** is fitted over and welded to the rearward end portion of each side rail **27**, and extends back and is bolted to rear fender **31**.

The aforementioned sloping hood plate **25** is securely fastened at its rearward edge, as shown best in FIG 11, to the forward edge of the top wall of cowl **24**. The front edge of hood plate **25** marginally overlies the upper, rearward edge portion of the previously mentioned cylindric bumper plate **26**, which is of heavy gauge sheet steel, e.g., 14 gauge. This bumper plate curves forwardly, downwardly, and under the frame head **40**, and its under inner edge portion **146** is bolted to the underside of frame reinforcement plate **41** (FIG. 11). It will be observed from FIG. 11 that the arc of bumper plate **26** is spaced within the perimeter of the front tires **21** of the vehicle. The upper edge portion of the bumper plate **26** is sturdily braced from the frame head **40**. as here shown, a front fender yoke **150**, fabricated of angle iron, has a central angle member **151** whose upper flange underlies the overlapped edge portions of bumper plate **26** and hood plate **25**, and these three members are bolted togethed, as indicated (FIG. 11). To the back of the other flange of angle member **151** is welded a heavy brace **152**, whose lower end is welded to frame head **40**; and to the front of said other flange is welded a heavy brace **153** extending from a base flange **154** which is bolted to the aforementioned clamp bars **45**.

The bumper **26** is thus securely mounted to the frame at its two edges. It is heavy and rugged, and somewhat resilient, such that, upon a collision, its generally cylindrical form will tend to flatten somewhat, and then elastically spring back. By being between and inside the perimeter of the tires, the bumper **26** functions only as regards objects between and in back of the front arcuate portions of the tires. The tires themselves take over the bumper function outside the two ends of the bumper **26**. The benefit arising in a rough-road vehicle of the character described from having the bumper inside the periphery of the tires has been amply stated hereinbefore.

Fender yoke **150** includes, in addition to the central cross member **151**, two upwardly and outwardly inclined angle members **158**, to which are joined two horizontal and outwardly extending angle members **159**. The front fenders **30** include flat top portions **160** which are flanged downwardly over and bolted to the outer yoke members **159**, and sloping portions **161** which are shaped along their inner edges to lie against the side walls **81** of the cowl. These fender portions **161** may be mounted on the cowl walls **81** in any suitable manner, as by use of angle strips (not shown) welded to the fender members and bolted to the cowl walls **81**. The front fender portions **160** also have connected thereto downwardly bent portions **166**, which are flanged under the hood plate **25** and bolted thereto, and are also flanged over the angle members **158** of yoke **150**, and bolted thereto.

A triangular sheet metal step **169** is fitted between the rearward edge of each front fender and the side rail **27** and is welded to the latter. It may be flanged and bolted to the fender.

Brackets **170** welded to yoke members **158** and connected to shock absorber brackets **63** afford sturdy support for the yoke **150** and front fenders from the chassis.

In the fabrication of the vehicle, the original frame is cut and shortened first, so that it appears as in FIG. 6. The cowl **24** is fabricated, and then added to the frame. Preferably, the box **28** is fabricated, and added to the frame next. The side rails **27** and new floor **97** are next added. Brace **152**, carrying yoke **150**, is mounted on the frame head, and the hood plate **25** is added. With the bumper plate **26** removed, the frame head is assembled with the torsion bar tubes **43** and **44** and the remainder of the conventional front suspension, axles, steering equipment, etc. With clamp bars **45** in place, brace **153** is bolted down. Brackets **170** are connected to the shock absorber brackets **63**, and the shock absorbers are installed. The conventional transmission assembly, rear torsion bars, spring plates, and rear shock absorbers are then added, as will be readily understood by those skilled in the art. The engine can then be installed. For necessary road clearance, the usual junction boxes and heater assemblies are removed, and new exhaust pipes (not shown) are installed at a higher level, as will be well within the skill of the art without further instruction.

The brake pedal cluster, master cylinder, hydraulic lines and emergency brakes are then installed in the usual manner, and no illustration of these conventional parts is required. It should be noted, however, that the throttle cable, clutch cable and emergency brake cable are necessarily shortened by twelve inches. The gearshift lever and rod, with the rod shortened twelve inches, are then installed, and, again, no illustration of these conventional parts will be required. Finally, seats **29** are installed in the conventional manner on the seat rails **56**.

It is believed that the off-the-road vehicle of the invention will now be understood, and its features and advantages will be appreciated, as will the process of the invention by which this unique vehicle may be constructed by conversion of a conventional Volkswagen chassis and addition of novel component parts thereto.

It will, of course, be understood that the presently disclosed embodiment of the invention, while now preferred, is for illustrative purposes only, and various changes in design, structure arrangement and procedure may be made without departing from the spirit and scope of the invention or of the appended claims.

What is claimed is:

1. In a vehicle, the combination of:

a vehicle frame comprised of a horizontally disposed, longitudinal sheet metal tunnel having at the front portion thereof a front cross member extending laterally in opposite directions therefrom and a fork at the rearward end thereof, a horizontal transverse cross tube carried by said fork, frame brackets on opposite ends of said cross tube, and floor plate members joined to a forward portion of said tunnel and to said front cross member, and including outer boundary edge portions,

a pair of longitudinally extending side rails rigidly secured to the rearward end portions of said outer boundary edge portions of said floor plate members,

and means supporting the rearward end portions of said side rails from said frame brackets, said means comprising a carry-all box including front, side and rear walls joined edge to edge, means rigidly joining said front wall of said box to the rearward end portions of said side rails, and means rigidly mounting said side walls of said box on said frame brackets, all in such manner that said front and side walls of said box function to assume a stress loading in support of said side rails on said frame brackets.

2. The subject matter of claim **1**, including rearward floor plate members disposed generally rearwardly of said first-mentioned floor plate members, said last-mentioned floor plate members being rigidly joined along the longitudinal edges thereof to said tunnel and said side rails, and each including an upwardly turned rear wall member meeting and welded to the lower portion of said front wall of said box.

3. In an automotive vehicle having a platform frame of the type comprised of a longitudinal substantially horizontal sheet metal tunnel having at the front portion thereof a front cross member extending laterally in opposite directions therefrom and a fork at the rearward end thereof, and having floor plate members joined at the inner edges thereof to said tunnel and at the front thereof to said front cross frame member at the bottom level of said tunnel, with stiff outer front end boundary frame members extending rearwardly along the outer boundary edges of a front portion of said floor plate members, the combination of:

a pair of upwardly and rearwardly inclined longitudinal side frame rails rigidly secured at their front ends to the rearward ends of said boundary frame members,

means supporting the rearward ends of said side frame rails from said fork,

portions of said floor plate members along said side rails being supported from said side rails, and

said floor plate members being upwardly and rearwardly inclined for a substantial distance along said tunnel for progressive increase of road clearance in the rearward direction on opposite sides of said tunnel.

4. In an automotive vehicle having a platform frame of the type comprised of a longitudinal sheet metal tunnel having at the front portion thereof, a front cross member extending laterally in opposite directions therefrom and a fork at the rearward end thereof, and having floor plate members joined at their inner edges to said tunnel and at the front to said front cross member, with stiff outer front end boundary frame members extending rearwardly along the outer boundary edges of the front portion of said floor plate members, the combination of:

a pair of longitudinally extending side rails rigidly secured to the rearward ends of said boundary frame members, said side rails being of channel section and having inwardly projecting upper flange formations, and

means supporting the rearward end portions of said side rails comprising a box structure including side walls supported from the frame fork and a front wall including tabs projecting laterally into the channels of said side rails and engaging and supporting said upper flange formations of said side rails.

5. In a vehicle, the combination of:

a platform frame comprised of a longitudinal sheet metal tunnel having at the front portion thereof a front cross member, a frame head projecting forwardly of said cross member, and floor plate members joined to said tunnel and to said front cross member and including, at the outer boundary edges thereof, raised box sections extending rearwardly from opposite ends of said front cross member,

a cowl structure mounted on said box sections on the outer edges of said floor plate portions, said cowl structure including side and top walls and a front wall, said side walls engaging and being rigidly secured to said box sections,

a hood plate secured to and extending forwardly from the top of said cowl structure,

a bumper plate having an upper edge portion secured to the front edge of said hood plate, and extending therefrom forwardly, down, around and under said frame head to a rearward edge portion secured to the underside of said frame head,

and means supporting the upper edge of said bumper plate from said frame head.

6. In a vehicle, the combination of:

a platform frame comprised of a longitudinal sheet metal tunnel having at the front portion thereof a front cross member, and a frame head projecting forwardly of said cross member,

a bumper plate having an upper edge mounting portion spaced above said frame head, and extending from said mounting portion forwardly, down, around and under said frame head to a rearward edge portion secured to the underside of said frame head, and

a bracing means erected on said frame head and supporting said upper edge mounting portion of said bumper plate from said frame head.

7. The subject matter of claim **6**, wherein said frame is a part of a vehicle chassis including front and rear wheels and tires, and

wherein said bumper plate is disposed between said front wheels and is substantially in the form of a semicylinder which curves around the frame head on an arc spaced inside the outer perimeter of the front tires.

8. In a vehicle, the combination of a frame including a central longitudinal sheet metal frame member having a front end portion;

a brace erected on the top of said front end portion of said frame member, and

a bumper plate having an upper edge mounting portion secured to the top of said brace and extending therefrom forwardly, down, around and under said front end portion of said longitudinal frame member to a rearward edge portion secured to the underside of said front end portion, of said longitudinal frame member.

9. The subject matter of claim **8**, wherein the frame is part of a vehicle chassis including front and rear wheels and tires, and

wherein said bumper plate is disposed between said front wheels and is positioned inside the outer perimeter of the front tires.

References Cited

UNITED STATES PATENTS

1,498,482	6/1924	Rutherford	293—60
2,251,970	8/1941	Avery et al.	296—28
2,288,978	7/1942	Talley	293—64
2,292,646	8/1942	McIntosh et al.	280—106
2,480,526	8/1949	Voltz	293—68
2,711,340	6/1955	Lindsay	296—28
2,733,096	11/1956	Waterhouse	296—28
2,841,439	7/1958	Schwenk	296—28
2,829,915	4/1958	Claveau	293—63 X
2,933,341	4/1960	Muller	296—28

FOREIGN PATENTS

143,367	11/1935	Austria.
742,977	6/1944	Germany,
972,170	5/1959	Germany.
805,577	12/1958	Great Britain.

BENJAMIN HERSH, *Primary Examiner.*

J. A. PEKAR, *Assistant Examiner.*

3 Sheets.
Sheet 1

S. Colt.

Impt in Fire Arms.

No 1304. Patented Aug 29. 1839.

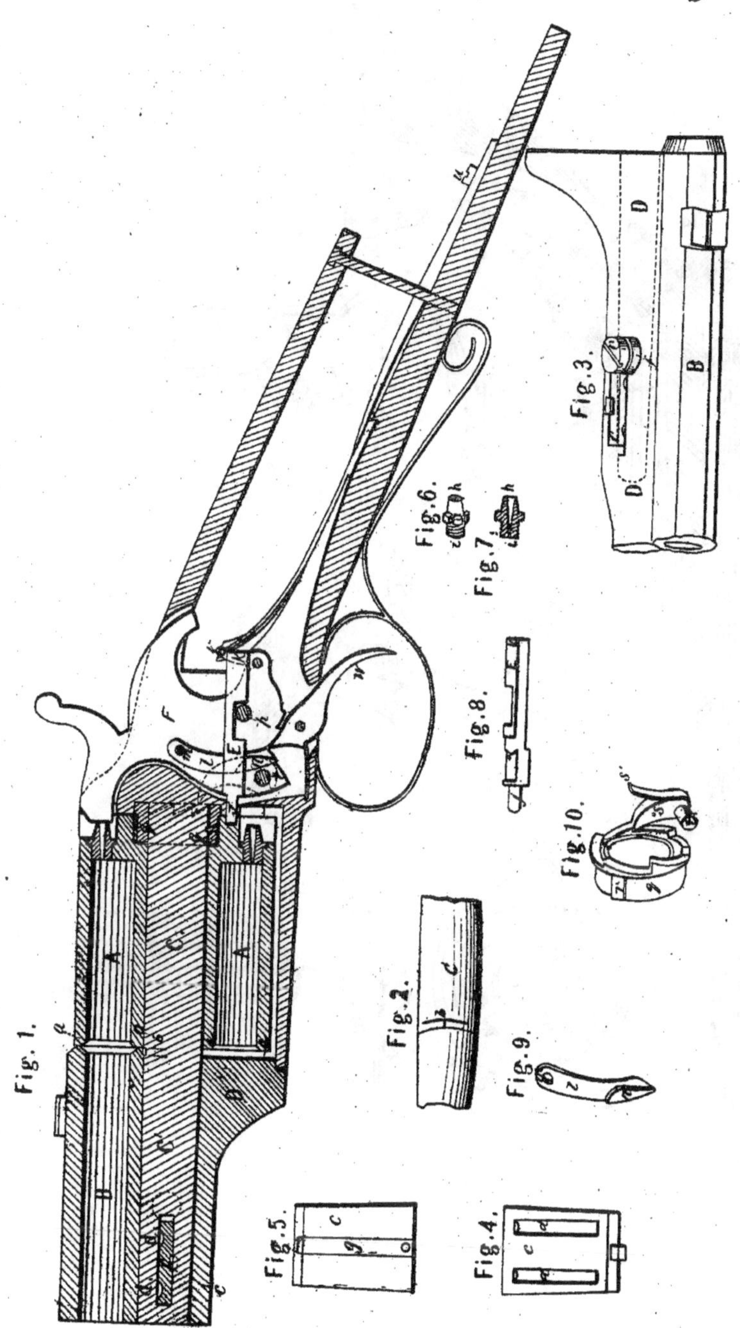

3 Sheets.
Sheet 2.

S. Colt.

Imp-t in Fire Arms.

No 1,304.

Patented Aug. 29, 1839.

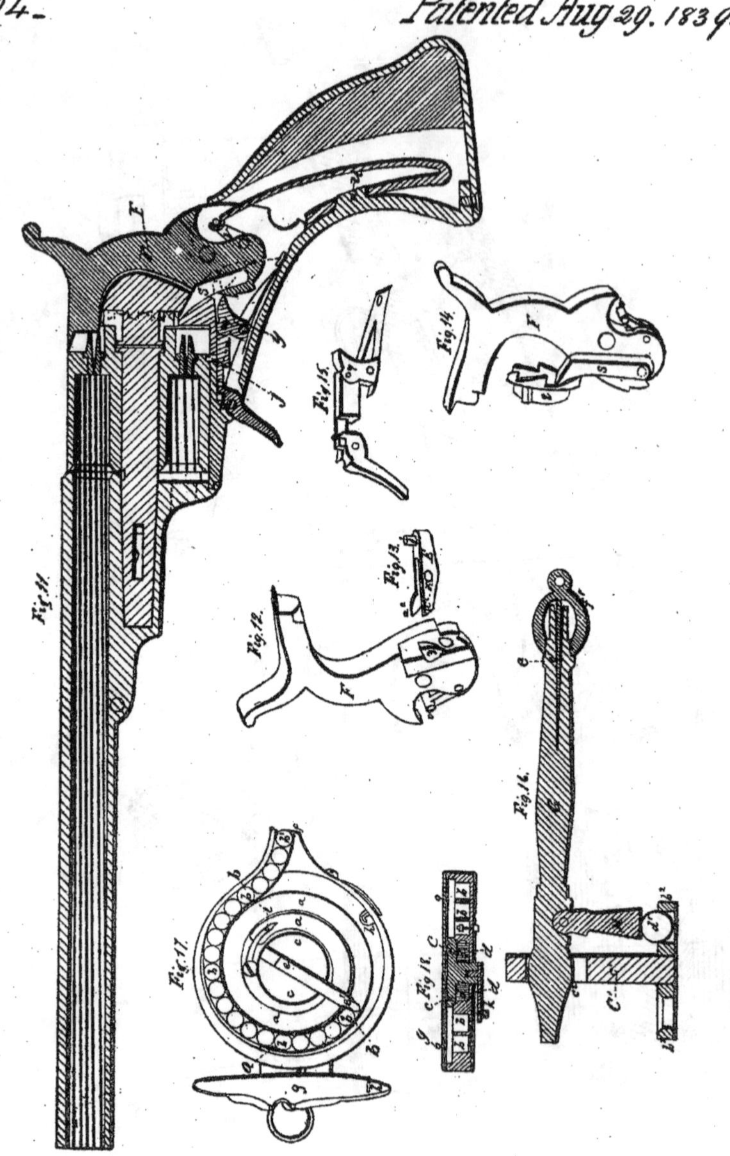

S. Colt.

3 Sheets. Sheet. 3.

Imp't in Fire Arms.

No 1304. Patented Aug 29. 1839.

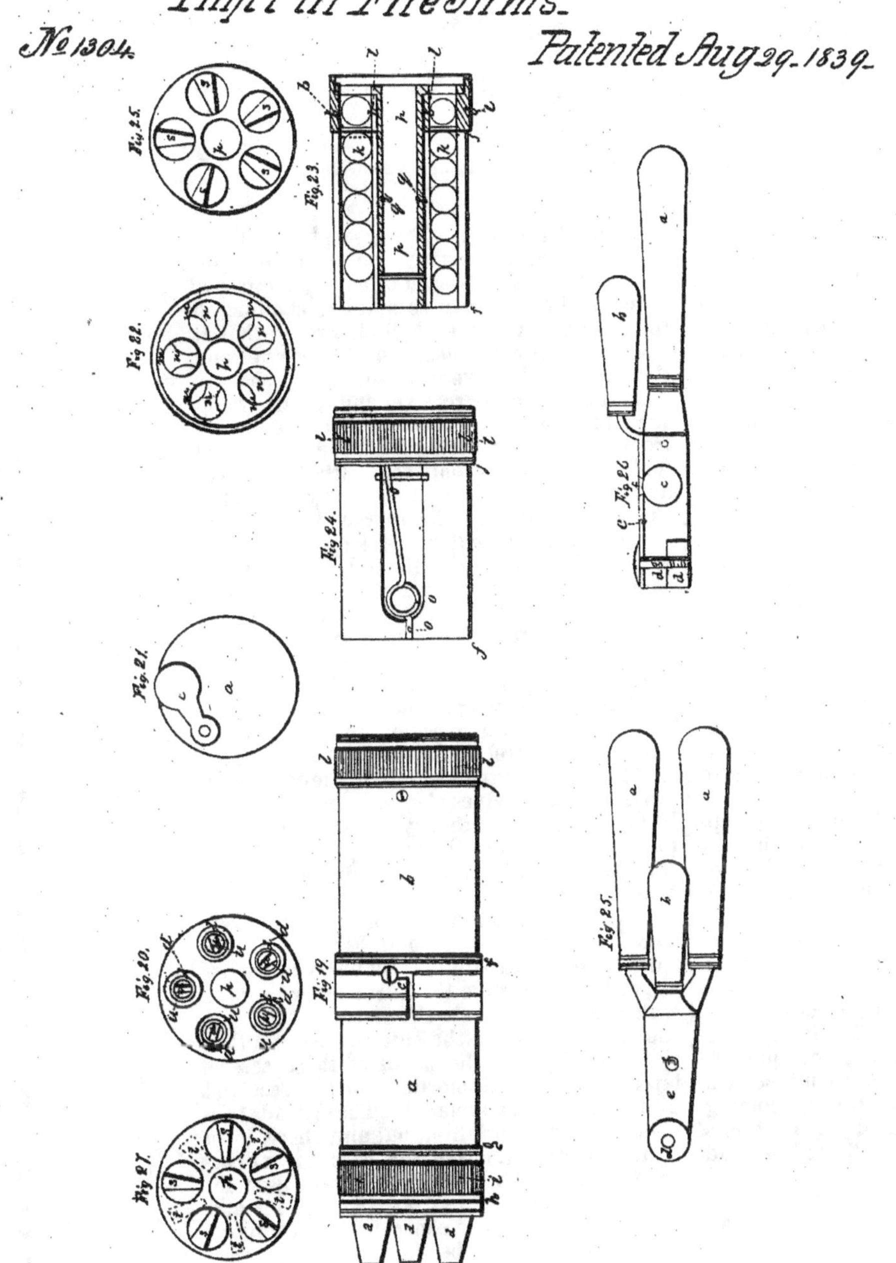

UNITED STATES PATENT OFFICE.

SAMUEL COLT, OF PATERSON, NEW JERSEY.

IMPROVEMENT IN FIRE-ARMS AND IN THE APPARATUS USED THEREWITH.

Specification forming part of Letters Patent No. 1,304, dated August 29, 1839.

To all whom it may concern:

Be it known that I, SAMUEL COLT, of Paterson, in the county of Passaic and State of New Jersey, did obtain Letters Patent of the United States for an Improvement in Fire-Arms, which Letters Patent bear date on the 25th day of February, in the year 1836, and that I have made certain improvements in the construction of the said fire-arms, and also in the apparatus for loading and priming the same; and I do hereby delare that the following is a full and exact description of my said improvements.

My first improvements appertain to rifles, guns, and pistols; my second to the construction of a cap-primer for containing the percussion-caps and placing the same upon the nipples, and my third to a flask and other apparatus for loading the rifle or gun.

For the general construction of my fire-arms, as originally patented I refer to the Letters Patent first above named, the same being necessary to a perfect understanding of the improvements thereon, which I am now about to describe.

Figure 1 in the accompanying drawings represents a section through the lock and breech of my rifle or gun and two of the chambers of the revolving receiver, B being a part of the barrel of the gun. The mouths of the chambers and the end of the barrel have their edges chamfered or beveled, as shown at *a a* in the drawings. In all guns of this description there is necessarily a lateral discharge between the receiver and the barrel, and this lateral discharge may endanger the ignition of the powder in the loaded chambers not in contact with the barrel; but the ignited matter, by coming into contact with the beveled edge as it crosses said chamber, is effectually reflected off, and does not enter them. The beveling of the end of the barrel is not a thing of importance, it being intended merely to prevent its scraping or cutting the ball in its passage from the chamber.

Fig. 2 shows a part of the arbor upon which the receiver turns. *b* is the portion thereof which is immediately below the chamber in contact with the barrel, and in this part a channel or groove is made descending from the point *b* in both directions, so as to form two inclined planes meeting at the point *b*. These planes or sections of the grooves may form an angle at forty-five degrees with each other. This groove or channel serves as a chimney to conduct off the smoke of the lateral discharge, so that it shall pass directly between the receiver and the barrel and prevent its spreading, so as to pass in between the receiver and the arbor and the barrel and the arbor, and consequently from condensing there and rendering them foul. The barrel is connected with the receiver and with the stock of the gun by the aid of the arbor which at the rear end, is a solid piece with the shield or solid piece of metal which receives the recoil and constitutes a component part of the metallic frame-work or foundation of the lock and its appendages. The part C of the arbor is that upon which the receiver revolves, and the part C′ enters a cylindrical cavity in a mass of iron, D D, to which the barrel is brazed or otherwise attached.

Fig. 3 shows the piece of iron D D and a part of the barrel B which is attached to it. The dotted lines in this represent the cylindrical cavity which receives the outer end of the arbor C′. The two are keyed together by the passing of a suitable key through a mortise in the piece D D and a corresponding one in the arbor.

c, Figs. 1, 3, 4, and 5, represents the key. Fig. 4 shows its upper and Fig. 5 its under side. *d d* are two fluted channels along its upper side to receive the heads of two screws which are screwed into the piece D D close to the mortise and on the side opposite to that shown in Fig. 3. The heads of these screws lap a little over the mortise and are received into the fluted channels *d d*. These heads prevent the key from falling out and check it in either direction, and must be withdrawn in order to remove the key. As the key *c* is to act laterally as a wedge to draw the receiver and the barrel into proper contact, it is of importance that it should be checked when forced sufficiently far in, or the receiver might be wedged up and prevented from turning. For this purpose I insert a screw, *e*, Fig. 3, into the steel button *f*, which is attached to D D, to strengthen the end of the mortise and prevent the bruising of it by the recoil. The head of this screw, overlapping the end of the mortise, receives the wedge and checks it. By turning this screw the force of the wedge may be tempered. In Fig. 5, *g* is a spring-latch on the under side of the key, which catches upon D when the key is forced in and prevents its accidental removal.

Fig. 6 represents a percussion tube or nipple, through which the fire from the percussion-cap is to be conducted to the chamber. Fig. 7 represents the same in section. The outer end, *h*, of the tube has the opening made as large as convenience will allow, and it goes tapering or conical until at the inner end, *i*, it is as small as a proper entrance of the flame from the percussion-powder will warrant. By giving the conical or funnel-formed opening to the tube the effect of the percussion-powder is greatly increased.

E, Figs. 1 and 8, is a bolt for locking the receiver when a loaded chamber is brought to coincide with the barrel, the rounded end *j* being forced into a hole in the receiver by the action of a spring on its outer end *k*. This bolt is drawn back by the hammer F in the act of cocking.

l, Figs. 1 and 9, is a spring-cam, which is screwed to the hammer at *m*. It is made of spring-steel, so that its cam end *n* may recede from or approach the hammer F. The cam *n* bears against the projection *o* on the bolt E on the side which lies against the hammer, and as the hammer is drawn back causes the bolt to recede. The bolt is notched to enable it to be withdrawn without interfering with the joint-pin *q* of the hammer. The lateral springing of the cam-piece *l* is necessary to admit of its passing by the projection *o* of the bolt when the hammer is made to strike upon a percussion-cap. To enable the spring-cam to pass the bolt E, the lower end of it, *n*, is made wedge-shaped, diminishing to a point or edge at its extremity, and as it is made to spring laterally it is received into a recess in the hammer as the latter passes the bolt in making the discharge.

Fig. 10 shows the ratchet-wheel and hand or pawl by which the receiver is made to revolve to the distance from one chamber to another in the act of cocking. The cylindrical periphery *q* of the ratchet-wheel fits into a corresponding cavity on the back end of the receiver, as shown at *q q*, Fig. 1. *r* is a projection to prevent its turning round, this being adapted to a notch made to receive it. S is a hand or pawl, which falls into the teeth of the ratchet-wheel, said pawl being forced forward by the spring S′. The arbor *t* on which the hand turns is received into the opening *t*. In the hammer, Fig. 1, the hand itself being on the opposite side of said hammer from that shown, its position is shown by the dotted lines surrounding its arbor *t*. The cocking of the gun causes it to act upon the ratchet-wheel, and when turned to the proper distance the bolt E is forced by its spring into the proper opening in the receiver. The mainspring is connected to the lock-plate at *u*, and to the hammer by a stirrup at V. The trigger is shown at *w*. These parts, not differing in their construction and operation from analogous parts in other gun-locks, need no particular description, and from the description above given of the structure and operation of those parts of the rifle or gun which are new, the action of the whole will, it is believed, be clearly understood.

Fig. 11 is a sectional view of a pistol, the general construction of which is the same with that of the rifle or gun already described, such modifications only being made as are rendered necessary by its size and other considerations. F is the hammer carrying the hand or pawl S, which operates on the ratchet-wheel, which wheel and hand are arranged in the same way with the same parts in the rifle; but the hand is as here represented on the reverse side. The bolt which holds the receiver is, however, differently constructed to enable it to act in the space which it must occupy.

Fig. 12 is a view of the hammer on the side the reverse of that shown in Fig. 11; and E, Fig. 13, is the bolt adapted thereto. *j* is the pin on the bolt, which holds the receiver by falling into openings on its periphery instead of in its end. The pin *j* is shown in place in Fig. 11. The bolt E vibrates on a joint-pin at *x*, which is nearly in the same line with the joint-pin *y* on the trigger, Figs. 11 and 15, by which it is hidden in those figures. *z* is a cam formed in a recess in the hammer, Fig. 12, which cam is to act upon the bolt E and to disengage it from the receiver. The ends a' a^2 of this bolt are capable of receding from or approaching toward each other, as they constitute two spring-cheeks formed by splitting or forking the bolts, as shown in the drawings. The end a^2 lies above the cam *z* on the hammer when the pistol is not cocked, and the lower end of a^2, as well as the upper end of *z*, being flat, the bolt E is lifted in the act of cocking until the pin *j* is disengaged, and the ends of a^2 and *z* then pass each other. The cam *z* is made wedge shape by sloping from its upper to its lower end, and the end a^2 of the bolt is similarly formed, but in the reverse direction, so that when the piece is discharged the end a^2 will be made to spring in, allowing the hammer to pass readily, when the end a^2 again rests upon *z* as before.

Fig. 14 shows the hammer with the hand S and ratchet *q*, which need no further description.

Fig. 15 shows the trigger and its appendages, which are formed in a way not presenting any claim to novelty. *u* is the attachment of the mainspring to the lock-plate, and *v* its attachment to the stirrup and hammer.

Fig. 16 shows the apparatus which I employ for forcing the balls into the chambers. b^2 b^2 are two of the chambers, shown in section; and C, the arbor by which the barrel is attached to the stock and upon which the receiver turns, as already explained. G is a lever carrying a rammer, H, by which the balls are to be forced into the chambers. The forward end of the lever H passes into the mortise c^2, which receives the key by which the barrel is attached. The operation of the rammer H upon the ball d' will be apparent. In using this lever the receiver is to be turned upon the arbor, and the chambers brought in succession under the rammer. This lever, at its end e', constitutes a wrench for screwing and unscrewing the percussion-tubes, and also contains a picker attached to a screw-cap, f'. A fulcrum for the lever H may be formed on the barrel or otherwise, instead of using the mortise c^2, if preferred.

Figs. 17 and 18 are a top and sectional view of my improved cap-primer, which differs in some important particulars from the English and other cap-primers now in use. I make a spiral groove, *a a a*, in a plate of brass or other metal, which groove is of such depth and width as to receive the percussion-caps, and to allow them to move freely therein. *b b b* are caps within said groove. In the center of the primer, under the plate *c c*, is a spiral spring, *d d*, operating like the mainspring of a watch upon its barrels and turning the plate *c c*. This plate has a groove across it which carries a sliding arm, *e e'*, having under its end *e'* a projecting piece which enters the groove, draws the arm out, and presses upon the row of caps. At the mouth of the spiral groove, where the the cap *b'* is seen, a steel spring, *f*, checks the cap and counteracts the pressure of the spiral spring *d d*; but when the cap *b'* is placed upon the tube or nipple the spring *f* will recede by the withdrawal of the cap, and a new one will be made to occupy its place, and so on until the whole are exhausted. The cover *g*, which in Fig. 17 is shown as raised, is held down by a spring-catch at *h*. There is a spring-catch at *i*, which holds the sliding arm *e* at its end *e'*, when it is brought round to the inner end of the spiral groove, its use being to detain the arm while the groove is being filled with caps, when it is to be raised, and the arm left at liberty to operate. The spring *d d* may be wound up by a small key, *k*, or by inserting a screw-driver in a notch made for that purpose, or simply by forcing the arm *e e* round until it is caught by the catch *i*.

Fig. 19 is a representation of my ammunition-flask, by means of which all the chambers in my receiver may be simultaneously charged with powder and with balls. It consists of two separate chambers, one of which is a powder and the other a bullet magazine, *a* being the former and *b* the latter, the two being connected together by a bayonet-joint at *c*. *d d d* are charging-tubes adapted in number and position to the mouths of the chambers of the receiver which they are to enter.

Fig. 21 shows the closed top of the powder-magazine, with a valve or turn cover, *e*, which closes a hole through which the magazine is to be filled. This magazine occupies the space from *f* to *g*, Fig. 19, where the powder is contained in bulk. The space from *g* to *h* is a receptacle which is divided by partitions into separate chambers, the same in number with the tubes *d d*, each of which chambers contains the quantity of powder required for the charge of a single chamber. *i i* is the rim of this chambered receptacle, which is capable of being turned round to a short distance by the thumb and finger for the purpose of charging the chambers with powder. This turning round brings openings *u u*, Fig. 20, in the lower end of the chambered box to coincide with the openings in the tubes *d d*, so that the powder contained in the chambers in *i i* may pass out therefrom into the chambers of the receiver. There are openings also in the upper plate or top of the receptacle *i i* corresponding with openings in the bottom of the magazine *a*, which are closed by turning the rim *i i*, so as to prevent powder from falling through from the magazine while the receiver is being filled.

Fig. 27 is a section through the middle of the chambered receptacle *i i*, the circles *r r* representing the chambers for containing the powder. *s s* are the openings in the top plate of these chambers, through which the powder is admitted into them from the magazine *a*. The dotted lines *t t* show the plan of the openings in the bottom plate of the chamber *a*, the chambered receiver being shown in the position in which those openings are covered.

In Fig. 28 the same parts are represented; but the chambered receiver is supposed to be turned round or standing in its ordinary position, so that the openings S S and *t t* coincide. The chambered receptacle is restored to its place by means of a spring of any suitable form. The whole operation of this part will be more clearly made known by the sectional representation of the magazine for balls, which I am now about to describe.

The end of the flask, Fig. 19, is, I have said, the magazine for balls. Fig. 22 is an end view of this magazine, and Fig. 23 a section along its axis. The portion from *f* to *j*, Figs. 19, 23, and 24, is divided into as many tubular chambers as there are chambers in the receiver—say five. These are open at top and are to be filled with balls, as shown at *k k* in the section Fig. 23. These tubes are also open at their lower ends, so that the balls may pass from them into a chambered receptacle, *l l*, similar to that for the powder. From this chambered receptacle they are to fall into the chambers of the receiver when the lower end of the flask, Fig. 22, is applied thereto for that purpose, the openings *m m* in the lower end of the flask being adapted thereto. The rim of the chambered receiver *l l* is to be turned round to allow the balls to escape through *m m*, as already described in the charging with powder. In Fig. 22, *n n* are the divisions between the chambers of the chambered receiver, and which retain one set or tier of balls until the rim is turned around so as to cause the chambers to coincide with the openings *m m*. The balls will then pass through. The same motion of the chambered receiver causes the divisions between the tubes and the chambered receiver to pass under and sustain the balls in the magazine. In the case of the powder-magazine the action is the same; but the powder being in fine particles, the apertures at one end of the receptacles must be perfectly closed before those at the other begin to be opened, which is not necessary with the balls. In Fig. 24 a portion of the exterior of the magazine is removed to show how a spring, *o o*, may be placed within it so as to act upon *l l*; but spiral or other springs may be placed in many ways to answer the same purpose. The central part of both the magazines is tubular, as shown at *p p*, said tube fitting onto the arbor C′, Fig. 16, when the barrel is removed therefrom, and the receiver left on for the purpose of being charged, which operation does not require to be further explained. Upon the barrel of this tubular part the chambered receptacles are received and revolve.

Figs. 25 and 26 represent a top view and a side view of a part of an improved bullet-mold, which I describe without intending to make any claim thereto, but merely for the purpose of showing the whole of the apparatus employed in a complete and connected series. *a a* are the two handles of the mold, and *b* the handle of the knife by which the sprue is cut off. *c* is one-half of the mold, of which *d d* is the hinge-joint. *e* is a plate of steel, through which there is a hole, *f*, for pouring in the lead, the lower edges of which constitute a knife by which the sprue is cut off and the ball left perfect. This knife turns on the joint-pin *g*.

Having thus fully described the manner in which I construct and use my improved firearms, and the respective articles of apparatus appertaining thereto, it has been necessary in so doing to mention many parts which I do not claim as new, the same being similar to what has been before used and patented by me, or which are common property. I do hereby declare, therefore, that I limit my claim to the following particulars.

I claim—

1. The making of a groove or channel on the arbor, as represented at *b*, Fig. 2, for the purpose of conducting off the smoke from the lateral discharge, and thus preserving the arbor clean within the receiver, and the tube by which the barrel is connected.

2. The particular manner of forming and governing the key by which the barrel is attached to the stock by making the same with grooves in which the heads of overlapping screw-heads are received, and with a tempering-screw to check and regulate its action as a wedge, as set forth.

3. The making the aperture through the tubes or nipples (which receive the percussion-caps) conical or funnel-shaped, for the purpose of freely admitting the fire from the percussion-cap and concentrating it as it enters the chamber.

4. The manner of arranging the bolt E of the rifle and its spring cam *l n* for locking and unlocking the receiver, the same being constructed and operating as herein described.

5. The manner of constructing and arranging the bolt E and its spring-cam, operated upon by the cam or projecting piece *z* under that modification thereof adopted in the pistol, and herein fully made known.

6. The improved manner of arranging the ratchet-wheel and hand, as set forth, by which the hinge-joint to allow of the lateral motion of this hand, as described by me in my former patent, is dispensed with in consequence of the placing of the ratchet-teeth on the face instead of on the side of the wheel, and operating the same in the manner described, as applied to the rifle and to the pistol.

7. The combination of the lever with its rammer for forcing the balls into the chambers of the receiver, as described.

8. In the improved cap-primer, the making thereof with a spiral groove to receive the caps, and with the sliding arm acted upon by the spiral spring elongating itself and forcing the percussion-caps forward in the manner set forth.

9. The manner of constructing and arranging the respective parts of the magazines for powder and balls, in the flask, by means of which the powder and the balls are in turn supplied to all the chambers in the receiver at the same time, the whole being made with the chambered receptacles and other parts, as set forth.

SAML. COLT.

Witnesses:
THOS. B. JONES,
GEORGE WEST.

Disclaimer forming part of Letters Patent No. 1,304, dated August 29, 1839.

To the Honorable Commissioner of Patents:

The petition of SAMUEL COLT, of Hartford, in the State of Connecticut, respectfully represents that he is the sole patentee and owner of Letters Patent granted to him on the 29th day of August, 1839, for an improvement in fire-arms and in the apparatus used therewith; that he has reason to believe that through inadvertence and mistake the claim made in the specification of said Letters Patent is too broad, including that of which the said patentee was not the first inventor, although he avers that he was an original inventor thereof, and had no knowledge when he applied for Letters Patent therefor that any other person had ever used the said improvement before that time.

Your petitioner therefore hereby enters his disclaimer to that part of the claim in the before-mentioned specification which is in the following words, viz:

"I claim making the aperture through the tubes or nipples (which receive the percussion-caps) conical or funnel shaped, for the purpose of freely admitting the fire from the percussion-cap and concentrating it as it enters the chamber," which disclaimer is to operate to the extent of the interest in said Letters Patent vested in your petitioner, the same being the whole right, title, and interest thereby granted to him, as aforesaid, he having paid ten dollars into the Treasury of the United States agreeably to the provisions of the act of Congress in that case made and provided.

Dated at Hartford this 5th day of August, A. D. 1853.

SAM. COLT.

In presence of—
L. P. SARGEANT.

(No Model.)

J. H. & H. H. DOW.

ROTARY STEAM ENGINE.

No. 403,335. Patented May 14, 1889.

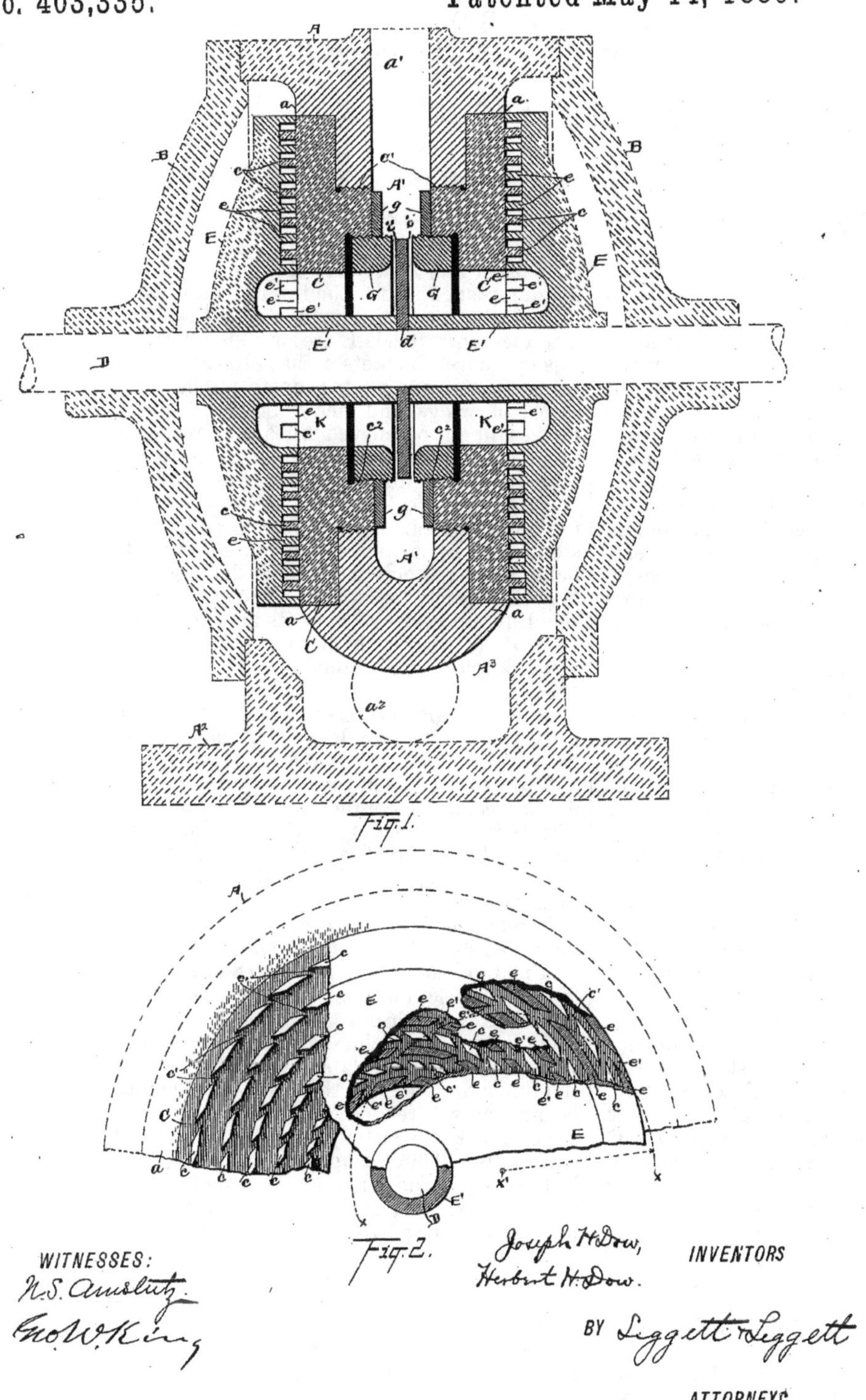

Fig. 1.

Fig. 2.

WITNESSES:

N. S. Amstutz.

Geo. W. King

Joseph H. Dow,

Herbert H. Dow.

INVENTORS

BY Liggett & Liggett

ATTORNEYS

N. PETERS, Photo-Lithographer, Washington, D. C.

UNITED STATES PATENT OFFICE.

JOSEPH H. DOW AND HERBERT H. DOW, BOTH OF CLEVELAND, OHIO; SAID HERBERT H. DOW ASSIGNOR TO WILLIAM CHISHOLM, OF SAME PLACE.

ROTARY STEAM-ENGINE.

SPECIFICATION forming part of Letters Patent No. 403,335, dated May 14, 1889.

Application filed August 25, 1888. Serial No. 283,751. (No model.)

To all whom it may concern:

Be it known that we, JOSEPH H. DOW and HERBERT H. DOW, of Cleveland, in the county of Cuyahoga and State of Ohio, have invented certain new and useful Improvements in Rotary Steam-Engines; and we do hereby declare the following to be a full, clear, and exact description of the invention, such as will enable others skilled in the art to which it pertains to make and use the same.

Our invention relates to improvements in the rotary engine shown and described in Letters Patent No. 392,545, granted November 6, 1888; and it consists in certain features of construction and in combination of parts hereinafter described, and pointed out in the claims.

In the accompanying drawings, Figure 1 is an elevation in longitudinal section through the center of the engine. Fig. 2 is an end elevation, partly in section, the cylinder-head of the foreground being removed and portions being broken away to show the construction.

A represents the body or shell of the engine, to which are detachably secured heads B, the latter having suitable boxes, in which is journaled the engine-shaft D. The shell A is provided with internal flanges, a, these flanges being "faced off" and screw-threaded at the internal periphery to receive disks C, the latter being screw-threaded at C' for engaging the aforesaid flanges. The two flanges and the backs of these disks are separated to form steam-chamber A', the latter connecting with induction-pipes a'. The shell A terminates in a supporting-base, A^2, to which exhaust-pipe a^2 is attached, the latter connecting with exhaust-chamber A^3. The disks C, on the outer faces thereof, are provided, respectively, with series of curved wings or chutes c, each series being arranged in concentric order with the axis of the shaft D. The different wings c of a series are separated the one from the other to form ports c', the latter being preferably arranged in the curved order shown in Fig. 2. (See dotted line $x\ x$ with center x'.)

E are rotating disks or wheels, the same being mounted on and secured to shaft D. These disks have long hubs E', projecting toward each other and nearly meeting on the shaft, leaving room only for the intermediate disk, d, the latter also being made fast to the shaft. The faces of wheels E are provided with series of curved buckets e, that alternate with wings c, a series of buckets operating between series of wings, and vice versa, the vents e' being laid out on curved lines the reverse of line $x\ x$, and these vents e' being less in aggregate area of discharge than the aforesaid ports c', the object being to utilize the reactive force of steam rather than the impact of steam, and by means of the ports and vents being arranged on curved lines, deflecting in opposite directions, as shown, the openings of the different series of buckets and wings can be made to cross each other at substantially equal angles throughout the series and at such angles as will give the greatest effectiveness, such angles at present advised being approximately right angles. This class of engines is intended to run at a very high speed, and to avoid friction the disk C and wings c do not come in contact with the opposing wheels E and buckets e, although the clearance at the ends of the wings and buckets is very small—say a two-hundredth part of an inch, more or less.

Heretofore the great difficulty has been to hold the shaft D endwise. The high speed attained was such that ordinary collars caused too much friction, and were consequently worthless for such purpose.

Our improved mechanism is as follows: The aforesaid disk d overlaps in radial direction the inner periphery of the disks C. The latter are screw-threaded at c^2 to engage rings or collars G, the latter being screw-threaded externally. These rings are provided with lock-nuts g. The rings are adjusted toward or from disk d to leave suitable and equal passage ways or ports, i and i', for the passage of steam from chamber A' to chamber K, from which latter the steam passes between the disks C and opposing wheels E *via* ports c and vents e' to the exhaust-chamber. Rings G having been adjusted and secured by the respective lock-nuts, so long as there is no disturbing cause spaces i and i' will remain equal and an equal volume of steam will issue to the two sides of the engine; but suppose from any cause the shaft were moved toward, for instance, what is the right-hand side, as shown in Fig. 1, such a movement would partially close the passage-way or port *i*, and would consequently shut off a portion of the steam to the right-hand side of the engine, and would at the same time further separate disk C and wheel E at this side of the engine. The reduction of steam-volume and the widening of steam-space between the disk and wheel would tend to diminish the endwise pressure on the shaft, while on the other side of the engine the reverse would take place—that is to say, steam-space *i'* would enlarge, thereby giving greater volume of steam on this the left-hand side of the engine, and the disk and wheel on the left-hand side of the engine would approach each other, thereby decreasing the steam-space between the two, and consequently giving greater steam-pressure to act in the opposite direction endwise of the shaft, all of which would tend to return the shaft to its central or normal position lengthwise of the engine. The apparatus whereby a steam-balance is thus had is very sensitive, by reason of which the buckets may run in close proximity to their seats without absolute contact, thus avoiding all friction except at the journal-bearings of the engine-shaft. In place of a single disk, *d*, rigidly secured to the shaft, two such disks might be employed, and these might be adjusted lengthwise of the shaft instead of the adjusting-rings G; also, the curved lines on which the vents and ports are laid out need not necessarily be on true circles.

What we claim is—

1. In a rotary steam-engine, the combination, with stationary disks with steam-chamber located between the disks and opposing rotating wheels located outside the disks, the wheels being mounted on the engine-shaft, substantially as indicated, of an engine-shaft, and a disk mounted on the engine-shaft between the line of the stationary disks, said rotating disk extending into and dividing the eduction of the said steam-chamber, substantially as set forth.

2. In a rotary engine, the combination of shaft carrying with it two wheels and a disk between them, these wheels being opposed inwardly by stationary disks, the central or revolving disks being opposed on opposite sides by rings connected with the stationary disks, leaving annular spaces between the central disk and rings that act as steam-ports, substantially as set forth.

3. The combination, with a rotary engine of the variety indicated, having stationary disks and movable disk, the latter being mounted on the engine-shaft and extending into the steam-space between the stationary disks, of rings mounted on such stationary disks and adjustable toward and from the rotating disk, substantially as set forth.

4. The combination, with stationary disks and intervening rotating disk, the latter being mounted on the engine-shaft, substantially as indicated, of screw-threaded rings engaging the stationary disks, such rings being adjustable toward and from the rotating disk, and lock-nuts mounted on the respective rings, substantially as set forth.

5. In a rotary engine, the combination of stationary disks and rotating wheels, the former bearing wings separated so as to form intervening ports and the latter bearing buckets separated so as to form intervening vents, the system of ports and vents being arranged in curved lines deflecting in opposite directions, the aggregate area of the vents being less than the aggregate area of the ports, substantially as set forth.

6. In a rotary engine, the combination, with stationary disks provided with a curved series of ports, and revolving wheels provided with a curved series of vents, the latter being arranged in curved lines deflecting in opposite directions to those of the ports, said revolving wheels being provided with inwardly-projecting limbs, of a revolving disk located on the shaft between the adjacent ends of said limbs and extending into and dividing the steam-outlet of the steam-chamber, substantially as set forth.

In testimony whereof we sign this specification, in the presence of two witnesses, this 15th day of May, 1888.

JOSEPH H. DOW.
HERBERT H. DOW.

Witnesses:
CHAS. H. DORER,
ALBERT E. LYNCH.

No. 608,845. **Patented Aug. 9, 1898.**

R. DIESEL.

INTERNAL COMBUSTION ENGINE.

(Application filed July 15, 1895.)

(No Model.) 2 Sheets—Sheet 1.

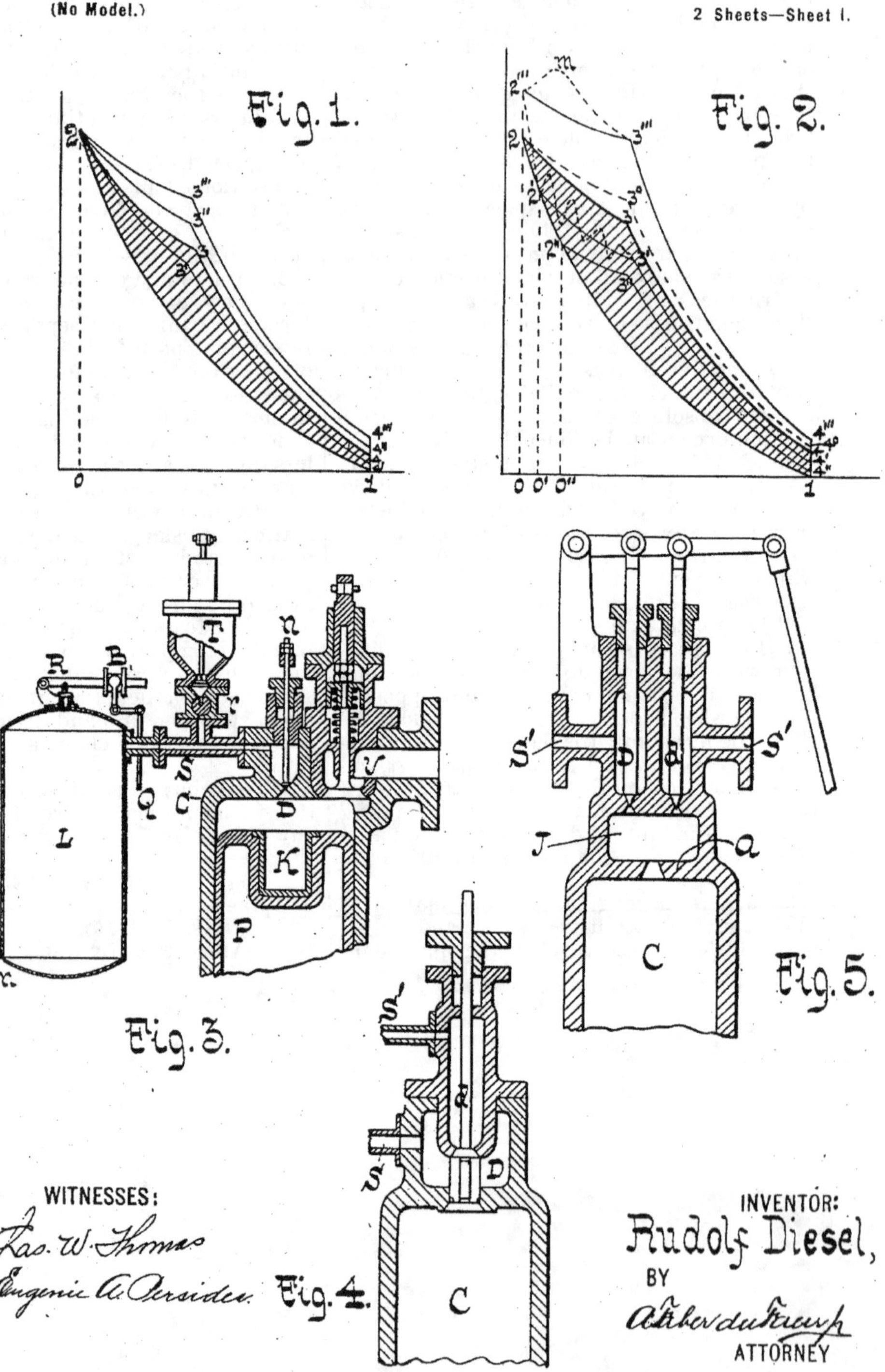

No. 608,845. Patented Aug. 9, 1898.

R. DIESEL.

INTERNAL COMBUSTION ENGINE.

(Application filed July 15, 1895.)

(No Model.) 2 Sheets—Sheet 2.

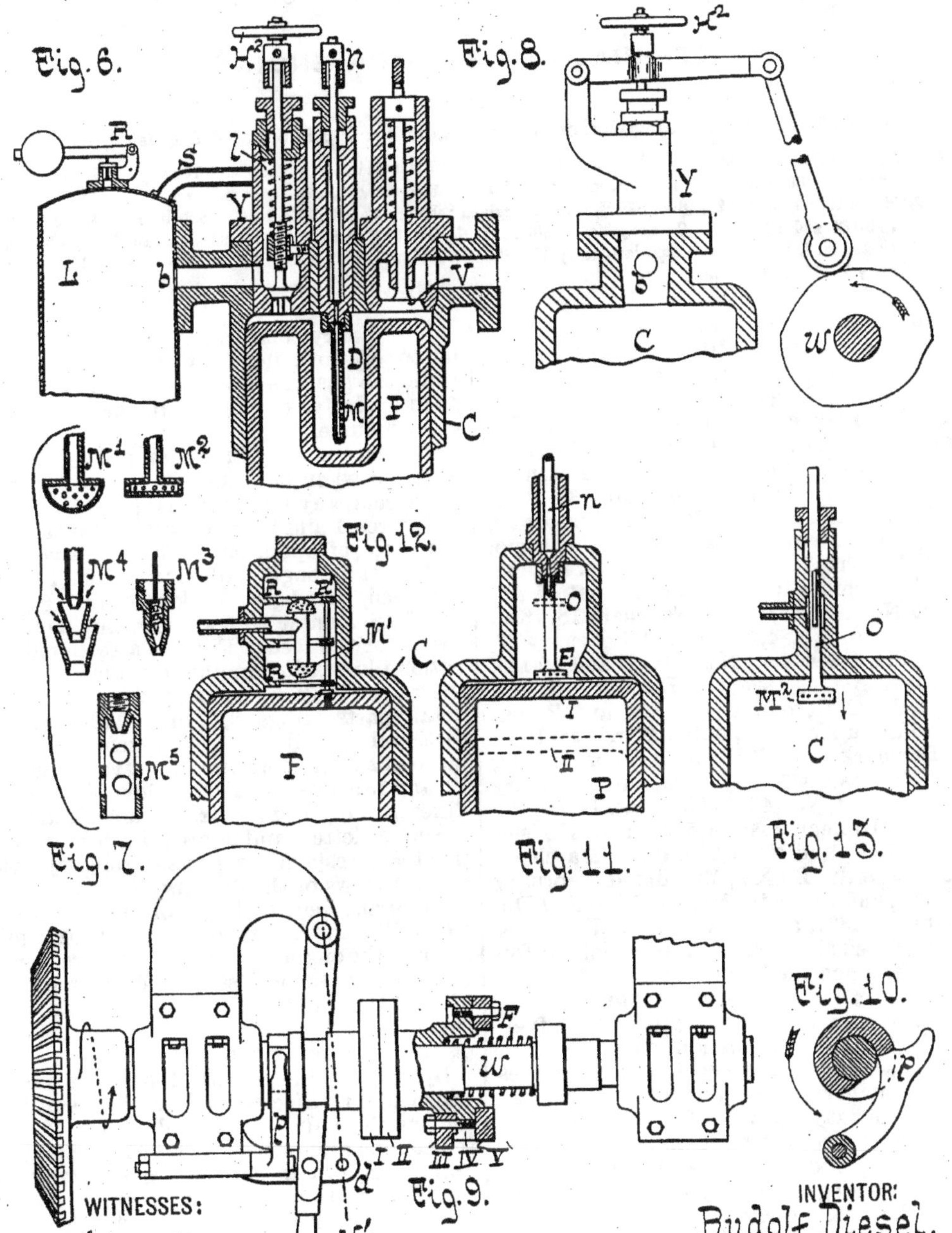

Fig. 6. Fig. 8. Fig. 12. Fig. 7. Fig. 11. Fig. 13. Fig. 10. Fig. 9.

WITNESSES:

Jas. W. Thomas.

Eugenie A. Persides.

INVENTOR:

Rudolf Diesel,

BY

A. Faber du Faur Jr.

ATTORNEY

UNITED STATES PATENT OFFICE.

RUDOLF DIESEL, OF BERLIN, GERMANY, ASSIGNOR, BY MESNE ASSIGNMENTS, TO THE DIESEL MOTOR COMPANY OF AMERICA, OF NEW YORK.

INTERNAL-COMBUSTION ENGINE.

SPECIFICATION forming part of Letters Patent No. 608,845, dated August 9, 1898.

Application filed July 15, 1895. Serial No. 556,059. (No model.) Patented in Spain December 3, 1894, No. 16,654; in France December 10, 1894, No. 243,531; in Belgium December 10, 1894, No. 113,139; in Luxemburg December 10, 1894, No. 2,192; in Italy February 21, 1895, LXXV, 132; in England February 27, 1895, No. 4,243; in Switzerland March 5, 1895, Nos. 10,134 and 10,135; in Germany March 30, 1895, No. 86,633; in Hungary November 23, 1895, No. 4,539, and March 20, 1897, No. 7,876; in Austria January 18, 1896, No. 46/203, and May 22, 1896, No. 46/2,038, and in Denmark February 12, 1896, No. 393.

To all whom it may concern:

Be it known that I, RUDOLF DIESEL, a subject of the King of Bavaria, and a resident of Berlin, in the Kingdom of Prussia, Germany, have invented certain new and useful Improvements in Internal-Combustion Engines, (for which I have obtained Letters Patent in Germany, No. 86,633, dated March 30, 1895; in France, No. 243,531, dated December 10, 1894, and patent of addition to the same, dated March 1, 1895; in Belgium, No. 113,139, dated December 10, 1894, and Patent of Addition No. 114,346, dated February 18, 1895; in England, No. 4,243, dated February 27, 1895; in Switzerland, Nos. 10,134 and 10,135, dated March 5, 1895; in Luxemburg, No. 2,192, dated December 10, 1894, and Patent of Addition No. 2,265, dated March 22, 1895; in Denmark, No. 393, dated February 12, 1896; in Austria, No. 46/203, dated January 18, 1896, and No. 46/2,038, dated May 22, 1896; in Hungary, No. 4,539, dated November 23, 1895, and No. 7,876, dated March 20, 1897; in Italy, LXXV, 132, dated February 21, 1895, and in Spain, No. 16,654, dated December 3, 1894, and Patent of Addition No. 17,085, dated March 4, 1895,) of which the following is a specification.

My invention has reference to improvements in apparatus for regulating the fuel-supply in slow-combustion motors, and in particular to internal-combustion engines adapted for carrying out the process described in my prior patent, No. 542,846, dated July 16, 1895, which process consists in first compressing air or a mixture of air and neutral gas or vapor to a degree producing a temperature above the igniting-point of the fuel to be consumed, then gradually introducing the fuel for combustion into the compressed air while expanding against resistance sufficiently to prevent an essential increase of temperature and pressure, then discontinuing the supply of fuel and further expanding without transfer of heat.

In ordinary combustion-engines the regulation of work done was performed either while the gas was at a constant pressure or, as in explosive engines, with the gas at constant volume.

The nature of my invention will best be understood when described in connection with the accompanying drawings, in which—

Figures 1 and 2 are diagrams illustrating the cycle of operation. Fig. 3 is a vertical section of an engine, illustrating one form of fuel-feed, part being broken away. Figs. 4 and 5 are similar views illustrating modified forms for the feed. Fig. 6 is a sectional elevation illustrating another modified form for the same. Fig. 7 shows sectional views of detail parts. Figs. 8, 9, and 10 illustrate in sectional elevation the arrangement of the mechanism for operating the valve. Figs. 11, 12, and 13 are sectional elevations illustrating different devices for mixing the air and fuel.

Similar letters and figures of reference designate corresponding parts throughout the several views of the drawings.

Referring now to Fig. 1 of the drawings, which illustrates a theoretical indicator-diagram of the engine, the curve 2 3 corresponds to the period of admission and consumption of fuel, the fuel being injected under a pressure greater than the pressure 0 2 at the point of highest compression.

By varying the excess of pressure under which fuel is injected and in the meantime the length or duration of admission of fuel the combustion-curve 2 3, Fig. 1, is changed both in its form or position, as in its length 2 3′, 2 3², &c., thus producing diagrams, such as 1 2 3 4 or 1 2 3′ 4′, &c. In all the diagrams shown in Fig. 1 the fuel is admitted at the point 2 of highest compression. In Fig. 2 the beginning of admission is variable, as will be hereinafter explained.

Referring now to Fig. 3 for a description of an apparatus for carrying out the regulation of the supply of fuel, the letter C designates a cylinder provided with a piston P and with an air-valve V. D is a nozzle for regulating the supply of fuel, by means of which the periods of admission and cut-off, and consequently the length of the curve 2 3 or 2 3′, &c., are determined with the use of a needle-valve *n*, actuated by any well-known mechanism. Pulverulent solid fuel is contained in a hopper T, provided with a rotary distributing-valve *r*. L is a reservoir which is supplied with suppressed gas through a pipe *m*. The gas may be air, a combustible gas, or a mixture of combustible gas and air. The air or gas or the mixture of the same is held under a pressure (by means of a pump or other well-known means) in excess of the highest pressure in the cylinder C. Said reservoir L is connected with the cylinder C by a pipe S and with the hopper T by a suitable branch pipe in communication with the pipe S.

When the valve *n* is lifted to open the nozzle D, the excess of pressure in the reservoir L causes the gas to flow through the pipe S and the nozzle D into the cylinder C, carrying with it the pulverulent fuel discharged by the turning of the valve *r*. In this manner an intimate mixture of gas and fuel is obtained and injected into the cylinder and rapid and complete combustion is insured.

If the pressure in the reservoir L were fixed and constant, the same combustion-curve 2 3 would always result for a predetermined and fixed admission and cut-off and a predetermined or fixed highest compression in the cylinder C; but if under these conditions of admission and cut-off the curve of combustion is to be altered or varied then the pressure in the reservoir L must be changed. This change is effected by means of the pressure-regulating valve R, Fig. 3, the weight B of which can be shifted by means of the rod Q, suitably connected with the governor of the engine. (Not shown.) By the coöperation of the needle-valve *n*, which regulates fuel admission and cut-off and the adjustable excess of gas-pressure, the form of the working diagram is determined. In other words, both variations conjointly produce the variable form of the combustion-curve, distinctly marking the new method of regulating. Both can be effected by the governor, or one by the governor and the other by hand, according to the degree of sensitiveness required. The valve for regulating the pressure in the reservoir L may of course be of any other construction which will answer the purpose and may in the usual manner regulate the supply through the pipe *m*. The pressure regulation can also be applied, as desired, to the pump feeding the tube *m*. This latter method would be adopted should fluid fuel be exclusively used, in which case the reservoir L would act as the pressure vessel of the pump. The fuel-supply apparatus might be placed directly on the reservoir L, as the motion of the gas therein would keep the dust in suspension. The hopper T might also contain fluid fuel.

The mixture of fuel and gas may take place in the interior of the cylinder or a prolongation thereof, as shown in Fig. 4. In this case the reservoir L contains pure compressed air, and in addition to the nozzle D for pulverulent fuel I provide a nozzle *d* for liquid or gaseous fuel for the purpose of intensifying combustion. In this instance the nozzle *d* is arranged concentric with the nozzle D, the liquid or gaseous fuel being supplied to said nozzle *d* through the lateral pipe S′, while the air for combustion and the solid fuel are supplied to the nozzle D through the pipe S, leading from the reservoir.

The nozzles D and *d*, instead of being arranged concentrically, may be arranged side by side, as shown in Fig. 5, and caused to discharge into a common combustion-chamber J, forming a prolongation of the cylinder and separated from the bore proper of the same by a perforated partition *a*. The regulation may be rendered still more sensitive by changing the fixed point 2 of the diagram, for instance, to 2′ or 2^2, Fig. 2, thus varying at the same time the height of the ordinate 0 2, 0′ 2′, 0^2 2^2, &c., and the length 0 1, 0′ 1, 0^2 1, &c., as well as the expansion-curve 3 4, 3′ 4′, 3^2 4^2, &c. This regulation is easily effected by opening the fuel-valve *n* not when the piston is at the commencement of its return stroke, but somewhat later, in which case compression takes place from 1 to 2, Fig. 2, as before; but the compressed gas first expands on the return stroke from 2 to 2′ or 2 2^2, &c., before the commencement of the period of combustion 2′ 3′, 2^2 3^2, &c.

Of course in practice the lines of the diagram are not so regular as shown, but about as indicated in dotted lines between 2′ and 3′, Fig. 2. It is also evident that the lead may be given to the fuel-valve *n* on the compression stroke, whereby the upper end of the compression-line 2′ 2 is made steeper and the combustion-curve changed to 2^3 3^3, taking, under circumstances, even a form like $2^3 m 3^3$. This lead of the valve may be effected by changing the position of the cam actuating said valve, so that the fuel will be introduced somewhat in advance of the end of the compression stroke of the piston and the valve kept open during part of the working stroke of the piston.

It is of particular importance that the fuel entering at the mouth should be thoroughly consumed and without the formation of soot. For this purpose all of the above-described devices for the admission of fuel may be provided within the cylinder with an additional burner similar in construction to those used for the same purpose for gas-burners—that is to say, the jet is not permitted to enter in solid cylindrical form, but is subdivided into thin sheets or jets. The construction may be similar to the Bunsen burner, which, as well known, gives a smokeless non-luminous frame. Such burners, located within the compression-space, are shown in Figs. 6, 7, 11, 12, and 13.

Fig. 6 shows a burner which subdivides the flame into a large number of very small tongue-shaped slow-burning jets. M' and M^2 show other forms of the same. The principle of the Bunsen burner is embodied in M^4 and M^5, the jet leaving the lower end while burning slowly and without discoloration. A similar effect is produced by the use of the twyer M^3.

The uniformity of diffusion of the heat throughout the whole mass of air in the compression-space is further increased by the peculiar arrangement of the burners—as, for instance, in Fig. 11, where owing to the lengthening of the twyer-pipe the burner is attached at E, so that while the piston is receding from I to II the greater part of the air is compelled to pass across the burner E. A second burner O may be provided.

Fig. 12 shows an arrangement for introducing the fuel laterally. The ribs R R on the left force the air on its way from the chamber to the cylinder and while expanding over the burners. The ribs R' R' to the right may be attached to the piston, so that the motion of the latter causes considerable agitation of the air. Finally, as shown in Fig. 13, the burner itself may be made movable for the purpose of obtaining more perfect distribution of heat. In this case the burner may be attached to the piston and the fuel supplied through a hollow piston-rod.

Figs. 6, 8, 9, and 10 show another way of carrying out the above-described method of regulation, the use of a special air-pump being dispensed with. In this instance the piston itself compresses the air necessary, not, however, in the usual way—by the momentum of the fly-wheel after cessation of combustion—but during the normal process of working without interrupting combustion and as an integral part of the working process itself.

In Figs. 6 and 8 the letter Y designates a valve through which during the regular working a small quantity of compressed air escapes at the end of each compressing stroke of the piston and passes by the tube *b*, Fig. 6, into the reservoir L. The air-pressure in the reservoir, therefore, equals the highest compression-pressure in the cylinder; but according to the previous description of the process an excess of pressure is required for the injection of the fuel. To obtain this result, the fuel-nozzle is not opened until the piston has slightly receded from the dead-point—that is to say, until the pressure in the cylinder has become somewhat lessened. As the opening of the nozzle by the governor occurs sooner or later, so the excess of pressure in the reservoir L varies. The injection of the fuel takes place, as previously described, S, Fig. 6, being the connecting-tube between the reservoir L and the nozzle, as in Fig. 3. The valve Y can also be arranged to be opened at the end of the stroke by the piston itself, or it might be a self-acting relief-valve, or for it might be substituted a cock or slide-valve.

Figs. 9 and 10 show the details of the gear for positively operating the valve Y. W is a cam-shaft provided with a number of cams I to V. Cam II works valve Y in normal working. Cam III works the fuel-valve for nozzle D, and cam IV operates the main valve V of the motor. This gear serves also in reverse order to start the motor, compressed air passing through valve Y from the reservoir L into the cylinder to drive the piston and then escapes through the valve V. During this very short starting period the lever H, Fig. 9, takes the dotted position H', so that the valve Y is moved by cam I instead of cam II, the valve V by cam IV instead of cam V, while the fuel-cam III is disengaged. After a few revolutions made in this manner the motor obtains its normal speed. At this moment the detent *p*, which retains the lever H in position, is removed. The lever is automatically pushed by spring F to the normal working position H, carrying with it the system of cams to continue the normal working without interruption. As the moving of the cams has to occur at the exact moment, it can only take place when a specially-arranged notch in the hub of the cams receives the detent *p*.

The valve Y (shown in Fig. 7) serves three purposes: first, to start the motor with compressed air; secondly, to fill the reservoir L during normal working, and, thirdly, to operate as a safety-valve, it being loaded by a spring *l*, so that on explosion in the cylinder the gases can pass to the reservoir L and thence through the safety-valve R.

To determine the maximum pressure in the cylinder, and consequently that in the reservoir L, a hand-wheel H^2 is applied, by means of which the spring *l* can be compressed more or less either while the engine is stopped or in motion.

It is evident that by adjusting the position of the cam III on the shaft W the time of opening of the fuel-supply valve can be varied—that is to say, by turning the cam either to the right or the left on the cam-shaft the time of admission will be made earlier or later. It is also evident that by interchanging cams a different timing of admission can be obtained.

What I claim as new is—

1. In an internal-combustion engine, the combination of a cylinder and piston constructed and arranged to compress air to a degree producing a temperature above the igniting-point of the fuel, a supply for compressed air or gas; a fuel-supply; a distributing-valve for fuel, a passage from the air-supply to the cylinder in communication with the fuel-distributing valve, an inlet to the cylinder in communication with the air-supply and with the fuel-valve, and a cut-off, substantially as described.

2. In an internal-combustion engine, the combination of a cylinder and piston constructed and arranged to compress air to a degree producing a temperature above the igniting-point of the fuel; a distributing-valve for fuel; a cut-off for varying the time and duration of the supply of fuel, and a burner placed in the combustion-space and constructed for slow and perfect combustion of the gradually-introduced stream of fuel, substantially as shown and described.

3. In an internal-combustion engine, the combination of a cylinder and piston constructed and arranged to compress air to a degree producing a temperature above the igniting-point of the fuel, a supply for compressed air or gas, a hopper, a distributing-valve for pulverulent fuel, a passage from the air-supply to the cylinder in communication with the fuel-distributing valve, an inlet-valve to the cylinder in communication with the air-supply and with the valve for pulverulent fuel, and a cut-off for the fuel-supply, substantially as shown and described.

4. In an internal slow-combustion engine, the combination of a cylinder and piston constructed and arranged to compress air to a degree producing a temperature above the igniting-point of the fuel, a supply for compressed air, a hopper and distributing-valve for pulverulent fuel, a supply-pipe for liquid fuel, a valve or valves leading to the cylinder and communicating with the pulverulent-fuel-distributing valve and the liquid-fuel-supply pipe, and a cut-off for the fuel-supply, substantially as specified.

5. In an internal-combustion engine, the combination of a supply for compressed air, a feed for pulverulent fuel placed in communication with the air-supply and with the cylinder, and an auxiliary feed for liquid fuel communicating with the cylinder, substantially as specified.

6. In an internal-combustion engine, the combination of a cylinder and piston, a supply for compressed air, a distributing-valve communicating with the air-supply and with a fuel-supply for gradually introducing a unitary, or mixed fuel, into the combustion-space, a valve placed between the air-supply and the cylinder, and a reversing-gear in cooperation with said valve for starting the motor with the compressed air from the air-supply, substantially as described.

7. In an internal-combustion engine, the combination with a cylinder and a piston constructed to compress air to a degree producing a temperature above the igniting-point of the fuel, of a fuel-feed, and a valve mechanism adapted to open the fuel-feed somewhat in advance of the end of the compression stroke of the piston and to keep it open during part of the working stroke, substantially as and for the purpose specified.

8. In an internal-combustion engine, the combination of a cylinder and piston constructed to compress air or a mixture of air and neutral gas, a storage-reservoir in communication with the combustion-space of the cylinder, a valve controlling this communication and opening to admit compressed air from the cylinder to the reservoir, and a fuel-feed in communication with said reservoir for the introduction of fuel to the combustion-space under the pressure of the compressed air or gas in the reservoir, substantially as described.

9. In an internal-combustion engine, the combination of a cylinder and piston constructed and arranged to compress air to a degree producing a temperature above the igniting-point of the fuel, a distributing-valve for fuel, and a cut-off for varying the time and duration of the supply of fuel by said valve, substantially as described.

In testimony that I claim the foregoing as my invention I have signed my name in presence of two witnesses.

RUDOLF DIESEL.

Witnesses:
WM. HAUPT,
CHR. KRÜGER.

United States Patent Office.

ALFRED NOBEL, OF HAMBURG, GERMANY, ASSIGNOR TO JULIUS BANDMANN, OF SAN FRANCISCO, CALIFORNIA.

Letters Patent No. 78,317, *dated May* 26, 1868.

IMPROVED EXPLOSIVE COMPOUND.

The Schedule referred to in these Letters Patent and making part of the same.

TO ALL WHOM IT MAY CONCERN:

Be it known that I, ALFRED NOBEL, of the city of Hamburg, Germany, have invented a new and useful Composition of Matter, to wit, an Explosive Powder.

The nature of the invention consists in forming out of two ingredients long known, viz, the explosive substance nitro-glycerine, and an inexplosive porous substance, hereafter specified, a composition which, without losing the great explosive power of nitro-glycerine, is very much altered as to its explosive and other properties, being far more safe and convenient for transportation, storage, and use, than nitro-glycerine.

In general terms, my invention consists in mixing with nitro-glycerine a substance which possesses a very great absorbent capacity, and which, at the same time, is free from any quality which will decompose, destroy, or injure the nitro-glycerine, or its explosiveness.

It is undoubtedly true, as a general rule, that nitro-glycerine, when mixed with another substance, possesses less concentration of power than when used alone; but while the safety of the miner (to prevent leakage into seams in the rock) prohibits the use of nitro-glycerine without cartridges, which latter must of course be somewhat less in diameter than the bore-holes which are to contain them, the powder herein described can be made to form a semi-pasty mass, which yields to the slightest pressure, and thus can be made to fill up the bore-hole entirely. Practically, therefore, the miner will have as much nitro-glycerine in the same height of bore-hole with this powder as with nitro-glycerine in its pure state.

This is the real character and purpose of my invention; and in order to enable others skilled in the art to which it appertains (or with which it is most nearly connected) to make, compound, and use the same, I will proceed to describe the same, and also the manner and process of making, compounding, and using it, in full, clear, and exact terms.

The substance which most fully meets the requirements above mentioned, so far as I know or have been able to ascertain from numerous experiments, is a certain kind of silicious earth or silicic acid, found in various parts of the globe, and known under the several names of silicious marl, tripoli, rotten-stone, &c. The particular variety of this material which is best for my compound is homogeneous, has a low specific gravity, great absorbent capacity, and is generally composed of the remains of *infusoria*.

So great is the absorbent capacity of this earth, that it will take up about three times its own weight of nitro-glycerine and still retain its powder-form, thus leaving the nitro-glycerine so compact and concentrated as to have very nearly its original explosive power; whereas, if another substance, having a less absorbent capacity, is used, a correspondingly less proportion of nitro-glycerine will be absorbed, and the powder be correspondingly weak or wholly inexplosive.

For example, most chalk will take but about fifteen per cent. of nitro-glycerine and retain its powder-form. Twenty per cent. will reduce it to a paste.

Porous charcoal has also a considerable absorbent capacity, but it has the defect of being itself a combustible material, and also of less elasticity of its particles, which renders it easy to squeeze out a part of its nitro-glycerine.

The two materials are combined in the following manner:

The earth, thoroughly dried and pulverized, is placed in a wooden vessel. To it is introduced the nitro-glycerine in a steady stream so small that the two ingredients can be kept thoroughly mixed.

The mixing may be effected by the naked hand, or by any proper wooden instrument used in the hand, or by wooden machinery

Sufficient of nitro-glycerine should be used to render the compound explosive, but not so much as to change its form of powder to a liquid or pasty consistency.

Practically, about sixty parts, by weight, of nitro-glycerine to forty of earth, forms the useful minimum, and seventy-eight parts, by weight, of nitro-glycerine to twenty-two of earth, the useful maximum of explosive power. The former has a perfectly dry appearance, the latter is pasty.

Between these two extremes the composition will be explosive powder, and it will be more easily exploded, and its explosive power greater, as the relative proportion of the nitro-glycerine is greater.

The proportions, by weight, of seventy-five of nitro-glycerine to twenty-five of earth, gives a powder as well adapted to ordinary practical purposes as that from any proportions I am now able to name, and can be easily compressed to a specific gravity nearly equal to that of pure nitro-glycerine.

When the mass has been intimately mixed and thoroughly incorporated by stirring and kneading, it is rubbed through a hair, silk, or brass-wire sieve, (iron corrodes,) and any lumps which may remain are rubbed with a stiff-bristle brush till they are reduced and made to pass through the sieve.

The powder is then finished and ready for use.

The fineness desired for the powder will determine the fineness of the sieve to be used.

The chief characteristic of this powder is its nearly perfect exemption from liability to accidental or involuntary explosion.

It is far less sensitive than nitro-glycerine to concussion or percussion, and contained in its usual packing, (a wooden cask or box,) the latter may be smashed completely to pieces without any danger of an explosion.

Unlike gunpowder, in the open air or in ordinary packing, (a wooden cask or box,) it burns up, when set fire to, without exploding. It can, therefore, be handled, stored, and transported with less danger than ordinary gunpowder.

When confined in a tight and strong enclosure it explodes by heat applied in any form when above the temperature of 360° Fahrenheit. Under all other circumstances it may be exploded by some other explosion in it or into it.

The most simple and certain method known to me of exploding it is as follows:

The end of a common blasting-fuse is inserted into a percussion-cap, and the rim of the cap crimped tightly and firmly about the fuse by nippers, or other means, so as to leave the fulminating-powder of the cap and the end of the fuse tightly and firmly enclosed together. The end of the fuse, with the cap attached, is then embedded in the powder—the more firmly, the more certain the explosion.

In blasting, the powder is pressed tightly about the cap and fuse, and tamping, of sand or other proper material, added, and pressed but not pounded in. A tamping firmly pressed is as good as if rammed in the most solid manner.

The fuse explodes the cap, and this explosion explodes the powder.

I will add here that by carefully packing the end of a good fuse amidst the powder of a charge enclosed, like a blasting charge, in a tight place, the fuse alone will explode the powder, especially if the powder is strongly charged with nitro-glycerine. But this method of explosion requires too much care, and is too uncertain to be depended upon or generally used.

As before stated, the more strongly the powder is charged with nitro-glycerine the more easily it explodes. If, therefore, the powder contains a low proportion of nitro-glycerine, it is necessary to employ in its explosion a correspondingly long, strong, and heavily-charged percussion-cap, made especially for the purpose. For the sake of certainty of explosion it is better to use such a cap in all cases.

If the fire from the fuse comes in contact with the powder before the cap is exploded, which is liable to occur if the fuse is leaky and the cap extends too far into the powder, a portion of the powder will be burned before the explosion takes place. To guard against this, the cap should only be fairly inserted into the powder, and poor fuses wound next to the cap firmly with strong glued paper or hemp, or otherwise secured.

The bore-holes, as a practical but not absolute rule, should be about one-half the size, and the charge should be from one-fifth to one-tenth the quantity ordinarily used in gunpowder-blasting.

A very convenient form in which to use the powder is to pack it firmly in cartridges of strong paper.

Having thus described my invention, what I claim as new, and desire to secure by Letters Patent, is—

The composition of matter, made substantially of the ingredients and in the manner and for the purposes set forth.

ALFRED NOBEL.

Witnesses:

FR. T. PROHME,

HEINR. BARTELTSSEN

W. Kelly.

Manuf. of Iron & Steel.

No. 17.628. Patented June 23. 1857.

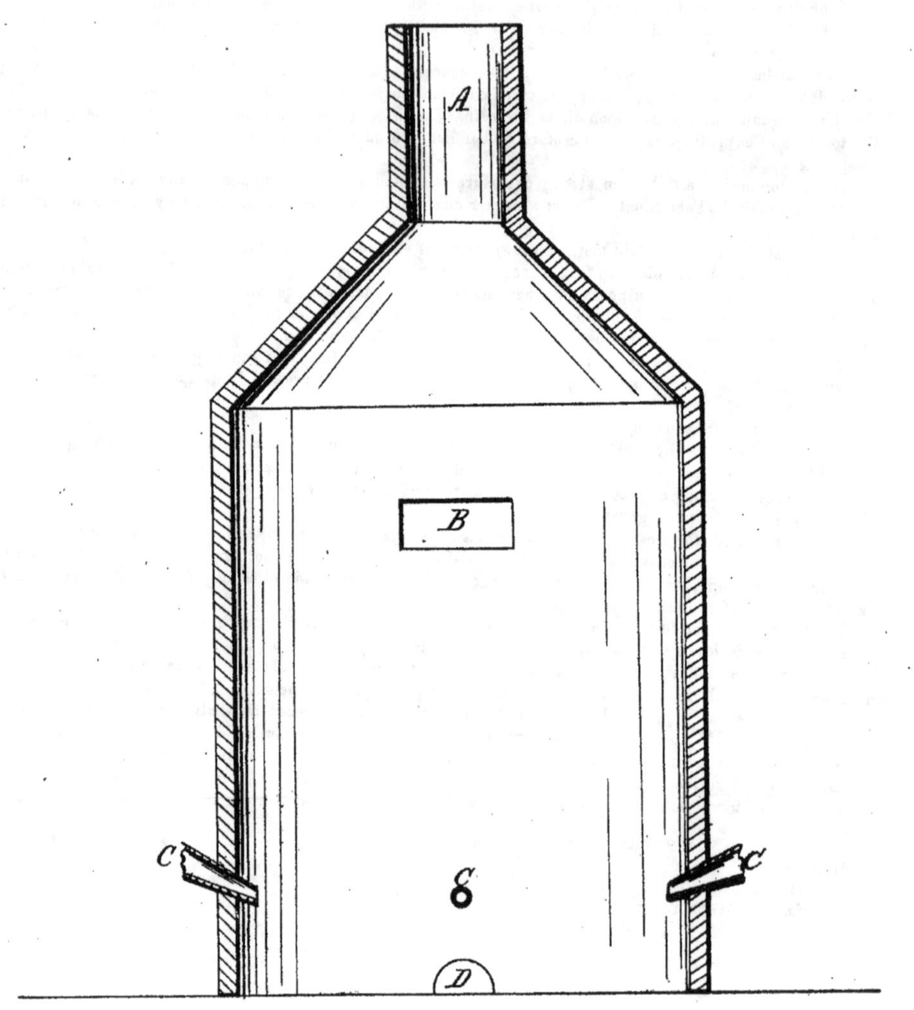

UNITED STATES PATENT OFFICE.

WILLIAM KELLY, OF LYON COUNTY, KENTUCKY.

IMPROVEMENT IN THE MANUFACTURE OF IRON.

Specification forming part of Letters Patent No. **17,628**, dated June 23, 1857.

To all whom it may concern:

Be it known that I, WILLIAM KELLY, of Lyon county, Kentucky, have discovered a new and Improved Method of Treating Iron, by which I am enabled to refine and decarbonize crude pig metal or iron in a fluid state without the use of fuel.

The nature of my invention consists in the discovery that the carbon mechanically combined with iron, and which is burned from the fuel while in the process of smelting in the blast-furnace, is of itself (the carbon) sufficient, when combined with the oxygen of the air, to create heat enough and of sufficient intensity to keep melted pig iron or metal in a fluid or lively state long enough to carry the metal through, without chilling, all the various manipulations of refining without the aid of any other heat than that obtained by the above-described chemical union of oxygen and carbon.

I am aware that it is well known that oxygen and carbon, when combined or brought together, produce heat; but it is not known that the amount of these chemical properties in air and iron is the required quantity necessary to produce heat sufficient to carry out the practical refining of crude pig-iron; hence the prevailing opinion among iron-workers that a blast of cold air driven into a body of liquid iron would chill it. Therefore, when iron is worked in the finery or run-out fire the presence of heat from other sources is deemed indispensable to prevent the chilling of the iron. The finery or run-out fire is usually open on three sides, sometimes closed except at top to receive the charge of coal and iron.

A furnace or cupola to work iron under my new process must be constructed as close as possible to prevent a loss of iron which would occur on account of its violent boiling, during which particles are thrown up and adhere to the sides and top of the chamber, but which during the process are remelted and flow down to the mass in the bottom. In the finery or run-out this loss is prevented by the iron being covered by fuel. It is also first charged with metal in a solid state. In my process the metal is taken in a fluid state from the blast-furnace and put in the cupola or furnace. In the finery or run-out the iron is brought to a fluid state by mixing it with large quantities of fuel, and when melted falls to the bottom of the finery, where it is decarbonized by strong blasts of air in connection with the fuel. In my process no fuel of any kind is used or required, as I rely exclusively on the heat created or generated by the chemical union of oxygen in the air and carbon in the iron.

In the accompanying drawings, Figure 1 represents a vertical section of cupola or furnace used in my process, being a close cylindrical chamber with a flue, A, at top to carry off the carbonic-acid gas formed in decarbonizing the iron.

B is a small opening to receive the charge of fluid iron.

C C C are the tuyeres placed around the sides of the furnace, pointing downward at an angle such that they sweep about three-quarters of the bottom of the chamber, the muzzles of the tuyeres being about six inches above the bottom of the chamber.

D is a tap-hole for letting out the metal when refined. The chamber should not exceed three or four times the space occupied by the fluid iron. The blast is first let on into this chamber or cupola; then the fluid iron is poured in, which, by the cause hereinbefore described, commences a violent ebullition or boiling, which continues until the iron is sufficiently refined, when the tap-hole is opened and the metal let out.

What I claim as my invention or discovery, and desire to secure by Letters Patent, is—

Blowing blasts of air, either hot or cold, up and through a mass of liquid iron, the oxygen in the air combining with the carbon in the iron, causing a greatly increased heat and boiling commotion in the fluid mass and decarbonizing and refining the iron.

WILLIAM KELLY.

Witnesses:

W. B. MACHEW,

JAS. N. GRACY.

(No Model.) 4 Sheets—Sheet 1.

N. TESLA.

ELECTRO MAGNETIC MOTOR.

No. 381,968. Patented May 1, 1888.

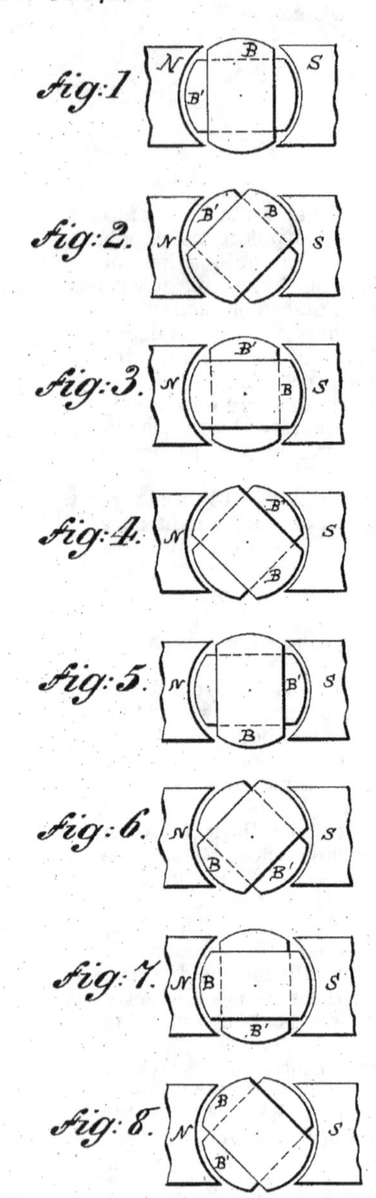

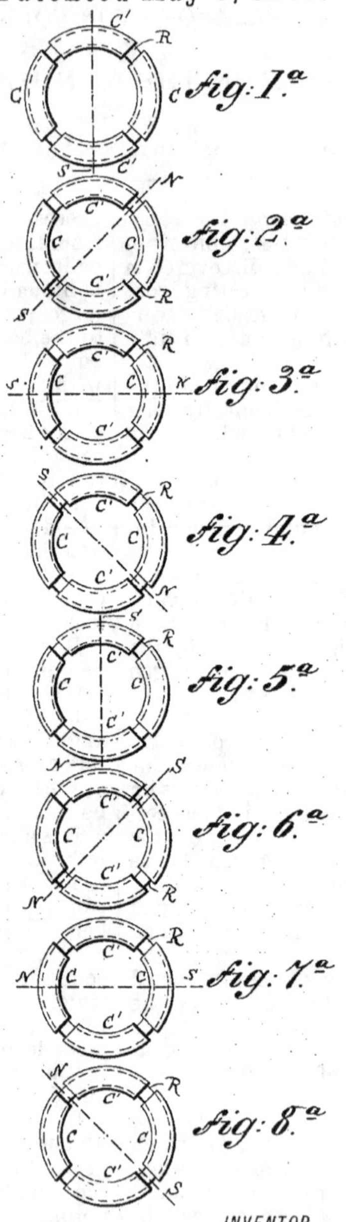

WITNESSES:

Frank E. Hartley

Frank B. Murphy

INVENTOR.

Nikola Tesla.

BY

Duncan, Curtis & Page

ATTORNEYS.

(No Model.) 4 Sheets—Sheet 2.

N. TESLA.

ELECTRO MAGNETIC MOTOR.

No. 381,968. Patented May 1, 1888.

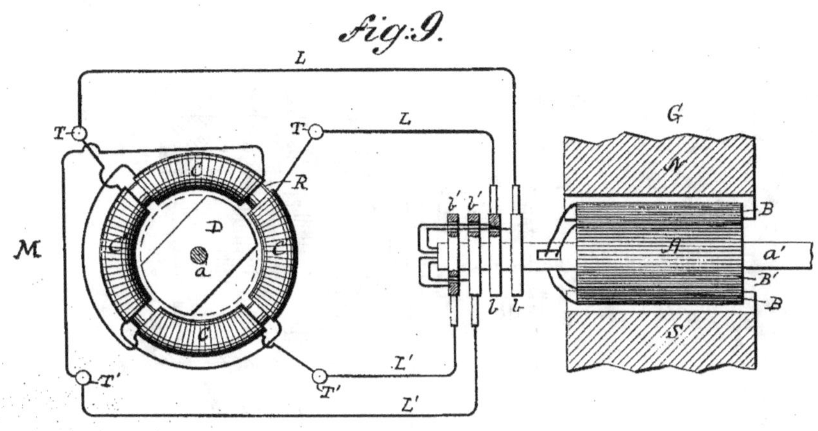

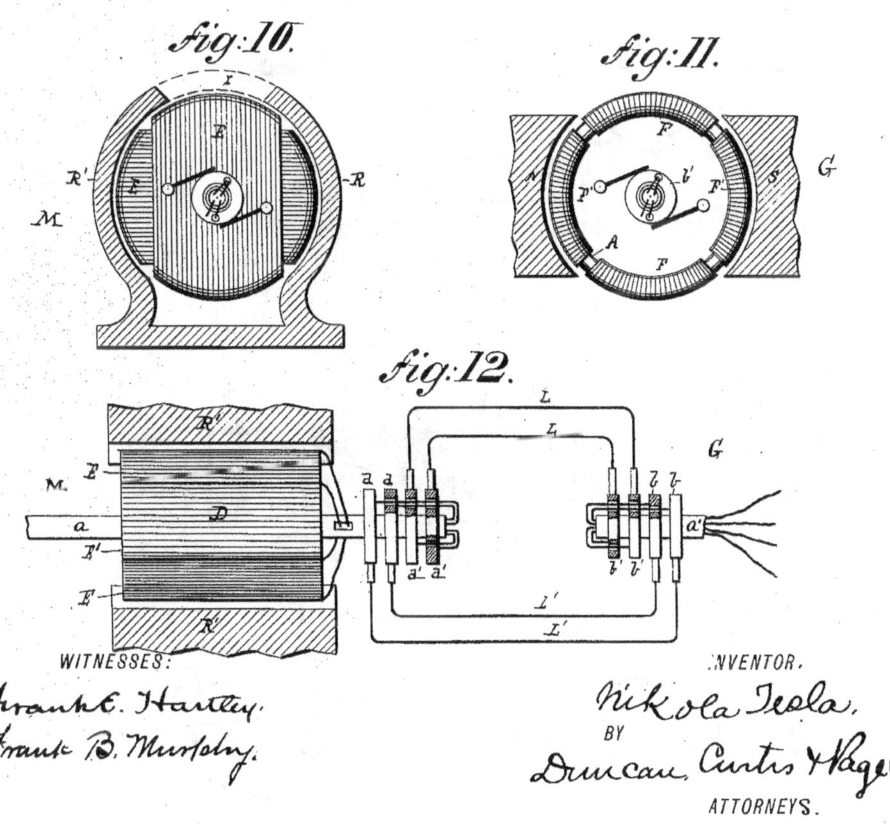

WITNESSES:

Frank E. Hartley.

Frank B. Murphy.

INVENTOR.

Nikola Tesla.

BY

Duncan, Curtis & Page

ATTORNEYS.

(No Model.) 4 Sheets—Sheet 3.

N. TESLA.

ELECTRO MAGNETIC MOTOR.

No. 381,968. Patented May 1, 1888.

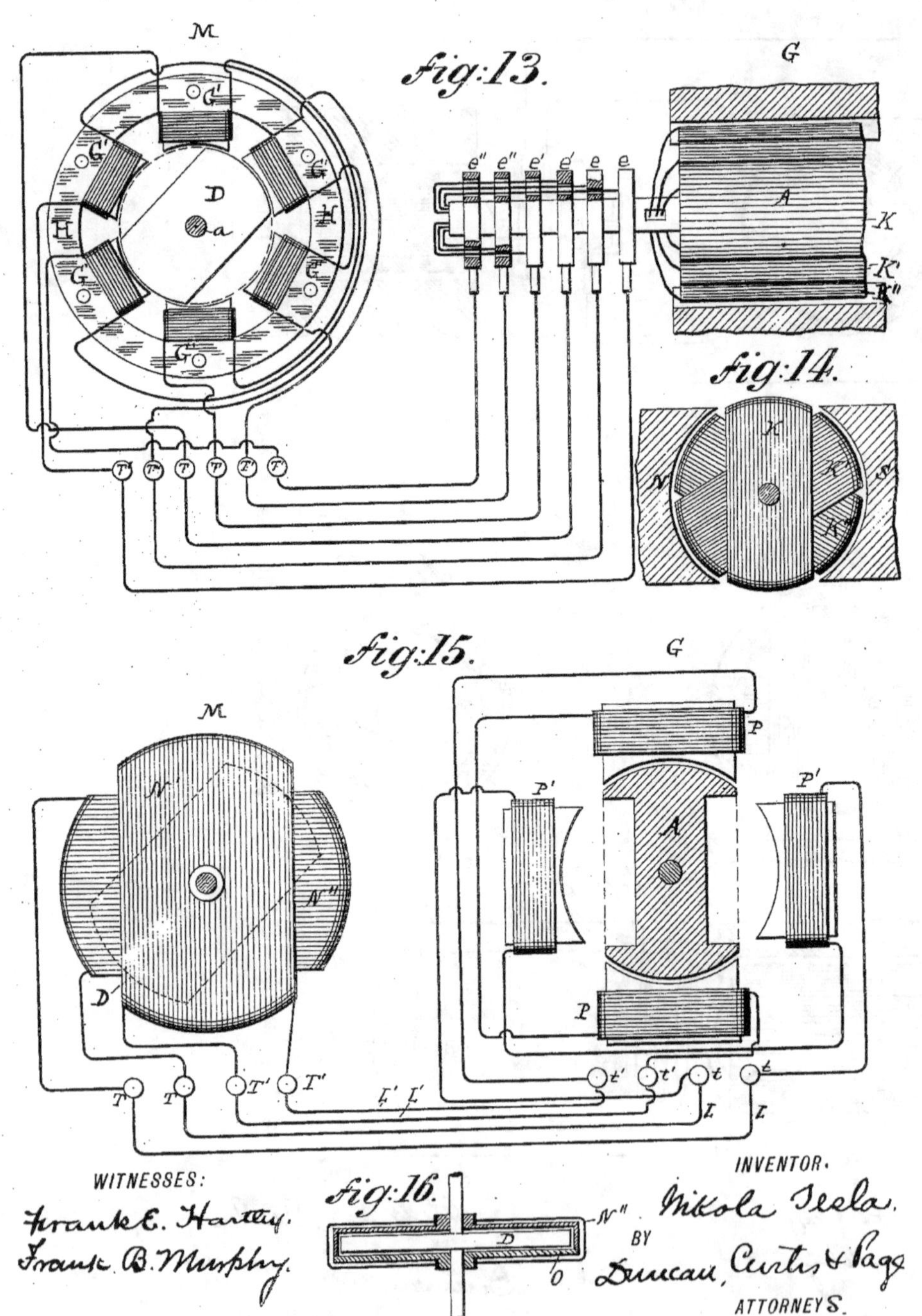

(No Model.) 4 Sheets—Sheet 4.

N. TESLA.

ELECTRO MAGNETIC MOTOR.

No. 381,968. Patented May 1, 1888.

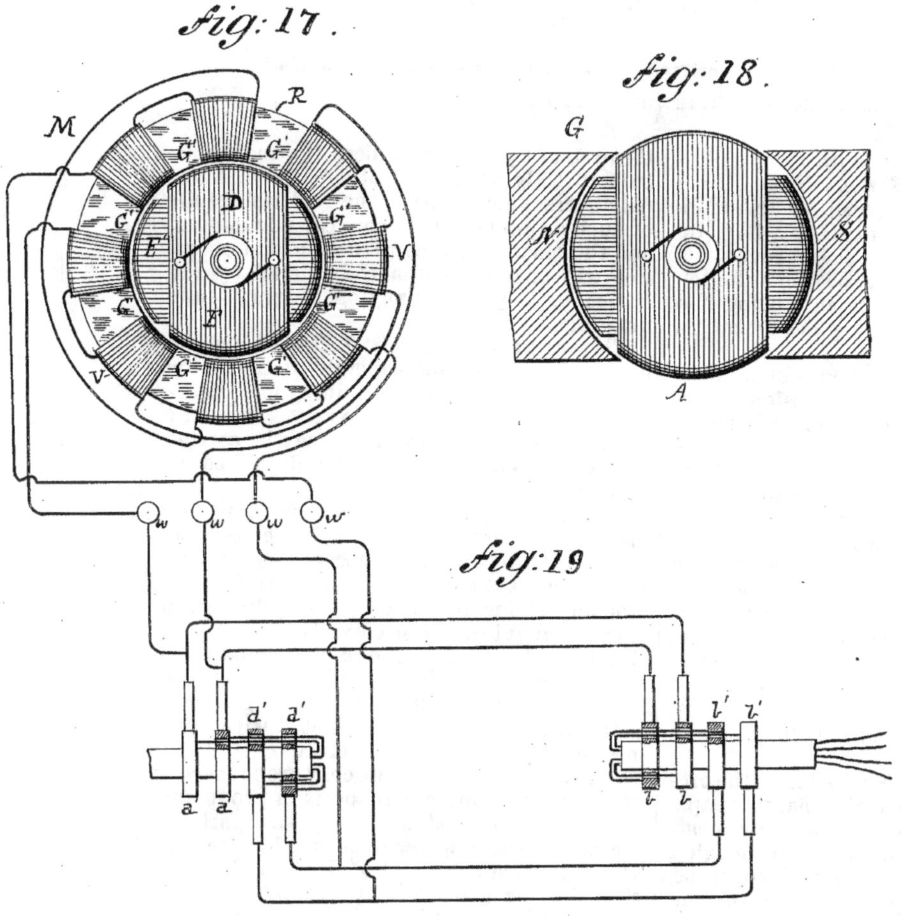

WITNESSES:

Frank E. Hartley.
Frank B. Murphy.

INVENTOR.

Nikola Tesla,

BY

Duncan, Curtis & Page

ATTORNEYS.

UNITED STATES PATENT OFFICE.

NIKOLA TESLA, OF NEW YORK, N. Y., ASSIGNOR OF ONE-HALF TO CHARLES F. PECK, OF ENGLEWOOD, NEW JERSEY.

ELECTRO-MAGNETIC MOTOR.

SPECIFICATION forming part of Letters Patent No. 381,968, dated May 1, 1888.

Application filed October 12, 1887. Serial No. 252,132. (No model.)

To all whom it may concern:

Be it known that I, NIKOLA TESLA, from Smiljan Lika, border country of Austria-Hungary, residing at New York, N. Y., have invented certain new and useful Improvements in Electro-Magnetic Motors, of which the following is a specification, reference being had to the drawings accompanying and forming a part of the same.

The practical solution of the problem of the electrical conversion and transmission of mechanical energy involves certain requirements which the apparatus and systems heretofore employed have not been capable of fulfilling. Such a solution, primarily, demands a uniformity of speed in the motor irrespective of its load within its normal working limits. On the other hand, it is necessary, to attain a greater economy of conversion than has heretofore existed, to construct cheaper and more reliable and simple apparatus, and, lastly, the apparatus must be capable of easy management, and such that all danger from the use of currents of high tension, which are necessary to an economical transmission, may be avoided.

My present invention is directed to the production and improvement of apparatus capable of more nearly meeting these requirements than those heretofore available, and though I have described various means for the purpose, they involve the same main principles of construction and mode of operation, which may be described as follows: A motor is employed in which there are two or more independent circuits through which alternate currents are passed at proper intervals, in the manner hereinafter described, for the purpose of effecting a progressive shifting of the magnetism or of the "lines of force" in accordance with the well-known theory, and a consequent action of the motor. It is obvious that a proper progressive shifting of the lines of force may be utilized to set up a movement or rotation of either element of the motor, the armature, or the field-magnet, and that if the currents directed through the several circuits of the motor are in the proper direction no commutator for the motor will be required; but to avoid all the usual commutating appliances in the system I prefer to connect the motor-circuits directly with those of a suitable alternate-current generator. The practical results of such a system, its economical advantages, and the mode of its construction and operation will be described more in detail by reference to the accompanying diagrams and drawings.

Figures 1 to 8 and 1ª to 8ª, inclusive, are diagrams illustrating the principle of the action of my invention. The remaining figures are views of the apparatus in various forms by means of which the invention may be carried into effect, and which will be described in their order.

Referring first to Fig. 9, which is a diagrammatic representation of a motor, a generator, and connecting-circuits in accordance with my invention, M is the motor, and G the generator for driving it. The motor comprises a ring or annulus, R, preferably built up of thin insulated iron rings or annular plates, so as to be as susceptible as possible to variations in its magnetic condition. This ring is surrounded by four coils of insulated wire symmetrically placed, and designated by C C C′ C′. The diametrically-opposite coils are connected up so as to co-operate in pairs in producing free poles on diametrically-opposite parts of the ring. The four free ends thus left are connected to terminals T T T′ T′, as indicated. Near the ring, and preferably inside of it, there is mounted on an axis or shaft, *a*, a magnetic disk, D, generally circular in shape, but having two segments cut away, as shown. This disk is mounted so as to turn freely within the ring R. The generator G is of any ordinary type, that shown in the present instance having field-magnets N S and a cylindrical armature-core, A, wound with the two coils B B′. The free ends of each coil are carried through the shaft *a′* and connected, respectively, to insulated contact-rings *b b b′ b′*. Any convenient form of collector or brush bears on each ring and forms a terminal by which the current to and from a ring is conveyed. These terminals are connected to the terminals of the motor by the wires L and L′ in the manner indicated, whereby two complete circuits are formed—one including, say, the coils B of the generator C′ C′ of the motor, and the other the remaining coils B′ and C C of the generator and the motor.

It remains now to explain the mode of operation of this system, and for this purpose I refer to the diagrams, Figs. 1 to 8, and 1ª to 8ª, for an illustration of the various phases through which the coils of the generator pass when in operation, and the corresponding and resultant magnetic changes produced in the motor. The revolution of the armature of the generator between the field-magnets N S obviously produces in the coils B B′ alternating currents, the intensity and direction of which depend upon well-known laws. In the position of the coils indicated in Fig. 1 the current in the coil B is practically *nil*, whereas the coil B′ at the same time is developing its maximum current, and by the means indicated in the description of Fig. 9 the circuit including this coil B′ may also include, say, the coils C C of the motor, Fig. 1ª. The result, with the proper connections, would be the magnetization of the ring R′, the poles being on the line N S. The same order of connections being observed between the coil B and the coils C′, the latter, when traversed by a current, tend to fix the poles at right angles to the line N S of Fig. 1ª. It results, therefore, that when the generator-coils have made one eighth of a revolution, reaching the position shown in Fig. 2, both pairs of coils C and C′ will be traversed by currents and act in opposition, in so far as the location of the poles is concerned. The position of the poles will therefore be the resultant of the magnetizing forces of the coils—that is to say, it will advance along the ring to a position corresponding to one-eighth of the revolution of the armature of the generator. In Fig. 3 the armature of the generator has progressed to one-quarter of a revolution. At the point indicated the current in the coil B is maximum, while in B′ it is *nil*, the latter coil being in its neutral position. The poles of the ring R in Fig. 3ª will, in consequence, be shifted to a position ninety degrees from that at the start, as shown. I have in like manner shown the conditions existing at each successive eighth of one revolution in the remaining figures. A short reference to these figures will suffice for an understanding of their significance. Figs. 4 and 4ª illustrate the conditions which exist when the generator-armature has completed three eighths of a revolution. Here both coils are generating current; but the coil B′, having now entered the opposite field, is generating a current in the opposite direction, having the opposite magnetizing effect; hence the resultant pole will be on the line N S, as shown. In Fig. 5 one-half of one revolution of the armature of the generator has been completed, and the resulting magnetic condition of the ring is shown in Fig. 5ª. In this phase coil B is in the neutral position while coil B′ is generating its maximum current, which is in the same direction as in Fig. 4. The poles will consequently be shifted through one half of the ring. In Fig. 6 the armature has completed five-eighths of a revolution. In this position coil B′ develops a less powerful current, but in the same direction as before. The coil B, on the other hand, having entered a field of opposite polarity, generates a current of opposite direction. The resultant poles will therefore be in the line N S, Fig. 6ª, or, in other words, the poles of the ring will be shifted along five-eighths of its periphery. Figs. 7 and 7ª in the same manner illustrate the phases of the generator and ring at three-quarters of a revolution, and Figs. 8 and 8ª the same at seven-eighths of a revolution of the generator-armature. These figures will be readily understood from the foregoing. When a complete revolution is accomplished, the conditions existing at the start are re-established and the same action is repeated for the next and all subsequent revolutions, and, in general, it will now be seen that every revolution of the armature of the generator produces a corresponding shifting of the poles or lines of force around the ring. This effect I utilize in producing the rotation of a body or armature in a variety of ways—for example, applying the principle above described to the apparatus shown in Fig. 9. The disk D, owing to its tendency to assume that position in which it embraces the greatest possible number of the magnetic lines, is set in rotation, following the motion of the lines or the points of greatest attraction.

The disk D in Fig. 9 is shown as cut away on opposite sides; but this, I have found, is not essential to effecting its rotation, as a circular disk, as indicated by dotted lines, is also set in rotation. This phenomenon I attribute to a certain inertia or resistance inherent in the metal to the rapid shifting of the lines of force through the same, which results in a continuous tangential pull upon the disk, causing its rotation. This seems to be confirmed by the fact that a circular disk of steel is more effectively rotated than one of soft iron, for the reason that the former is assumed to possess a greater resistance to the shifting of the magnetic lines.

In illustration of other forms of my invention, I shall now describe the remaining figures of the drawings.

Fig. 10 is a view in elevation and part vertical section of a motor. Fig. 12 is a top view of the same with the field in section and a diagram of connections. Fig. 11 is an end or side view of a generator with the fields in section. This form of motor may be used in place of that shown above. D is a cylindrical or drum-armature core, which, for obvious reasons, should be split up as far as practicable to prevent the circulation within it of currents of induction. The core is wound longitudinally with two coils, E and E′, the ends of which are respectively connected to insulated contact-rings *d d d′ d′*, carried by the shaft *a*, upon which the armature is mounted. The armature is set to revolve within an iron shell, R′, which constitutes the field-magnet, or other element of the motor. This shell is preferably formed with a slot or opening, *r*, but it may be continuous, as shown by the dotted lines, and in this event it is preferably made of steel. It is also desirable that this shell should be divided up similarly to the armature and for similar reasons. As a generator for driving this motor I may use the device shown in Fig. 11. This represents an annular or ring armature, A, surrounded by four coils, F F F′ F′, of which those diametrically opposite are connected in series, so that four free ends are left, which are connected to the insulated contact-rings *b b b′ b′*. The ring is suitably mounted on a shaft, *a′*, between the poles N S. The contact-rings of each pair of generator-coils are connected to those of the motor, respectively, by means of contact-brushes and the two pairs of conductors L L and L′ L′, as indicated diagrammatically in Fig. 12. Now it is obvious from a consideration of the preceding figures that the rotation of the generator-ring produces currents in the coils F F′, which, being transmitted to the motor-coils, impart to the core of the latter magnetic poles constantly shifting or whirling around the core. This effect sets up a rotation of the armature owing to the attractive force between the shell and the poles of the armature, but inasmuch as the coils in this case move relative to the shell or field-magnet the movement of the coils is in the opposite direction to the progressive shifting of the poles.

Other arrangements of the coils of both generator and motor are possible, and a greater number of circuits may be used, as will be seen in the two succeeding figures.

Fig. 13 is a diagrammatic illustration of a motor and a generator constructed and connected in accordance with my invention. Fig. 14 is an end view of the generator with its field-magnets in section. The field of the motor M is produced by six magnetic poles, G′ G′, secured to or projecting from a ring or frame, H. These magnets or poles are wound with insulated coils, those diametrically opposite to each other being connected in pairs so as to produce opposite poles in each pair. This leaves six free ends, which are connected to the terminals T T T′ T′ T″ T″. The armature, which is mounted to rotate between the poles, is a cylinder or disk, D, of wrought-iron, mounted on the shaft *a*. Two segments of the same are cut away, as shown. The generator for this motor has in this instance an armature, A, wound with three coils, K K′ K″, at sixty degrees apart. The ends of these coils are connected, respectively, to insulated contact-rings *e e e′ e′ e″ e″*. These rings are connected to those of the motor in proper order by means of collecting-brushes and six wires, forming three independent circuits. The variations in the strength and direction of the currents transmitted through these circuits and traversing the coils of the motor produce a steadily-progressive shifting of the resultant attractive force exerted by the poles G′ upon the armature D, and consequently keep the armature rapidly rotating. The peculiar advantage of this disposition is in obtaining a more concentrated and powerful field. The application of this principle to systems involving multiple circuits generally will be understood from this apparatus.

Referring, now, to Figs. 15 and 16, Fig. 15 is a diagrammatic representation of a modified disposition of my invention. Fig. 16 is a horizontal cross-section of the motor. In this case a disk, D, of magnetic metal, preferably cut away at opposite edges, as shown in dotted lines in Fig. 15, is mounted so as to turn freely inside two stationary coils, N′ N″, placed at right angles to one another. The coils are preferably wound on a frame, O, of insulating material, and their ends are connected to the fixed terminals T T T′ T′. The generator G is a representative of that class of alternating-current machines in which a stationary induced element is employed. That shown consists of a revolving permanent or electro magnet, A, and four independent stationary magnets, P P′, wound with coils, those diametrically opposite to each other being connected in series and having their ends secured to the terminals *t t t′ t′*. From these terminals the currents are led to the terminals of the motor, as shown in the drawings. The mode of operation is substantially the same as in the previous cases, the currents traversing the coils of the motor having the effect to turn the disk D. This mode of carrying out the invention has the advantage of dispensing with the sliding contacts in the system.

In the forms of motor above described only one of the elements, the armature or the field-magnet, is provided with energizing-coils. It remains, then, to show how both elements may be wound with coils. Reference is therefore had to Figs. 17, 18, and 19. Fig. 17 is an end view of such a motor. Fig. 18 is a similar view of the generator with the field-magnets in section, and Fig. 19 is a diagram of the circuit-connections. In Fig. 17 the field-magnet of the motor consists of a ring, R, preferably of thin insulated iron sheets or bands with eight pole pieces, G′, and corresponding recesses, in which four pairs of coils, V, are wound. The diametrically-opposite pairs of coils are connected in series and the free ends connected to four terminals, *w*, the rule to be followed in connecting being the same as hereinbefore explained. An armature, D, with two coils, E E′, at right angles to each other, is mounted to rotate inside of the field-magnet R. The ends of the armature-coils are connected to two pairs of contact-rings, *d d d′ d′*, Fig. 19. The generator for this motor may be of any suitable kind to produce currents of the desired character. In the present instance it consists of a field-magnet, N S, and an armature, A, with two coils at right angles, the ends of which are connected to four contact-rings, *b b b′ b′*, carried by its shaft. The circuit-connections are established between the rings on the generator-shaft and those on the motor-shaft by collecting brushes and wires, as previously explained. In order to properly energize the field-magnet of the motor, however, the connections are so made with the armature coils or wires leading thereto that while the points of greatest attraction or greatest density of magnetic lines of force upon the armature are shifted in one direction those upon the field-magnet are made to progress in an opposite direction. In other respects the operation is identically the same as in the other cases cited. This arrangement results in an increased speed of rotation. In Figs. 17 and 19, for example, the terminals of each set of field-coils are connected with the wires to the two armature-coils in such way that the field-coils will maintain opposite poles in advance of the poles of the armature.

In the drawings the field-coils are in shunts to the armature, but they may be in series or in independent circuits.

It is obvious that the same principle may be applied to the various typical forms of motor hereinbefore described.

Having now described the nature of my invention and some of the various ways in which it is or may be carried into effect, I would call attention to certain characteristics which the applications of the invention possess and the advantages which the invention secures.

In my motor, considering for convenience that represented in Fig. 9, it will be observed that since the disk D has a tendency to follow continuously the points of greatest attraction, and since these points are shifted around the ring once for each revolution of the armature of the generator, it follows that the movement of the disk D will be synchronous with that of the armature A. This feature by practical demonstrations I have found to exist in all other forms in which one revolution of the armature of the generator produces a shifting of the poles of the motor through three hundred and sixty degrees.

In the particular construction shown in Fig. 15, or in others constructed on a similar plan, the number of alternating impulses resulting from one revolution of the generator armature is double as compared with the preceding cases, and the polarities in the motor are shifted around twice by one revolution of the generator-armature. The speed of the motor will, therefore, be twice that of the generator. The same result is evidently obtained by such a disposition as that shown in Fig. 17, where the poles of both elements are shifted in opposite directions.

Again, considering the apparatus illustrated by Fig. 9 as typic... of the invention, it is obvious that since the attractive effect upon the disk D is greatest when the disk is in its proper relative position to the poles developed in the ring R—that is to say, when its ends or poles immediately follow those of the ring—the speed of the motor for all the loads within the normal working limits of the motor will be practically constant. It is clearly apparent that the speed can never exceed the arbitrary limit as determined by the generator, and also that within certain limits at least the speed of the motor will be independent of the strength of the current.

It will now be more readily seen from the above description how far the requirements of a practical system of electrical transmission of power are realized in my invention. I secure, first, a uniform speed under all loads within the normal working limits of the motor without the use of any auxiliary regulator; second, synchronism between the motor and generator; third, greater efficiency by the more direct application of the current, no commutating devices being required on either the motor or generator; fourth, cheapness and simplicity of mechanical construction and economy in maintenance; fifth, the capability of being very easily managed or controlled; and, sixth, diminution of danger from injury to persons and apparatus.

These motors may be run in series, multiple arc or multiple series, under conditions well understood by those skilled in the art.

The means or devices for carrying out the principle may be varied to a far greater extent than I have been able to indicate; but I regard as within my invention, and I desire to secure by Letters Patent in general, motors containing two or more independent circuits through which the operating-currents are led in the manner described. By "independent" I do not mean to imply that the circuits are necessarily isolated from one another, for in some instances there might be electrical connections between them to regulate or modify the action of the motor without necessarily producing a new or different action.

I am aware that the rotation of the armature of a motor wound with two energizing-coils at right angles to each other has been effected by an intermittent shifting of the energizing effect of both coils through which a direct current by means of mechanical devices has been transmitted in alternately-opposite directions; but this method or plan I regard as absolutely impracticable for the purposes for which my invention is designed—at least on any extended scale—for the reasons, mainly, that a great waste of energy is necessarily involved unless the number of energizing-circuits is very great, and that the interruption and reversal of a current of any considerable strength by means of any known mechanical devices is a matter of the greatest difficulty and expense.

In this application I do not claim the method of operating motors which is herein involved, having made separate application for such method.

I therefore claim the following:

1. The combination, with a motor containing separate or independent circuits on the armature or field-magnet, or both, of an alternating-current generator containing induced

circuits connected independently to corresponding circuits in the motor, whereby a rotation of the generator produces a progressive shifting of the poles of the motor, as herein described.

2. In a system for the electrical transmission of power, the combination of a motor provided with two or more independent magnetizing-coils and an alternating-current generator containing induced coils corresponding to the motor-coils, and circuits connecting directly the motor and generator coils in such order that the currents developed by the generator will be passed through the corresponding motor-coils, and thereby produce a progressive shifting of the poles of the motor, as herein set forth.

3. The combination, with a motor having an annular or ring-shaped field-magnet and a cylindrical or equivalent armature, and independent coils on the field-magnet or armature, or both, of an alternating-current generator having correspondingly independent coils, and circuits including the generator-coils and corresponding motor-coils in such manner that the rotation of the generator causes a progressive shifting of the poles of the motor in the manner set forth.

4. In a system for the electrical transmission of power, the combination of the following instrumentalities, to wit: a motor composed of a disk or its equivalent mounted within a ring or annular field-magnet, which is provided with magnetizing-coils connected in diametrically-opposite pairs or groups to independent terminals, a generator having induced coils or groups of coils equal in number to the pairs or groups of motor-coils, and circuits connecting the terminals of said coils to the terminals of the motor, respectively, and in such order that the rotation of the generator and the consequent production of alternating currents in the respective circuits produces a progressive shifting of the poles of the motor, as hereinbefore described.

NIKOLA TESLA.

Witnesses:
FRANK E. HARTLEY,
FRANK B. MURPHY.

P. LALLEMENT.

VELOCIPEDE.

No. 59,915. Patented Nov. 20, 1866.

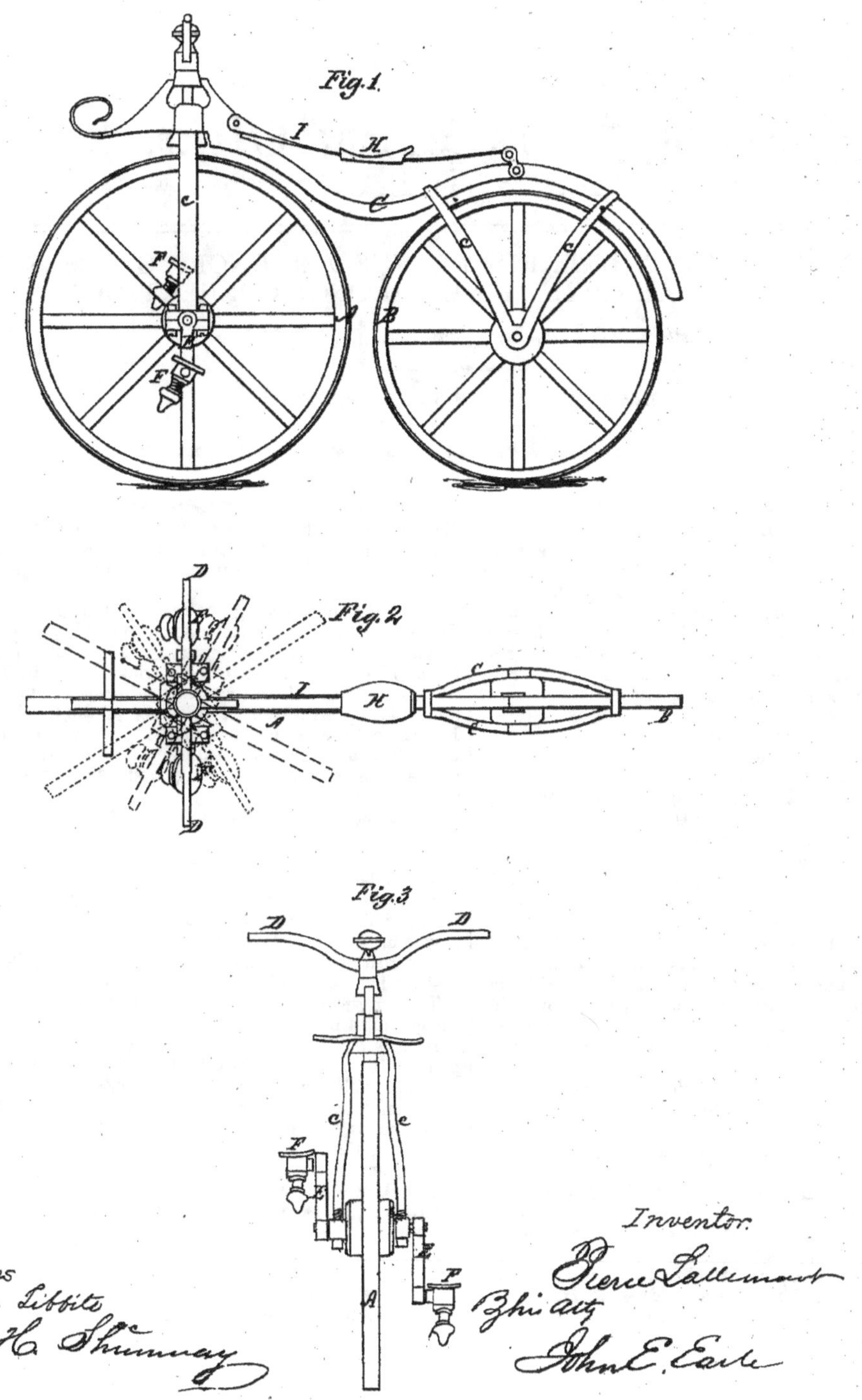

Witnesses
Albert J. Tibbits
John H. Shumway

Inventor.
Pierre Lallement
By his atty
John E. Earle

THE NORRIS PETERS CO., PHOTO-LITHO., WASHINGTON, D. C.

PIERRE LALLEMENT, OF PARIS, FRANCE, ASSIGNOR TO HIMSELF AND JAMES CARROLL, OF NEW HAVEN, CONNECTICUT.

Letters Patent No. 59,915, dated Nove[illegible] 1866.

IMPROVEMENT IN VELOCIPEDES.

The Schedule referred to in these Letters Patent and making part of the same

To all whom it may concern:

Be it known that I, PIERRE LALLEMENT, of Paris, temporarily residing at New Haven, in the county of New Haven, and State of Connecticut, have invented a new Improvement in Velocipedes; and I do hereby declare the following, when taken in connection with the accompanying drawings, and the letters of reference marked thereon, to be a full, clear, and exact description of the same, and which said drawings constitute part of this specification, and represent, in—

Figure 1, side view,

Figure 2, top view, and in

Figure 3, a front end view.

My invention consists in the arrangement of two wheels, the one directly in front of the other, combined with a mechanism for driving the wheels, and an arrangement for guiding; which arrangement also enables the rider to balance himself upon the two wheels.

To enable others to construct and operate my velocipede, I will proceed to describe the same, as illustrated in the accompanying drawings.

A and B are two wheels of common construction, each arranged upon separate axles, and placed, one directly in front of the other, as seen in figs. 1 and 2, the two connected together by a bar, C, passing over the two, as seen in fig. 1, with arms, *c*, extending down and supported on the axles of each wheel, as seen in fig. 3.

The arms of the forward wheel, A, are arranged upon a pivot on the bar C, so that, by means of handles, D D, the forward wheel may be turned to the right or left, as denoted in red and blue, fig. 2.

To the axle of the forward wheel A, I fix cranks E, to each of which I also fix a rocking-treadle, F, the same treadle being balanced by an extension below the crank-pin, so that the flat surface, as seen in fig. 3, will always be uppermost.

Above the bar C, and attached thereto in any convenient manner, I arrange a saddle-seat, H, upon a spring, I, as seen in figs. 1 and 2.

It is evident that, if left to its natural inclination, this carriage could not be made to stand upright. I will, therefore, proceed to describe how the carriage is put in motion, and, when in motion, an upright position maintained.

The rider, first setting the carriage upright, as in figs. 1 and 3, seats himself upon the saddle, in like manner as upon other carriages of this character, giving a forward movement to the carriage, either by his feet in contact with the earth or otherwise, immediately placing his feet, each, upon one of the treadles F, and each hand upon one of the guiding-arms, D, by his feet causing the forward wheel A to revolve, and by the hands guiding the carriage and maintaining his upright position.

If the carriage is inclined to lean to the right, turn the wheel as denoted in red, which throws the carriage over to the left; or, if inclined to the left, turn the wheel as denoted in blue.

Thus the carriage is maintained in an upright position, and driven with great velocity by means of the cranks in the forward wheel.

The greater the velocity, the more easily the upright position is maintained.

To turn the carriage either to the right or left, turn the guiding-wheel accordingly.

By this construction of a velocipede, after a little practice the rider is enabled to drive the same at an incredible velocity, with the greatest ease.

Having, therefore, thus fully described my invention,

What I claim as new and useful, and desire to secure by Letters Patent, is—

The combination and arrangement of the two wheels, A and B, provided with the treadles F, and the guiding-arms D, so as to operate substantially as and for the purpose herein set forth.

PIERRE LALLEMENT.

Witnesses:

JOHN E. EARL,

ALTSIE J. TIBBITS.

E. G. OTIS.

HOISTING APPARATUS.

No. 31,128. Patented Jan. 15, 1861.

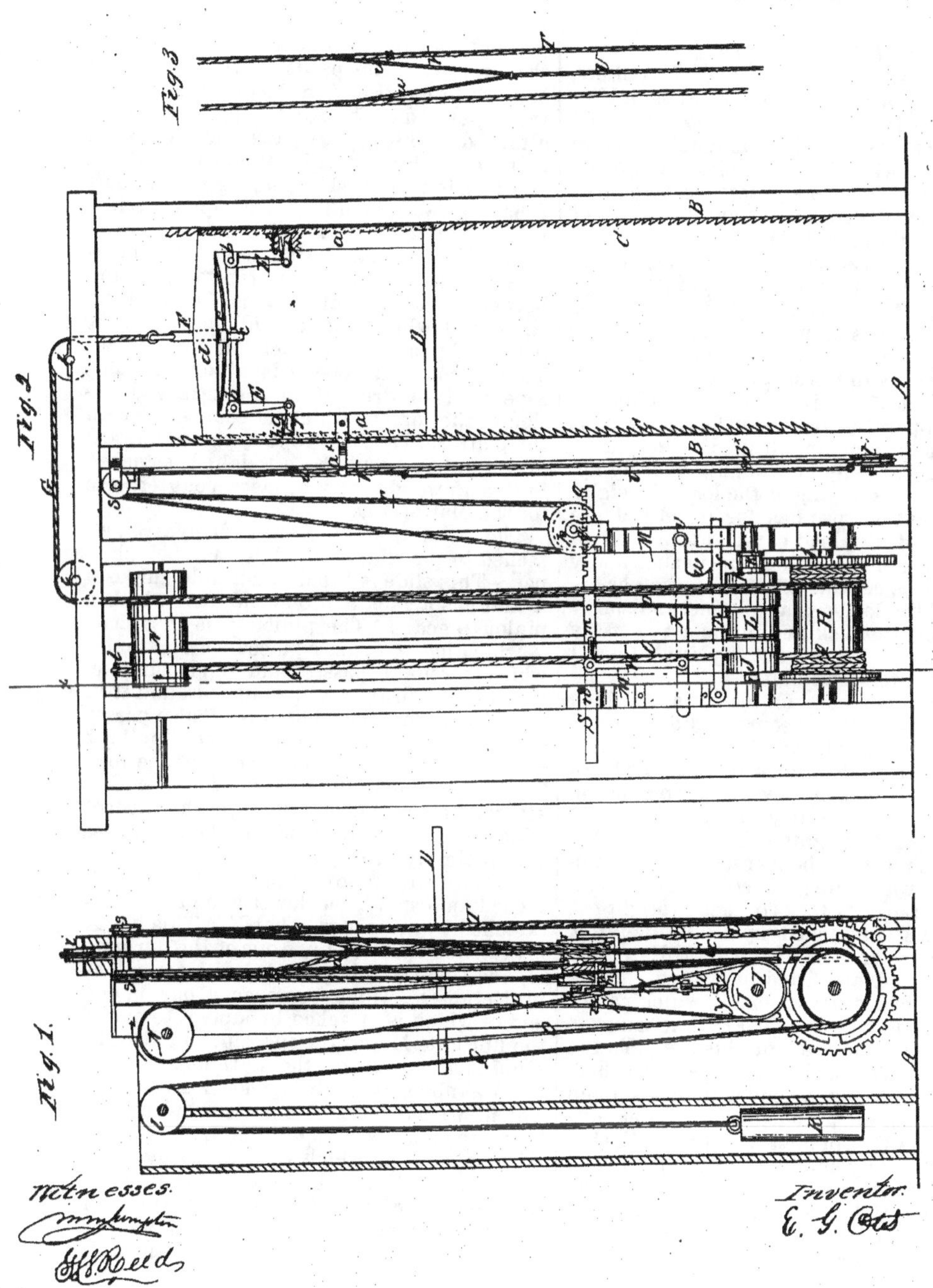

Witnesses.

Inventor.
E. G. Otis

THE N. PETERS CO., WASHINGTON, D. C.

UNITED STATES PATENT OFFICE.

E. G. OTIS, OF YONKERS, NEW YORK.

IMPROVEMENT IN HOISTING APPARATUS.

Specification forming part of Letters Patent No. **31,128**, dated January 15, 1861.

To all whom it may concern:

Be it known that I, E. G. OTIS, of Yonkers, in the county of Westchester and State of New York, have invented a new and Improved Hoisting Apparatus; and I do hereby declare that the following is a full, clear, and exact description of the same, reference being had to the annexed drawings, making a part of this specification, in which—

Figure 1 is a vertical section of my invention, taken in the line *x x*, Fig. 2; Fig. 2, a front view of the same; Fig. 3, a detached side view of the stop mechanism.

Similar letters of reference indicate corresponding parts in the several figures.

The object of this invention is to obtain a hoisting apparatus which may have its weight or load stopped at any desired point and a brake automatically and simultaneously applied with the stopping of the load or weight.

The invention also has for its object the sustaining of the load or weight in case of the breaking of the lifting-rope in such a way as to insure a certain effectual action or operation of the load-sustaining mechanism.

To enable those skilled in the art to fully understand and construct my invention, I will proceed to describe it.

A represents a base or platform, to which two uprights B B are secured, said uprights having each a rack C at its inner side. These racks C have teeth of hook form, or the teeth may be described as having an inclination upward, as shown clearly in Fig. 2.

Between the uprights B B a platform D is placed, the platform being secured to two uprights *a a*, which are grooved vertically to receive the racks C C. To each upright *a* a bent lever E is attached by a fulcrum-pin *b*, and the inner ends of the levers E E overlap each other and are fitted in an eye *c* at the lower end of a vertical bar F, which passes loosely through a rail or bar *d*, that connects the upper ends of the uprights *a a*. To the lower end of the bar F a spring *e* is attached, said spring having a tendency to keep the pawls *f*, which are attached to the lower ends of the levers E E, in gear with the racks C C. This will be fully understood by referring to Fig. 2, in which it will be seen that the pawls *f* are connected to the ends of the levers E by pivots, and have springs *g* attached, which springs have a tendency to keep the pawls pressed down into or between the teeth of the racks C. The pawls, it will be seen, fit or work in mortises *h* in the uprights *a a*. To the upper end of the bar F there is a rope G attached. This rope G passes over pulleys *i i*, and extending down is attached to a drum H, which is connected by gearing *j k* to a shaft I, having two idle-pulleys J K upon it and a working-pulley L between them.

The drum H and shaft I have their bearings attached to suitable uprights M M, and between these uprights there is placed a drum N, around which and the idle-pulleys J K belts O P pass, one of which P is a cross-belt. To the drum H a rope Q is attached. This rope winds on drum H in a contrary direction to the rope G, which is connected with the platform D. The rope Q passes upward over a pulley *l* and has a weight R attached to it, said weight serving as a counterpoise for the platform D.

The belts O P pass through eyes *m*, attached to the slide S, which forms a belt-shipper. This slide is fitted in suitable guides *n n* and has a rack *o* at one end, into which a pinion *p* gears. The pinion *p* is on a shaft *q*, which has a drum *r* placed on it, around which a rope T passes, said rope being secured to the drum *r* and wound around it in opposite directions. The rope T also passes over pulleys *s s* and down around a pulley *t* near the base A. To the portion of the rope T between the pulleys *s s* and *t* a rope U is attached by a branched end V, each part *u* of which is attached to a side of the rope T, as shown clearly in Fig. 3.

To the slide or belt-shipper S there is attached an arm W, the lower end of which is attached by a pivot to a bar X. This bar X is attached by a pivot *v* to one of the uprights M, and the bar X is provided with a pendent projection *w*, which bears on a bar Y, one end of which is attached by a pivot to one of the uprights M and the opposite end fitted in a guide *a'* on one of the uprights. To the bar Y at about its center a shoe Z is attached, which, when the bar Y is pressed downward, bears upon the working-pulley L.

The operation, which will be readily seen, is as follows: When the drum N is turned in the direction of the arrow and the belt P on the working-pulley L, the rope G will be wound on the drum H and the platform D elevated, and in order to lower the platform the cross-belt P is moved on the working-pulley L, the belt O being moved on the idle-pulley J. The shifting of these belts is effected by actuating the rope T by hand, the movement of which turns the drum *r* so that the pinion *p* will, in consequence of gearing into the rack *o*, move the slide S. The rope U forms the stop, and when pulled down both parts *u u* of the branched end V of the rope U have their upper ends brought in the same horizontal plane, and the slide S will be so actuated that the belt O will be on the idle-pulley J and belt P on the idle-pulley K, the shoe Z being at the same time pressed down on the working-pulley L and serving as a brake. The branched end V of the rope U, it will be seen, actuates the rope T when the machine is in operation, but will have no effect on said rope when the brake is applied, as the upper ends of both parts *u u* of the end V will be in a horizontal line with each other. In order to raise the platform D, the rope T is moved by hand so as to throw the belt O on the working-pulley L, the shoe Z being simultaneously raised, and in order to reverse the movement of the platform D and allow it to descend, the rope T is moved so as to shift the cross-belt P on the working-pulley L.

In case the rope G should break in hoisting the loaded platform D, the pawls *f f*, in consequence of being released from the pull of said rope, will immediately be thrown in connection with the racks C C by the springs *e g g*, and in consequence of the teeth of the racks being of hook form or pointed upward the pawls *f f*, under the weight of the load on the platform, will have a tendency to draw the uprights B B toward each other instead of forcing them apart, and the pawls lock themselves with the racks, so that casual disengagement is impossible. By having the counterpoise R attached to the drum H instead of to the platform D the platform or load-sustaining mechanism is not at all interfered with, as would be the case were the rope Q attached directly to the cross-piece *d*. To one of the uprights *a* an arm $a^{\times}$ is attached, said arm having an eye at its outer end, through which the rope T passes, and said rope has a knot or projection $b^{\times}$ on it, against which the arm $a^{\times}$ acts when the platform reaches its lowest point of descent, and thereby throws the belt O off the working-pulley L and stops the descent of the platform, while the brake Z is simultaneously applied.

Having thus described my invention, what I claim as new, and desire to secure by Letters Patent, is—

1. Having the pawls *f f* and the teeth of the racks C C hook formed, essentially as shown, so that the weight of the platform will, in case of the breaking of the rope G, cause the pawls and teeth to lock together and prevent the contingency of a separation of the same, as herein set forth.

2. The arrangement of the ropes T, U, and V, combined and operating substantially as and for the purpose set forth.

3. The arrangement of the slide or belt-shipper S with the shoe or brake Z and rope T, substantially as shown, to admit of the simultaneous application of the brake and the shifting of the belts O P on the idle-pulleys J K, as set forth.

4. Attaching the rope Q of the counterpoise R to the drum H on the opposite side from the lifting-rope G, substantially as shown, so as to counterpoise the platform D without preventing or interfering with the action of the safety mechanism E *e f*.

E. G. OTIS.

Witnesses:
M. M. LIVINGSTON,
G. H. REED.

July 3, 1962 F. PORSCHE ET AL 3,042,444

MOTOR VEHICLE BODY CONSTRUCTION

Filed April 22, 1960 2 Sheets-Sheet 1

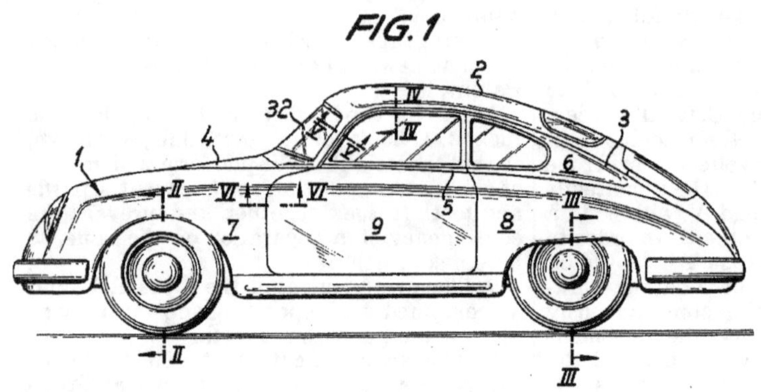

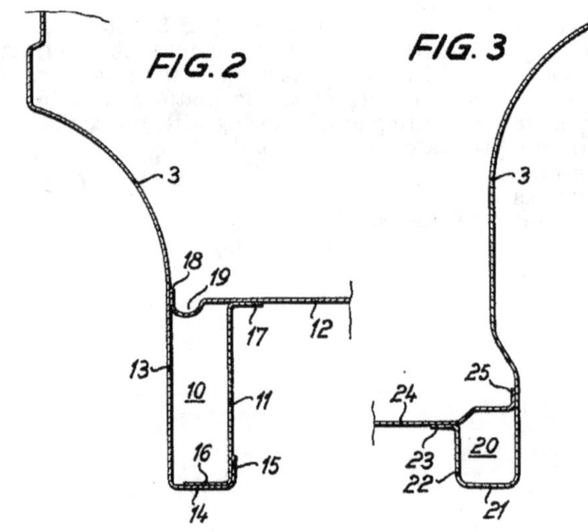

INVENTORS

FERDINAND PORSCHE

ERWIN KOMENDA

BY Dicke, Craig & Freudenberg

ATTORNEYS

July 3, 1962 F. PORSCHE ET AL 3,042,444

MOTOR VEHICLE BODY CONSTRUCTION

Filed April 22, 1960 2 Sheets-Sheet 2

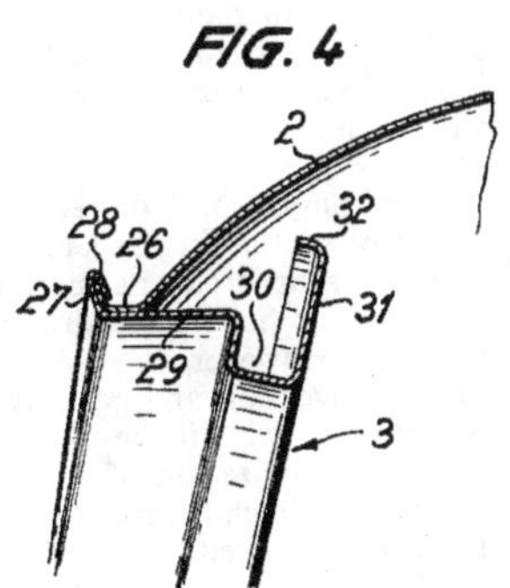

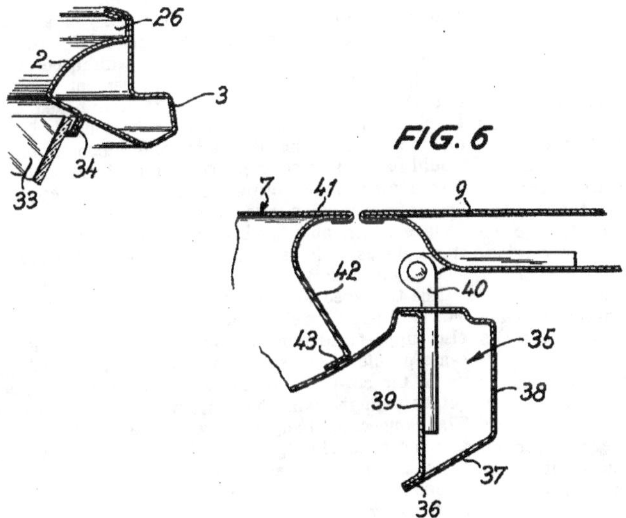

INVENTORS
FERDINAND PORSCHE
ERWIN KOMENDA
BY Dicke, Craig & Freudenberg
ATTORNEYS

United States Patent Office

3,042,444
Patented July 3, 1962

3,042,444

MOTOR VEHICLE BODY CONSTRUCTION

Ferdinand Porsche, Stuttgart, and Erwin Komenda, Korntal, Germany, assignors to Dr. Ing. h.c. F. Porsche KG, Stuttgart-Zuffenhausen, Germany

Filed Apr. 22, 1960, Ser. No. 24,080
Claims priority, application Germany Apr. 25, 1959
2 Claims. (Cl. 296—28)

The present invention relates to a side-wall panel construction for a self-supporting type vehicle body of motor vehicles which extends in one piece over the entire length of the vehicle, includes the door aperture, and is provided with flanges for securing thereto adjoining pressed or stamped parts and/or for forming therewith bearer members of box-shaped cross-section.

Motor vehicle body constructions are known in the prior art which consists essentially of unitary side-walls and of a stamped roof part. The individual side wall parts are thereby provided along the edges thereof with flanges which form, in most sections thereof, box-shaped hollow bearer members together with further stamped parts as well as with the angularly-bent portions of the adjoining wall parts. The relatively great number of welding seams necessary in the prior art constructions not only requires high expenditures in connection with such constructions but also involves a danger of corrosion of the body which is very great corresponding to the large number of welded seams. Particularly, the welded seams of the box-shaped longitudinal bearer members which are exposed to or face the vehicle road surface entail an accumulation of dirt and are thereby exposed over relatively long periods of time, in an interrupted manner, to the humidity and dampness normally encountered.

A further difficulty usually occurs in connection with the handling of relatively large vehicle wall parts or body panels. It is known already in the prior art to provide the wall parts or panels with edge flanges for purposes of increasing the inherent rigidity thereof which edge flanges also serve during final assembly as securing flanges with the adjacent wall parts or body panels or with the additional stamped parts of the body. These flanges are relatively narrow so that the reinforcement of the stamped wall parts or panels obtained thereby is relatively slight.

The present invention aims at improving the handling of the side wall parts or panels by means of a far-reaching reinforcement thereof, at reducing the costs involved in the formation of the hollow longitudinal bearer member as well as at a physical relocation and displacement of the absolutely necessary welding seams out of the soiling danger zone.

This is achieved in accordance with the present invention in that sections of the side-wall edge or rim portions are stamped out in a U-shaped manner whereby the free leg portion of the stamping extends at a distance from the side-wall panel and is disposed essentially parallel thereto. By the use of such an arrangement, it is possible to achieve a construction which requires fewer stamped parts for the formation of the hollow, longitudinal bearer members and in which the welding seams are disposed exclusively along one side of the box-shaped bearer member. The reinforcement of the side-wall part or panel takes place in a manner corresponding to the size thereof whereby the largest part of the bearer member is already stamped out together with the wall part or panel of the body. The side-walls forming within the outer regions thereof the wheel casings are provided inwardly thereof with U-shaped, angularly-bent edge portions which form channels extending in the vehicle longitudinal direction. The open side of these channels is

located on the side of the bearer member opposite the road surface and is closed off by means of a stamped part extending essentially horizontally and preferably forming the vehicle floor. As a result of such an arrangement, the otherwise necessary connecting flanges provided at the side of the longitudinal bearer member facing the road surface and therewith any corners that would otherwise collect dirt are effectively eliminated. The corrosion danger is also lessened thereby and the cost for the welding installations is reduced since the longitudinal bearer member in accordance with the present invention is composed exclusively of two stamped parts.

The stamped part thereby abuts against an angularly-bent rim of the free leg portion, bridges the open side of the U-shaped channel part and is arranged at the side wall part or body panel by means of an angularly-bent portion. However, it is also within the purview of the present invention and possibly also of advantage in connection therewith if the free leg portion of the stamping is extended by means of a separate stamped part the rim or edge of which is provided with a flange for supporting thereon a stamped part closing the channel.

The door aperture of the side-wall panel, preferably the hinge post or column is defined or limited by a stamping of channel-shaped configuration, open on one side only which is formed by the rim portion of the stamped wall part or panel. The opening of the channel-shaped stamping is closed off by another stamped part. The use of such an arrangement entails, in addition to the advantages already generally mentioned hereinabove, the elimination of connecting flanges within the region of the door aperture. Such an arrangement produces a smooth door-frame without any sharp-edged corners which otherwise might possibly be the source of damage to the clothes of the passengers and which in any event would have to be covered by additional covering means. The stamped part closing the channel is thereby formed by a web portion provided with angularly-bent rim flanges which abut against the free leg portion of the stamping as well as against the side wall body panel. The stamped part may thereby form the hinge support.

The U-shaped stamping consisting of one piece with the side wall body panel which bridges the door aperture and also supports thereon the roof extends with basically the same profile or cross-section along the windshield frame up to the cowl and forms within the region thereof together with the windshield frame the box-type column. The connecting flanges between the windshield frame and the channel-shaped rim part of the side wall body panel form, in an advantageous manner, the securing flange for the windshield and a rain water drain channel.

Accordingly, it is an object of the present invention to provide a vehicle body construction for a self-supporting-type body which avoids the disadvantages encountered in the prior art and which is simple and relatively inexpensive in manufacture and assembly, yet relatively sturdy so as to be capable to readily withstand all forces that many occur therein.

Another object of the present invention is the provision of a vehicle body construction which eliminates the necessity of numerous welded seams.

A further object of the present invention is the provision of a vehicle body construction which effectively eliminates the danger of corrosion thereof, particularly in those places where moist dirt thrown up from the road surface is likely to accumulate.

Still a further object of the present invention is the provision of a motor vehicle construction which facilitates handling of the relatively large body stampings by imparting thereto a sufficient rigidity.

Another object of the present invention is the provision

of body stampings of relatively large dimensions which are stamped out in such a manner as to provide an increased inherent rigidity in these stampings.

Still another object of the present invention resides in the provision of a motor vehicle body construction in which a minimum of individual parts are used to form the self-supporting frame-like members of the body.

Another object of the present invention resides in the provision of a door aperture provided in the vehicle side wall parts which exhibits only smooth edges so as to avoid the danger of tearing the passengers' clothes and therewith also eliminates the requirement for a separate covering of such edges.

These and other objects, features and advantages of the present invention will become more obvious from the following description when taken in connection with the accompanying drawing which shows, for purposes of illustration only, one embodiment in accordance with the present invention, and wherein

FIGURE 1 is a side elevation view of a passenger motor vehicle in accordance with the present invention,

FIGURE 2 is an enlarged partial cross-sectional view taken along line II—II of FIGURE 1,

FIGURE 3 is an enlarged partial cross-sectional view taken along line III—III of FIGURE 1,

FIGURE 4 is an enlarged partial cross-sectional view taken along line IV—IV of FIGURE 1,

FIGURE 5 is an enlarged partial cross-sectional view taken along line V—V of FIGURE 1, and

FIGURE 6 is an enlarged cross-sectional view taken along line VI—VI of FIGURE 1.

Referring now to the drawing wherein like reference numerals are used throughout the various views to designate like parts, and more particularly to FIGURE 1, reference numeral **1** generally designates therein a passenger motor vehicle which includes a vehicle body consisting of several stamped body parts or body panels. In particular, the motor vehicle **1** is composed essentially of a stamped roof part **2**, of stamped side wall parts **3** which are essentially of mirror image-like construction and of which only one is therefore shown in the drawing, and of a front hood **4**. The side wall stamping **3** forms above the so-called belt line **5** of the motor vehicle the outer wall **6** thereof, as is known in the prior art, whereby the forward fender **7** and the rear fender **8** are arranged below the line **5** in the front and rear, respectively, of the motor vehicle. Within the regions of the fenders **7** and **8**, the wall part **3** forms the inner walls of the wheel casings.

The side wall **3** is provided, among others, with a door aperture for the door **9** which door aperture is suitably cut out of the stamping. As a result thereof, additional beams or bearers which would have to be welded to the body and which would bridge the door aperture are obviated thereby.

The side wall body panel stamping **3** is reinforced in itself in accordance with the present invention by a plurality of edge flanges which are arranged within the edge regions of the stamping **3** and always have essentially the same basic shape or configuration. The edge flanges are constructed as a U-shaped stamping whereby the free leg portion of the channel formed thereby extends at a distance from the stamped part or body panel and also essentially parallel with respect thereto.

FIGURES 2 through 6 illustrate, on an enlarged scale, some examples of the construction of the edge or rim regions of the side wall panel **3** as well as the connections thereof with the adjoining stamped parts of the body.

FIGURE 2 illustrates the construction of the side wall as well as the formation of a hollow bearer member within the forward region of the vehicle. The hollow bearer member generally designated by reference numeral **10** (FIGURE 2) is limited or defined by the body panel of the side wall **3**, by a web portion **11** and by a horizontal stamped part **12** representing a partition wall. The vertical wall portion **13** of the side wall panel **3** which forms the wheel casing inner wall is thereby provided with a horizontally extending angularly-bent portion **14** which terminates in an upright flange portion **15** spaced at a distance from the side wall portion **13**. The flange **15** is extended by a web portion **11** which is supported with the flange **16** thereof against the angularly-bent flange portion **15**. Additionally, the web portion **11** is provided with an angularly-bent flange portion **17** by means of which the web portion **11** is connected with the stamped part **12**. The stamped part **12** thereby bridges the gap between the web portion **11** and the wall portion **13**, and is secured at the latter by means of a flange **18**. It is thereby advantageous to provide simultaneously a water drainage channel **19** within the stamped part **12** forming the partition wall.

The longitudinal bearer member generally designated by reference numeral **20** (FIGURE 3) is provided within the rear portion of the vehicle and is formed exclusively by two stamped parts. The side wall body panel **3** is thereby provided, in a manner similar to FIGURE 2, with a horizontal, angularly-bent portion **21** which is adjoined by an essentially vertical or upright web portion **22**. The web portion **22** in turn is provided with a flange portion **23**. The channel formed by the wall **3**, by the angularly-bent portion **21** as well as by the web portion **22** is covered by means of a horizontal stamped part **24** whereby a box-shaped hollow bearer member **20** is produced thereby. The provision of a further stamped part, such as web portion **11** of FIGURE 2, is thereby obviated since the stamped part **24** extends at a slight distance from the angularly-bent portion **21**. The stamped part **24** is also secured at the wall body panel **3** by means of a flange **25**.

The connection of the side wall body panel **3** with the stamped roof part or panel **2** is illustrated in FIGURES 4 and 5. FIGURE 4 thereby illustrates the cross-section, on an enlarged scale, within the region of the door **9** whereby the connecting places between the stamped parts serve for the formation of a water drainage channel **26**. The rim flange **27** of the stamped roof part **2** is thereby angularly-bent upwardly and is retained within the flanged rim **28** bent back upon itself of the side wall **3**. The door folding **29** is provided with a U-shaped indentation **30** the inwardly-disposed leg portion **31** of which terminates in a reinforcing flange **32**.

The basic profile of the door folding **29** is also present within the region of the windshield **33** (FIGURE 5) whereby the door folding **29** forms a box-type column together with the stamped roof part **2** which constitutes within this area the outer stamping of the windshield frame. The connecing piece between the stamped parts **2** and **3** forms, on the one hand, a drainage channel **26** and, on the other, the securing flange **34** for the windshield **33**.

The hinge post or column generally designated by reference numeral **35** (FIGURE 6) is also formed by a U-shaped upright rim portion of the side wall body panel **3** whereby an essentially U-shaped web portion **39** extends from the edge **36** of the free leg portion **37** of the channel-shaped stamping **38** toward the oppositely disposed wall of the stamped part **3**. The web portion **39** thereby complements the stamping **38** into a box-shaped bearer member. The web portion **39** also serves preferably for supporting thereon the hinges **40** of the door **9**.

The fender generally designated by reference numeral **7** is also secured within the region of the hinge column **35** which fender **7** essentially consists of an outer wall **41** to which is secured by means of bent-back flanges a stamped part **42**. The stamped part **42** is connected by means of flange **43** with the side wall **3** and is preferably detachably secured thereto.

While we have shown and described an embodiment in accordance with the present invention, it is understood that the same is not limited thereto but is susceptible of many changes and modifications within the spirit and scope of the present invention. We, therefore, do not wish to be limited to the details shown and described herein but intend to cover all such changes and modifications as are encompassed by the scope of the appended claims.

We claim:

1. A side-wall construction for a vehicle body, especially of the self-supporting type, for motor vehicles in which the side-wall body panel extends in one piece essentially over the entire length of the vehicle and is provided with a door aperture, said side-wall body panel including rim portions having flanges for connection with flanges of adjoining stamped body parts, at least some of the sections of the rim portions of said side-wall body panel being stamped out in an essentially U-shaped manner to form substantially parallel leg portions, one of which is free, said free leg portion extending at a distance from the side wall and essentially parallel thereto, said side-wall panel constituting within the outer regions thereof at least a part of the wheel casings, said wall panel comprising inwardly U-shaped angularly bent rim portions thereof, said last-mentioned rim portions extending essentially in the vehicle longitudinal direction and constituting channels, said channels being open on the side thereof opposite the road surface, and an essentially horizontally-extending stamped body part closing off the open side of said channel.

2. A side-wall construction according to claim 1, wherein said essentially horizontally-extending body part forms the vehicle floor.

References Cited in the file of this patent

UNITED STATES PATENTS

2,202,859	Ledwinka	June 4, 1940
2,254,458	Swallow	Sept. 2, 1941

FOREIGN PATENTS

812,475	France	Feb. 1, 1937
964,469	Germany	May 23, 1957
626,305	Great Britain	July 13, 1949
209,058	Switzerland	June 1, 1940

Aug. 26, 1930. P. T. FARNSWORTH 1,773,980

TELEVISION SYSTEM

Filed Jan. 7, 1927 4 Sheets-Sheet 1

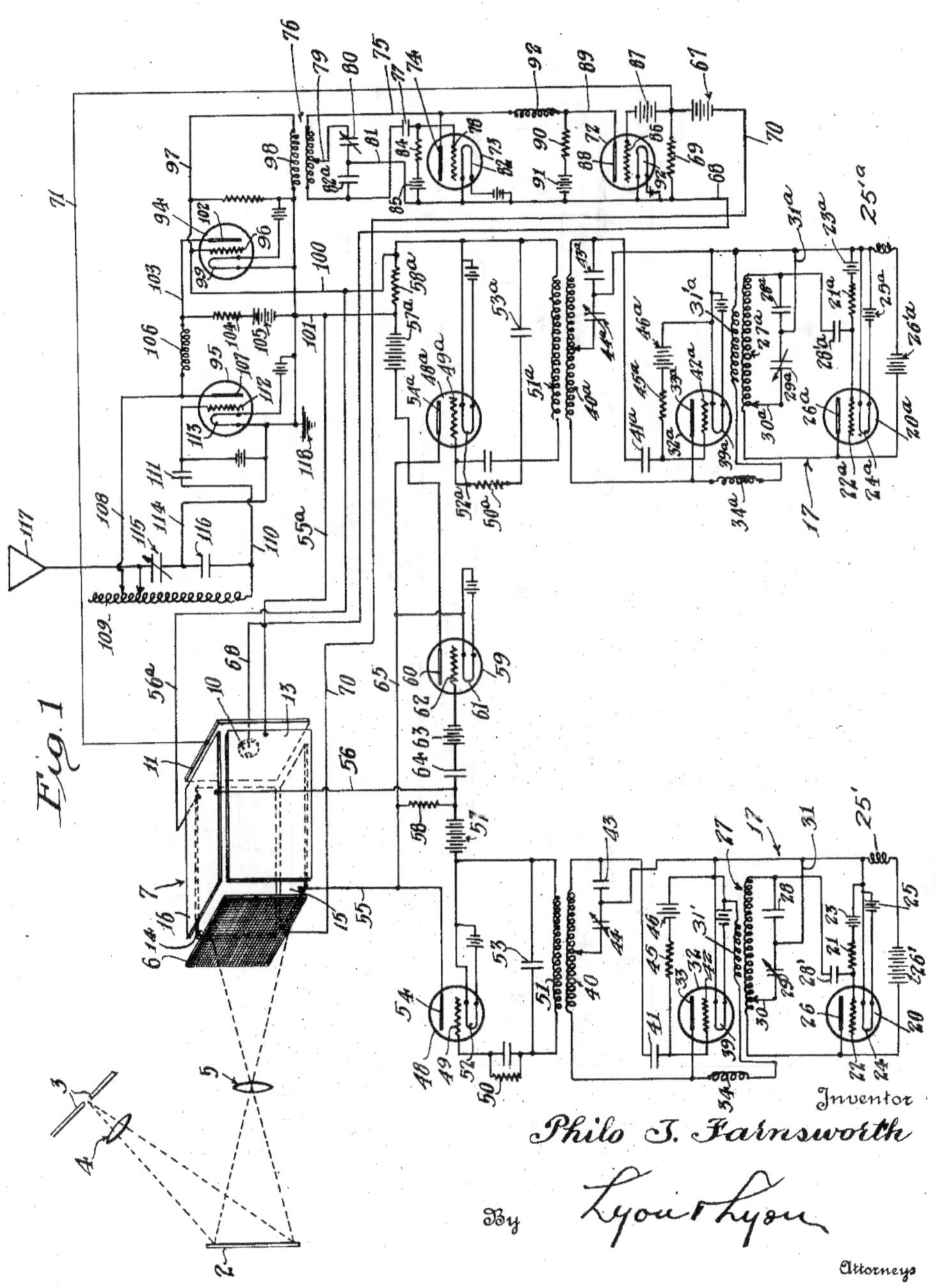

Aug. 26, 1930. P. T. FARNSWORTH 1,773,980

TELEVISION SYSTEM

Filed Jan. 7, 1927 4 Sheets-Sheet 2

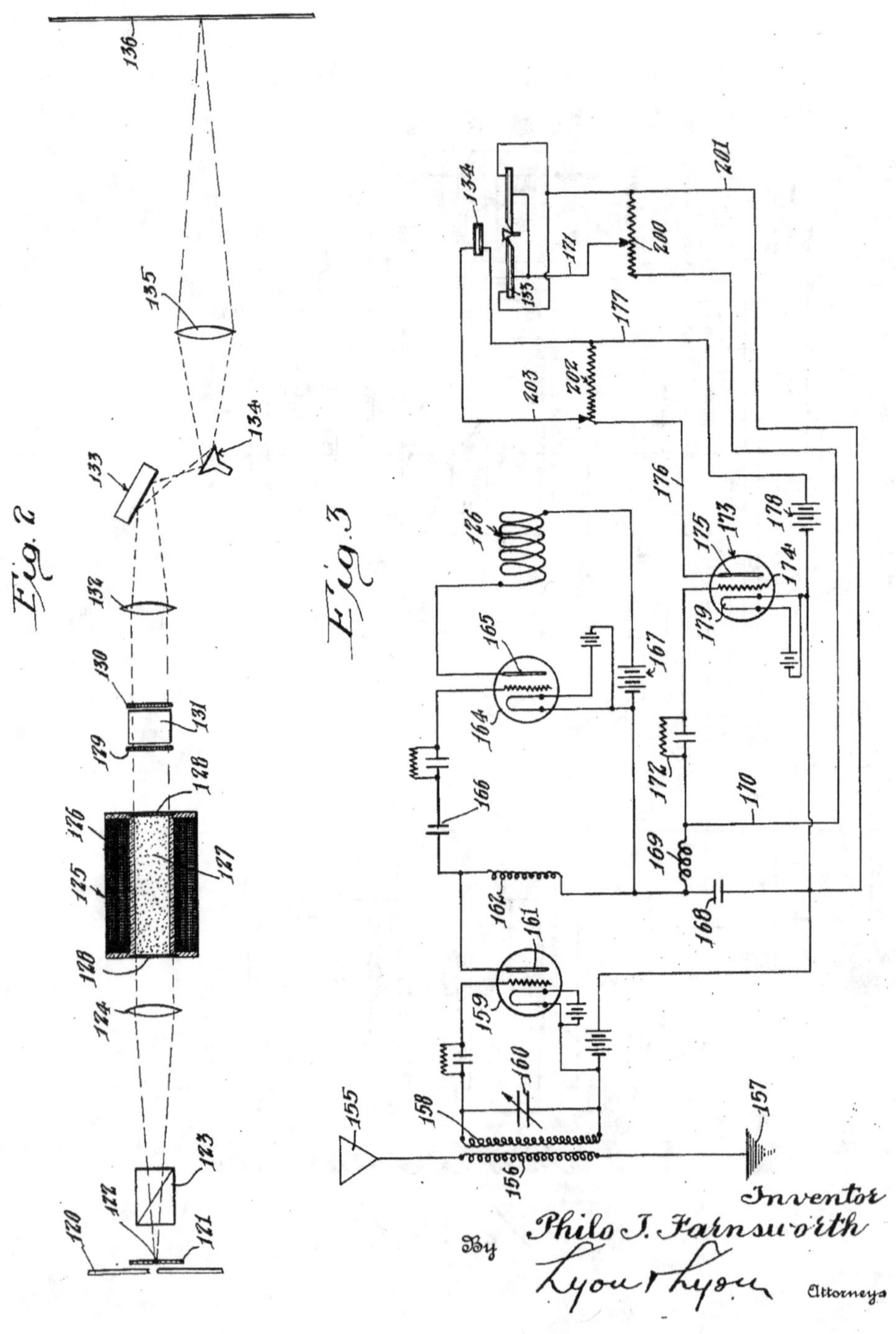

Aug. 26, 1930. P. T. FARNSWORTH 1,773,980

TELEVISION SYSTEM

Filed Jan. 7, 1927 4 Sheets-Sheet 3

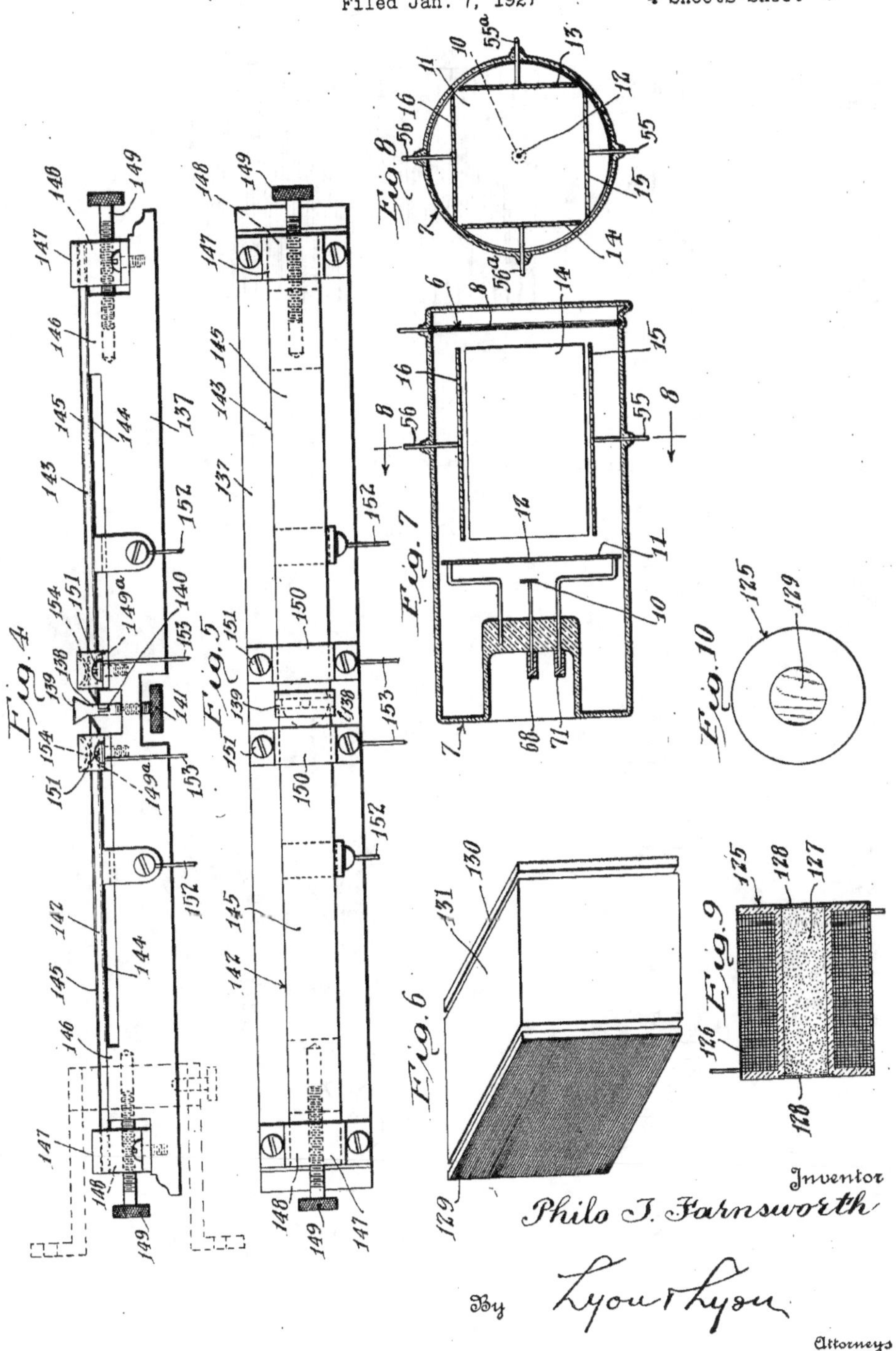

Inventor

Philo T. Farnsworth

By Lyon & Lyon

Attorneys

TELEVISION SYSTEM

Filed Jan. 7, 1927 4 Sheets-Sheet 4

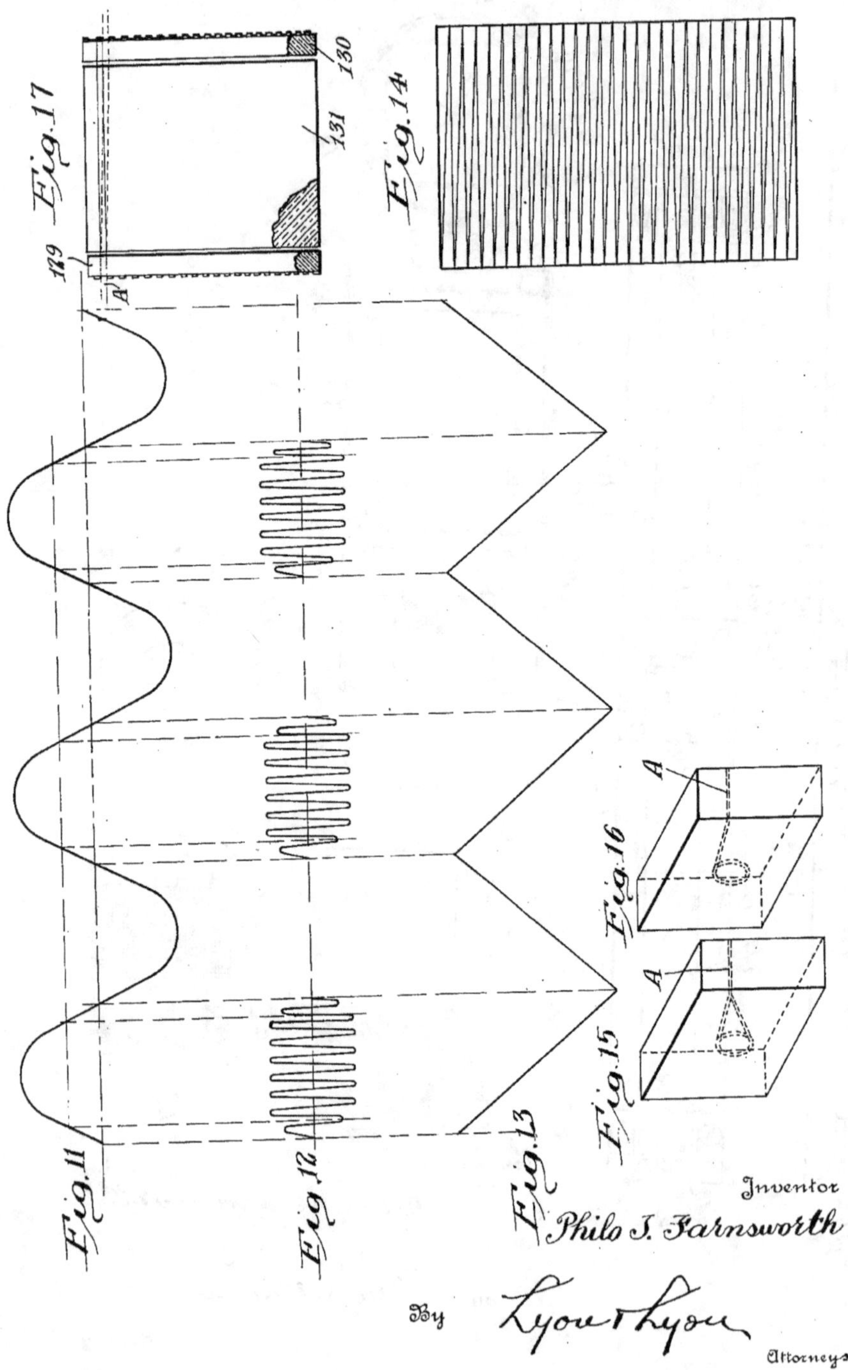

Inventor
Philo T. Farnsworth

By Lyon & Lyon
Attorneys

Patented Aug. 26, 1930 **1,773,980**

UNITED STATES PATENT OFFICE

PHILO T. FARNSWORTH, OF BERKELEY, CALIFORNIA, ASSIGNOR, BY MESNE ASSIGNMENTS, TO TELEVISION LABORATORIES, INC., OF SAN FRANCISCO, CALIFORNIA, A CORPORATION OF CALIFORNIA

TELEVISION SYSTEM

Application filed January 7, 1927. Serial No. 159,540.

This invention relates to a television apparatus and process, that is, it is directed to an apparatus and process for the instantaneous transmission of a scene or moving image of an object located at a distance in which the transmission is by electricity.

Heretofore attempts have been made to transmit an image of an object by electricity so that the image of the object will instantaneously appear at a distance. These prior attempts at television have generally embodied an apparatus and method in which each particular elementary area of the image of the object is successively converted into an electrical current, the intensity of which is proportional to the intensity of the light at that particular elementary area; all the elementary areas of the image being covered in that fraction of a second during which the eye will retain a picture, hereafter referred to as the optical period. This is followed by a transmission of such current and a conversion of such current to light corresponding in intensity to the intensities of the light of the individual areas of the original image; the reconversion process likewise being performed within the optical period so that, by a proper coordination of the developed light, an image of the object to be transmitted appears as instantly formed at the receiving end of the apparatus and method.

The time during which the human eye will retain a picture is of such short duration that the conversion of the light shades of the original image of the object to electricity and the reconversion of said electricity to light and the proper coordination of such light must be performed at a very tremendous speed. All prior attempts at television have attempted to employ some mechanically moving part for dissecting the image of the original object during the process of forming an electrical current which varies in intensity in accordance with the light shades of the respective elementary areas of the image. None of these prior attempts at television have proven successful. They have resulted at best in the production of a crude moving silhouette of the object to be transmitted. This has generally been due to the fact that the mechanically moving parts of the prior apparatus have not been able to travel at the necessary speed requirements with the synchronism required in a television apparatus.

An object of the present invention is to provide a method and apparatus for television, which is adapted to transmit electrically a true moving image in full light shades of the object to be transmitted.

Another object of the present invention is to provide a method and apparatus for television in which the conversion and dissecting of the light shades of the object to be transmitted, to electricity and the reconversion of such electricity to form an image is accomplished in the following manner:

In the process and apparatus of the present invention, light from all portions of the object whose image is to be transmitted, is focused at one time upon a light sensitive plate of a photo-electrical cell to thereby develop an electronic discharge from said plate, in which each portion of the cross section of such electronic discharge will correspond in electrical intensity with the intensity of light imposed on that portion of the sensitive plate from which the electrical discharge originated. Such a discharge is herein termed an electrical image. An electrical shutter is then interposed between said sensitive plate and the anode of the photo-electrical cell, the shutter having a small aperture therein so that there can be received upon said anode at one instant, only the electrons which originate from one elementary area of the light sensitive plate. There is then imposed upon the electrical discharge a plurality of electrical potentials of different frequencies for causing the electrical discharge to bend in two directions, whereby the electrons from each elementary portion of the sensitive plate are successively directed through said shutter, this action taking place so as to completely cover the area of the sensitive plate within the optical period. The scene to be transmitted is thus analyzed or dissected to produce an electrical current or "light" current having variations in intensity in accordance with the light shades of the object to be transmitted and this is accomplished within the optical period without the necessity of employing any mechanically moving parts.

The produced electrical current or "light" current may be transmitted to the receiving end of the apparatus by either wires or may be superimposed upon a wireless carrier wave. There is also transmitted at the same time and preferably superimposed upon the same carrier wave, the two electric potentials of different frequencies which are employed in analyzing the image so that such currents may be employed to synchronize the receiving apparatus and process.

At the receiving end of the apparatus and process, the "light" current is reconverted to light and the light coordinated to form an image of the object transmitted in accordance with the following apparatus and process.

Preferably a constant source of light is utilized which is directed, first, through a polarizing prism and hence through an apparatus or means by which the plane of polarization of the light may be rotated by the "light" current. In this manner an instantaneous response to the variations of such light current is obtained in the rotation of the plane of polarization of the light. The light is then directed through a suitable screen capable of shutting off the light in accordance with the rotation of its plane of polarization. In this manner, a beam of light is developed fluctuating in intensity to the variations of intensity of the "light" current transmitted without the necessity of employing any mechanically moving parts. This said beam of light is then projected by means of two cooperating oscillographs upon the screen where the image is to be transmitted, said oscillographs being operated by the synchronizing frequencies transmitted with the "light" current to correctly coordinate the light upon the screen to form a correct image.

The present invention, together with various objects and advantages thereof will best be understood from a description of a preferred form or example of a process and apparatus for television embodying the invention. For this purpose, I have hereinafter set forth one form of example of a method and apparatus for producing television in accordance with the present invention, and have illustrated said apparatus and method as it is adapted for television by wireless. It is to be understood, however, that the invention is capable of various and numerous modifications, changes, and substitutions, and is not necessarily limited to the transmission by wireless or radio.

The apparatus and method will best be understood from a description of the accompanying drawings, in which:

Figure 1 is a diagrammatic view of a complete television transmitter, including a circuit diagram therefor,

Figure 2 is a diagrammatic view of the television receiver,

Figure 3 is a circuit diagram of the electrical connections for the television receiver,

Figure 4 is an elevation of one of the oscillographs,

Figure 5 is a plan view of one of the oscillographs,

Figure 6 is a perspective view of the light diverting means,

Figure 7 is a sectional view of the photo-electric cell,

Figure 8 is a section on the line 8—8 of Figure 7,

Figure 9 is a section of the light rotator,

Figure 10 is an end view thereof,

Figure 11 is a representation of the form of electric current of the first oscillator employed in developing a potential for the photo-electric cell,

Figure 12 is a representation of the form of electric current produced in the second oscillator,

Figure 13 is a representation of the resulting straight lined potential,

Figure 14 is a view of the scanning path and also a view of the path of the light beam over the receiving screen,

Figure 15 is a perspective view of a bi-axial crystal showing the conical refraction of unpolarized light,

Figure 16 is a perspective view of a bi-axial crystal showing the refraction of polarized light, and,

Figure 17 is a diagrammatic illustration of the path of light through the gratings.

Referring to the drawings, 2 represents an object, an image of which is to be transmitted. Said object may be an actual scene or a photograph, a projection of a motion picture film, or any other object. The object 2 is preferably illuminated, for example, by means of an arc light 3 focused thereon by a lens 4. 5 indicates a lens for focusing an image of the object 2, upon the light sensitive plate 6 of a photo electric cell 7.

The photo-electric cell is preferably constructed as follows:

The light sensitive plate 6 or cathode of the cell is preferably made flat and is formed of a fine mesh screen 8, and said screen 8 is covered or coated with a light sensitive material such as sodium, potassium, or rubidium. 10 is the anode of the photo-electric cell positioned at the other end of the cell. Between the sensitive plate 6 and anode 10 and closely adjacent to anode 10 is placed an electric shutter 11 formed by a metallic plate in which there is a small aperture 12. Between the shutter 11 and light sensitive plate 6, four plates 13, 14, 15, and 16 are placed at right angles to each other and outside the path of electrons from the plate 6 to the shutter 11. Each opposed pair of the plates are connected to a source of electrical potential of a different frequency. The photo-electric cell should be highly evacuated, such for example as to 10^{-7} cm. mercury to permit a high potential across the cell without ionization.

The necessity for employing a high potential across the cell arises from the fact that the photo electrons emitted from the cathode 6 have a small emission velocity which depends upon the color of the light causing their emission. This emission velocity is always small, of the order of that which an electron would acquire by falling through a volt or two, but it may have nearly any direction. This haphazard motion tends to distort the electric image and is only prevented from doing so by making the potential between the cathode 6 and the anode 10 high enough to insure that the time taken for an electron to traverse the distance between cathode 6 and anode 10 is so small that the small velocity transverse to this path produces no appreciable distortion. Hence the vacuum in the photo-electric cell 7 should be the highest obtainable.

The electrical potentials are provided by an oscillator 17, capable of developing two different high frequency electrical currents. Said oscillator 17 not only is required to provide a source of oscillating energy but is required to provide a form of oscillating energy, the wave form of which is composed of substantially straight lines, as will be hereinafter pointed out. Such a wave form is essential to accomplish a uniform lighting of all portions of the image which is to be produced.

The oscillator comprises a tri-electrode valve 20 connected in a circuit acting as an oscillator to produce an oscillating energy of low frequency, such for example as 10 cycles per second. It is understood that any customary or preferred form of circuit for this purpose may be employed, the particular circuit described being provided with a grid leak 21 connected with the grid 22 of the tube 20, and hence through a negative bias battery 23 to the filament 24. The filament 24 is indicated as heated by a battery 25. The plate 26 of the tube is connected through a battery 26′ and the choke coil 25′ to the filament 24. The plate 26 also connects through an inductance 27 and capacity 28′ with the grid. The inductance 27 is shunted by a fixed capacity 28 and a variable capacity 29 in series, one end of the series being connected to the end of the inductance 27 and the other end having a variable connection with said inductance. Between these capacities 28 and 29, a lead 31 is connected which connects with the filament 24 of the tube 20.

By this connection, the constants of the oscillating circuit may be any value of inductance and capacity to bring the oscillating circuit in resonance with the frequency of the desired circuit. Said oscillator in turn provides a source of potential for a second oscillating circuit of similar design, the second oscillator operating at a higher frequency such, for example, as 500 kilo-cycles. The second oscillator comprises the tube 32, the plate 33 of which is charged with the oscillatory energy of the first oscillator. The first oscillator is coupled thru the secondary coil 31′ to plate 33, the inductance 34 being included in series therewith. The inductance 34 may be any suitable radio frequency choke to prevent the high frequencies in the second oscillating circuit from being imposed on the first oscillating circuit. The plate 33 is connected through the primary 40 of a radio frequency transformer and hence through the capacity 41 with the grid 42. Capacities 43 and 44 are shunted around all or part of the primary 40 and a lead is connected from their midpoint to the filament 39 of the tube 32. The grid 42 of the tube is connected through a suitable leak 45 and negative bias battery 46 with filament 39. It is understood that the second oscillating circuit thus described is only one example of a circuit adapted for this purpose and the various constants of the circuit may be of any value suitable for bringing the circuit into resonance with the frequency of the oscillations (500 kilo-cycles) desired to be produced therein.

The voltage of the first oscillator is adjusted to be well above the value required for maximum plate current of the second oscillator. Hence, since the second oscillator will generate oscillations only when the plate voltage is positive, the current generated by the second oscillator will be similar to that shown in Figure 12. The harmonic oscillating current developed by the first oscillator is represented in Figure 11. This current, when imposed upon the second oscillator, develops a current such as illustrated in Figure 12, in which it will be seen that each positive cycle of the first harmonic current produces a series of harmonic oscillations in the second oscillator of substantially equal intensity, while during the negative period of the first harmonic current, substantially no oscillations are developed in the second oscillator.

The output from the second oscillator is then imposed upon an audion circuit having a tube 48 with its grid 49 connected by a line through the grid leak and grid condenser 50 to an inductance 51 inductively coupled to the inductance 40. Said secondary 51 is connected to the filament 52 of the audion 48. Shunted across the secondary 52 is a condenser 53 of value suitable to produce resonance with the oscillations developed in the second oscillator. The plate 54 and the audion 48 is connected by the lead 55 with the plate 15 of the photo-electric cell, and the opposed plate 16 of the photo-electric cell is connected by a lead 56 through the battery 57 to the filament 52. The resistance 58 is shunted across the leads 55 and 56 to provide a potential for the plates 15 and 16.

The action of the audion circuit including the tube 48 is to produce an alternating current equal to the frequency developed in the first oscillator but the wave form of said frequency is of substantially straight lines such, for example, as indicated in Figure 13. In producing this wave form, the audion tube 48 operates due to the bias of the grid leak and condenser 50 to accumulate a charge during the passage of each wave train indicated in Figure 12, and such accumulated charge leaks off during the interval between successive trains, so that the output of the audion 40 into the plate circuit, indicated by the leads 55 and 56 (passing to the plates 15 and 16 of the photo-electric cells) assumes the straight line form of Figure 13.

There is also a duplicate form of audion circuit for supplying a similar wave form of electrical oscillations for the plates 13 and 14 of the photo-electric cell, said oscillations being, however, at a higher frequency such, for example, as 5000 cycles per second. Inasmuch as this circuit is identical except in value of constants to the circuit just described, the parts corresponding to those numbered 20 to 54 are numbered 20^a to 54^a. It is understood that the oscillating tube 20^a develops a harmonic oscillating current of 5000 cycles which will be imposed upon the oscillator including the tube 32^a, operating at 500 kilo-cycles producing a straight line alternating current in tube 48^a of a frequency of 5000 cycles per second. The output from tube 48^a to the plates 13 and 14 is from filament 52^a, through resistance 58^a, battery 57^a, and hence through a modulating tube 59 through the plate 60 thereof, and to the filament 61 thereof, and hence to the plate 54^a of the tube 48^a. The potential drop across resistance 58^a is utilized to provide the potential for plates 13 and 14 through leads 55^a and 56^a. The modulated tube 59 has its grid 62 connected through the negative bias battery 63 and condenser 64 with lead 56 while the filament 61 is connected to lead 65 with the lead 55. In this way, the tube 61 acts to modulate the low frequency from the first oscillator circuit upon the higher frequency of the second oscillating circuit.

The potential for the photo-electric cell is provided by a battery 67. The negative terminal of the battery 67 is connected by a line 70 with the light sensitive plate 6 of the photo-electric cell and the positive terminal of the battery 67 is connected through a resistance 69 to a lead 68 connecting with the anode 10 of the photo-electric cell. The battery 67 has preferably a high potential, such as the order of 1000 volts and the resistance 69 is of high resistance such, for example, as one megohm, in order that the drop across such resistance induced by the fluctuations of light in the photo-electric cell may be amplified before being transmitted. The shutter 11 of the photo-electric cell is connected by line 71 to the positive terminal of the battery 67 between the resistance 69 and the battery 67 so that it operates at the same potential as the anode 10 of the cell but its current supply does not pass through the resistance 69.

The effect of the potential applied to the plates 13 and 14 is to cause the electric discharge from the light sensitive plate 6 to be bent back and forth between the plates 13 and 14 at a frequency corresponding to the frequency of the electric potential imposed on the plates 13 and 14 (for example, 10 cycles per second). The effect of the potential applied to the plates 15 and 16 is to cause the electric discharge from the light sensitive plate to be bent back and forth between the plates 15 and 16 at a frequency corresponding to the frequency of the electric potential imposed on the plates 13 and 14, (for example, 5000 cycles per second). The resulting effect is the same as if the opening 12 of the shutter 11 was mechanically moved over the light sensitive plate in accordance with the line shown in Figure 14, in which the substantially parallel lines indicate the movement caused by the potential on the plates 15 and 16. The oscillations of the electric discharge in the direction at right angles to the lines of Figure 14 is caused by the potential on plates 13 and 14, causing the image on the plate 6 to be traversed once every 1/20th of a second with a 10 cycle per second potential. During this period of time, the 5000 cycle per second frequency imposed on plates 15 and 16 will have caused five hundred passages across the image as contrasted with the other television attempts which have succeeded in securing only about thirty-five lines across the image during the optical period. Moreover, it is understood that the frequencies imposed on the plates 13 to 16 inclusive may be increased without limit (up to at least ten thousand kilo-cycles per second), giving any desired number of passages over the image within the optical period, or to make the optical period as short as desired.

There will now be described the apparatus utilized for amplifying the light current and for transmitting such current on a wireless carrier wave, together with the two analyzing oscillator currents or potentials employed on the plates 13 to 16 inclusive, of the photo-electric cell. The transmitting means comprises the tube 72, said tube operating both as an amplifier of the light current and as a modulator of a further tube 73, it being illustrated as in a Heising modulating circuit. The tube 73 produces a first carrier wave of suitable frequency such, for example, as of about 500 kilo-cycles. For this purpose, the tube is illustrated as having its plate 74 connected by lead 75 with an inductance 76, the opposite end of which is connected through the condenser 77 to the grid 78 of the tube.

The inductance 76 is tapped in the center by a variable tap 79 which connects to a variable condenser 80 and hence by a line 81 to the filament 82. The condenser 80 and the coil 76 may have any values provided that the condenser 80 and the inductance 76 are adapted to bring the circuit in resonance with the carrier wave to be produced. The line 81 is also connected with the line 77 by a condenser 82[a]. The grid 78 is also connected with the filament 82 through a grid leak 84 and negative battery 85. The potential for the tube 73 is provided by the battery 91, through the resistance or choke 90. The tube 72 acts as a variable resistance across 90 and 91, increasing or decreasing the potential drop and thereby modulating the potential on plate 74 of the tube 73. The tube 72 has its grid 86 connected by a negative bias battery 87 with the resistance 69, across which there is imposed the "light" potential whereby said "light" potential is amplified in the tube 73. The plate 88 of the amplifying and modulating tube 72 is connected by a line 89 through a choke or resistance 90 and a battery 91, the negative side of which is connected with the filament 92 of the tube 72 and also with the filament 82 of the oscillating tube 73.

The choke 90 operates to fluctuate the potential supply to the plate of the oscillating tube in accordance with the amplified light current. In the lead between the choke 90 and plate 74 is provided a choke 92 which prevents the carrier wave produced in the oscillator 73 from being imposed upon the amplifying and modulating tube 72 by the circuit thus described. The carrier wave produced in the oscillator 73 is modulated by the amplified light current. This potential is then imposed upon a double modulating tube 94 which operates to modulate an oscillator 95 producing a second carrier wave of higher frequency, such for example as 1500 kilo-cycles, or the wave length to be transmitted.

Said double modulator tube 94 not only modulates the second carrier wave with the modulated first carrier wave from oscillator 73, but also modulates said carrier wave with the analyzing potentials from the modulator tube 59. The double modulating tube 94 has its grid 96 connected by lead 97 with a coil 98, the coil 98 being connected to the filament 99 of the double modulating tube. By this means, the output from the oscillator 73 is imposed upon the double modulating grid. The analyzing potentials are imposed upon the grid 96 by a lead 100 which connects across the resistance 58[a] and hence by a lead 101 to the filament 99. The tube 94 is part of a Heising modulator that has its plate 102 connected by a lead 103 through a radio frequency choke or resistance 104 to the positive terminal of battery 105, the negative terminal of which is connected with the filament 99. The lead 103 also connects with the radio frequency choke 106 to the plate 107 of the oscillator tube 95. The choke 106 prevents the second carrier wave from being imposed upon the double modulating tube 94 while the choke or resistance 104 fluctuates the potential supply to the plate 107 of the oscillator 95 in accordance with the output of the double modulating tube 94. The plate 107 connects with the lead 108 to an inductance 109 producing the second carrier wave, said inductance being connected with the lead 110 through condenser 111 with the grid 112 of the oscillator tube 95. The filament 113 of the tube is connected by lead 114 through a variable condenser 115 to the inductance 109. There is also a condenser 116 between the lead 114 and the grid leak 110. The inductance is also connected with an antenna 117 or other means for radiating the output from the transmitter. The filament 113 is grounded as indicated at 118.

The receiver of the television apparatus and process is constructed and operates as follows: Preferably there is employed a source of light of constant intensity, such as an arc light 120 and to obtain a pencil of light therefrom, there is placed a shutter 121 with a small aperture 122 in front of the arc light. The light from said shutter is then passed through a polarizer 123. The polarizer is indicated as preferably in the form of a Nicol prism. The polarized light from the Nicol prism 123 is then passed through a lens 124 which parallels the polarized light and the paralleled light is then passed through a device 125 for rotating the plane of the polarized light. The device 125 may be any device suitable for rotating the plane of the polarized light in accordance with the fluctuations of the light current received at the receiver. The method of receiving and separating this light current from the transmitted wave will be hereinafter pointed out. The preferred form of such device is illustrated as comprising a means for producing a magnetic field fluctuating in accordance with the light current, such as the coil 126, surrounding an electrically optically active medium 127, such for example as a thin film of iron, cobalt, or nickel, or carbon disulfide, glass, or any other material in which a beam of polarized light rotates considerably when subjected to a magnetic field. I prefer to employ carbon disulfide and said carbon disulfide is held in the core of the coil 126 by glass plates 128.

The light from the light rotator 125 is then passed through a device adapted for restricting the passage of light in accordance with its degree of rotation. I preferably employ a combination of a pair of gratings 129 and 130 and a bi-axial crystal 131. The gratings 129 and 130 may be any usual form of light gratings, for example, ruled upon a silvered transparent surface, and are placed at opposite ends or sides of the bi-axial crystal with their gratings opposed. The bi-axial crystal employed between the gratings is adapted to produce a conical refraction of the light. As an example of a suitable crystal of this kind, I have employed a crystal of arragonite one centimeter thick between the gratings ruled with 100 lines per millimeter. With this combination, the rotation between complete extinction and complete restoration is of the order of two degrees. Thus with this analyzer, very small currents may be employed upon the rotator, permitting the use of a coil of very high natural period.

The operation of this analyzer will best be understood from Figures 15, 16 and 17, in which Figures 15, 16 and 17 there is disclosed how a rotation of a few degrees will change complete extinction to complete restoration. A indicates a beam of light passing through the first grating 129 and hence through the bi-axial crystal 131 to the second grating 130, the lines of which are opposed to the lines of the grating 129. If the beam of light passes directly through the bi-axial crystal, it is completely extinguished by the lines of the grating 130 but if the plane of polarization of the beam A is rotated slightly, the ray A will take the direction of the dotted lines through the crystal and pass between the lines of the grating 130, a slight difference in refraction of the light in the bi-axial crystal 131 being sufficient for this purpose.

In explanation of the action of the bi-axial crystal 131, it is understood that the light is directed on said crystal along one of its optic axes. When this is done, the light is refracted to an extent depending on the position of the plane of polarization. When unpolarized light from an aperture is directed on such a crystal along one of its axes, said light will appear as a circle from the other side of the crystal, but when a beam of polarized light is directed along one of the axes of the crystal, it appears as a point of light lying in the circle produced by the unpolarized light, but its position is dependent on the position of the plane of polarization of the beam of light. A 90 degree rotation of the plane of polarization of the beam of light will rotate the light from the crystal from one side of the circle 8 to the opposite side. The two extreme positions of a polarized beam of light are indicated in Figure 15, by the two branches of the beam of light A. During the passage of the light through the bi-axial crystal, the wave front of the beam of light remains parallel and the wave front of the beam passes through perpendicularly to the optic axis of the crystal.

By means of the polarizer 123, light rotator 125, and analyzer comprising the gratings 129 and 130 and the bi-axial crystal 131, the constant supply of light through arc light 120 is caused to produce a light of varying intensity, varying in accordance with the intensity of the light current supplied to the coil 126. Thereby, without the employment of any mechanical moving apparatus, the light current is reconverted into light.

Such light is then passed through a lens 132 by which it is focused upon a pair of cooperating oscillographs 133 and 134. Said cooperating oscillographs 133 and 134 are positioned at right angles one to the other and so that the light from one strikes the other oscillograph. Said oscillographs are operated at different frequencies with the result that the light is by said oscillographs projected in horizontal vibrations, which are successively lowered or raised vertically so that the light can pass through a lens 135 upon a screen 136 and covers successively an entire rectangular area of said screen. The oscillographs 133 and 134 are operated by electrical currents of the frequencies of the two analyzer currents applied to the plates 13 to 16 of the photo-electric cell so that the passage of the beam of light over the screen 136 is in synchronism with the bending of the electrical discharge from the sensitive plate 6 of the photo-electric cell and thereby each portion of light is properly coordinated to produce a correct image of the object being transmitted.

The details of the construction of the oscillographs 133 and 134 are shown in Figures 4 and 5, only one of the oscillographs being illustrated since they are of similar construction. The oscillographs comprise a base or body 137 of any suitable material. In the center thereof, is mounted a quartz strip 138 having a silvered mirror surface 139 at its top. Said quartz strip vibrator 138 is held in a holder 140 which is vertically adjustable by a set-screw 141. The quartz strip vibrator is engaged at opposite sides and at points spaced apart slightly vertically by a pair of quartz strips 142 and 143 laid horizontally and plated at the tops and bottoms by a metallic plating, such as copper, as indicated at 144 and 145. The outer ends of such quartz strips 142 and 143 engage guides 146 on the body, and hence engage clamps 147 by which they are held to carriers 148. The clamps 147 are connected by adjusting screws 149 to the body 137 by means of which the quartz strips 142 and 143 may have their pressure against the quartz strip vibrator 138 adjusted. At the inner ends of the quartz strips 142 and 143 are placed rests 149[a] over which are placed a resilient material, such as rubber, and thereabove is placed a further quantity of rubber.

Clamps 150 are placed over the top of the inner ends of the quartz strips and connected with adjusting screws 151 by means of which the vertical positions of the ends of the quartz strips may be adjusted. It is understood that in the showing of Figures 4 and 5, the quartz strips are greatly exaggerated in thickness inasmuch as in practice such strips are very thin, approximating the thickness of a sheet of paper, and are cut with their thickness in the direction of the electric axis, their length in the direction of the axis of extension and their width along the optic axis of the crystal. The bottom sides of the strips 142 and 143 are connected by conductors 152 while the top plating on the strips is connected by conductors 153 connected with springs 154 at the top of the clamps 150.

Referring to Figure 3, the electrical apparatus for receiving the transmitted wave in the transmitter and correctly applying the light current and analyzing currents to the light rotator 125 and oscillographs 133 and 134 is as follows: 155 indicates a receiving antenna or other means for collecting wireless waves which antenna is connected through an inductance 156 to a ground indicated at 157. Inductance 156 forms a primary of a transformer in which the secondary 158 is in the grid circuit of a detector 159. 160 indicates a tuning condenser for bringing the receiver in resonance with the carrier wave of the transmitter. The plate 161 is indicated as connected to a plurality of filters, the first of which comprises the inductance 162, the voltage across which is applied to the grid of a second detector 164. The first filter comprising the inductance 162 should be in resonance with the first carrier wave developed in the transmitter or tube 159 thereof. There is thus imposed upon the grid of a detector tube 164 a current comprising the light current modulated upon the first carrier wave formed in the transmitter. In the detector 164, such carrier wave is detected to produce a current output from the plate 165, which is equivalent to the light current developed in the transmitter. In the second detector circuit 164, 166 indicates a condenser for passing the high frequency and blocking the low frequency currents, and 167 indicates a battery for supplying the plate potential. The plate 165 is indicated as connected with the coil 126 of the light rotator.

The complete circuit of the detector tube 159 also includes a condenser 168 of a capacity suitable for by-passing the high frequency of the first carrier wave which is detected by the tube 164 and of a capacity to block the frequency of the analyzing currents. Such analyzing currents are therefrom passed through a choke 169 and line 170 to one of the oscillographs 133, connecting for example with the top platings of both of the quartz strips thereof, the bottom plating of the quartz strips of said oscillographs 133 being connected by a line 171 with a resistance 200 shunted across line 170, and line 201 which line connects with the opposite side of the condenser 168. By this connection, the oscillograph 133 is operated by the higher analyzing frequency, i. e., the 500 cycles per second frequency. Said frequency also passes through the grid leak 172 to a grid 174 of a detector tube 173 wherein said frequency is detected to deliver from its plate 175 a potential of the frequency of the first analyzing current, or 10 cycles per second. The plate 175 is indicated as connected by the line 176 to the resistance 202 which is connected by a tap 203 to the top plating of the oscillograph 134 and the bottom plating of the oscillograph 134 is indicated as connected by line 177 through the battery 178 to the filament 179 of the detector 173. The filament 179 is also connected by the lead 180 with the condenser 168. The resistance 200 and 202 provide a means for controlling the potential of the currents applied to the oscillographs.

It will be readily apparent from the description of the apparatus and operation thereof, how the detected light current imposed upon the coil 126 modulates the light in accordance with the intensity of light at the particular point from which said light current originated from the light sensitive plate 6. It will also be seen that said light is projected upon the screen 136 by the oscillations of the oscillographs 133 and 134 to form a correct image of the object transmitted, the light being caused to travel back and forth across the screen similar to the action of the shutter 11 of the transmitter, making the example given 500 reciprocations across the screen in covering the complete area thereof, and said reciprocations are made within a period of 1/20th of a second. It is understood, however, that the process and apparatus of the present invention is not necessarily limited to the use of the particular frequencies given for the purpose of facilitating the description of a preferred process and apparatus.

The process and apparatus of the present invention permit the selection of such small elementary areas of the image to be transmitted that the produced image on the screen 136 follows all of the light shades of the object, producing a correct image thereof. This is accomplished without the employment of mechanically moving parts, excepting the vibrating strips of the oscillographs. The apparatus is thus free from mechanical problems.

While the process and apparatus for producing television herein described is well adapted for carrying out the objects of the present invention, it is understood that various modifications and changes may be made without departing from the invention, and the invention includes all such modifications and changes as come within the scope of the following appended claims.

I claim:

1. The method of television which includes forming an electrical image, and traversing each elementary area of the electrical image by an electric shutter at a velocity sufficient to cover the entire image within the optical period.

2. The process of television which comprises forming an electrical image, moving said electrical image in more than one direction by an analyzing potential, and varying the intensity of an electric current in accordance with the position of the electrical image.

3. The method of television which comprises focusing an image of an object upon the sensitive plate of a photo-electric cell, imposing a shutter in the path of the electrical discharge from said plate, and forming transverse to the electrical discharge two electrical potentials of different frequencies.

4. An apparatus for picture dissecting comprising a cell having a plate of photo sensitive material, an anode, a plurality of plates positioned between the photo sensitive plate and anode, and means for imposing upon said plates a plurality of electrical potentials of different frequencies.

5. An apparatus for dissecting an image comprising a cell having a photo sensitive plate, an anode, a shutter between the anode and plate, and electrical means for bending the electrical discharge from said plate.

6. The method of television which comprises forming an electrical discharge which corresponds in cross section in electrical intensity to the light intensity of an image to be transmitted, transmitting successive portions of said electric discharge, and modulating light thereby.

7. A method of television which comprises analyzing an image into elementary areas, producing a train of energy varying according to the intensity of light of said areas, all of the elementary areas being covered within the optical period, causing said train of energy to modulate a source of light of constant intensity according to the light of said areas, and correlating successive portions of said light to reform said image, said latter operation being completed within the optical period.

8. A method of television which comprises producing an electrical oscillation having a substantially straight line wave form, utilizing said electrical potential to analyze an image into elementary areas, producing a train of energy varying according to the intensity of light of said areas, and converting said train of energy into light varying according to the light of said areas.

9. A method of television which comprises producing an electrical oscillation having a substantially straight line wave form, utilizing said electrical potential to analyze an image into elementary areas, producing a train of energy varying according to the intensity of light of said areas, converting said train of energy into light varying according to the light of said areas, and utilizing said electric potential of substantially straight line wave form to correlate successive portions of said light.

10. A method of television which comprises producing two electrical potentials of different frequencies, each of said electrical potentials having substantially straight line wave forms, causing said electrical potentials to analyze an image into elementary areas, producing a train of energy varying according to the intensity of light of said areas, and converting said train of energy into light varying according to the light of said areas.

11. A method of television which comprises producing two electrical potentials of different frequencies, each of said electrical potentials having substantially straight line wave forms, causing said electrical potentials to analyze an image into elementary areas, producing a train of energy varying according to the intensity of light of said areas, converting said train of energy into light varying according to the light of said areas, and causing said electrical potentials of different frequencies to correlate successive portions of said light to reform said image.

12. In a system of television, analyzing an image into elementary areas by causing a scanning device to scan all elements of said image successively at a substantially uniform velocity, over a continuous path reciprocating transversely of the image and the reciprocations having a slow motion transverse thereto.

13. A method of television which comprises forming an electrical image, moving the image in two directions over an electrical shutter having a small aperture, thus forming an electrical current which is a function of the intensity of the portion of the electrical image at said aperture.

14. A method of television which comprises forming an electrical image, impressing upon said image two electrical potentials of different frequencies, thereby causing said image to move in two directions respecting an electrical shutter and forming an electric current from the portion of the electrical image registered with the electrical shutter.

15. An apparatus for television which comprises means for forming an electrical image, and means for scanning each elementary area of the electrical image, and means for producing a train of electrical energy in accordance with the intensity of the elementary area of the electrical image being scanned.

16. An apparatus for television which comprises means for forming an electric image,

means for moving said electric image in more than one direction by an analyzing potential, and means for varying the intensity of an electrical current in accordance with the position of the electrical image.

17. An apparatus for television which comprises means for focusing an image of an object upon the sensitive plate of a photo-electric cell, said photo-electric cell having an anode therein to receive an electrical discharge from said plate, said cell having a shutter in the path of the electrical discharge from the sensitive plate, said cell having plates positioned transverse to the electrical discharge, and means for imposing upon said plates electrical potentials of different frequencies.

18. An apparatus of the class described, including an oscillator, an oscillator of higher frequency operated by the oscillations from the first oscillator, thereby producing successive trains of oscillations during the positive cycle of oscillations of the first oscillator, a device for accumulating and discharging said oscillations thereby producing oscillations having substantially straight lined wave form, similar means producing an alternating potential of straight lined wave form and higher frequency, means for utilizing said potentials to scan an image in two directions, means for modulating the lower frequency upon the higher frequency, means for producing a train of energy varying in intensity in accordance with the area scanned, means for modulating a carrier wave with said train of energy and said scanning potentials, means for receiving and detecting said train of energy and said analyzing potentials, means for modulating the light in accordance with said analyzing potentials, and means for correlating said light to form an image actuated by said potentials having straight line wave forms.

Signed at San Francisco, California, this 21st day of December, 1926.

PHILO T. FARNSWORTH.

(No Model.)

3 Sheets—Sheet 1.

G. EASTMAN.

CAMERA.

No. 388,850. Patented Sept. 4, 1888.

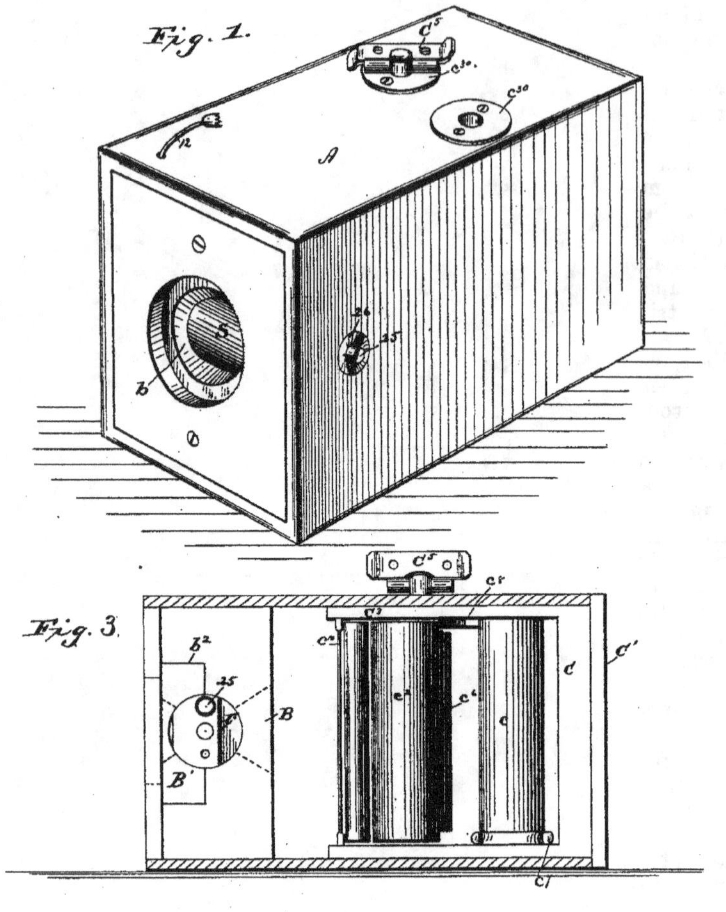

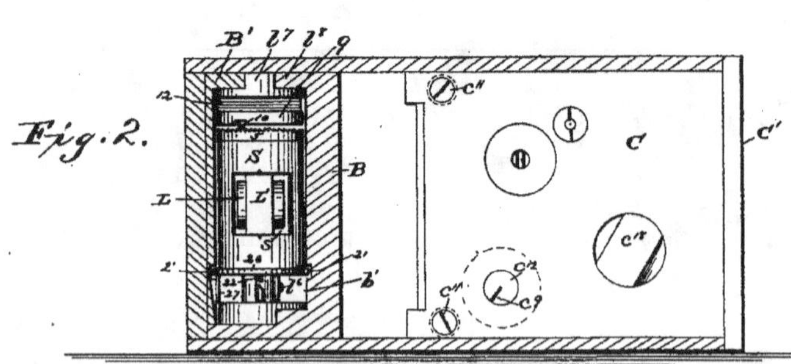

Witnesses.
Chas. R. Burr.
A. J. Stewart.

Inventor.
George Eastman.
by Church & Church
his Attorneys.

N. PETERS. Photo-Lithographer, Washington, D. C.

(No Model.)

3 Sheets—Sheet 2.

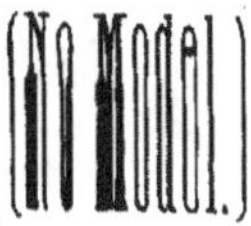

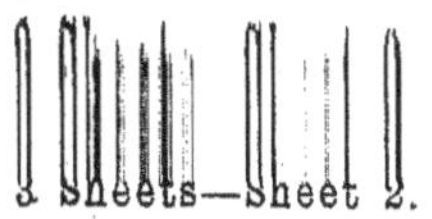

G. EASTMAN.

CAMERA.

No. 388,850. Patented Sept. 4, 1888.

Fig. 4.

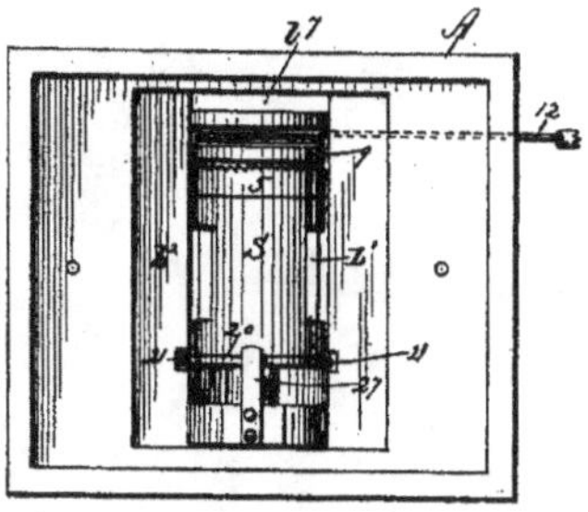

Fig. 5.

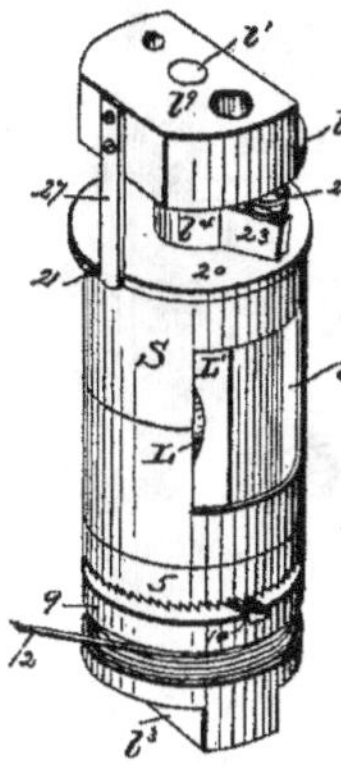

Fig. 6.

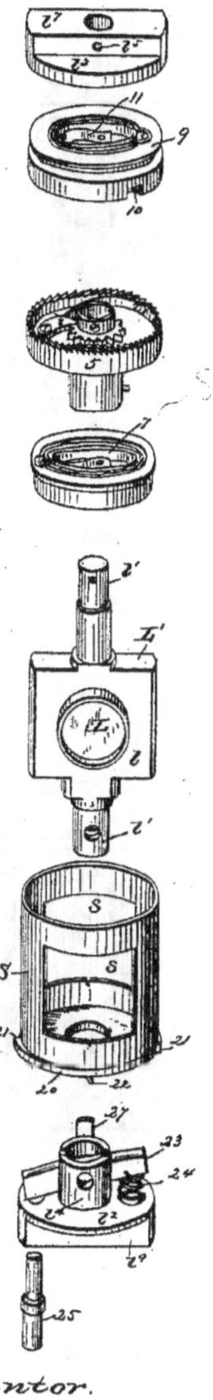

Fig. 7.

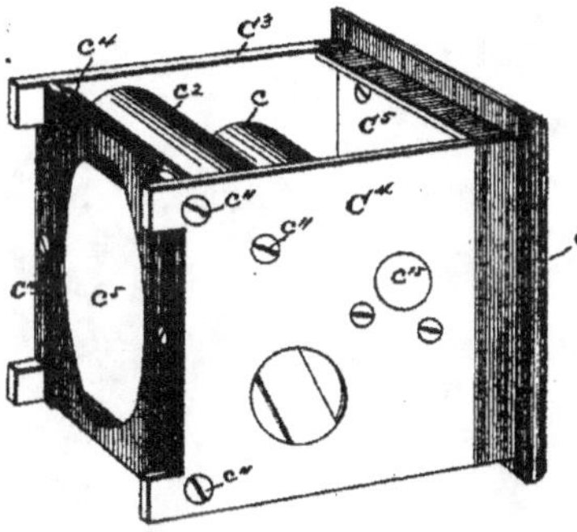

Fig. 8.

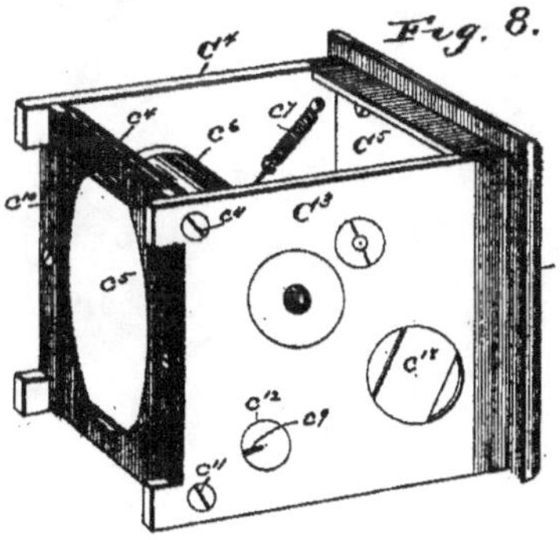

Fig. 11.

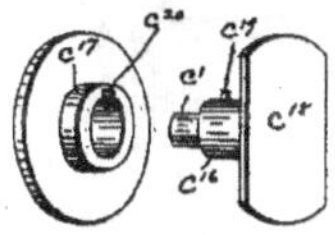

Witnesses.

Chas. R. Burr.

A. J. Stewart.

Inventor.

George Eastman,

by Church & Church

his Attorneys.

N. PETERS, Photo-Lithographer, Washington, D. C.

(No Model.)

3 Sheets—Sheet 3.

G. EASTMAN.

CAMERA.

No. 388,850.

Patented Sept. 4, 1888.

Fig. 9.

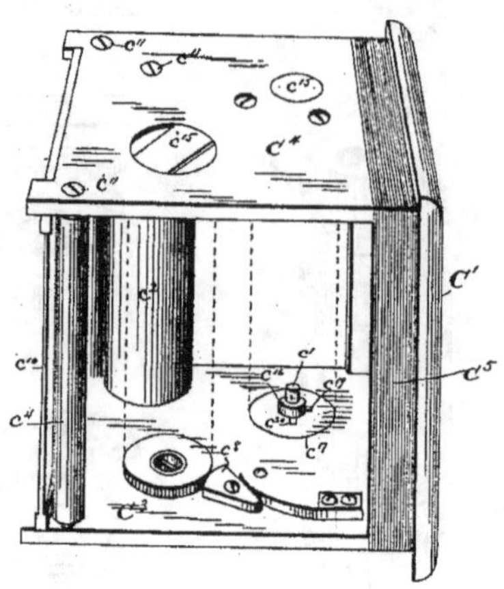

Fig. 10.

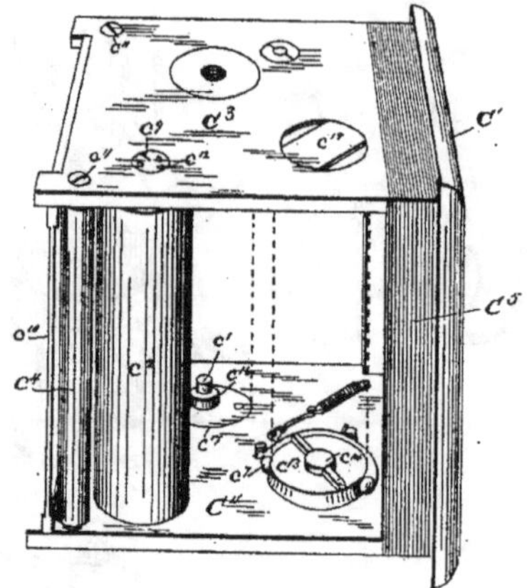

Witnesses.
Chas. R. Burr.
Thomas Durant.

Inventor.
George Eastman.
by Church & Church
his Attorneys.

N. PETERS, Photo-Lithographer, Washington, D. C.

UNITED STATES PATENT OFFICE.

GEORGE EASTMAN, OF ROCHESTER, NEW YORK.

CAMERA.

SPECIFICATION forming part of Letters Patent No. 388,850, dated September 4, 1888.

Application filed March 30, 1888. Serial No. 268,964. (No model.)

To all whom it may concern:

Be it known that I, GEORGE EASTMAN, of Rochester, in the county of Monroe and State of New York, have invented certain new and useful Improvements in Cameras; and I do hereby declare the following to be a full, clear, and exact description of the same, reference being had to the accompanying drawings, forming a part of this specification, and to the figures and letters of reference marked thereon.

This invention relates more particularly to improvements in that class of photographic apparatus known as "detective cameras;" and said invention consists in the novel and improved form, construction, and arrangement of parts constituting the case or body, the lens-support and shutter, and the film-holder, together with the various combinations of such instrumentalities as are hereinafter described, and set forth in the claims.

In the accompanying drawings, wherein I have illustrated one embodiment of my present improvements, Figure 1 is a view in perspective of the complete instrument. Fig. 2 is a side and Fig. 3 a top view of the camera, the side and top of the box being removed to disclose the interior arrangement. Fig. 4 is a front view with cap or end of box removed. Fig. 5 is a view in perspective of the lens-holder and shutter. Fig. 6 is a view in perspective of the lens-holder and shutter, several of the parts being detached. Figs. 7, 8, 9, and 10 are views in perspective of the roller-holder, looking from opposite sides. Fig. 11 is a detail view illustrating the manner of mounting and supporting the spool.

Similar letters of reference in the several figures indicate the same parts.

The letter A designates the camera box or case, preferably constructed in the form of a rectangular tube, at or near one end of which is fastened a block, B, recessed for the reception or accommodation of the lens-holder and shutter and perforated, as at *b*. In this block B, or between it and the end piece or cover, B′, is located the lens and shutter mechanism, and in rear of said block is located the roller-holder C, the latter being inserted within and closing the rear end of the box or tube A.

The roller-holder C is secured to or formed integral with the rear cap plate or cover, C′, and the sides of said holder upon which the operating devices are supported is formed to fit snugly within the tube A, the cover C′ overlapping or otherwise co-operating with the end of the box A to form a light-tight connection, so that when said roller-holder is inserted and held within the end of the box the sensitized film or plate will be entirely protected from light in rear and at the sides, and will only be exposed to light entering through the perforation *b* in block or diaphragm B.

The block or end piece, B, is designed not only to close the front end of case A and protect the sensitized plate or film, but also to receive and encompass the lens holder and shutter; and to this end said block, whether formed in one, two, or more sections, is provided with a chamber or recess, *b*′, and coincident openings *b*. In the illustration given the block B is formed in two sections, the rear section or that one permanently secured within the case A being provided with a rectangular or other shaped recess, b^2, into which the front section enters and fits snugly, the recess or chamber *b*′, for the reception of the lens-support and shutter, being formed in the proximate surfaces of the two sections in the rear of the walls of recess b^2, so that light will be entirely excluded from the interior of the case and from the chamber *b*′, with the exception of such as may be permitted to enter through openings *b*. To render the exclusion of light more certain, the front section is attached to the cap or cover B′, the latter making contact with the front face of the rear section and entering the mouth or front end of case A. The principal object and function of this construction is to furnish a convenient light-tight end piece for the camera-box and a receptacle for the lens and shutter—one which will furnish ample support for the lens and shutter mechanism and permit access to the latter when desired; hence any device or contrivance which, when located in front of the camera box or tube, and whether composed of one, two, or more parts or sections, will furnish a light-tight casing or chamber between the apertures through which light is admitted to the interior of the camera through the lens. Any contrivance possessing these capacities and functions will be the equivalent of the block B in its generic relation to the lens-support and shutter mechanism about to be described.

The lens-support and shutter mechanism employed, although specially adapted for use in connection with this camera, illustrates one of the many forms or embodiments of my invention—that is to say, it illustrates a new type rather than a new species of an old type—its principal distinguishing feature being a lens mounted or held within a chamber—such as b'—in front of the exposing-chamber, and a cylindrical case or shutter provided with coincident slots or openings on opposite sides, said case or shutter being rotated or reciprocated intermittingly to cause the slots or openings therein to pass across the opposite ends or faces of the lens, the solid opaque portions or body of the shutter standing normally and when at rest across both faces of the lens and between the exposing-aperture and the plate or film within the camera.

Suitable devices are to be employed for reciprocating or intermittingly rotating the cylindrical shutter and for holding it at rest.

Referring to the preferred form or embodiment of the principle, as illustrated in the drawings, S represents the cylindrical shutter, L the lens, and L′ the lens-support, the latter also constituting the support for the shutter and its actuating mechanism, as will hereinafter appear.

The lens-support, with its attachments, is located within the recess or chamber b' in block B; and in order that the whole of said mechanism may be readily inserted or removed, and at the same time avoid all danger of possible leakage of light, it is seated and held in position within the chamber b' between the fixed and removable sections of the block B. To this end the lens-support proper, consisting of a frame or plate, l, in which the lens L is detachably secured, is mounted upon or provided with posts l' at opposite ends, these posts being formed or provided at or near their outer extremities with heads l^2 l^3, of cylindrical or other shape, adapted to fit the interior of chamber b' on opposite sides of the opening b.

For convenience of manufacture and simplicity of construction, the head l^2 is formed with a collar or socket, l^4, in which the post l' of the lens-support is secured by a pin or screw, l^5, while the head l^3 at the opposite end is removably attached to its post l' by a screw or pin, l^6. The head l^3 is also formed or provided with a cross-piece or shoulder, l^7, which engages with a shoulder or abutment within the chamber b'—such as the shoulder l^8 on the two sections of block B—the purpose of said shoulder l^7 and a similar shoulder, l^9, on the opposite head l^2 being to sustain the lens-support in position and prevent rotation.

The shutter consists of a hollow cylinder, S, provided with coincident openings s, and preferably closed at both ends.

The lens is supported within this cylinder S, and the latter is mounted to rotate or reciprocate about an axis transverse to the axis of the lens, the openings s moving in a plane intersecting the axis of the lens on opposite faces thereof, to which end said cylinder is journaled at one or, preferably, both ends upon the posts l' of the lens support, so that it has a motion about said posts and around the lens supported between them, and at each half-revolution of the cylinder the openings s will be caused to approach and cross the opening or axis of the lens at opposite ends and from opposite sides or edges.

It is of course desirable that some competent means be provided for holding and actuating the cylindrical shutter. A mechanism of this character is shown, comprising a motor mechanism for impelling or moving the shutter and a stop and release mechanism.

The motor mechanism consists of a ratchet-wheel, 5, rotating loosely upon the pin l', attached to one head of the cylinder S, and connected to the latter by a spring, 7, inclosed within the end of the cylinder S and between it and the wheel 5. A ratchet-and-pawl connection, 8, between the wheel 5 and pin l', permits the said wheel to rotate freely in a direction to wind up the spring 7, but prevents a retrograde movement.

On the pin l', between the head l^3 and wheel 5, is mounted a drum or pulley, 9, carrying a spring-pawl, 10, for engaging the teeth on wheel 5 to rotate the latter in a direction to wind up the spring 7, a spring, 11, serving to retract said drum 9. A cord or other flexible connection, 12, is attached at one end to this drum, so that by alternately pulling upon said cord the drum will be reciprocated upon its axis, and through the ratchet-connection the wheel 5 will be rotated to wind up the spring 7, one end of which is secured to the shutter, and hence tends to rotate the latter. This constitutes the motor mechanism, the spring serving to drive or advance the shutter and the drum 9 to wind up the spring, and in the improved structure shown all of said mechanism is mounted upon the lens-support.

The winding-cord 12 is passed through a suitable orifice in the block B with a knot or knots, and is manipulated from the outside of the case.

The devices for controlling the movements of the shutter, also mounted upon the lens-support, comprise a plate, 20, formed upon or secured to one end of the hollow cylindrical shutter, and provided with peripheral shoulders 21 and surface shoulders or abutments 22.

Above the plate 20 and mounted upon a transverse pin or pivot is a latch, 23, the lower edge on opposite sides of the pivot being inclined or beveled in the same direction. One end of this latch 23 is held pressed down by a spring, 24, while a push-pin, 25, passing through head l^2, bears upon the opposite end of the latch. The pin 25 is preferably formed with a collar fitting an enlarged recess in the head l^2, and is held in place by a collar or thimble, 26, screwed or otherwise detachably secured in the case or box A. A spring, 27, attached to head l^2 bears against the periphery of the plate 20.

The operation of the controlling devices, arranged as described, is as follows: The rotation of the shutter, when acted upon by the motor devices, is arrested by one of the abutments, 22, resting in contact with the beveled side of the latch, the latter being held in place by its spring 24. When in this position the opposite end of the latch stands with its beveled edge above the opposite abutment 22, so that when the latch is oscillated by pressure upon the push-pin 23 the abutment on one side will ride down the beveled face of the latch and the shutter be correspondingly advanced before the elevated end of the latch reaches the abutment which is beneath it, so that when the engaging end of the latch passes off from the abutment on that side the other abutment will have passed beyond or cleared the edge of the latch on the opposite side, and the shutter impelled by its motor or spring will be driven forward until the previously-engaged abutment comes in contact with the previously-elevated but now depressed end of the latch. The push-pin being released and the latch permitted to resume its first position, the abutment beneath the push-pin, and now held against the vertical face of the latch, is released, and at the same time the beveled face on the opposite end of the latch is interposed in front of the other abutment, a very slight forward movement of the shutter being permitted, when it is again arrested and held until the push-pin is depressed. One edge of the latch is at all times projected in the path traversed by the abutments 22, so that the shutter can, under no circumstances, perform more than a half-revolution. It will thus be observed that each time the push-pin is depressed the shutter is permitted to make a half-revolution about the lens, is arrested at this point, and upon removing the pressure from the push-pin the devices are automatically set in position to repeat the same operations. The abutments 22 are so disposed relatively to the openings in the shutter that when the latter is arrested by the latch the solid or opaque portions of the cylinder will stand in front of or across the axis of the lens, and when released, as in passing from one abutment to the next, the lens will be uncovered simultaneously at opposite ends and from opposite edges. The spring 27, bearing upon plate 20, operates as a brake to control or regulate the speed at which the cylindrical shutter revolves, and, in conjunction with the shoulders 21, it acts as a detent or stop for preventing a retrograde movement of the shutter.

To provide for holding the shutter at an intermediate point, so that the lens will be wholly exposed, radial grooves 29 are formed in the face of the plate 20 to receive the edges of the latch, so that the shutter can be retained in position with the openings in line with the axis of the lens, and by pressing the push-pin can be at once set for action, as before described.

It will be observed that as thus constructed and arranged not only is a simple, cheap, compact, and withal effective and accurate shutter and lens-support formed, the whole constituting in effect a complete article distinct from its case, and which is not only complete in itself and can readily be applied with but slight changes and modifications to almost any form of cameras, but when applied and used in the manner indicated—that is to say, when located within a closed chamber at the end of the camera box or case—it can be quickly inserted or removed without the aid of special skill, can be adapted and held in position wholly by the walls of the case or chamber, and when so applied it forms an efficient means for preventing the entrance of light into the exposing-chamber until such time as the shutter is actuated to uncover the aperture in the lens.

The next, or what may be termed the third, constituent element of my improved instrument comprises a competent means for holding and presenting the plate or film within the camera box or case in position to receive the rays of light passing through the lens. As is obvious, any holder competent to perform the well-understood operation may be employed for this purpose; but I prefer, for obvious reasons, to make use of prepared films, and to this end have devised the following improved form of roll-holder: The operating mechanism of this holder C comprises a supply-spool, c, detachably mounted on supporting pins or journals c', a measuring roller, c^2, furnished with suitable puncturing-pins or marking devices, guide-rollers c^4, a platen or support, c^5, for the film, a winding-roller, c^6, provided with a detaining or film-attaching device, a tension device, c^7, applied to the supply-spool, a ratchet or detaining device, c^8, applied to the winding spool, and an indicator, c^9, connected to or actuated by the measuring-roller, all as set forth and described, in part or in whole, in Patents Nos. 317,049, 317,050, and 316,933, and in my prior application Serial No. 199,329, filed April 19, 1886.

The improvements herein claimed relate, mainly, to the construction and arrangement of parts whereby the holder is adapted for use in connection with a tubular case—such as A—forming part of the camera proper.

The several co-operating elements are mounted and supported wholly upon and between the two side pieces, C^3 C^4, attached to a base, C^5, which forms or is secured to the cap C', for closing the rear end of the case or box A, the several attaching devices being so arranged and constructed as not to project beyond the outer surfaces of the side pieces, leaving the latter smooth and unobstructed, so that the holder can be fitted and slid within the opening in the rear end of the box A. To secure this result, the platen c^5 is interposed between and secured to the side pieces at or near the outer end thereof, as is also the covering-plate c^{10}, which serves to retain the film and hold the edges down upon the platen. The pins or journals supporting rollers c^2 c^4 pass through the side plates, their outer ends being flush with or standing below the outer faces of the side pieces. The rollers c^4 and one end of the roller c^2 are preferably supported upon the ends of pins c^{11}, screwed through the side pieces, while the opposite end of roller c^2 carries a collar bearing against the inner surface of the side piece, C^3, and is supported upon a stud, c^{12}, screwed into the end of said roller and passing through a hole in the side piece, the outer end of said stud being formed or provided with a head resting in a countersunk recess and bearing a mark, c^9, which latter, when brought into proper relation to a fixed point, serves to indicate the quantity of film wound on the roller c^6.

The head c^{13}, for receiving and supporting one end of the spool, and to which the tension device c^7 is applied, is held in place and supported upon a detachable pin, c^{14}, passing through the side piece, C^4, and provided with a flat head or disk, c^{15}, countersunk in the outer face of side piece, C^4, while the opposite end of the spool is supported upon a pin, c^{16}, resting in a socket or bearing, c^{17}, secured in position on the inner face of side piece, C^3. The pin c^{16} is also provided with a head or cross-piece, c^{18}, and a stud, c^{19}, the former secured in a countersink on the outer face of the side piece, and the latter adapted to be inserted or withdrawn through a notch or way, c^{20}, in the bearing c^{17}, serving, when inserted, to hold the pin in place by bearing against the inner face of the socket. The pin c^{16} is thus held in place within its bearing by the head on one side and the stud on the other; and when it is desired to detach or insert a spool the pin can readily be withdrawn by turning it until the stud registers with the notch in the bearing. When, however, the holder is inserted in the camera, the accidental withdrawal of the pin is prevented, should the stud stand in line with the notch, by the head of the pin resting and bearing against the smooth inner face of the box A.

The supports for the winding-spool are constructed substantially the same as those for the supply-spool, with this exception, however, that the pin corresponding to c^{14}, supporting the head to which the ratchet or detaining devices are applied, is provided with a central screw-threaded aperture or equivalent connecting means for the reception of the winding-key C^5.

The construction described is more especially designed for use when the side pieces are made of wood; but constructed of metal or other material corresponding modifications might be made in the fittings without involving any material change in the invention.

The roller-holder as thus constructed is adapted for insertion within any tube or box A, of the proper dimensions, and without other fitting than such as will accommodate the winding-key C^5 and render visible the indicator, and at the same time prevent access of light to the interior of the camera; for, as is well understood, all light except such as is transmitted through the lens must be excluded from the exposing-chamber in rear of the lens. Ample provision of this nature is made by forming two holes or openings in the case A opposite the axes of the winding roller and spool, respectively; and when, as is proposed, the box A is constructed of wood, leather, or similar material, thimbles c^{30} may be inserted and secured in these openings and a mark made on that one next the indicator.

Having thus described my invention, what I claim as new is—

1. The combination, to form a camera such as described, of a tubular case or box closed at one end by a detachable film-holder and at the other by a block or end piece, and a removable shutter and lens-support held within a chamber in said block or end piece.

2. In a camera such as described, the combination, with the box or case, of a block or partition closing the front end and forming a chamber with coincident openings, a fixed lens-support, and rotary shutter mechanism connected in fixed relation and supported within said chamber, as and for the purposes set forth.

3. In a camera, and in combination with a tubular box or case, a block or diaphragm fixed in position within the front end and provided with an aperture, a cap or end piece provided with an aperture and co-operating with the said block or diaphragm to form a chamber within the box or case, a lens supported within said chamber between the said apertures, and a shutter mounted upon the lens support, substantially as described.

4. The combination, in a camera and with the lens, of a shutter surrounding the latter and provided with coincident apertures, substantially as described.

5. In a camera, and in combination with the lens, a shutter projected on opposite faces of the lens and provided with coincident apertures, said shutter being operated to simultaneously uncover both faces of the lens from opposite edges thereof, substantially as described.

6. In a camera, and in combination with a lens and its support, a shutter provided with coincident apertures or spaces and intermediate covering-plates projected on opposite sides of the lens, and devices for operating said shutter to alternately uncover and cover both faces of the lens simultaneously, as and for the purpose set forth.

7. In a camera, and in combination with the lens, a double shutter embracing the lens and mounted upon a pivotal support, about which it is rotated intermittingly to uncover both faces of the lens simultaneously, substantially as described.

8. In combination with the lens and its fixed support, a shutter embracing the lens and mounted to rotate intermittingly about the

latter on an axis transverse to the axis of the lens, substantially as described.

9. In combination with a lens and its support, a hollow shutter encircling the lens and provided with coincident apertures, said shutter being pivotally mounted upon the lens-support, substantially as described.

10. The combination, with a lens, of a hollow case or shutter surrounding said lens and provided with coincident apertures, said shutter being pivotally attached to the lens-support, the whole constituting a complete lens and shutter attachment adapted for application to a camera, substantially as described.

11. A combined lens-holder and shutter, consisting, essentially, of a lens-support and an intermittingly-rotating inclosing case or shutter provided with coincident apertures, said shutter being sustained wholly upon the lens-support, substantially as described.

12. In combination with the lens-support and the embracing-shutter provided with coincident apertures and mounted to rotate about the lens, a motor or impelling device connected to the shutter, and devices for releasing and arresting the shutter, substantially as described.

13. A pivotal hollow shutter provided with coincident apertures, a motor, and stopping and releasing devices, in combination with a lens fixedly supported within the shutter, substantially as described.

14. A lens-support and shutter mechanism connected together and adapted to be inserted within a chamber or recess in the front end of the camera box or tube, said lens and its support being held fixedly in position by the inclosing-walls thereof, substantially as described.

15. In combination with a hollow shutter closed at both ends and provided with coincident apertures, a lens supported and held within said hollow shutter, substantially as described.

16. In combination with a lens and its support, a shutter surrounding the lens and pivotally mounted upon the support of the latter, a motor device applied to one end and a releasing device on the opposite end of the shutter, substantially as described.

17. In combination with the lens-support and the shutter mounted thereon and inclosing the lens, the motor and releasing devices for controlling the movements of the shutter, also mounted upon the lens support, substantially as described.

18. In combination with the lens support and the inclosing-shutter mounted thereon and provided with apertures rotating in the plane of the axis of the lens, a chamber formed in the front end of the camera-tube, provided with coincident apertures and adapted to receive and hold the lens and shutter mechanism by engaging the lens support, substantially as described.

19. In combination with the lens-support and a shutter mounted thereon and provided with coincident apertures, a wheel, a spring interposed between said wheel and the shutter, a ratchet interposed between said wheel and its support, and a drum connected to said wheel by a ratchet or equivalent device for communicating motion to the wheel to wind up the spring, substantially as described.

20. In combination with the lens-support and its encircling shutter pivotally mounted thereon, a motor device comprising ratchet, spring, and winding-drum mounted on the lens-support at one end of the shutter, and a releasing device mounted upon the lens-support and engaging the opposite end of the shutter, substantially as described.

21. In combination with a rotary shutter and a motor device applied thereto, abutments secured to said shutter on opposite sides of its axis, and a pivoted latch provided with an inclined or beveled edge co-operating with said abutments, substantially as and for the purpose set forth.

22. In combination with the rotary shutter, its motor and releasing devices mounted upon the lens-support and between the heads by which said support is held in place within the camera, substantially as described.

23. In combination with the lens-support, provided with posts on opposite sides of the lens and sustaining-heads on said posts, a hollow shutter journaled on said posts and inclosing the lens, a latch for engaging one end of the shutter, and a motor device applied to the opposite end, substantially as described.

24. In combination with a pivoted tubular shutter, and a motor acting thereon to rotate it continuously in one direction, a pivoted latch for alternately engaging and releasing the shutter, and a tension device, also engaging the shutter to regulate its movements, substantially as described.

25. In combination with the rotary shutter mounted upon the lens-support, a cam-plate secured to the shutter and provided with shoulders or abutments co-operating with a latch pivoted in the post of the lens-support and actuated by a pin passing through the head of said support, substantially as described.

26. In combination with a rotating shutter, provided with a cam-plate carrying abutments, a latch engaging said abutments, and a spring bearing against the cam-plate and serving both as a brake and stop, substantially as described.

27. In a camera such as described, and in combination with the box or case, the apertured block secured within the front end of the box and provided with a transverse groove for the reception of the combined lens-support and shutter, substantially as described.

28. In a camera such as described, the combination, with the box or case, of the sectional block secured within the front end, said block being provided with coincident apertures, and a transverse groove or chamber for the reception of a lens-support and shutter mechanism, substantially as described.

29. In a camera such as described, and in combination with a box, and a combined lens-support and shutter mechanism located within a chamber in the forward end of the camera-box, a flexible connection extending through the walls of the box and connected to the motor devices of the shutter, and a push-pin connected to the shutter-releasing devices and guided by the lens support, substantially as described.

30. An improved detachable lens and shutter device such as hereinbefore described, the same comprising a lens and inclosing-shutter, both mounted upon a single supporting-frame adapted to be inserted within the camera tube or case, substantially as described.

31. A detachable lens and shutter device substantially such as herein described, consisting, essentially, of a supporting-frame provided with a transverse aperture for the reception of the lens, a hollow shutter pivotally attached to said frame and surrounding the lens and devices carried by the supporting-frame for actuating the shutter, substantially as described.

32. In a camera such as described, and in combination with the tubular box or case, a film carrying and feeding mechanism, substantially such as indicated, for insertion longitudinally of and within the rear end of the said box or case to close the rear end thereof, and consisting, essentially, of the supply-spool, guiding-rolls, platen, and winding-roller, the whole mounted and supported between the outer faces of side pieces, so that no part projects beyond the said outer faces, substantially as described.

33. In a roller-holder adapted for insertion within a tubular camera box or case, substantially as described, the combination, with the side pieces, of the removable supports for the spool passed through the side pieces and having their outer edges flush with or below the outer faces of said side pieces, as and for the purpose set forth.

34. In a roller-holder such as described, the combination, with the supporting-frame or side pieces fitted to the interior walls of the camera box or case, of the socket c^{17} and pin c^{16}, provided with cross-pieces or head c^{18}, resting in a countersunk recess in the outer face of the side piece, and a stud, c^{19}, adapted to enter through a slot or notch in the socket c^{17}, as and for the purpose specified.

35. In a roller-holder such as described, as a means for detachably supporting the film-carrying roller, a head for engaging one end of the roller supported upon a pin passing through one side piece, with its head countersunk in the outer face of the latter, a socket-plate applied to the inner face of the opposite side piece, and a pin for supporting the end of the roller passing through said socket, its head countersunk in the outer face of the side piece, said pin being detachably held in position by the wall of the camera-box, and a stud projecting on the inner side of the socket and adapted to pass through a slot or notch in the latter when the pin is withdrawn.

36. The combination, in a camera such as described, and with the tubular box or case thereof provided with sockets or openings for the indicator and winding-key, of a roller-holder provided with closed side pieces sustaining the film holding and carrying mechanism and projected, when in place, within the box or case beyond the sockets or openings therein, an indicator flush with or slightly below the outer face of the side piece and in line with one of the openings in the case or box, and a removable key inserted through the other opening and engaging the winding mechanism below the outer face of the side piece, whereby the roller-holder is adapted to be inserted and withdrawn, and to fit snugly within the camera box or case, and the openings for the indicator and winding-key are closed to prevent the entrance of light, as set forth.

37. In a camera such as described, the combination, with the box or case open at the rear end, of a roller-holder adapted for insertion within the rear end of said case or box and fitting snugly against the walls of the latter, and a detachable key passing through said box or case and engaging the film-winding devices, said key serving to lock and hold the roller-holder against longitudinal movement within the case, substantially as described.

38. In a camera, and in combination with the lens thereof, a shutter mechanism operating on opposite faces of the lens to simultaneously uncover and cover the front and rear of the lens tube or aperture, substantially as described.

39. In combination with the lens and its inclosing case or tube, through which light is conducted to the interior of the camera, a shutter mechanism adapted to interpose and withdraw light-excluding media simultaneously in front and in rear of the lens, to cover and uncover both faces of the lens, substantially as described.

40. In a camera such as described, and in combination with the lens, an inclosing casing or tube, shutter mechanism provided with two light-excluding media, the one movable in front and the other in rear of the lens, with devices for actuating said light-excluding media to simultaneously uncover and cover both faces of the lens, substantially as described.

41. In a camera such as described, the combination, with the box or case open at both ends, of the lens and diaphragm closing the front end, and the roller-holder fitted to the walls of the case, inserted from the rear end and removably held in position therein, substantially as described.

GEO. EASTMAN.

Witnesses:
EDWIN O. SAGO,
GEO. W. DEMING.

Nº 6732  A.D. 1904

(Under International Convention.)

Date claimed for Patent under Patents Act, 1901, being date of first Foreign Application (in United States), } *23rd Mar., 1903*

Date of Application (in the United Kingdom), 19th Mar., 1904

Accepted, 12th May, 1904

COMPLETE SPECIFICATION.

Improvements in Aeronautical Machines.

We, Orville Wright and Wilbur Wright, both of 1127 W. Third Street, Dayton, County of Montgomery,, State of Ohio, United States of America, Manufacturers, do hereby declare the nature of our invention and in what manner the same is to be performed, to be particularly described and ascertained in and by the following statement:—

Our invention relates to improvements in that class of aeronautical machines in which the weight is sustained by the reactions resulting when thin surfaces, or wings, are moved horizontally almost edgewise through the air at a small angle of incidence, either by the application of mechanical power, or by the utilization of the force of gravity.

The objects of our invention are, first, to provide a structure combining lightness, strength, convenience of construction, and the least possible edge resistance; second, to provide means for maintaining or restoring the equilibrium of the apparatus; and third, to provide efficient means of guiding the machine in both vertical and horizontal directions. We obtain these objects by the mechanism shown in the accompanying drawing, in which Fig. 1 is a view in perspective of the machine, Fig. 2 a side elevation, and Fig. 3 a top plan view.

The superposed horizontal surfaces 1, formed by stretching cloth upon frames of wood and wire, constitute the "wings," or supporting part of the apparatus. They are connected to each other through hinge joints by the upright standards 2 and the lateral stay wires 3, which together with the lateral spars 4 of the wing framing, form truss systems giving the whole machine great transverse rigidity and strength. The hinge joints admit of both flexing and twisting movements, and may be either ball and socket joints, or any joint of sufficiently loose construction to admit of the movements specified. The object of joints having both flexing and twisting movements is to permit superposed wing surfaces, or parts thereof, when joined together by upright standards, to be twisted or bent out of their normal planes for the purpose hereafter specified. We do not restrict outselves to the use of any particular form of joint, nor to its use at any particular number of places.

One end of the rope 5 is attached near the rear corner of the upper surface, passes diagonally downward around the pulleys 6, and diagonally upward to the corresponding corner at the opposite end of the machine. The rope 8 is attached to the front corner of the upper surface, passes around the pulleys 7 and back to the opposite upper corner. The movable cradle 9 is attached to

[*Price 8d.*]

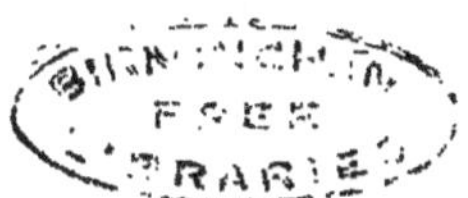

the rope 5 at the point where the operator's body rests, and provides a means of imparting movement to the ropes 5 and 8. The operator lies prone on the lower surface, his hips resting in the cradle, and his hands grasping the roller 10, which actuates the front rudder. The ropes 5 and 8 maintain the fore and aft positions of the two surfaces 1 with respect to each other, and by their movement impart a twist to the entire structure, including the wings 1, as will be more fully described hereafter. We have shown the operating system by means of ropes, which we now prefer to use, but we do not restrict ourselves to the use of any particular method of imparting this twist to a structure formed in the manner specified.

The main frames of the wings 1 are formed by uniting the lateral spars 4 (Fig. 3) by means of end bows 11. The cloth for each wing, previous to its attachment to the frame, is cut on the bias and made up into a single piece approximately the size and shape of the wing, having the threads of the cloth 12 (Fig. 3) diagonal to the lateral spars 4 and the longitudinal ribs 13, with which they form truss systems. A wide hem is sewed in the rear edge to form a pocket for the insertion of the wire 14. By the combination of a frame work with a cloth covering, each formed in the manner described, we secure a surface of very great strength to withstand lateral and longitudinal strains, but capable of some twisting movement.

When the two surfaces 1 are joined together by the wire stays 3, the ropes 5 and 8, and the upright standards 2, as already described, a system is formed capable of sustaining great weight without distortion. But when the cradle 9 is moved to right or left by the operator, the motion is communicated through the ropes 5 and 8 and the upright standards 2 in such a manner that the wing surfaces are twisted, the rear edge of the wing tips being drawn downward at one end of the machine and drawn upward at the other; thus presenting the left set of wing tips to the wind at a greater or a less angle than the right. When in flight, the end having the greater angle will necessarily rise and the other end will sink, so that the lateral balance of the machine is under control through twisting movements of the wing tips by the operator, by means of the cradle 9.

The struts 15, together with the struts 16 (Fig. 2) in combination with the main frame, form trussed skids which prevent the machine from rolling over forward when it lands, and also relieves the jerk on the rope 8. They are also utilized as a part of the front rudder steering system.

The flexible front rudder 17 consists of the stiff cross sticks 18, 19, 20 and the thin ribs 21, over which is stretched a cloth covering. The rudder is mounted upon the struts 15 by attachment to the cross stick 18, which is located near the centre of pressure, so as to form a balanced rudder. The up and down motion of the front edge of the rudder is in part restrained by the springs 23. The rear edge is raised and lowered by means of the axles 10, 22, the bands 24 and the arms 25 and 26, or by any other suitable means. The restraining action of the springs 23 causes the ribs 21 to bend when the rear edge is raised or lowered, thus presenting a concave surface to the action of the wind, and very greatly increasing its power as compared with a plane of equal area. By regulating the pressure on the upper and lower sides of the rudder, through changes of angle and curvature, a turning movement is communicated to the main structure and the course of the machine is directed upward or downward at the will of the operator, and the longitudinal balance maintained.

Contrary to the usual custom, we place the horizontal rudder in front of the main surfaces or "wings" at a negative angle, and use no horizontal tail at all. By this arrangement we obtain a forward surface which is almost free from pressure under ordinary conditions of flight, but which, even if not moved at all, becomes an efficient lifting surface whenever the speed of the machine is accidentally reduced very much below the normal, and thus largely counteracts that backward travel of the centre of pressure on the main surfaces or wings which has frequently been productive of serious injuries by causing the machine to turn downward and strike the ground head on. We are aware that a forward horizontal rudder of different construction has been used in combination with a supporting surface and a rear horizontal rudder, but this combination was not intended to effect and did not effect the object which we obtain by the arrangement of surfaces here described.

The vertical tail or rudder 27 is attached through universal joints to the two pairs of struts 28, which lie in parallel horizontal planes, and are connected to the rear edges of the main surfaces 1 by hinged joints. This combination secures the tail rigidly in a vertical position, but enables it to turn on a vertical axis, and also to rise bodily in case it strikes the ground, and thus escapes breakage. The cords 29 are tiller ropes which connect the rudder wheel 30 to the rope 8, which in conjunction with the rope 5 imparts the twisting motion to the wing tips as heretofore described. By this method of attachment the same motion of the ropes 8 and 5 which actuates the wing tips also presents to the wind that side of the vertical rear rudder which is toward the tip having the smaller angle of incidence. The wing tip presented to the wind at the greater angle, under the usual conditions of flight, has both greater lift and greater drift, or resistance, than the other. The wing with the greater angle therefore, tends to rise and drop behind, while the other sinks and moves ahead. Under these circumstances the longitudinal axis of the machine tends to turn toward the wing having the greater angle, while the general course of the machine through the air tends toward that wing which is the lowest with the result that a wide divergence soon arises between the direction which the machine faces and its actual direction of travel. By the use of a rear movable vertical rudder, so operated as to present to the wind that side which is toward the wing having the least angle, we obtain a turning force opposite to and greater than that arising from the difference in the resistance of the two wings, and thus are able to keep the longitudinal axis of the machine approximately in coincidence with the line of flight. We do not confine ourselves to the particular construction and attachment of the rear rudder hereinbefore described, nor to this particular construction of surfaces or wings, but may employ this combination in the use of any movable vertical rear rudder operated in conjunction with any wings capable of being presented to the wind at respectively differing angles at their opposite tips for the purpose of restoring the lateral balance of a flying machine and guiding the machine to right or left.

We are aware that prior to our invention flying machines have been constructed having superposed wings in combination with horizontal and vertical rudders; we therefore do not claim such combination broadly.

Having now particularly described and ascertained the nature of our said invention and in what manner the same is to be performed, we declare that what we claim is:

1. In a flying machine, the combination of superposed surfaces or "wings", with upright connecting standards one or more of which has its attachment by means of hinges or flexible joints, substantially as described and for the purpose specified.

2. In a flying machine, the combination of superposed surfaces or wings with upright connecting standards attached through flexible joints, and laterally extending stay wires, substantially as described.

3. In a flying machine, the combination of one or more supporting surfaces or wings with a device for imparting a twist to the said surfaces or wings for the purpose stated.

4. In a flying machine, the combination of superposed wings, upright standards attached by flexible joints, and laterally extending stay wires, with a device for imparting a twisting to the wings for the purpose specified.

5. In a flying machine, the combination of superposed wings, upright standards attached by means of flexible joints, and laterally extending stay wires, with actuating ropes attached and operated substantially as described.

6. In a flying machine, the combination of wings having their right and left tips capable of being adjusted so as to be presented to the wind at respectively differing angles, with a vertical adjustable rear rudder operating in conjunction therewith in the manner and for the purpose specified.

7. In a flying machine having wings capable of being twisted by actuating ropes, the combination therewith of a movable vertical rear rudder having tiller cords attached to said actuating ropes, substantially as described.

8. In a flying machine, the combination of superposed surfaces with a vertical rear rudder, and hinged connecting arms in parallel planes substantially as described.

9. In a flying machine having surfaces or wings composed of a cloth covered frame, the combination of laterally extending spars and longitudinal ribs, with a covering having the threads of the cloth diagonal to the main lines of the framing, substantially as set forth.

10. In a flying machine the combination of superposed surfaces with forwardly extending struts arranged in the manner and for the purpose specified.

11. In a flying machine, the combination of supporting wings with a smaller inert surface which becomes a supporting surface when the speed of the machine is greatly diminished, substantially as described and for the purpose specified.

12. In a flying machine, the combination of supporting wings and a horizontal rudder, having stiff lateral sticks, thin longitudinal ribs, and cloth covering, and a device for imparting a slight curvature to the rudder in the manner and for the purpose specified.

13. In a flying machine, the combination of supporting wings with a flexible horizontal rudder and a device for simultaneously regulating the angle of the rudder with the wind and imparting to it a slight curvature, substantially as described and for the purpose specified.

14. In a flying machine, the combination of superposed surfaces capable of being twisted with a forward horizontal rudder and an adjustable vertical rear rudder, substantially as described and for the purposes specified.

Dated this 19th day of March 1904.

HERBERT HADDAN & Co.,
Agents to Applicants.
18 Buckingham Street, Strand, W.C. London.

Redhill: Printed for His Majesty's Stationery Office, by Love & Malcomson, Ltd.—1904.

A.D. 1904. MARCH 19. N°. 6732.

WRIGHT & *another's* COMPLETE SPECIFICATION.

(1 SHEET)

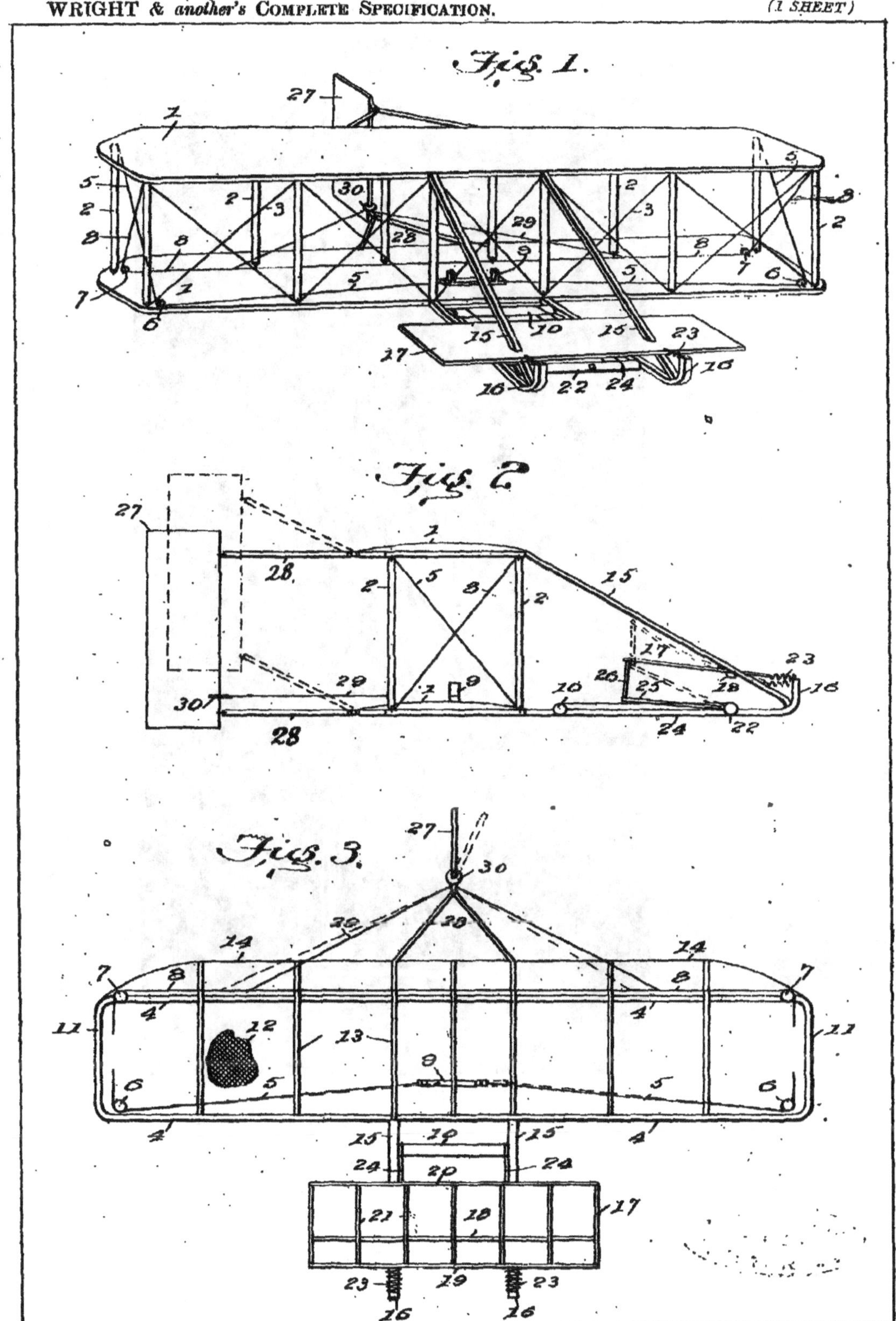

[This Drawing is a reproduction of the Original on a reduced scale.]

Malby & Sons, Photo-Litho.

DESIGN.

A. BARTHOLDI.

Statue.

No. 11,023. Patented Feb. 18, 1879.

LIBERTY ENLIGHTENING THE WORLD.

E. A. Dick
J. E. Carpenter.

Auguste Bartholdi
by A. Pollok
atty.

UNITED STATES PATENT OFFICE.

AUGUSTE BARTHOLDI, OF PARIS, FRANCE.

DESIGN FOR A STATUE.

Specification forming part of Design No. **11,023**, dated February 18, 1879; application filed January 2, 1879.
[Term of patent 14 years.]

To all whom it may concern:

Be it known that I, AUGUSTE BARTHOLDI, of Paris, in the Republic of France, have originated and produced a Design of a Monumental Statue, representing "Liberty enlightening the world," being intended as a commenorative monument of the independence of the United States; and I hereby declare the following to be a full, clear, and exact description of the same, reference being had to the accompanying illustration, which I submit as part of this specification.

The statue is that of a female figure standing erect upon a pedestal or block, the body being thrown slightly over to the left, so as to gravitate upon the left leg, the whole figure being thus in equilibrium, and symmetrically arranged with respect to a perpendicular line or axis passing through the head and left foot. The right leg, with its lower limb thrown back, is bent, resting upon the bent toe, thus giving grace to the general attitude of the figure. The body is clothed in the classical drapery, being a stola, or mantle gathered in upon the left shoulder and thrown over the skirt or tunic or under-garment, which drops in voluminous folds upon the feet. The right arm is thrown up and stretched out, with a flamboyant torch grasped in the hand. The flame of the torch is thus held high up above the figure. The arm is nude; the drapery of the sleeve is dropping down upon the shoulder in voluminous folds. In the left arm, which is falling against the body, is held a tablet, upon which is inscribed "4th July, 1776." This tablet is made to rest against the side of the body, above the hip, and so as to occupy an inclined position with relation thereto, exhibiting the inscription. The left hand clasps the tablet so as to bring the four fingers onto the face thereof. The head, with its classical, yet severe and calm, features, is surmounted by a crown or diadem, from which radiate divergingly seven rays, tapering from the crown, and representing a halo. The feet are bare and sandal-strapped.

This design may be carried out in any manner known to the glyptic art in the form of a statue or statuette, or in alto-relievo or bass-relief, in metal, stone, terra-cotta, plaster-of-paris, or other plastic composition. It may also be carried out pictorially in print from engravings on metal, wood, or stone, or by photographing or otherwise.

What I claim as my invention is—

The herein-described design of a statue representing Liberty enlightening the world, the same consisting, essentially, of the draped female figure, with one arm upraised, bearing a torch, while the other holds an inscribed tablet, and having upon the head a diadem, substantially as set forth.

In testimony whereof I have signed this specification in the presence of two subscribing witnesses.

A. BARTHOLDI.

Witnesses:
C. TERINIER,
COTTIN.

Aug. 2, 1955 T. M. McCARTY 2,714,326

STRINGED MUSICAL INSTRUMENT OF THE GUITAR TYPE AND COMBINED BRIDGE AND TAILPIECE THEREFOR

Filed Jan. 21, 1953 2 Sheets-Sheet 1

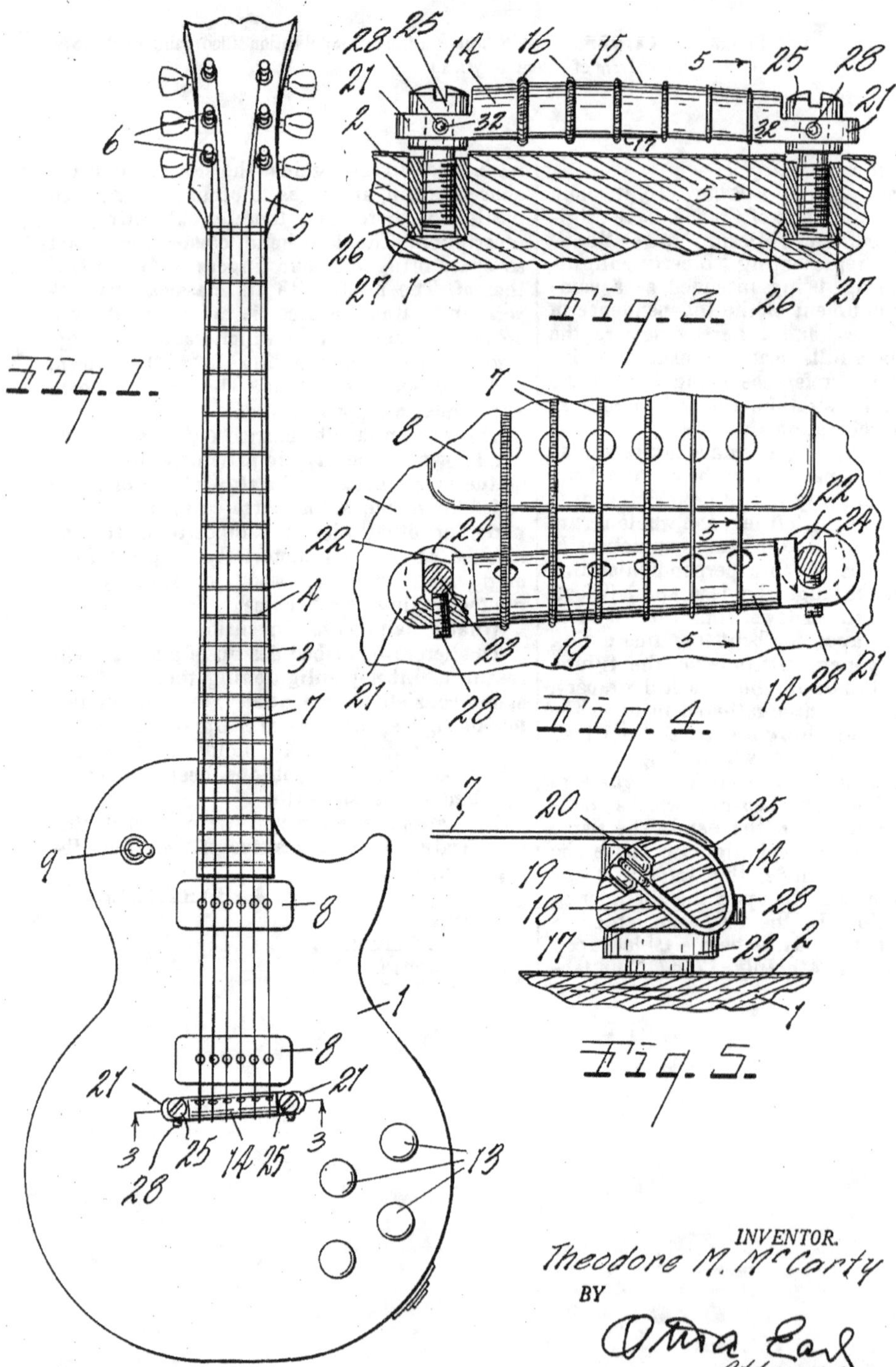

INVENTOR.
Theodore M. McCarty
BY
Attorney.

Aug. 2, 1955 T. M. McCARTY 2,714,326

STRINGED MUSICAL INSTRUMENT OF THE GUITAR TYPE
AND COMBINED BRIDGE AND TAILPIECE THEREFOR

Filed Jan. 21, 1953 2 Sheets-Sheet 2

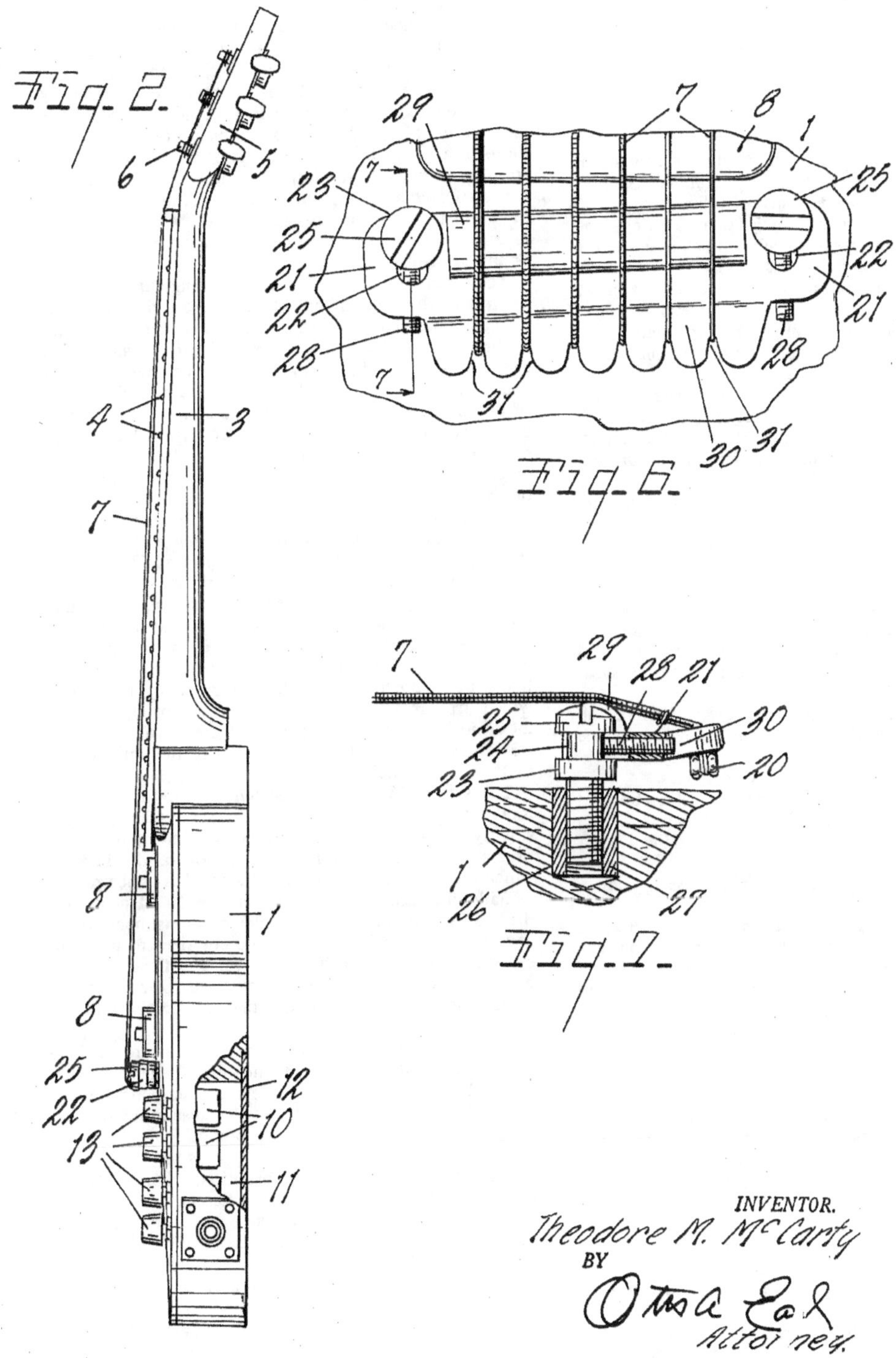

United States Patent Office

2,714,326

Patented Aug. 2, 1955

2,714,326

STRINGED MUSICAL INSTRUMENT OF THE GUITAR TYPE AND COMBINED BRIDGE AND TAILPIECE THEREFOR

Theodore M. McCarty, Kalamazoo, Mich., assignor to Gibson, Inc., Kalamazoo, Mich.

Application January 21, 1953, Serial No. 332,374
12 Claims. (Cl. 84—299)

This invention relates to improvements in a stringed musical instrument of the guitar type and combined bridge and tailpiece therefor.

The main objects of this invention are:

First, to provide a bridge assembly for stringed musical instruments of the electrically amplified Spanish guitar type which has a wide range of adjustment to accommodate different sized sets of graduated strings and one which has a wide range of adjustment to meet the requirements of particular players.

Second, to provide a combined bridge and tailpiece for electrically amplified stringed instruments which is tiltably adjustable both vertically and horizontally, thus adapting it for a wide range of strings and player requirements.

Third, to provide a stringed musical instrument of the class described having these advantages and one which is attractive in appearance and one in which the adjusting elements for controlling the electrical pickup are unobstructed by the bridge.

Objects relating to details and economies of the invention will appear from the description to follow. The invention is defined and pointed out in the claims.

A preferred embodiment of the invention is illustrated in the accompanying drawings, in which:

Fig. 1 is a plan view of a stringed musical instrument of the Spanish guitar type embodying my invention.

Fig. 2 is a side elevational view thereof looking from the right of Fig. 1 and partly broken away to show structural details.

Fig. 3 is an enlarged fragmentary view partially in vertical transverse section on a line corresponding to line 3—3 of Fig. 1.

Fig. 4 is an enlarged fragmentary plan view.

Fig. 5 is an enlarged fragmentary vertical section on a line corresponding to line 5—5 of Figs. 3 and 4.

Fig. 6 is a fragmentary plan view of a modified form or embodiment of my invention.

Fig. 7 is a fragmentary view partially in section on a line corresponding to line 7—7 of Fig. 6.

In the accompanying drawing, **1** represents the body of the instrument which is desirably solid except for certain chambers or recesses formed therein to receive parts, as will be described. The top is flat and **2** indicates the finish. The neck **3** is provided with the usual frets **4** and the head **5** with tuning pins **6** for the graduated set of strings indicated at **7**.

The instrument illustrated is provided with two electrical pickups **8, 8** controlled by the switch **9**. The tone and volume control elements are conventionally indicated at **10** in Fig. 2, these being arranged in a chamber **11** provided in the body **1** and having a closure **12**. Finger pieces **13** are disposed or project from the top of the body.

The combined bridge and tailpiece **14** of Figs. 1 to 5, inclusive, is bar-like in form and has an upwardly and transversely curved string supporting surface or face **15** for a graduated set of strings as **16**. The under side of the bridge member **14** is flattened at **17**. The bridge member has a downwardly and rearwardly inclined bore **18** for each string terminating at the front side of the bridge member in an enlargement **19** receiving the string anchoring element **20**. The strings are wrapped around the transversely curved string supporting face of the bridge member and their ends inserted in the bores **18** and secured or anchored by the anchoring elements **20**. At each end the bridge member is provided with an ear-like extension **21** of flat section **9** having a slot **22** therein opening at its forward edge. The supporting posts **23** have annular slots or grooves **24** in the head portions **25** thereof receiving these ears.

The body member **1** is provided with vertical bores **26** receiving the sleeve-like socket members **27** which are internally threaded to threadedly receive the post. The post sockets are mounted on the body in transversely spaced relation and desirably in a plane inclined to the longitudinal plane of the neck, the angle being such that there is a gradual increase in the length of the strings from the smaller to the larger. The bridge is supported for vertically adjusting the strings relative to the frets of the instrument and also relative to the pickups. Adjustment of the bridge may be equal at both ends thereof or the bridge may be tilted longitudinally.

The bridge may also be tiltingly angularly adjusted relative to the longitudinal plane of the strings and this is accomplished by the adjusting and string thrust supporting screws **28** which are threaded into the supporting ears **21** of the bridge from the rear side thereof to engage the posts as is illustrated, and sustain the thrust of the bridge on the posts under the stress of the strings. The ears slidably fit within the grooves or slots on the post heads with the side edges of the slots in the ears in engagement with the posts and sustaining end thrust on the bridge.

It will be understood that the load on the bridge as the result of the tensioning of the strings is a very substantial load. This mounting of the post and the mounting of the bridge on the post effectively sustains such load and the parts are so associated and supported that there is no vibration between the bridge and its supports.

In the modification shown in Figs. 6 and 7, the bridge member **29** is provided with a rearwardly projecting web-like portion **30** having notches **31** receiving the strings which are provided with the string anchoring elements **20** disposed on the under side of this extension. Otherwise the structure is the same as that of the embodiment of Figs. 1 to 5 inclusive.

The screws **28** are of the Allen setscrew type having tool receiving sockets **32**. These screws project but slightly so they are not likely to be engaged by the hands or clothing of the player. The portion of the body in which the control elements **13** art located is unobstructed.

With the parts arranged as illustrated and described, the instrument has a wide range of adjustment to accommodate sets of strings of different sizes to meet the requirements of the particular player and the instrument may be quickly and accurately adjusted to meet the requirements of the strings and the requirements of the particular player.

I have illustrated and described my invention in a highly practical embodiment thereof. I have not attempted to illustrate or describe other embodiments or adaptations as it is believed that this disclosure will enable those skilled in the art to embody or adapt the invention as may be desired.

Having thus described my invention, what I claim as new and desire to secure by Letters Patent is:

1. A stringed musical instrument of the class described including a body and a neck, a bar-like bridge member having a longitudinally and transversely curved string supporting face and having inwardly and rearwardly inclined string bores provided with enlargements at their front ends opening on the front side of the bridge member and adapted to receive anchoring elements of strings disposed on the bridge and wrapped around the rear side thereof and inserted in said string bores, said bridge member having ears at its ends provided with slots opening at their front edges, post sockets mounted on

said body in transversely spaced relation and in a plane inclined to the longitudinal plane of the neck, posts threadedly engaging said sockets for vertical adjustment therein and provided with slots in which the ears of said bridge member are disposed with the side edges of the slots in supported engagement with the posts, and bridge adjusting and thrust sustaining screws threaded into said ears to engage said posts.

2. A stringed musical instrument of the class described including a body, a bar-like bridge member having a longitudinally and transversely curved string supporting face and having inwardly and rearwardly inclined string bore provided with enlargements at their front ends opening on the front side of the bridge member and adapted to receive anchoring elements of strings disposed on the bridge and wrapped around the rear side thereof and inserted in said string bores, said bridge member having ears at its ends provided with slots opening at their front edges, post sockets mounted on said body in transversely spaced relation, posts threadedly engaging said sockets for vertical adjustment therein and provided with slots in which the ears of said bridge member are disposed, and bridge adjusting and thrust sustaining screws threaded into said ears to engage said posts.

3. A stringed musical instrument of the class described including a body, a bridge member having a string supporting face and openings with which strings disposed on the bridge may be retainingly engaged, said bridge member having ear-like portions at the ends thereof provided with slots opening at their forward edges, post sockets mounted on said body in transversely spaced relation, posts threadedly engaging said sockets for vertical adjustment therein and provided with slots in which the ears of said bridge member are disposed, and bridge adjusting screws threaded into said ears to supportingly engage said posts.

4. A stringed musical instrument of the class described including a body, a bridge member having a transversely curved string supporting face, a rearwardly projecting web-like string anchoring element having spaced ring receiving notches in its edge, said bridge member having post engaging portions at its ends provided with forwardly opening slots, post sockets mounted on said body in transversely spaced relation, posts threadedly engaging said sockets for vertical adjustment and with which the said forwardly opening slots of said bridge member are engaged, and bridge adjusting screws carried by said bridge member and supportingly engaging said posts.

5. A stringed musical instrument of the class described including a body and a neck, a bridge member having an upwardly facing spring supporting face and provided with integral string connecting means for strings supported on said string supporting face, posts threadedly mounted on said body for vertical adjustment thereon and having heads provided with annular grooves, said bridge member being provided with horizontally flat end portions disposed in said grooves and having forwardly opening slots receiving said posts, and bridge adjusting screws carried by the bridge and coacting with said posts.

6. A stringed musical instrument of the class described including a body, a bridge member having a string supporting face and openings with which strings disposed on the bridge may be retainingly engaged, said bridge member having ear-like portions at the ends thereof provided with slots opening at their forward edges, posts mounted on said body for vertical adjustment and provided with slots in which the ears of said bridge member are engaged, and bridge adjusting screws threaded into said ears to supportingly engage said posts.

7. A stringed musical instrument of the class described including a body, a bridge member having a rearwardly projecting string anchoring element provided with string attaching notches in its rear edge, said bridge member having slotted post engaging portions, posts mounted on said body for vertical adjustment and receiving said slots of said bridge member, and bridge adjusting screws carried by said bridge member and supportingly engaging said posts.

8. A stringed musical instrument of the class described including a body, a bridge member having an upwardly facing transversely curved string supporting face and provided with means for attaching strings thereto, an independently adjustable supporting post for each end of the bridge member providing means for independent vertical adjustment thereof the bridge member being supported by the post for adjustment transversely thereof, and bridge adjusting and thrust sustaining screws carried by said bridge member and coacting with the posts for adjustment of the bridge member transversely relative to the posts.

9. A combined bridge and tailpiece assembly for stringed musical instruments comprising a bar-like bridge member having a longitudinally and transversely curved string supporting face, said bridge member being flattened on its under side and having ears at its ends provided with slots opening at their front edges and having inwardly and rearwardly inclined string bores provided with enlargements at their front ends opening on the front side of the bridge member and adapted to receive anchoring elements of strings disposed on the bridge and wrapped around the rear side thereof and inserted in said string bores, post sockets adapted to be mounted on the body of an instrument, posts threadedly engageable with said sockets for adjustment therein and provided with slots in which the ears of said bridge member are disposed with the side edges of the slots in supported engagement with the posts, and bridge adjusting and thrust sustaining screws threaded into said ears to engage said posts.

10. A combined bridge and tailpiece assembly for stringed musical instruments comprising a bridge member having a string supporting face, said bridge member having ears at its ends provided with slots opening at their front edges and having inwardly and rearwardly inclined string bores provided with enlargements at their front ends opening on the front side of the bridge member and adapted to receive anchoring elements of strings disposed on the bridge and wrapped around the rear side thereof and inserted in said string bores, post sockets adapted to be mounted on the body of an instrument, posts threadedly engageable with said sockets for adjustment therein and provided with slots in which the ears of said bridge member are disposed, and bridge adjusting and thrust sustaining screws threaded into said ears to engage said posts.

11. A combined bridge and tailpiece assembly for stringed musical instruments having a string supporting face and means with which strings disposed on the bridge may be retainingly engaged, said bridge member having ear-like portions at the ends thereof provided with slots opening at their forward edges, independently adjustable supporting posts provided with slots in which the ears of said bridge member are engaged, and bridge adjusting and thrust sustaining screws threaded into said ears to engage said posts.

12. A bridge member for stringed musical instruments of the class described provided with means for attaching strings thereto, independently and vertically adjustable supporting means for each end of the bridge member, and bridge member adjusting means carried by said bridge member and coacting with the posts for adjustment of the bridge member relative to the posts.

References Cited in the file of this patent

UNITED STATES PATENTS

455,221	Lorang	June 30, 1891
601,071	Borcur	Mar. 22, 1898
897,964	De Julio	Sept. 8, 1908
2,074,982	Di Marzio	Mar. 23, 1937
2,588,726	Hoover	Mar. 11, 1952

United States Patent [19]

Glock

[11] **Patent Number: 4,825,744**

[45] **Date of Patent: May 2, 1989**

[54] **AUTOMATIC PISTOL**

[76] Inventor: **Gaston Glock,** Siebenbürger Strasse 16-12, A-1220 Vienna, Austria

[21] Appl. No.: **227,514**

[22] Filed: **Aug. 2, 1988**

Related U.S. Application Data

[63] Continuation of Ser. No. 773,352, Sep. 6, 1985, abandoned, which is a continuation of Ser. No. 456,056, filed as PCT AT82/00015 on Apr. 29, 1982, published as WO82/03910 on Nov. 11, 1982, Pat. No. 4,539,889.

[30] **Foreign Application Priority Data**

Apr. 30, 1981 [AT] Austria 1944/81

[51] **Int. Cl.**[4] **F41C 19/14**
[52] **U.S. Cl.** **89/145;** 89/147
[58] **Field of Search** 42/65, 69.01, 69.02; 89/145, 147

[56] **References Cited**

U.S. PATENT DOCUMENTS

3,857,325 12/1974 Thomas 89/147
4,539,889 9/1985 Glock 89/147

FOREIGN PATENT DOCUMENTS

605729 11/1934 Fed. Rep. of Germany .

Primary Examiner—Stephen C. Bentley
Attorney, Agent, or Firm—Toren, McGeady & Associates

[57] **ABSTRACT**

A pistol has a frame, a barrel slidable on the frame and having a cartridge-receiving rear end, and a breech slidable on the frame and engageable over the rear end of the barrel to form a cartridge chamber. A standard slide carries the barrel and breech. A firing element and a firing pin operatively linked thereto are movable on the breech toward and away from the barrel between a rear position in which the firing pin is out of the cartridge chamber and a front position with the firing pin projecting forward into the cartridge chamber for firing a cartridge in the chamber when the firing element moves from the rear to the front position. A relatively strong firing spring braced against the firing element urges same into the front position and a relatively weak spring braced against the firing element urges same into the rear position. A trigger movable on the frame between an actuated and an unactuated position and an abutment engageable with the firing element and displaceable backward on the frame are linked together so as to displace the firing element back into the rear position on displacement of the trigger from the unactuated to the actuated position and to displace the abutment out of operative engagement with the firing element on displacement of the trigger into the actuated position for displacement of the firing element by the springs into the front position.

8 Claims, 9 Drawing Sheets

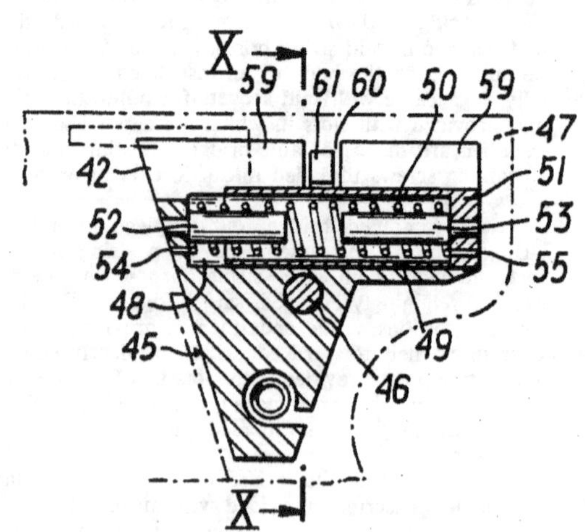

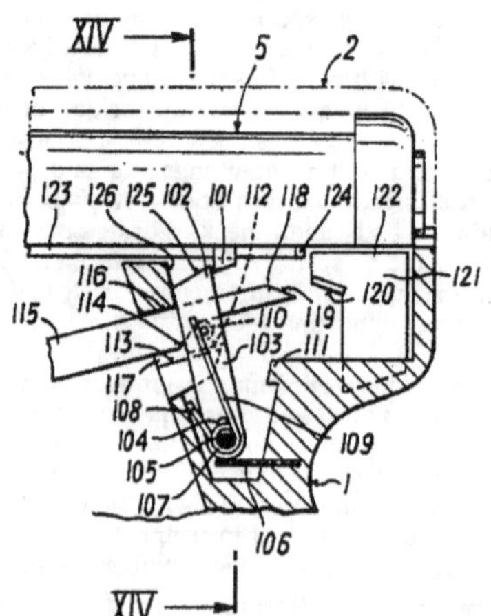

FIG. 1

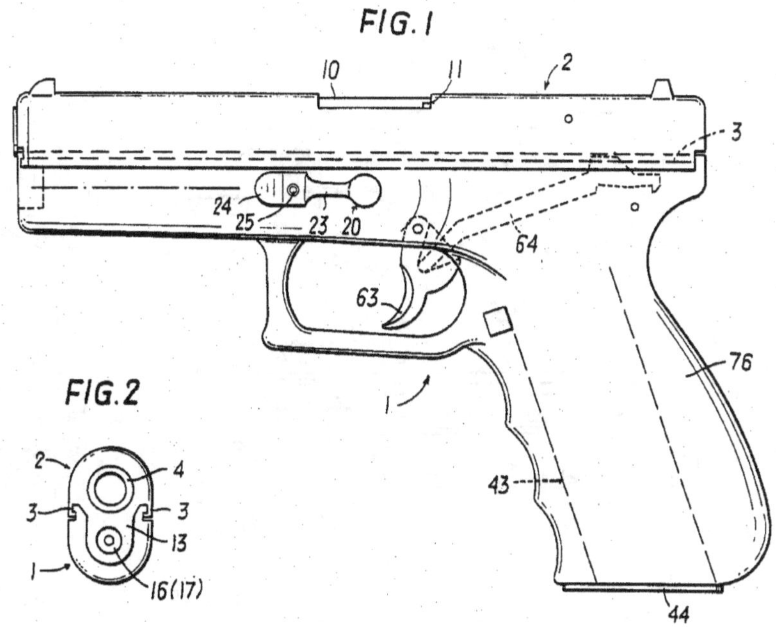

FIG. 3

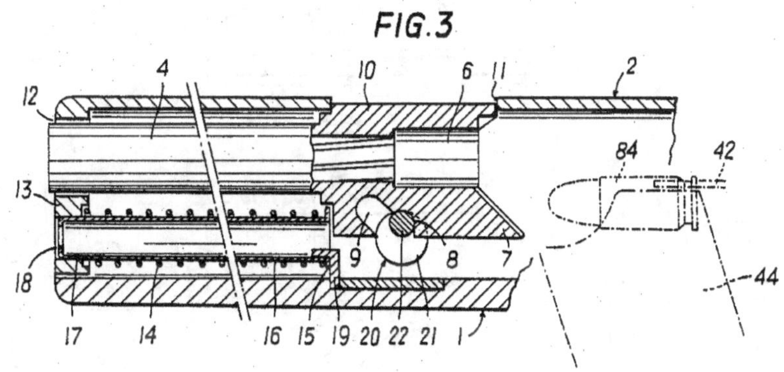

FIG. 4

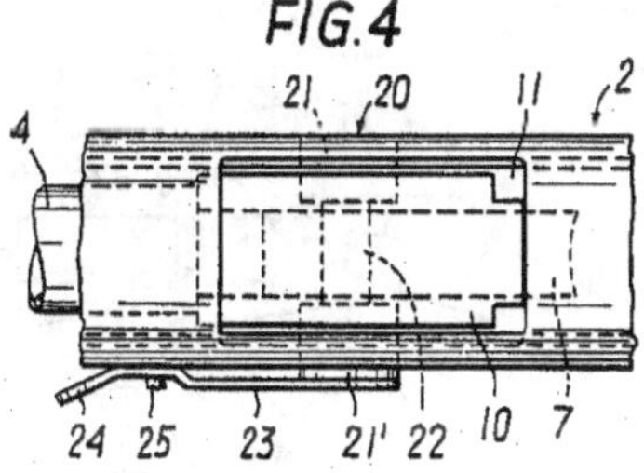

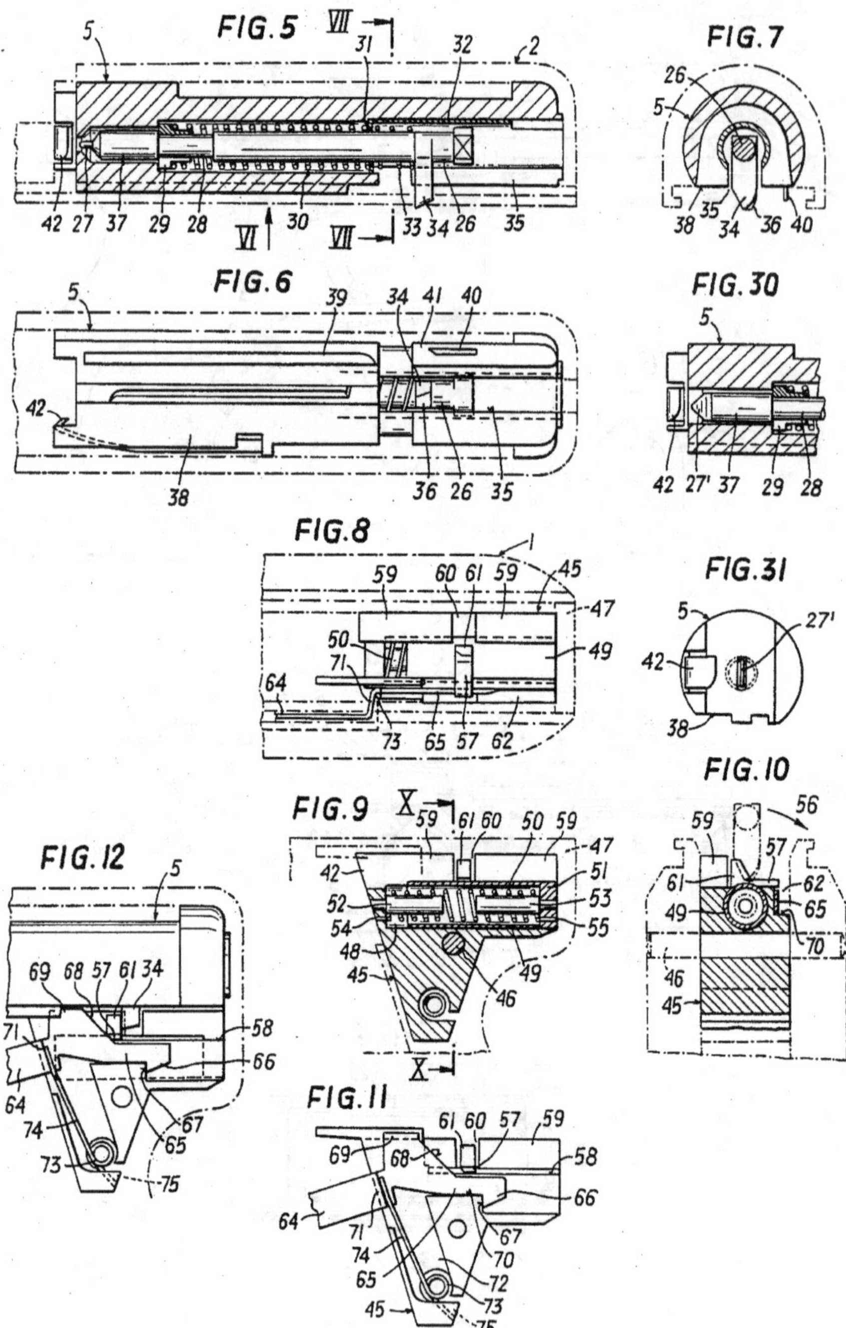
FIG.5
FIG.6
FIG.7
FIG.8
FIG.9
FIG.10
FIG.11
FIG.12
FIG.30
FIG.31

FIG. 13

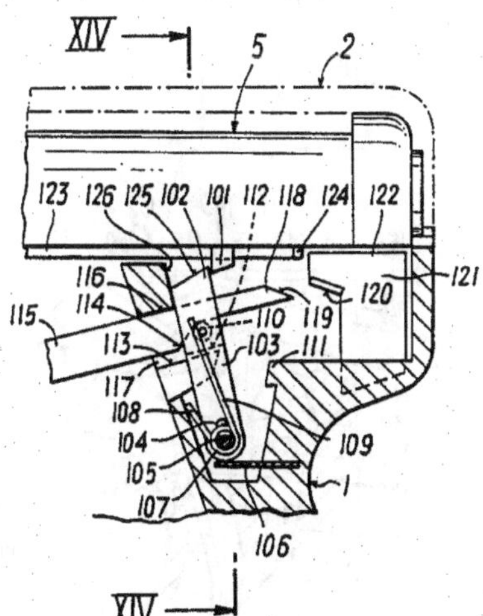

FIG. 14

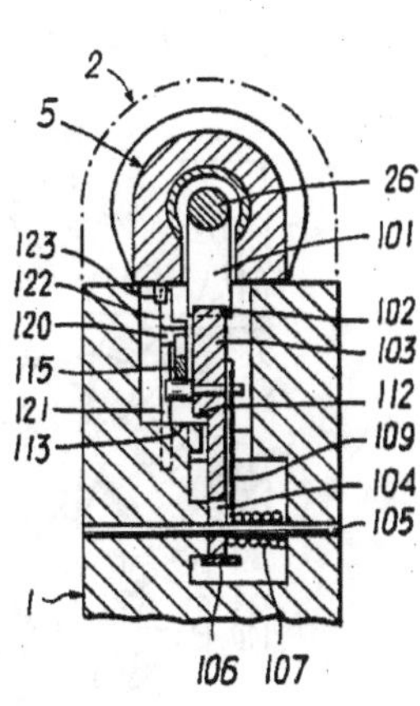

FIG. 15

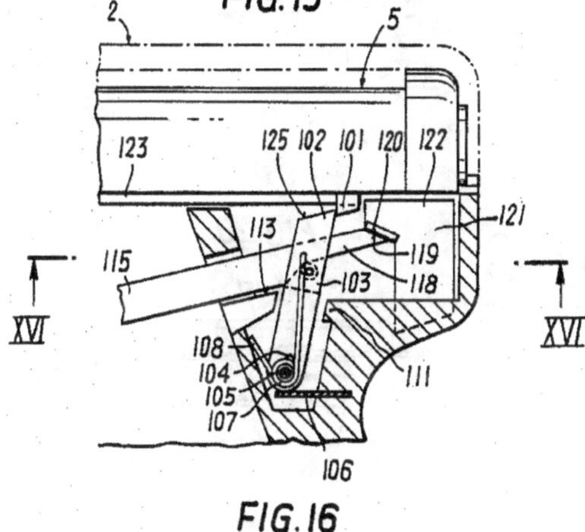

FIG. 16

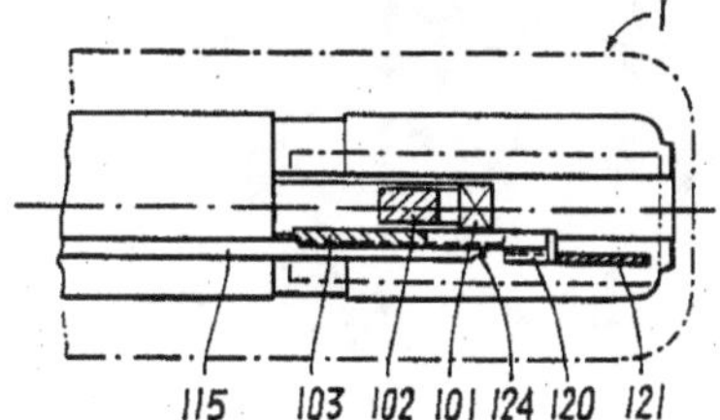

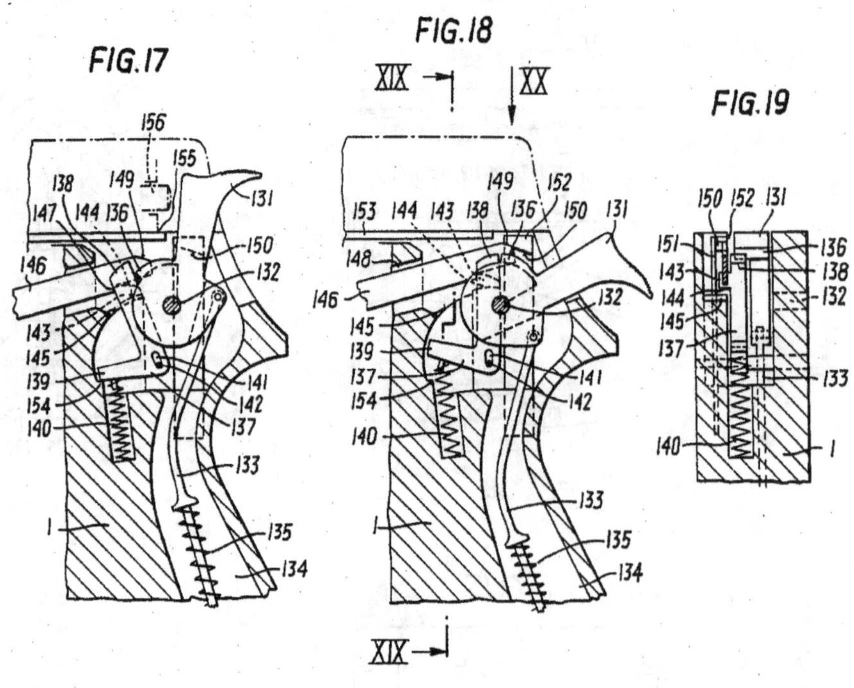
FIG.17
FIG.18
FIG.19
XIX
XX
131
132
133
134
135
136
137
138
139
140
141
142
143
144
145
146
147
148
149
150
151
152
153
154
155
156
1

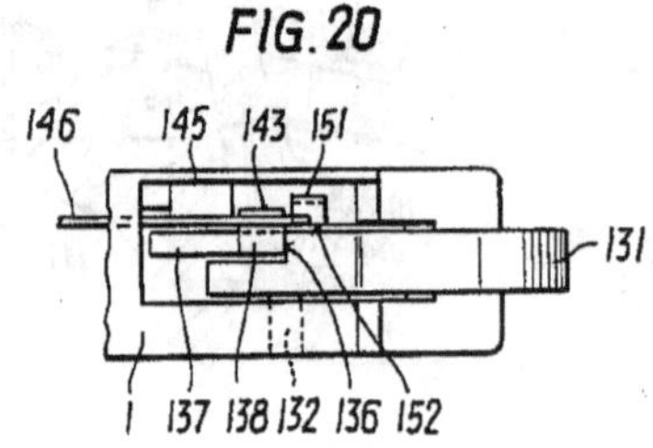
FIG.20
146
145
143
151
131
1
137
138
132
136
152

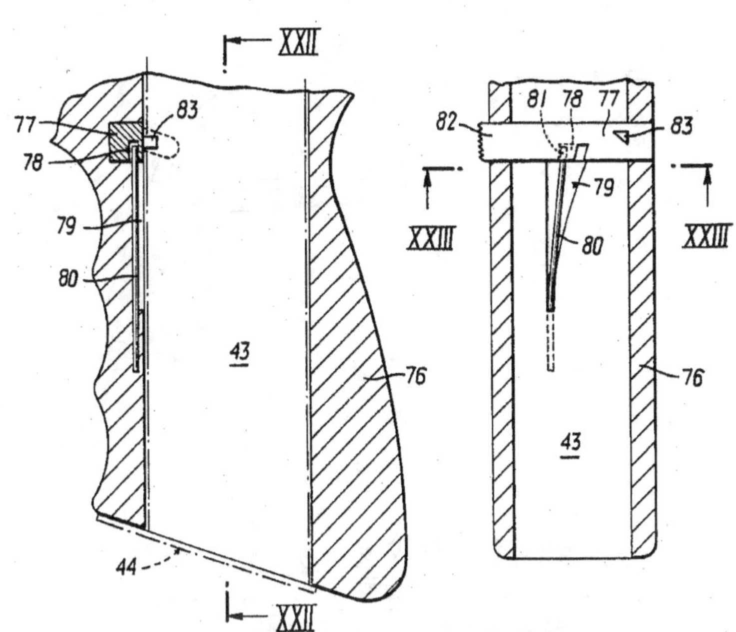
FIG. 21
FIG. 22
XXII
77
83
78
79
80
43
76
44
XXII
81 78 77
82
83
XXIII
79
80
XXIII
76
43

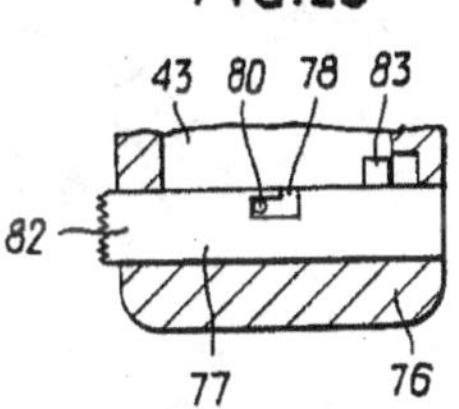
FIG. 23
43 80 78 83
82
77
76

FIG. 24

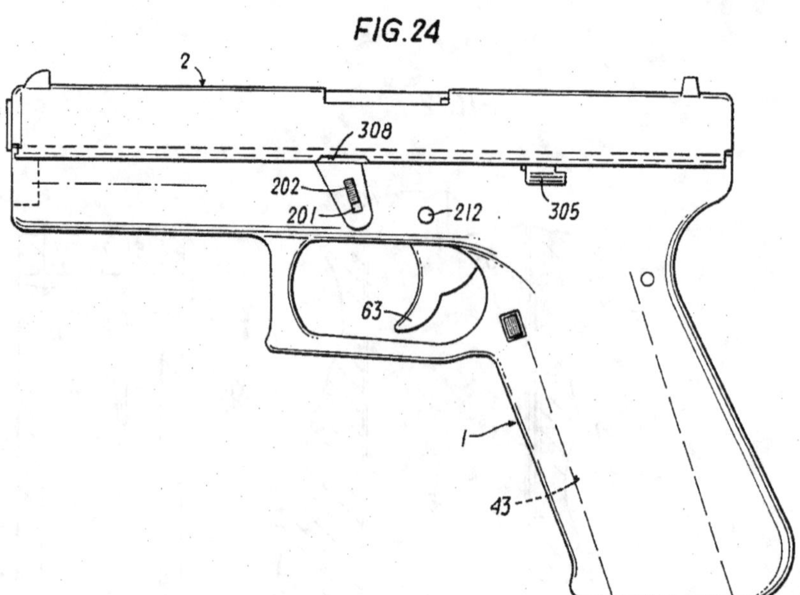

FIG. 25

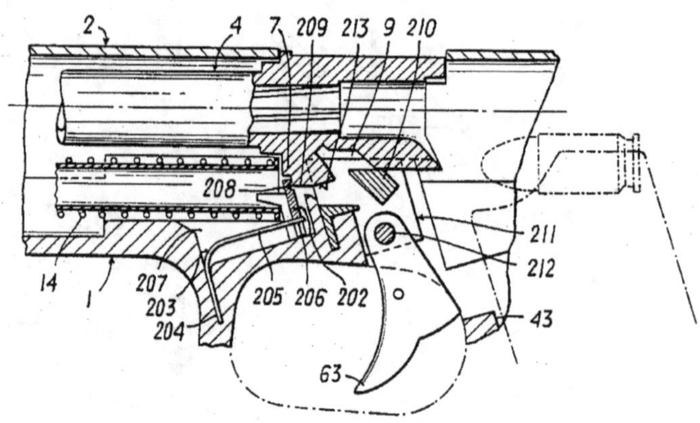

FIG. 26

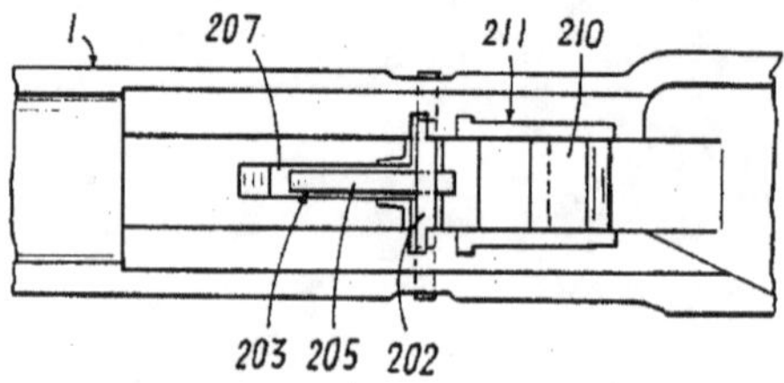

FIG. 27

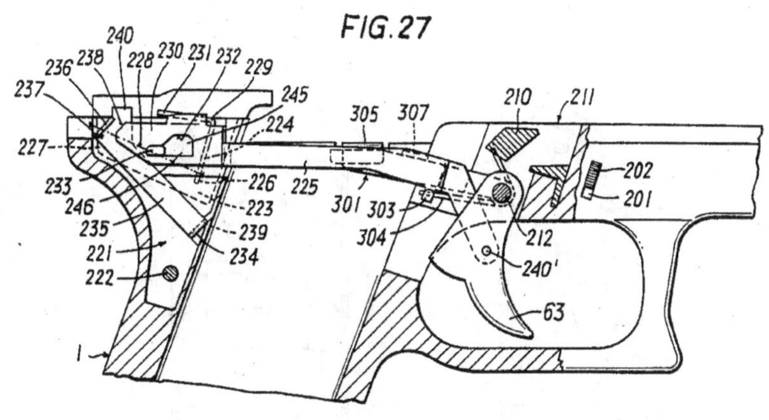

FIG. 28

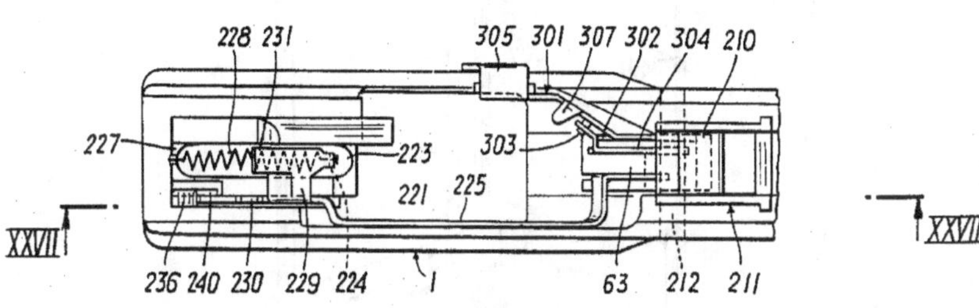

FIG. 29

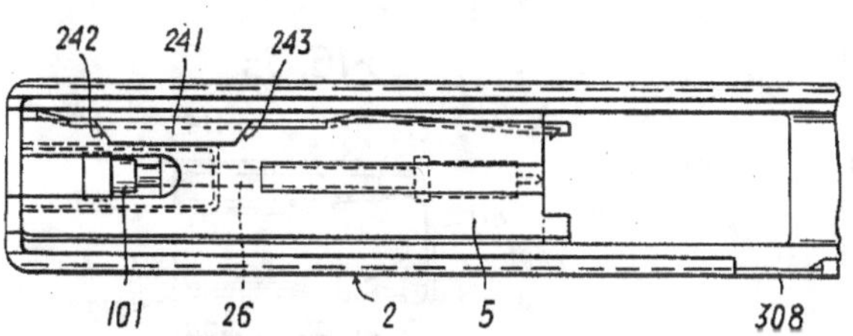

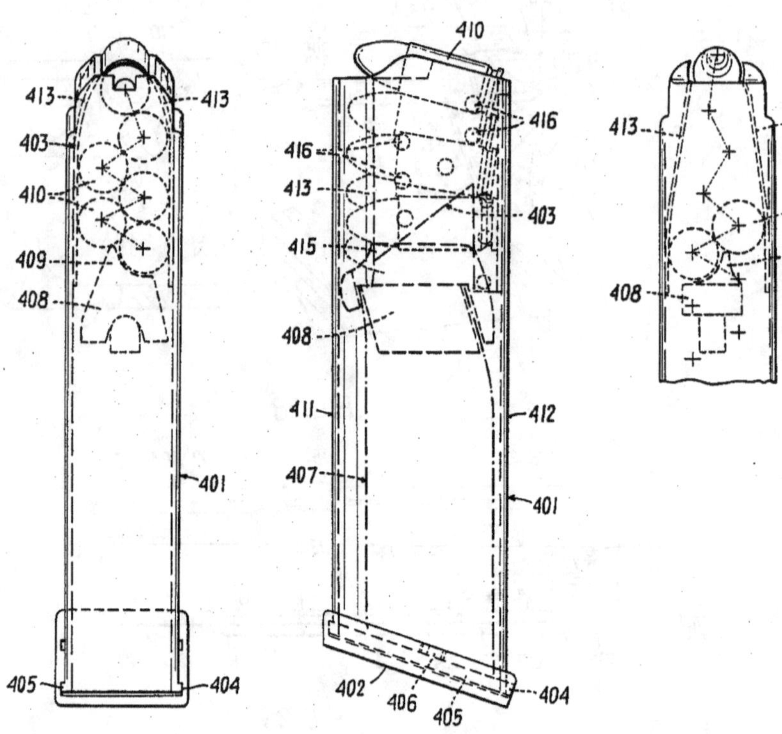
FIG.33
FIG.32
FIG.34
410
413
413
416
403
416
410
413
403
409
415
408
408
411
412
401
407
401
405
404
402
406
405
404
413
413
410
409
408

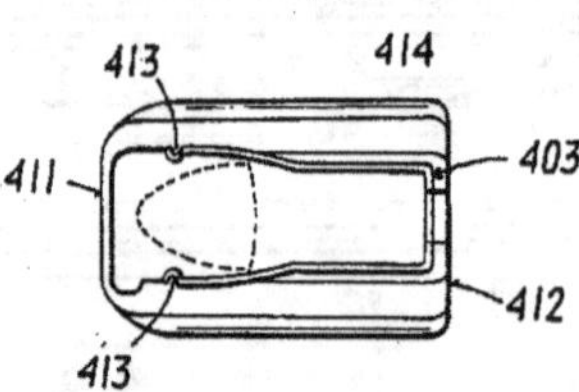
FIG.35
413
414
411
403
412
413

FIG. 36

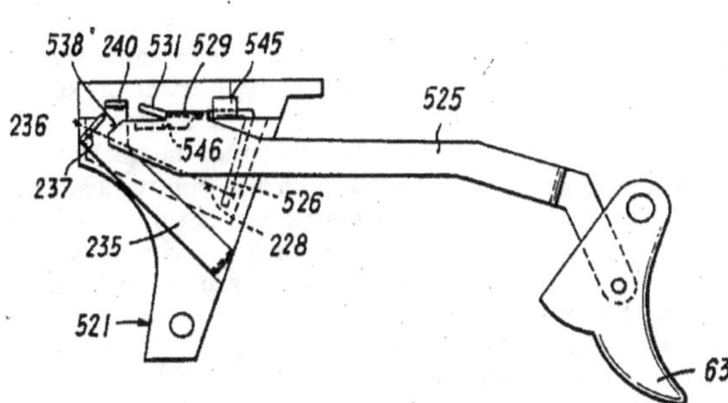

FIG. 38

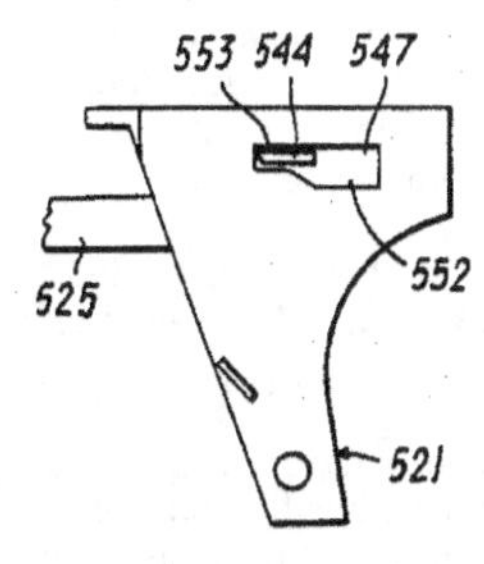

FIG. 37

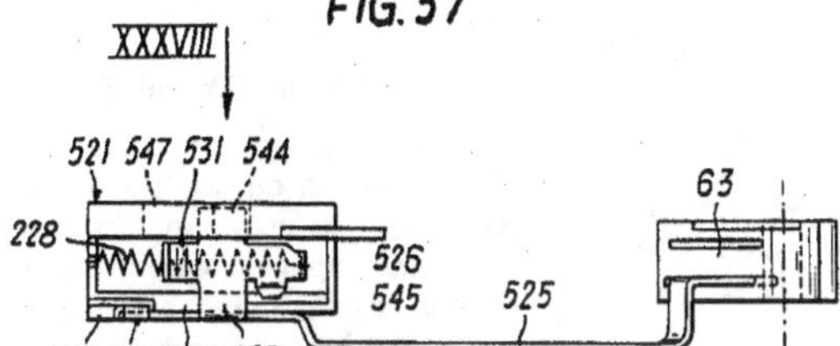

FIG. 39

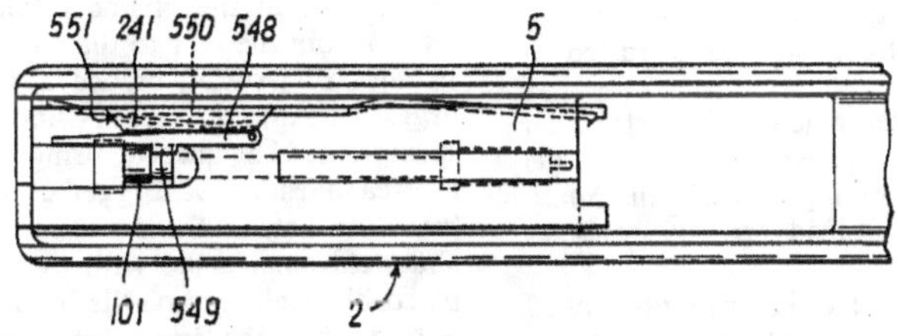

AUTOMATIC PISTOL

CROSS-REFERENCE TO RELATED APPLICATION

This is a continuation of application Ser. No. 773,352, filed Sept. 6, 1985, now abandoned, which is in turn a continuation of Ser. No. 456,056, filed as PCT AT82/00015 on Apr. 29, 1982, published as WO82/03910 on Nov. 11, 1982, now U.S. Pat. No. 4,539,889.

This application is related to my copending application Ser. No. 456,056 filed Dec. 30, 1982.

FIELD OF THE INVENTION

The present invention relates to a pistol. More particularly this invention concerns an automatic pistol of the type which automatically ejects the spent casing and chambers a new cartridge after each shot.

BACKGROUND OF THE INVENTION

A standard automatic pistol has a frame provided with a firing mechanism and carrying a slide comprising the barrel and breech block slidable on the frame. A recoil spring is braced between the frame and slide and the breech block is provided with a firing pin operable by a firing-pin spring or a hammer loaded by a hammer spring. The firing mechanism has an abutment guided generally parallel to and extending into the path of the firing pin or the hammer. This abutment is connected via a link with the trigger so that on actuation of same it is movable in a direction loading the firing pin or hammer.

Such pistols are relatively complicated to use. To chamber a cartridge it is necessary to pull back and then push forward the slide. Similarly when the cartridge clip is empty, the slide must be pulled back, a new clip inserted, then the slide released and moved forward to chamber the cartridge. All these actions must be carried out against heavy spring forces and in only one sequence, so that such a pistol can only be entrusted to experienced hands.

The shooter cannot often tell whether the pistol is on or off safety, especially after a pause in shooting. Thus it is possible for a shot to be attempted while the safety is on, or for a shot to be loosed inadvertently by someone thinking it is on when it is not.

Trigger-type automatics have a trigger that is cocked by the slide when a cartridge is chambered. In order to carry the loaded pistol with safety, the hammer must be uncocked. Subsequent shooting necessitates manually cocking the hammer by means of the trigger. This procedure requires that quite some force be exerted, necessitating a long trigger stroke without any noticeable critical point. The pistol is off safety after the shot, and subsequent shots only require limited force on and a limited stroke of the trigger, so that the danger of an unintentional shot is great. For safety against jarring and dropping, a particular latch for the firing pin is provided that is released on operation of the trigger before the hammer strikes the firing pin.

In addition, pistols are known with a separate safety lever which is actuated by the three fingers surrounding the pistol grip. It is, however, difficult to move these three fingers independently of the trigger finger so that mistakes in handling happen. In addition, with such a pistol whenever it is solidly gripped it is off safety, so that unintentional shots can be fired.

The known pistols have in common that the firing mechanism holds the firing element, that is, the firing pin or the hammer, in its cocked position and in this position the pistol is off safety and cocked so it is sensitive to jarring or dropping.

Another problem with the known automatic pistols is that removal of the barrel for servicing of the gun is fairly difficult, necessitating tools. In view of the need to maintain such complicated mechanisms carefully, such difficulty is extremely disadvantageous.

Yet another disadvantage of the known automatic pistols is that after the last shot in a clip the slide returns forward on the empty chamber. To chamber a new cartridge it is necessary to pull back the slide, insert the new clip, then advance the slide. In a situation where a pistol is used, such extra handling is very disadvantageous.

OBJECTS OF THE INVENTION

It is therefore an object of the present invention to provide an improved pistol.

Another object is the provision of such an improved pistol which overcomes the above-given disadvantages.

Yet another object is to provide an easy-to-use but very safe automatic pistol which can be produced at low cost.

SUMMARY OF THE INVENTION

A pistol according to the invention has a frame, a barrel slidable on the frame and having a cartridge-receiving rear end, and a breach block slidable on the frame and engageable over the rear end of the barrel to form a cartridge chamber. A standard slide carries the barrel and breech block. A firing element and a firing pin operatively linked thereto are movable on the breech block toward and away from the barrel between a rear position in which the firing pin is out of the cartridge chamber and a front position with the firing pin projecting forward into the cartridge chamber for firing a cartridge in the chamber when the firing element moves from the rear to the front position. A relatively strong firing spring braced against the firing element urges same into the front position and a relatively weak spring braced against the firing element urges same into the rear position. A trigger movable on the frame between an actuated and an unactuated position and an abutment engageable with the firing element and displaceable backward on the frame are linked together so as to displace the firing element back into the rear position on displacement of the trigger from the unactuated to the actuated position and to displace the abutment out of operative engagement with the firing element on displacement of the trigger into the actuated position for displacement of the firing element by the springs into the front position.

According to another feature of this invention, the abutment is engageable with the firing element in an intermediate position thereof in the unactuated position of the trigger and the linkage displaces the firing element backward from the intermediate position into the rear position by means of the abutment on displacement of the trigger from the unactuated to the actuated position.

Thus, the starting position of the abutment for the firing bolt or hammer is at an intermediate location in the travel path of same. In this manner, the firing mechanism can be such that the trigger force is substantially less than with the known pistols. Preferably, the starting position of the firing pin is in a noncritical region of its travel path or that of the hammer, in which region the force of the partially loaded firing-pin spring or hammer spring is insufficient to fire a shot.

According to this invention, the spring means includes a relatively strong firing spring braced against the firing element and urging same into the front position and a relatively weak spring braced against the firing element and urging same into the back position. The trigger or cocking force is the difference between these spring forces and can be set at a hair trigger or a relatively stiff novice level. In other words, the pistol is always uncocked or at least partially uncocked. The cocking for each shot is effected by the trigger and is assisted by a spring, so that the condition of the pistol is the same before the first shot as it is before the subsequent shots.

According to another feature of this invention, the firing means includes a guide holding the abutment in operative engagement with the firing element on displacement of the trigger from the unactuated to the actuated position. In addition, when the element is in the intermediate position the abutment prevents any displacement of the firing element relative to the abutment. Accidental discharge of the pistol therefore is impossible.

In accordance with another feature of the invention the abutment is displaceable laterally relative to the firing element between a position in the path of same and engageable therewith and a position out of the path and unengageable therewith, the link means displacing the abutment into the out-of-path position on displacement of the trigger into the actuated position. This movement of the abutment which frees the firing element—firing pin or hammer—in a direction perpendicular to one in which this element moves to fire the cartridge means that if the pistol is jarred, as for example by being dropped, it is virtually impossible for the necessary forces to be exerted on the mechanism to fire the pistol.

The abutment can, according to this invention, be rotatable between the in-path and the out-of-path positions. More particularly when the firing element is a longitudinally displaceable bolt and the abutment has a sleeve carried thereon, the firing means includes a torsion spring urging the abutment into the in-path position. The abutment is an arm projecting from the sleeve and the breech block is formed with a guide holding the abutment in operative engagement with the firing element on displacement of the trigger from the unactuated to the actuated position. In addition, in such an arrangement, the link means includes a trigger slide displaceable parallel to and transversely of the path of travel of the firing pin and having a spring urging it into engagement with the abutment.

The abutment can be a lever having one pivoted end and an opposite end engageable in the path of the firing element and deflectable thereby out of the path thereof to free the firing element for firing. More particularly, when the abutment is such a lever it can have one end engaging in the path of the firing element and another end formed with a slot. The frame has a pivot pin traversing the slot and the firing mechanism comprises a spring urging the one end away from the pivot pin. Once again, the displacement direction for the link is transverse of the firing pin. Therefore, the pistol is very jar-resistant.

In accordance with this invention, the firing element can also be a hammer pivotal on the frame, in which case the firing mechanism includes a firing pin carrying the firing pin tip and engageable with the hammer and the spring means and abutment are engageable with the hammer. The abutment can be a two-arm lever blockingly engageable with the hammer and laterally deflectable out of engagement therewith. It can also be a longitudinally slidable trigger slide.

In a particularly simple construction according to this invention, the firing element is a firing bolt carrying the firing pin tip and a firing-pin nose and the abutment is directly engageable with the nose. Furthermore, the link means includes a trigger slide and an inclined surface on the frame engageable with the trigger slide in the actuated position of the trigger. This link means includes a spring urging the slide into engagement with the inclined surface.

To eliminate the problem of firing-pin tip breakage which plagues automatic pistols, the firing pin tip is lance-shaped and the breech block is formed with an elongated throughgoing slot through which the lance-shaped pin tip engages. More particularly, the firing pin has a flattened triangular tip lying in the pistol plane, so it is very strong in this direction, which is the same as the shell-ejection direction.

The pistol according to this invention has latch means for releasably securing the barrel to the frame. This means includes a projection on the barrel movable along a path on sliding of the barrel and breech block on the frame, a frame abutment on the frame and normally in the path, and means for moving the frame abutment out of engagement with the barrel projection. The frame abutment can be an eccentric pivotal into and out of the in-path position, or can be a slide displaceable parallel to the clip hole in the frame. Either arrangement makes removal of the slide relatively easy.

In addition, the pistol of this invention has a clip removably engageable with the frame and holding a supply of cartridges displaceable by the breech into the chamber, and safety means engageable between the clip and the link means for permitting the abutment to move out of engagement with the firing element only when at least one cartridge remains in the clip. To this end, the clip has a cartridge follower and the safety means includes an element on the frame engageable through the clip with the follower. Thus, when the last shot is fired, the slide will not return forward, so that a new clip can be inserted with automatic chambering of the first cartridge.

The pistol according to this invention is set up so that the abutment cannot catch the firing element and recock and fire it when the trigger is held back. Instead the trigger must be released between shots to move the link forward into its forward position where it can engage the abutment. In other words, the abutment and link can only engage one another when the firing element is in the intermediate position and the link is in the trigger-unactuated position.

With the pistol of this invention, releasing and firing are done with the same element. Thus, the condition of the pistol is the same before the first shot as it is before the subsequent shots. This is attained when the guide which establishes the path of the abutment during the loading motion blocks projection of the abutment into the travel path of the firing pin or hammer. The pistol is therefore always uncocked or partially uncocked.

Handling of the pistol according to the instant invention is therefore as simple as possible. The pistol is ready after chambering of the cartridge in the barrel for shooting at any time and is nonetheless completely safe from unintentional shots. Similarly in this condition the pistol is fully drop- and jar-resistant. As a result of the unchanging trigger force, accuracy is increased. Simple and safe handling of the pistol is ensured even for the unpracticed user.

As a result of the small number of parts and the possible fitting of the firing mechanism into a small space, the frame can be of one piece, preferably of a synthetic resin, so that the overall weight is substantially less than that of comparable known pistols. In addition, manufacture is simplified and made inexpensive. In fact, the entire frame can be a synthetic-resin casting made with a simple two-piece mold, and the various elements like the firing element and guide for the slide can be formed by metallic inserts.

DESCRIPTION OF THE DRAWING

The above and other features and advantages will become more readily apparent from the following, reference being made to the accompanying drawing in which:

FIG. **1** is a side view of a pistol according to the present invention;

FIG. **2** is a front end view of the pistol of FIG. **1**;

FIG. **3** is a large-scale longitudinal section through a detail of the pistol of FIG. **1**;

FIG. **4** is a top view of a detail of FIG. **1**;

FIG. **5** is a longitudinal section through the breech of the pistol of FIG. **1**;

FIG. **6** is a top view of the breech of FIG. **5**;

FIG. **7** is a section taken along line VII—VII of FIG. **5**;

FIG. **8** is a top view of the rear portion of the firing mechanism;

FIG. **9** is a longitudinal section through the mechanism of FIG. **8**;

FIG. **10** is a section taken along line X—X of FIG. **9**;

FIG. **11** is a side view of the firing mechanism in the uncocked position;

FIG. **12** is a side view of the firing mechanism in the ready-to-fire position;

FIG. **13** is a longitudinal section like FIG. **9** through a second embodiment of the firing mechanism, in the uncocked position;

FIG. **14** is a section taken along line XIV—XIV of FIG. **13**;

FIG. **15** is a section like FIG. **13** but in the cocked position;

FIG. **16** is a section taken along line XVI—XVI of FIG. **15**;

FIG. **17** is a longitudinal section like FIG. **9** through a third embodiment of the firing mechanism, in the uncocked position;

FIG. **18** is a section like FIG. **17** but in the cocked position;

FIG. **19** is a section taken along line XIX—XIX of FIG. **18**;

FIG. **20** is a top view taken in the direction of arrow XX of FIG. **18**;

FIG. **21** is a large-scale longitudinal section through a detail of FIG. **1**;

FIG. **22** is a section taken along line XXII—XXII of FIG. **21**;

FIG. **23** is a section taken along line XXIII—XXIII of FIG. **22**;

FIG. **24** is a side view of another pistol according to the present invention;

FIG. **25** is a large-scale longitudinal section through a detail of the pistol of FIG. **24**;

FIG. **26** is a top view of the detail of FIG. **25**;

FIG. **27** is a large-scale longitudinal section through the firing mechanism of the pistol of FIG. **24**;

FIG. **28** is a top view of the detail of FIG. **27**, line XXVII—XXVII being the section plane for FIG. **27**;

FIG. **29** is a bottom view of the slide of the pistol of FIG. **24**;

FIG. **30** (same sheet as FIGS. **5–12**) is a longitudinal section through a firing-pin assembly according to this invention;

FIG. **31** (same sheet as FIGS. **5–12**) is a front end view of the assembly of FIG. **30**;

FIG. **32** is a side view of the clip for the pistol of FIG. **1**;

FIGS. **33**, **34**, and **35** are respectively back end, partial front, and top views of the clip of FIG. **32**;

FIG. **36** is a side view of yet another firing mechanism according to the invention;

FIG. **37** is a top view of the structure of FIG. **36**;

FIG. **38** is a view in the direction of arrow XXXVIII—XXXVIII of a detail of FIG. **37**; and

FIG. **39** is a bottom view of a slide that works with the mechanism of FIGS. **36–38**.

SPECIFIC DESCRIPTION

As seen in FIGS. **1–12**, the pistol in general is comprised of a frame **1** and a slide **2** which is displaceable on the frame **1** by means of a tongue-and-groove guide **3**. The barrel **4** and breech block **5** are mounted on the slide **2**. The barrel **4** has a breech block chamber **6** and a downwardly projecting extension **7** in which a recess **8** is formed from which a forwardly upwardly inclined groove **9** extends. The barrel **4** has an upwardly directed projection **10** extending into a window **11** of the slide **2** so that the barrel **4** and breech block **5** are locked together for joint movement along the axis A of the barrel **4**. The barrel **4** projects with play through an opening **12** at the front end of the slide **2** so that it is only held in the region of its breech block chamber **6**.

The slide **2** has an end wall **13** which serves as rest for a recoil spring **14**. The other rest is formed by the rim **15** of a sleeve **16** that is surrounded with play by the recoil spring **14**. The front end **17** of the sleeve **16** is received with play in a bore **18** of the end wall **13**. The sleeve **16** is fixed axially by a bracket **19** which has a flange fitted in a slot of the frame **1**.

Also mounted in the frame **1** is an eccentric **20** formed of a pair of disks **21** and **21'** rotatable in a bore of the frame **1** and connected together by a stem **22**. The disk **21'** projects out of the frame **1** and carries a spring lever **23** whose free end **24** can snap over a pin **25**. When thus engaged over the pin **25**, rotation of the eccentric **20** is impeded.

The breech block **5** is mounted axially unmovably in the slide **2**. It only has centering surfaces which correspond to complementary surfaces of the barrel **4**. As seen in FIGS. **5–7**, a firing pin **26** generally centered on the barrel axis A is slidable in the breech block **5** and carries at its front end a firing pin tip **27**. This firing bolt **26** has a section **28** of small diameter straddled by a spring rest **29** against which the firing-pin spring **30** is braced and which bears with its other end against a shoulder **31** of a spring tube **32** which is downwardly open and which flatly abuts the breech block **5** with its rear end. A weak return spring **33** mounted in the tube **32** bears in one direction against the shoulder **31** of the spring tube and in the other direction on a nose **34** of the firing pin **26**, which nose **34** projects with its tip down out of and is guided in a slot **35** of the breech block **5**. The nose **34** is formed with an inclined surface **36**. The position of the firing bolt **26** is determined by engagement of its section **37** on the spring rest **29**.

Ridges **39** and **40** are provided spaced one behind the other on the underside **38** of the breech block **5** so that a gap **41** is formed. The outer flank of the ridge **39** and the inner flank of the ridge **40** lie substantially in a plane. In addition, the breech block **5** has an extractor hook **42**.

Behind a magazine hole **43** for the cartridge clip **44** (see FIG. **1**), the frame **1** has a guide **45** fixed in place by a pin **46** as seen in FIGS. **8–12**. The guide **45** extends to the rear end wall **47** of the frame **1** and is formed with a bore **48** whose axis is parallel to the bolt/barrel axis A. This bore **48** receives a stop sleeve **49** which contains a stop spring **50** which is braced between the floor of the bore **48** and the end wall **51** of the stop sleeve **49**. This stop spring **50** is considerably weaker than the firing spring **30** that is oppositely braced against it as will become apparent hereinbelow. A pin **52** in the floor of the bore **48** and a pin **53** in the end wall **51** of the stop sleeve **49** guide it. The spring **50** is received with radial play in the sleeve **49**. The wire ends **54** and **55** of the spring **50** are bent axially and are received in little bores respectively in the floor of the bore **48** and in the end wall **51**. The stop spring **50** is not only axially, but also angularly loaded since the stop sleeve **49** is is turned a few times when it is mounted to torsionally or angularly load the spring **50**. In this manner, the sleeve **49** is urged clockwise according to arrow **56** (FIG. **10**). To prevent such rotation it has an arm **57** which bears against the upper edge surface **58** of the guide **45**. The guide **45** further has a shoulder **59** which is interrupted by a slot **60**. The width of the slot **60** is such that an abutment **61** on the sleeve **49** can engage in it.

The guide **45** is provided with a longitudinally open recess **62** in which is slidable a trigger slide **64** connected to the trigger **63**. This trigger slide **64** occupies only a portion of the recess **62** and has an arm **65** engaging underneath the arm **57** as shown in FIG. **12**. The arm **65** terminates in a hook **66** which coacts with a step **67** on the guide **45**. Furthermore, the slide **64** has a cam surface **68** which ends at a guide edge **69**. The slide **64** is thus on one side guided by the floor **70** of the recess **62** and on the other side with the edge **69** on the underside **38** of the breech block **5** so that it can move substantially only parallel to the axis A of the firing pin **26**. It is provided with a bend **71** on which bears the long leg **74** of a hairpin spring **73** received in an elongated cutout **72** of the guide **45** and having a short leg engaged in a slot **75** of the frame **1**. The spring is constructed such that the long leg **74** urges the trigger slide **64** forward and the arm **65** toward the spring sleeve **49**.

As seen in FIGS. **21–23**, inside the grip **76** of the frame **1** is formed with a square-section passage transverse to the hole **43** for the cartridge clip **44** and receiving a slide **77** having a downwardly open bayonet guide **78**. Near this slide **77** the hole **43** is formed with a wedge-shaped recess **79** into which a bore opens that receives a spring wire **80** that engages with its free end **81** in the bayonet guide **78**. In addition, the slide **77** projects with its end **82** out of the grip **76** so that it can be shifted against the force of the spring **80**. It carries a wedge nose **83** which engages in a groove of the clip **44**.

FIGS. **5** to **11** show the pistol uncocked. In order to chamber the first cartridge **84** in the barrel **4**, the slide **2** is slid back against the force of the recoil spring **14**. The stem **22** of the eccentric **20** engages in the groove **9**, swings the barrel **4** down, and holds it against axial movement relative to the slide **2** once it reaches the base of the groove **9**. This action pulls the projection **10** of the barrel **4** out of the window **11** so that the slide **2** with the breech block **5** can move further back. Meanwhile the nose **34** of the firing bolt **26** carries back the abutment **61** as can be understood from a comparison of FIGS. **7** and **10**. The nose **34** is prevented from rotating by the groove **35**, as is the abutment **61** which lies on the front part of the shoulder **59**. Once the abutment **61** and the stop sleeve **49** have moved back sufficiently, while unloading the stop spring **50**, the abutment **61** can move angularly into the slot **60**. Thus, the cam surface **36** of the nose **34** pushes the abutment **61** against the torsional force of the stop spring **50** into the slot **60** so that this nose **34** can move back past the abutment **61**. As soon as the nose **34** passes the abutment **61**, it is turned by the spring **50** back into the position of FIG. **10**. The sleeve **49** engages with its end wall **51** against the rear end wall **47** of the frame **1** and is solidly positioned here by the axially effective force of the stop spring **50**.

When the slide **2** is returned somewhat forward, the nose **34** of the firing pin **26** engages forward against the abutment **61**. The force of the firing-pin spring **30** overcomes that of the stop spring **50** so that the nose **34** slides the abutment **61** and the sleeve **49** forward until the end of same engages the floor of the bore **48** in the guide **45**. Further advance of the slide **2** is impeded as the abutment **61** holds the nose **34** and the firing pin **26**, thereby partially tensioning the firing-pin spring **30**. Simultaneously, the breech block **6** is closed by the breech block **5** and moved forward with the barrel **4**. This lifts the rear portion of the barrel **4** from the stem **22** so the projection **10** engages again in the window **11** of the slide **2** and the barrel **4** and breech block **5** are locked together. During this action, as is known, the extractor finger **42** engages the lip of the casing of a shell in the chamber **6**, pulls it axially backward therefrom, and flips this spent casing out the window **20** as same comes level with the extracted casing.

Pulling the trigger **63** moves the trigger slide **64** back so the cam surface **68** engages under the arm **57** and urges it up. Since the abutment **61** lies on the front shoulder **59**, further movement of the trigger **63** moves the arm **57**, the sleeve **49**, the abutment **61** and the firing-pin nose **34** entrained thereby back while unloading the stop spring **50** and loading the firing-pin spring **30**. Thus, it is only necessary to bring a force to bear on the trigger **63** equal to the difference between the forces of the springs **30** and **50** to move these elements of the mechanism. By choosing appropriate spring forces, the pressure for the trigger **63** can be set at any desired level.

As soon as the abutment **61** reaches the slot **60**, the cam surface **68** of the trigger slide **64** raises the arm **57** and swings the abutment **61** into this slot **60**. This frees the nose **34** and the firing pin **26** is propelled forward by the force of the spring **30**, overcoming the spring force of the weak return spring **33**. The firing pin tip **27** strikes as a result of the kinetic energy of the firing tip **26** with the necessary force on the primer of the cartridge, exploding it. As soon as the shot has left the barrel **4**, the powder gases drive the slide **2** in the above-described manner back, with the rail **39** on the underside **38** of the breech block **5** engaging the guide edge **69** of the trigger slide **64** and moving it outward against the force of the hairpin spring **73**.

Meanwhile the arm **57** disengages the cam surface **68** and returns with the abutment **61**, which moves out of the slot **60** under the force of its stop spring **50**, into the starting position in which the arm **57** lies on the surface **58** of the guide **45**. If the slide **2** is moved by the recoil spring **14** forward again, the nose **34** entrains the abutment **61** until the sleeve **49** engages with its end surface on the base of the bore **48**. Further advance of the slide partially compresses the firing-pin spring **30** again. Since the trigger slide **64** does not yet engage the arm **57**, it is prevented from shooting automatically when the trigger **63** remains depressed. Only when the trigger **63** is released does the guide edge **69** slide along the rail **40** forward to be pressed by the spring **73** into the gap **41** between the ridges **39** and **40** so the arm **65** of the trigger slide **64** engages under the arm **57**. Simultaneously, the return spring **33** slides the firing pin **26** until it lies with its front section **37** on the spring rest **29**, whereupon the firing pin tip **27** is withdrawn into the breech block **5**. Now the pistol is again ready to fire.

If the spring arm **23** is pulled out of the catch **25** it can rotate the eccentric **20** through 180°. The stem **22** disengages the barrel **4** and the entire slide **2** can be pulled off toward the front. The recoil spring **14** meanwhile remains compressed since the shoulder **15** of the sleeve **16** bears on the projection **7** of the barrel **4**. The slide **2** can therefore be shifted without exerting substantial force.

FIGS. **13** to **16** show a further embodiment of the firing mechanism according to this invention. Otherwise the pistol is identical to that shown in FIGS. **1** to **12** with identical reference numerals referring to identical structure.

A nose **101** formed unitarily with the firing pin **26** projects down and toward the front from the breech **5**. This firing pin **26** is mounted in the breech block **5** which is fitted jointly with the barrel **4** in the slide **2**. Engaged in the path of the nose **101** is the abutment end **102** of a lever **103** which is formed on its other end with a slot **104** that is traversed by a pivot pin **105**, the lever **103** therefore being limitedly displaceable transverse to the breech block **5** in the frame **1**. A spring **106** fixed in the frame **1** presses with its free end against the pivoted end of the lever **103** so that the outer end of the slot **104** bears on the pivot pin **105**, that is so that the lever **103** is moved up on the pin **105** towards the slide **2**. A hairpin spring **107** carried on the pin **105** has a short leg **108** anchored in the frame **1** and a long leg **109** bearing on a pin **110** which is fixed on the lever **103**. This spring **107** therefore urges the lever clockwise toward a position lying on a stop **111** in the frame **1** (FIG. **15**).

The lever **103** has a shoulder **112** which coacts with a shoulder **113** of the frame **1**. The shoulder **113** is so long that it only moves clear of the shoulder **112** when the lever **103** is in its rear end position, (FIG. **15**) lying against the stop **111**. Before reaching this end position the shoulders **112** and **113** prevent a shifting of the lever **103** against the force of the spring **106** and a simultaneous sliding of the slot **104** along the pin **105**.

The pin **110** of the lever **103** is engaged by an edge **114** of the trigger slide **115** which passes through a hole **117** ion the frame **1**. It extends with its arm **118** beyond the pin **110** and its end surface **119** coacts with another surface **120** formed by a bent-over end of a leaf spring **121** that is fixed in the frame **1**. Above the surface **120**, the leaf spring **121** is provided with a control edge **122** which coacts with a ridge **123** on the underside of the breech block **5**. The end surface **124** of the ridge **123** is included as seen in FIG. **16**.

In the uncocked position of the pistol the lever **103** is urged by the spring **107** against the stop **111**. If the slide **2** is shifted back in order to chamber the first cartridge from the clip, the nose **101** slides over the end surface **125** of the lever **103** and moves it down against the force of the spring **106**. As a result, the lever **103** can be passed by the nose **101**. Forward shifting of the slide **2** causes this firing-pin nose **101** to again entrain the lever **103** whose end **102** extends into its path until the lever **103** engages against a stop **126**, assuming the position, of FIG. **13**. In this position the nose **101** and the firing pin **26** are prevented from moving forward any further. Further forward sliding of the slide **2** compresses the firing-pin spring **30** partially. The pistol is ready to shoot in this condition.

When the trigger slide **115** is pushed back by the trigger, it slides on the surface **116** of the frame **1** and its edge **114** pushes the pin **110**, the lever **103**, and the nose **101** back also. This fully loads the firing-pin spring **30**. When the lever **103** reaches its rear end position and engages the abutment **111**, the end surface **119** of the trigger slide **115** engages the surface **120** and moves the trigger slide **115** down against the force of the spring **106** so that it moves away from the surface **116**, and the pin **110** as well as the lever **103** move down. This moves the free lever end **102** out of the path of the nose **101**, freeing it, so that the firing pin **26** is propelled forward under the force of the fully compressed firing-pin spring **30** and fires the cartridge.

When the slide **2** is then driven back by gases from the shot, the inclined surface **124** of the ridge **123** on the breech block **5** pushes the leaf spring out (FIG. **14**). The control edge **122** of the spring **121** thus lies against the ridge **123** until the slide **2** returns to its full-forward position. Deflection of the spring **121** makes the surface **120** slide off the end surface **119** of the trigger slide **115** and the spring **106** lifts the lever **103** and with it the arm **118** of the trigger slide **115**. The ridge **123** frees the control edge **122** when the breech block **5** moves into the full-forward position, but the spring **121** cannot return to its original position because the arm **118** lies at the same level as the bend where the surface **120** is. Only when the trigger is released and the trigger slide **115** slides forward is the leaf spring **121** released and takes its illustrated rest position. The trigger slide **115** need not be in its full-forward position shown in FIG. **13** for the pistol to be ready to fire; the surfaces **119** and **120** need merely be separated. This embodiment of the firing mechanism allows rapid fire in that a shot following another shot can be made without complete release and depression of the trigger. Nonetheless, some forward return of the trigger is essential for a second shot so the pistol does not fire automatically.

In this arrangement also the mechanism can be mounted in a block, such as the guide **45**, in the frame **1**.

Instead of a sliding of the level **103** in its longitudinal direction, the lever can also be moved laterally out of the path of the firing-pin nose **101**. Such an arrangement is not illustrated, because it is basically a combination of the two described embodiments.

FIGS. **17** to **20** show a part of a pistol that is provided with a hammer **131** that on shooting is pivoted by the force of a spring **135** on a pivot pin **132** and engages a firing pin **26** which fires the cartridge. Even in this type of pistol it is possible to use the principles of the instant invention.

The hammer **131** is pivotal in the frame **1** about a pivot pin **132** and the rod **133** is pivoted on the hammer **131**. This rod **133** is arranged in a cutout **134** of the frame **1** and is braced against the hammer compression spring **135**. The hammer **131** has a lateral shoulder **136** on which a two-arm lever **137** engages with its one arm **138**. Its other arm **139** is braced against an abutment spring **140** which is received in a bore of the frame **1**. The two-arm lever **137** has in its center between its two arms a slot **141** that is traversed by a pivot pin **142** and about which the lever **137** is pivotal. Furthermore, the level **137** is provided with a lateral projection **143** which forms a shoulder **144** that faces a shoulder **145** of the frame **1**. The trigger slide **146** engages with a shoulder **147** on the projection **143** and extends through an opening **148** in the frame **1**. Its free end is provided with an inclined surface **149** that coacts with a surface **150** formed by the bend of a leaf spring **151** whose end surface **152** acts as a control edge that coacts with a ridge **153** of the breech **5**. The leaf spring **151** is anchored with its lower end in the frame **1**. In addition a stop **154** is formed in the frame **1** to limit the forward pivot motion of the lever **137**.

FIG. **17** shows the firing mechanism in the uncocked condition. To cock it the trigger slide **146** is moved back by means of the trigger so that it pivots the projection **143** of the lever **137** back by means of the shoulder **147**. The free end of the lever **137** serving as stop lies meanwhile on the step **136** of the hammer **131** and pivots it clockwise against force of the hammer spring **135**. As soon as the inclined surface **149** of the trigger slide **146** reaches the surface **150** (FIG. **18**) the trigger slide **146** swings down, whereupon the lever **137** is shifted down against the force of the abutment spring **140** in the slot **141** and its abutment end frees the shoulder **136** of the hammer. The hammer is snapped forward by the hammer spring **134** to strike the protruding rear end **156** of the firing pin **26**, which in turn is propelled forward to fire the cartridge.

After the shot the gases drive the slide **2** with the breech block **5** back so that the inclined end face of the ridge **153** runs against the control surface **152** of the leaf spring **151** to pivot same outward (FIG. **19**). As a result the surfaces **149** and **150** disengage each other and the stop spring **140** slides the lever **137** together with the trigger slide **146** up so that the end of same comes to lie next to the bend of the spring **151**. Simultaneously as the slide reverses direction and follows the breech block **5** a rear edge **155** of the breech block **5** engages the hammer **131** and pivots it back counterclockwise against the force of its spring **135**. The hammer **131** entrains with its shoulder **136** the abutment end **138** of the lever **137** until this lies on the stop **154**. In this position the hammer **131** is a sufficiently safe distance behind the rear end **156** of the firing pin **26**. As soon as the trigger is released, the trigger slide **146** moves forward so that the firing mechanism again assumes the position of FIG. **17**. Thus the pistol cannot make automatic fire, that is a separate actuation of the trigger is need for each shot.

FIG. **24** shows a pistol in side view in which disassembly is effected by a slide catch **202** which takes the place of the eccentric **20** of FIG. **1**. The frame **1** has a slot **201** in which the catch **202** is shiftable and urged upward by a leaf spring **203** having has a short arm **204** fitted into the frame **1** and a long arm **205** engaged in a groove **206** of the catch **202**. Since this spring arm **205** is arranged in a slot in the frame **1**, the catch **202** is also prevented from moving longitudinally of itself, that is transversely of the central plane of the pistol.

The catch **202** is formed in its central region near the upper edge with a groove **208** in which a ridge **209** of the projection **7** of the barrel **4** can engage. This projection **7** in turn has an inclined groove **9** which coacts with a strut **210** of an anchor piece **211** fitted into the frame **1** and secured therein by a pin **212** which also serves as pivot for the trigger **63**.

The groove **207**, the groove receiving the spring arm **204**, the slot **201**, and the recess for holding the support piece **211** are inclined to the barrel axis A and preferably are parallel to the hole **43** for the cartridge clip **44** so that shaping of the core for the manufacture of these grooves and recesses can be quite easy. It is therefore possible to make the frame **1** in one piece of a synthetic resin in a mold whose halves are separated in a direction parallel to the oblique hole **43** and the parallel grooves mentioned above.

The foundation of the strut **210** corresponds to that of the stem **22** of the eccentric **20** of FIGS. **1** to **4**. In order to pull the slide **2** with the barrel **4** and recoil spring **14** off the front of the frame **1**, the slide **2** must be pulled back a little so that the ridge **209** of the projection **7** moves out of the groove **208** of the catch **202**. Then the slide **2** is moved down against the force of the spring **203** so as to free the barrel **4** and slide **2** from the catch **202** for unimpeded forward movement on the frame **1**. On replacing the slide **2** on the frame **1** a wedge surface **213** runs over the upper edge of the catch **202** and pushes it down against the force of the spring **203**. Once the projection **7** has passed the catch **202**, the ridge **209** is moved by the spring **203** up and latches in the groove **208** of the catch **202**, whereupon mutual tongue-and-groove locking is ensured.

This type of barrel locking can also be used in other types of pistols.

FIGS. **27** to **29** show another trigger mechanism which is very simple. A block **221** is fitted in the frame **1** and is held in place therein by a pin **222**. It has in the region of the longitudinal central plane of the pistol a cavity **223** in which a bent arm **224** of the trigger slide **225** engages. Stretched between a lower end **226** of the trigger slide **225** and the rear wall **227** of the cavity **223** is a tension spring **228** which pulls the trigger slide **225** up and back.

The arm **224** of the trigger slide **225** is connected by a web **229** with a plate-like end **230** of the trigger slide **225**. This web **229** has a backwardly projecting part **231** that forms an abutment for the firing pin **26**. The plate **230** is formed with a polygonal recess **232** which is traversed by a projection **233** of the block **221**. Near this projection **223** the block **221** has a groove **234** which extends upward and back and in which a leaf spring **235** is fitted which has an outwardly directed edge **236** which forms a control surface **237** for an inclined end surface **238** of the trigger slide **225**. The other end of the leaf spring **235** is bent in the opposite direction from the edge **236** and is fitted in a groove **239** of the block **221**. Near the edge **236** is a control edge **240** which is arranged in the path of a control ridge **241** on the underside of the slide **2** and formed on its ends with two cam or wedge surfaces **242** and **243**. The front end of the trigger slide **235** is pivoted on a pin **240'** on the trigger **63**.

In the uncocked condition of the pistol, the spring **228** urges the trigger slide **225** into its back position with the inclined surface **238** lying on the control surface **237** so that the projection **233** is in the upper part **245** of the recess **232**. In this position of the slide **225**, the abutment **231** lies underneath the path of the nose **101** of the pin **26**.

To chamber the first cartridge, the slide **2** is pulled back so that the control edge **240** is moved inward by the wedge surface **242** of the guide **241** on the slide **2** and the inclined surface **238** of the trigger slide **225** and the control surface **237** of the leaf spring **235** disengage each other. As a result, the spring **228** can pivot the trigger slide **225** up so that the projection **233** comes to lie in the lower region **246** of the recess **232** of the plate **230**. Meanwhile the end of the trigger slide **225** has laterally run past the edge **236** so that the surfaces **237** and **238** are out of alignment with each other and therefore without mutual effect. In this position, the abutment **231** moves into the path of the nose **101** of the firing pin **26** so that, as the slide **2** is moved forward, the nose **101** engages the abutment **231** and moves the trigger slide **225** forward until the projection **233** assumes the position in the recess **232** shown in FIG. **27**. The pistol is now cocked.

On pulling the trigger at first the slide **225** is guided by the projection **233** and the spring **228** against the force of the bolt spring **30**, which is hereby loaded, and slides back until the surfaces **237** and **238** engage each other. In this position, the projection **233** has reached the broad part of the recess **232** so that the trigger slide **225** can swing down as the inclined surface **238** slides on the control surface **237**. The abutment **231** then frees the nose **101** of the firing pin **26** and the shot is fired.

On backward displacement of the slide as a result of the recoil, the wedge surface **242** moves the control edge **240** inward so that the above-described interaction can repeat itself. The position of the control ridge **241** ensures that the wedge surface **242** of the control edge **240** is only freed when the barrel **4** and breech block **5** are locked together. If for any reason the slide **2** has not moved fully forward, the ridge **241** holds the leaf spring **235** in its inwardly bent position in which pulling of the trigger **63** is ineffective because the inclined surface **238** cannot engage the control surface **237** so that the trigger slide **225** does not swing down and the abutment **231** cannot free the firing-bolt nose **101**.

A device can also be provided to ensure quick preparation to fire on changing the cartridge clip, whose construction is described in detail below with reference to FIGS. **32–35**. To this end, a lever shown generally at **301** is provided which is pivoted on the axle pin **212** of the trigger **63** and received in a laterally open recess **302** of the trigger **63**. The lever **301** is provided on its underside with a hook **303** in which is hooked the end of a hairpin spring **304** which surrounds the pin **212** partially and is caught in a groove of the web **210**. This spring **304** tries to pivot the lever **301** into a lower end position in which a handle **305** lies in a recess of the frame **1**. In addition the lever **301** has on its upper side a nose **307** which engages in the path of a slide that is backed up by a spring in the clip, as will be described below and which urges the cartridge upward. When the last cartridge of the clip is inserted into the barrel **4**, the slide of the clip engages the nose **307** of the lever and tries to pivot same up. Such pivoting of the lever **301** is prevented by the lower edge of the slide **2**. After firing the cartridge, the lever **301** enters into a recess **308** (FIG. **29**) on the lower edge of the slide **2** when this is in its end position. The lever **301** latches the slide **2** against moving forward. Swinging of the lever **301** on the pin **212** (FIG. **24**) maintains this latching even when the clip is removed and replaced with a new clip. As soon as the handle **305** of the lever **301** is swung down, the slide **2** is moved by the force of the recoil spring **14** forward and pushes the first cartridge of the new clip into the barrel **4**. Thus it is no longer necessary to pull back the slide after changing the cartridge clip.

With the known pistols and center-fire cartridges it is possible to break off or damage the firing pin tip on the firing pin when this pin does not pull back into the end of the breech block **5** quickly enough. As shown in FIGS. **30** and **31**, according to the invention, the firing pin tip **27'** is pointed and lance-shaped, that is it is not of cylindrical shape as is standard, but is of rectangular section with a substantially greater vertical dimension in the pistol plane than transverse dimension. The pin tip **27'** can engage through an elongated cutout lying on the longitudinal middle plane of the pistol at the breech end. When a spent shell is ejected or a new cartridge is chambered, before the pin tip **27'** engages fully in the cutout in the end of the breech block **5**, the firing pin is not damaged but is pushed back by the shell or cartridge. By means of the elongated shape of the cutout according to this invention pistols with drop barrels do not develop the otherwise normal brass-chip deposits.

Lance-shaped according to this invention means any shape which varies from the round section and cylindrical shape of the known firing pins and which has a generally flat shape. Preferred is a triangular shape which is obtained, for example, from a pyramid with spherically rounded points that is formed on opposite sides with symmetrical shoulders so that the remaining parabolic flanks can be made planar, slightly convex, or concave.

The cartridge clip shown in FIGS. **32** to **35** has an elongated and generally parallelepipedal synthetic-resin body **401**, a base **402** inclined obliquely to the body **401**, and a metal insert **403** inside the body **401**. The metal insert **403** has holes **416** that flare inward so that the insert can be well anchored in the synthetic-resin body **401** of the clip. The base **402** has grooves **404** that fit on ridges **405** on the lower end of the clip body **401**, with a latch **406** blocking unintentional sliding-off of the base **402**. A spring **407** is braced at one end on the base **402** and at the other end on a follower slide **408** so as to urge same upward. Shoulders **409** at one end of the follower **408** support the cartridges **410** in the lower portion of the body **401** with their axes perpendicular to a front wall **411** and a rear wall **412** of the body **401**, in the upper portion of the body **401** the insert **403** forms, in the region of the slugs, guides **413** that narrow considerably upward and that push the tips of the cartridges **410** together toward the central plane of the clip, which coincides with that of the pistol when the clip is in place in the well **43** thereof, so that the cartridges **410** align and finally assume a position generally parallel to the clip base **402**. Portions **414** of the metal insert **403** at the back of the cartridges **410** converge first toward the end of the clip so that as shown in FIG. **33** the backs of the cartridges **410** remain longer in their original staggered positions. This can be seen by a comparison of the dot-dash zig-zag lines of FIGS. **33** and **34**.

As a result of the construction according to the invention, two considerable advantages are obtained. The position of the cartridge backs ensures a solid contact between them and the back wall **412** of the clip so that friction is reduced as is the resistance to displacement. Catching on the back wall is impossible. The second advantage is that the contact between the cartridges **410** as they are aligned in the upper section of the clip changes. It moves from line contact to point contact so that in the upper part of the clip there is no wedging-together of the cartridges **410**.

The front end **415** of the shoulder **409** of the slide **408** can coact with the lever **301** of FIG. **28** so that this part comes to lie on the projection **307** when the last cartridge **410** is chambered.

FIGS. **36** to **39** show a variant of the firing mechanism of FIGS. **27** to **29**, in which the guide is not in the trigger slide but in the frame **1** or in a block **521** set into the frame **1**. The block **521** is of a synthetic resin and has a leaf spring **235** whose end **236** forms a control surface **237** which coacts with an inclined surface **538** of the trigger slide **525**. Similarly the control edge **240** coacts with the ridge **241** of the slide **2**. The trigger slide **525** has a web **529** which carries on one side the abutment **531** and on the other side the bent-over part **526**. In addition it is provided with a wing-like projection **544** and has a control edge **545**.

A recess **546** on one side and on the other side a recess **547** serve as a guide in the block **521** in which the wing **544** engages. The underside of the breech **5** is provided with a simple pivotal lever **548** which has a projection **549** extending into the path of the firing-pin nose **101**. This lever **548** is braced against a leaf spring **550** which is braced with its free end **551** on the breech block **5** and which lies underneath the ridge **241** so as to pivot the lever **548** such that the projection **549** is in the path of the firing-pin nose **101**.

The operation of this firing mechanism corresponds mainly with that of the mechanism shown in FIGS. **27** to **29**. The abutment spring **228** draws the trigger slide **525** back so that its inclined surface **538** slides down along the control surface **538** and the web **529** enters the recess **546** and the wing **544** enters the lower region **552** of the hole **547**. In this position, the abutment **531** is below and out of the path of the firing-pin nose **101**. Meanwhile the control edge **545** is below the lever **548** so that it takes the position of FIG. **39** in which the firing pin is locked.

When the slide **2** is moved back, the ridge **241** pushes the spring **235** toward the center so that the control surface **237** is guided by the inclined surface **538** and the trigger slide **525** is freed. The abutment spring **228** draws this slide **525** up and back while the abutment **531** and the control edge **545** are drawn into the path of the firing-pin nose **101**.

When the slide **2** is moved forward, the firing-bolt nose **101** entrains the abutment **531** and the slide **525** as well as the trigger **63**. Meanwhile the web **529** moves out of the recess **546** and the wing **544** engages in the narrow region **553** of the recess **547**. As soon as the wing **544** gets to the end of the recess **547**, the abutment **531** is held solidly by the firing-pin nose **101**. The narrow section **553** of the hole **547** holds the abutment **531** tightly.

Pulling the trigger moves the slide **525** back and the abutment **531** pushes back the pin-bolt nose **101** while further compressing the firing-pin spring **30**. As soon as the inclined surface **538** engages the control surface **237**, further movement of the slide **525** pushes it down sot hat the web **529** engages in the recess **546** and the wing **544** in the wide region of the pin **547**. As a result the abutment **531** releases the firing-pin nose **101** and the shot is fired.

Moving the slide **525** moves the control edge **545** back into the region of the projection **549** and pushes the lever **548** to the side so that the projection **549** moves out of the path of the firing-pin nose **101** and does not impede forward travel of the firing pin.

Since the additional safety catch **548**, **549** is on the breech block **5**, jarring and inertial forces cannot open it. To fire it is therefore necessary to actuate the trigger **63**.

The invention is not limited to the described and shown embodiments. Its parts can be combined in other than the shown manner. It is common to the described embodiments that the abutment for the firing pin or hammer moves in two directions that are not parallel to each other. It is therefore possible to use the abutment for securing as well as for releasing the firing pin which not only reduces the stroke of the trigger but also substantially reduces the number of parts. This double movement can also be split up between the abutment and firing bolt or hammer.

The provision of stop spring whose effect is opposite to that of the firing-pin spring means that only the difference between these forces need be overcome by the force on the trigger to shoot.

In addition to these basic advantages the pistol according to this invention has several advantages. The rest position, for example, of the trigger is established by abutments **67** and **233** in the guide **45** or block **221**, respectively, so that no particular stops for the trigger are needed. This eliminates expensive adjustment procedures. Since the parts necessary for the release and latching of the firing bolt can fit in a limited space, it is possible make the guide **45** or the block **221** relatively small. This leaves room in the frame **1** behind the hole for the cartridge clip. As a result, it is possible to have the grip near the axis A of the barrel **4** so that on firing there is little torsion on the hand. The pistol kicks up less so it shoots more accurately.

With the new pistol it is possible to form the spaces for mounting the trigger, trigger slide, and guides in the frame **1** by molding in such a manner that the frame **1** is easily demolded. The same applies for the clip hole and the passage for the clip-safety slide. As a result, the frame **1** can be made in one piece.

The guides for the slide in the frame **1** are made of metal. To do this it is sufficient to mount two guide rails which are imbedded in the frame **1** in the resin. They have the tongue and groove construction of FIG. **2**.

Particularities of the described features can be used independently of each other in pistols of known construction without losing the described advantages.

I claim:

1. In a pistol, including

a frame having a longitudinal axis and a barrel mounted in the frame;

a slide mounted on the frame so as to be slidable forwardly toward the barrel into a closed position in which the frame is in contact with the barrel and rearwardly out of a closed position, a recoil spring for biasing the slide into the closed position;

the slide including a breech block, the breech block closing off a cartridge chamber in the closed position of the slide;

a firing pin being mounted longitudinally movable in the breech block and having a nose projecting toward the frame, a firing pin spring for biasing the firing pin, the tension of the firing pin spring being releasable in a direction toward the barrel;

the frame further including a trigger mechanism with a trigger means and an abutment, the abutment being movable by the trigger means from an initial position initially parallel to the firing pin so that the firing pin nose engages the abutment and is moved and the firing pin spring is tensioned, the abutment being further movable in release direction until the abutment and the nose are disengaged;

the slide defining control means which during firing cause the abutment to be moved from the released position into the path of movement of the nose;

the improvement comprising,

a stop spring having first and second ends, the first end acting on the frame and the second end acting on the trigger means, wherein the stop spring acts on the trigger means in a direction which is opposite the direction of action of the firing pin spring by the nose of the firing pin on the abutment, the tension of the firing pin spring being greater than the tension of the stop spring.

2. The pistol according to claim **1**,

wherein the frame further comprises positive guide means serving for maintaining the abutment in the path of motion of the nose of the firing pin at least in a portion of the initial movement of the abutment parallel to the firing pin.

3. The pistol according to claim **1**,

wherein the frame comprises abutment guide means for effecting the movement of the abutment in the release direction, and

wherein the control means define a path, a guide edge of the abutment guide means projecting into the path of the control means of the slide whereby, during the recoil of the slide, the control means dislodges the guide edge and thereby frees the abutment from the guide means for movement back into engagement with the firing pin nose.

4. The pistol according to claim **1**,

wherein the abutment is a projection on an abutment sleeve which is slidable in a longitudinal direction parallel to the firing pin and rotatable in a bore defined in the frame and is biased in a direction opposite the direction of biasing the firing pin spring by the stop spring;

an arm being attached to the abutment sleeve;

the trigger means being provided with a wedge surface engageable with the arm;

the trigger means being biased by a spring in the direction of the abutment;

the slide having longitudinal median plane; the control means including a forward guide ridge; the side further including a rearward guide ridge;

a guide edge arranged on the trigger means engageable with the rearward guide ridge lying outside of and adjacent the guide edge when the slide is in the closed position;

the distance between the outer surface of the forward guide ridge and the longitudinal median plane of the slide being greater than the distance between the radial end of the arm and the longitudinal median plane of the slide;

and a gap between the forward guide ridge and the rearward guide ridge, the gap being wider than the length of the guide edge.

5. The pistol according to claim **1**,

wherein the trigger means comprises a lever having a slot extending in longitudinal direction thereof, a pivot pin attached to the frame extending through the slot, the lever being biased by an additional spring toward the slide;

the lever forming the abutment, the stop spring acting on the lever;

the lever having a shoulder and the frame being provided with a counter shoulder, the counter shoulder supporting the shoulder at least in the initial position of the lever to prevent movement of the lever away from the slide;

the trigger means having an end with a wedge surface;

a leaf spring attached to the frame, the leaf spring being displaceable transversely of the trigger means and having a guide edge engageable by the control means of the slide;

a lever guide attached to the leaf spring and engageable by the wedge surface of the trigger means;

the slide having a longitudinal median plane, wherein the distance in a direction perpendicular to the median plane between the control means and the inner surface of the leaf spring is at least equal to the distance between the outer surface of the trigger means and the inwardly facing edge of the lever guide when the slide is in the closed position.

6. The pistol according to claim **1**,

the abutment being rigidly attached to the trigger means;

the trigger means having a recess, a projection attached to the frame projecting into the recess;

the recess having a narrow portion for effecting the movement of the abutment parallel to the firing pin and a wide portion for permitting the movement of the abutment in release direction;

a leaf spring attached to the frame, the leaf spring being displaceable transversely of the trigger means and having a guide edge;

the leaf spring further having a wedge surface, the trigger means having an inclined end surface engageable by the wedge surface of the leaf spring;

the stop spring being inclined relative to the longitudinal axis, so that the trigger means is biased by the stop spring toward the slide;

the guide edge engageable by the control means of the slide.

7. The pistol according to claim **1**,

the abutment being rigidly attached to the trigger means;

the frame having a recess, a projection attached to the trigger means projecting into the recess;

the recess having a narrow portion for effecting the movement of the abutment parallel to the firing in and a wide portion for permitting the movement of the abutment in release direction;

a leaf spring attached to the frame, the leaf spring being displaceable transversely of the trigger means and having a guide edge;

the leaf spring further having wedge surface, the trigger means having an inclined end surface engageable by the wedge surface of the leaf spring;

the stop spring being inclined relative to the longitudinal axis, so that the trigger means is biased by the stop spring toward the slide;

the guide edge engageable by the control means of the slide.

8. The pistol according to claim **1**, wherein

the slide has a bore extending parallel to the longitudinal axis,

a first stop means in the bore;

the bore receiving the firing pin, the firing pin having a second step means;

a spring rest mounted slidably on the firing pin;

the firing pin spring having first and second ends, the first end resting against the slide, and the second end resting against the spring rest;

the first stop means serving to limit the forward movement of the spring rest;

the second stop means serving to move the spring rest in the rearward direction.

the first stop means and the spring rest are spaced from each other when the abutment is in the initial position.

* * * * *

T. A. EDISON.
Electric-Lamp.

No. 223,898. Patented Jan. 27, 1880.

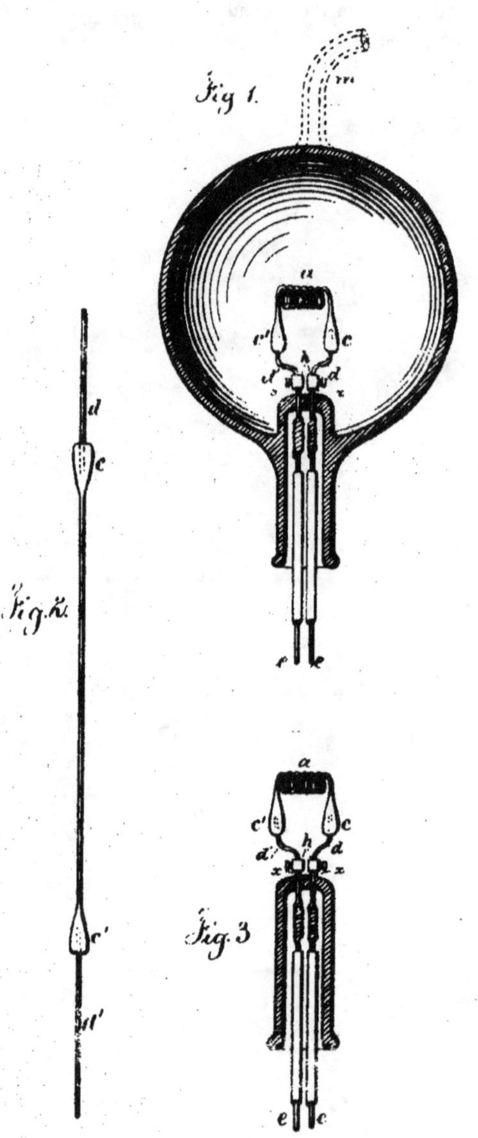

Witnesses
Chas H Smith
Geo T Pinckney

Inventor
Thomas A. Edison
per Lemuel W. Serrell
atty.

UNITED STATES PATENT OFFICE.

THOMAS A. EDISON, OF MENLO PARK, NEW JERSEY

ELECTRIC LAMP.

SPECIFICATION forming part of Letters Patent No. 223,898, dated January 27, 1880.

Application filed November 4, 1879.

To all whom it may concern:

Be it known that I, THOMAS ALVA EDISON, of Menlo Park, in the State of New Jersey, United States of America, have invented an Improvement in Electric Lamps, and in the method of manufacturing the same, (Case No. 186,) of which the following is a specification.

The object of this invention is to produce electric lamps giving light by incandescence, which lamps shall have high resistance, so as to allow of the practical subdivision of the electric light.

The invention consists in a light-giving body of carbon wire or sheets coiled or arranged in such a manner as to offer great resistance to the passage of the electric current, and at the same time present but a slight surface from which radiation can take place.

The invention further consists in placing such burner of great resistance in a nearly-perfect vacuum, to prevent oxidation and injury to the conductor by the atmosphere. The current is conducted into the vacuum-bulb through platina wires sealed into the glass.

The invention further consists in the method of manufacturing carbon conductors of high resistance, so as to be suitable for giving light by incandescence, and in the manner of securing perfect contact between the metallic conductors or leading-wires and the carbon conductor.

Heretofore light by incandescence has been obtained from rods of carbon of one to four ohms resistance, placed in closed vessels, in which the atmospheric air has been replaced by gases that do not combine chemically with the carbon. The vessel holding the burner has been composed of glass cemented to a metallic base. The connection between the leading wires and the carbon has been obtained by clamping the carbon to the metal. The leading-wires have always been large, so that their resistance shall be many times less than the burner, and, in general, the attempts of previous persons have been to reduce the resistance of the carbon rod. The disadvantages of following this practice are, that a lamp having but one to four ohms resistance cannot be worked in great numbers in multiple arc without the employment of main conductors of enormous dimensions; that, owing to the low resistance of the lamp, the leading-wires must be of large dimensions and good conductors, and a glass globe cannot be kept tight at the place where the wires pass in and are cemented; hence the carbon is consumed, because there must be almost a perfect vacuum to render the carbon stable, especially when such carbon is small in mass and high in electrical resistance.

The use of a gas in the receiver at the atmospheric pressure, although not attacking the carbon, serves to destroy it in time by "air-washing," or the attrition produced by the rapid passage of the air over the slightly-coherent highly-heated surface of the carbon. I have reversed this practice. I have discovered that even a cotton thread properly carbonized and placed in a sealed glass bulb exhausted to one-millionth of an atmosphere offers from one hundred to five hundred ohms resistance to the passage of the current, and that it is absolutely stable at very high temperatures; that if the thread be coiled as a spiral and carbonized, or if any fibrous vegetable substance which will leave a carbon residue after heating in a closed chamber be so coiled, as much as two thousand ohms resistance may be obtained without presenting a radiating-surface greater than three-sixteenths of an inch; that if such fibrous material be rubbed with a plastic composed of lamp-black and tar, its resistance may be made high or low, according to the amount of lamp-black placed upon it; that carbon filaments may be made by a combination of tar and lamp-black, the latter being previously ignited in a closed crucible for several hours and afterward moistened and kneaded until it assumes the consistency of thick putty. Small pieces of this material may be rolled out in the form of wire as small as seven one-thousandths of a inch in diameter and over a foot in length, and the same may be coated with a non-conducting non-carbonizing substance and wound on a bobbin, or as a spiral, and the tar carbonized in a closed chamber by subjecting it to high heat, the spiral after carbonization retaining its form.

All these forms are fragile and cannot be clamped to the leading wires with sufficient force to insure good contact and prevent heating. I have discovered that if platinum wires are used and the plastic lamp-black and tar material be molded around it in the act of carbonization there is an intimate union by com-

material be molded around it in the act of carbonization there is an intimate union by combination and by pressure between the carbon and platina, and nearly perfect contact is obtained without the necessity of clamps; hence the burner and the leading-wires are connected to the carbon ready to be placed in the vacuum-bulb.

When fibrous material is used the plastic lamp-black and tar are used to secure it to the platina before carbonizing.

By using the carbon wire of such high resistance I am enabled to use fine platinum wires for leading-wires, as they will have a small resistance compared to the burner, and hence will not heat and crack the sealed vacuum-bulb. Platina can only be used, as its expansion is nearly the same as that of glass.

By using a considerable length of carbon wire and coiling it the exterior, which is only a small portion of its entire surface, will form the principal radiating-surface; hence I am able to raise the specific heat of the whole of the carbon, and thus prevent the rapid reception and disappearance of the light, which on a plain wire is prejudicial, as it shows the least unsteadiness of the current by the flickering of the light; but if the current is steady the defect does not show.

I have carbonized and used cotton and linen thread, wood splints, papers coiled in various ways, also lamp-black, plumbago, and carbon in various forms, mixed with tar and kneaded so that the same may be rolled out into wires of various lengths and diameters. Each wire, however, is to be uniform in size throughout.

If the carbon thread is liable to be distorted during carbonization it is to be coiled between a helix of copper wire. The ends of the carbon or filament are secured to the platina leading-wires by plastic carbonizable material, and the whole placed in the carbonizing-chamber. The copper, which has served to prevent distortion of the carbon thread, is afterward eaten away by nitric acid, and the spiral soaked in water, and then dried and placed on the glass holder, and a glass bulb blown over the whole, with a leading-tube for exhaustion by a mercury-pump. This tube, when a high vacuum has been reached, is hermetically sealed.

With substances which are not greatly distorted in carbonizing, they may be coated with a non-conducting non-carbonizable substance, which allows one coil or turn of the carbon to rest upon and be supported by the other.

In the drawings, Figure 1 shows the lamp sectionally. *a* is the carbon spiral or thread. *c c'* are the thickened ends of the spiral, formed of the plastic compound of lamp-black and tar. *d d'* are the platina wires. *h h* are the clamps, which serve to connect the platina wires, cemented in the carbon, with the leading-wires *x x*, sealed in the glass vacuum-bulb. *e e* are copper wires, connected just outside the bulb to the wires *x x*. *m* is the tube (shown by dotted lines) leading to the vacuum-pump, which, after exhaustion, is hermetically sealed and the surplus removed.

Fig. 2 represents the plastic material before being wound into a spiral.

Fig. 3 shows the spiral after carbonization, ready to have a bulb blown over it.

I claim as my invention—

1. An electric lamp for giving light by incandescence, consisting of a filament of carbon of high resistance, made as described, and secured to metallic wires, as set forth.

2. The combination of carbon filaments with a receiver made entirely of glass and conductors passing through the glass, and from which receiver the air is exhausted, for the purposes set forth.

3. A carbon filament or strip coiled and connected to electric conductors so that only a portion of the surface of such carbon conductors shall be exposed for radiating light, as set forth.

4. The method herein described of securing the platina contact-wires to the carbon filament and carbonizing of the whole in a closed chamber, substantially as set forth.

Signed by me this 1st day of November, A. D. 1879.

THOMAS A. EDISON.

Witnesses:

S. L. GRIFFIN,

JOHN F. RANDOLPH.

Order of Cancellation of Certificate of Correction of Letters Patent No. 223,898.

It is found that the following certificate has been attached to Letters Patent granted to Thomas A. Edison for improvement in "Electric Lamps," No. 223,898, dated January 27, 1880:

DEPARTMENT OF THE INTERIOR,

UNITED STATES PATENT OFFICE,

WASHINGTON, D. C., *December 18, 1883.*

In compliance with the request of the party in interest Letters Patent No. 223,898, granted January 27, 1880, to Thomas A. Edison, of Menlo Park, New Jersey, for an improvement in "Electric Lamps," is hereby limited so as to expire at the same time with the patent of the following-named, having the shortest time to run, viz.: British patent, dated November 10, 1879, No. 4,576; Canadian patent, dated November 17, 1879, No. 10,654; Belgian patent, dated November 29, 1879, No. 49,884; Italian patent, dated December 6, 1879, and French patent, dated January 20, 1880, No. 133,756.

It is hereby certified that the proper entries and corrections have been made in the files and records of the Patent Office.

This amendment is made that the United States patent may conform to the provisions of section 4887 of the Revised Statutes.

[SEAL.]

BENJ. BUTTERWORTH,
Commissioner of Patents.

Approved:
M. L. JOSLYN,
Acting Secretary of the Interior.

Now, in compliance with the request of the parties in interest, said certificate is hereby *canceled* and proper entries and corrections have been made in the files and records of the Patent Office.

In testimony whereof I have hereunto set my hand and caused the seal of the Patent Office to be affixed, this 15th day of March, 1893.

W. E. SIMONDS,
Commissioner of Patents.

Approved:
CYRUS BUSSEY,
Assistant Secretary of the Interior.

Correction in Letters Patent No. 223,898.

DEPARTMENT OF THE INTERIOR,

UNITED STATES PATENT OFFICE,

WASHINGTON, D. C., *December 18, 1883.*

In compliance with the request of the party in interest, Letters Patent No. 223,898, granted January 27, 1880, to Thomas A. Edison, of Menlo Park, New Jersey, for an improvement in "Electric-Lamps," is hereby limited so as to expire at the same time with the patent of the following named, having the shortest time to run, viz: British Patent dated November 10, 1879, No. 4,576; Canadian Patent dated November 17, 1879, No. 10,654; Belgian Patent dated November 29, 1879, No. 49,884; Italian Patent dated December 6, 1879; and French Patent dated January 20, 1880, No. 133,756;

It is hereby certified that the proper entries and corrections have been made in the files and records of the Patent Office.

This amendment is made that the United States Patent may conform to the provisions of Section 4887 of the Revised Statutes.

BENJ. BUTTERWORTH,
Commissioner of Patents.

Approved:
M. L. JOSLYN,
Acting Secretary of the Interior.

(No Model.)

E. BERLINER.

GRAMOPHONE.

No. 372,786. Patented Nov. 8, 1887.

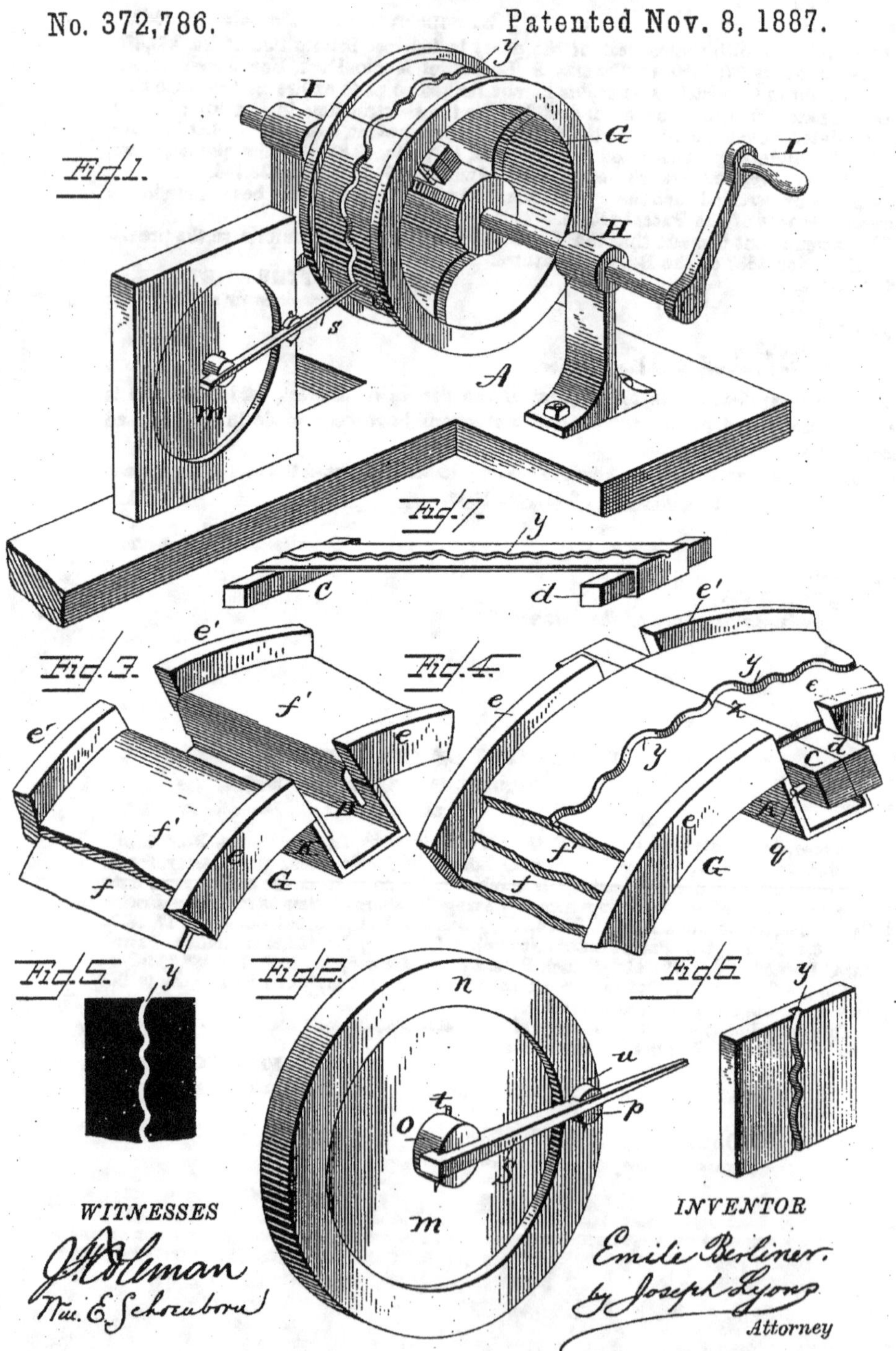

UNITED STATES PATENT OFFICE.

EMILE BERLINER, OF WASHINGTON, DISTRICT OF COLUMBIA.

GRAMOPHONE.

SPECIFICATION forming part of Letters Patent No. 372,786, dated November 8, 1887.

Original application filed May 4, 1887, Serial No. 237,060. Divided and this application filed September 26, 1887. Serial No. 250,721. (No model.)

To all whom it may concern:

Be it known that I, EMILE BERLINER, a citizen of the United States, residing at Washington, in the District of Columbia, have invented certain new and useful Improvements in Gramophones, of which the following is a specification.

This invention has reference to a novel method of and apparatus for recording and reproducing all kinds of sounds, including spoken words, and is designed to overcome the defects inherent in that art as now practiced and in the apparatus used therefor.

By the ordinary method of recording spoken words or other sounds for reproduction it is attempted to cause a stylus attached to a vibratory diaphragm to indent a traveling sheet of tin-foil or other like substance to a depth varying in accordance with the amplitudes of the sound-waves to be recorded. This attempt is necessarily more or less ineffective, for the reason that the force of a diaphragm vibrating under the impact of sound-waves is very weak, and that in the act of overcoming the resistance of the tin-foil or other material the vibrations of the diaphragm are not only weakened, but are also modified. Thus while the record contains as many undulations as the sounds which produce it, and in the same order of succession, the character of the recorded undulations is more or less different from those of the sounds uttered against the diaphragm. There is, then, a true record of the pitch, but a distorted record of the quality of the sounds obtained. The simple statement that the material upon which the record is made resists the movement of the diaphragm is not sufficient to explain the distortion of the character of the undulations, for if that resistance were uniform, or even proportional to the displacement of the stylus, the record would be simply weakened, but not distorted; but it is a fact that the resistance of any material to indentation increases faster than the depth of indentation, so that a vibration of greater amplitude of the stylus meets with a disproportionately greater resistance than a vibration of smaller amplitude. For this reason loud sounds are even less accurately recorded than faint sounds, and the individual voice of a loud speaker recorded and then reproduced by the phonograph cannot be recognized. With a view of overcoming this defect it has been attempted to engrave instead of indent a record of the vibrations of the diaphragm by employing a stylus shaped and operating like a chisel upon a suitably-prepared surface; but even in this case the disturbing causes above referred to are still present. In addition to this, if in the apparatus of the phonograph or graphophone type it is attempted to avoid the disturbing influence of the increase of resistance of the record-surface with the depth of indentation or cut as much as possible by primarily adjusting the stylus so as to touch the record-surface only lightly, then another disturbing influence is brought into existence by the fact that with such adjustment, when the diaphragm moves outwardly, the stylus will leave the record-surface entirely, so that part of each vibration will not be recorded at all. This is more particularly the case when loud sounds are recorded, and it manifests itself in the reproduction, which then yields quite unintelligible sounds.

It is the object of my invention to overcome these difficulties by recording spoken words or other sounds without perceptible friction between the recording-surface and the recording-stylus, and by maintaining the unavoidable friction uniform for all vibrations of the diaphragm. The record thus obtained, almost frictionless, I copy in a solid resisting material by any of the methods hereinafter described, and I employ such copy of the original record for the reproduction of the recorded sounds.

Instead of moving the recording-stylus at right angles to and against the record-surface, I cause the same to move under the influence of sound-waves parallel with and barely in contact with such surface, which latter is covered with a layer of any material that offers a minimum resistance to the action of a stylus operating to displace the same, all substantially in the manner of the well-known phonautograph by Leon Scott. All this will more fully appear from the following detailed description, in which reference is made to the accompanying drawings, which illustrate one of the numerous forms which my improved apparatus may assume, and in which—

Figure 1 is a perspective view of my recording and reproducing apparatus; Fig. 2, a like view of the recording and reproducing diaphragm with its stylus; Fig. 3, a similar view of a portion of the support for the record-surface; Fig. 4, the same view with the record-surface applied; Fig. 5, a plan view of a phonautographic record; Fig. 6, a perspective of a phonautographic record copied in solid resisting material; and Fig. 7, the copied record mounted, ready for application to the support.

The general arrangement of the parts is best illustrated in Fig. 1, in which a T-shaped base-plate, A, is shown, upon which two standards, H I, serving as journal-bearings for the shaft of drum G, are mounted. The drum G may be constructed with flanges *e e'*, which project beyond the cylinder-surface *f*, and from the edges of a gap, B, left upon the cylinder-surface extend the side walls of box K, as shown. A thin layer of felt or other yielding elastic substance is placed upon the cylinder-surface and is bent over the edges of the gap and secured to the side walls of the box K. This layer of elastic material is designed to serve as the support for the record-surface both in recording and reproducing.

For recording I employ a thin strip of paper, parchment, metal, or any other suitable substance, which is secured at both ends to bars *c d*, in the manner shown in Fig. 7, with reference to a copy of a record, and is then placed upon the elastic support *f'*, with the bars *c d* entering into but projecting at both ends beyond the box K, as illustrated in Figs. 1 and 4, with reference to an engraved copy of a record. Bolts *q*, passing through the projecting ends of bars *c d*, are employed to draw the record-strip tightly about the drum, and the length of the strip is such that the ends of the same meet as nearly as practicable upon a straight line, *z*. The record-sheet is then prepared to receive the record by covering its surface with a thin layer of any substance which is easily removed by the action of the recording-stylus. I may use lamp-black, which is deposited by placing a smoky flame under the record-strip and by slowly turning the drum until all parts of the strip are covered with the deposit. It is well known that a layer of lamp-black thus deposited, while it adheres well to the surface of a solid body, is nevertheless easily removed from the same. It requires only an exceedingly small force to draw a plainly-visible line upon such surface, owing to the fact that the spicules of carbon of which lamp-black is composed are only loosely superimposed upon each other, and are exceedingly light. All this has long since been recognized and utilized in the production of phonautographic records, and I take advantage of these facts in my improved method of recording and reproducing sounds.

The diaphragm *m* is mounted in a frame, *n*, with its plane at right angles to the axis of drum G. A post, O, is fixed to the center of the diaphragm, and a slot in said post receives one end of stylus S, which is pivoted in the post by a pin, *t*. The stylus extends over and beyond the frame, with its free end barely in contact with the record-surface, and is also pivotally supported in a slot in a post, *p*, secured to the frame by means of a pin, *u*, as shown in Figs. 1 and 2. It will now be seen that the stylus is in effect a lever having its fulcrum in the pin *u*, and that its free end can only move in lines practically parallel to the record-surface. If it is now desired to produce a record of sounds the drum is slowly and uniformly rotated by means of crank L, or by any other suitable means, and sounds are uttered or directed against the diaphragm. Under the impact of the sound-waves the diaphragm is set into vibrations, whereby the free end of the stylus is also caused to vibrate to the right and left of its normal position, removing at the same time an undulating line, *y*, of lamp-black from the record-surface, as indicated, greatly exaggerated, in Fig. 5. Since in this operation the stylus only penetrates a uniform layer of loosely-heaped carbon spicules and barely touches the record-surface, it is clear that the slight friction at the free end of the stylus will be uniform, whatever be the amplitude of vibration. Consequently the vibrations of the diaphragm will not be modified or changed by the reaction upon the same of a sensible and varying resistance, as is the case in all other mechanical sound-recorders.

Having thus obtained an accurate phonautographic record, the same may be fixed by applying a thin solution of varnish of any kind which dries very rapidly and which does not obliterate or change the record.

If in this process the deposit of lamp-black be made thick enough, the line drawn by the stylus would represent a groove of even depth, preserving all the characteristics of the sounds which produced it and which may be handled and touched with impunity. The latter is then removed from the drum and may be preserved any length of time without danger of its being disfigured. This record I then copy in solid resisting material, preferably metal, either by the purely mechanical process of engraving, or by chemical deposition, or by photo-engraving. I prefer the last-named process, which enables me to produce the most accurate copy of the original record in copper, nickel, or any other metal without in any way or manner affecting the original record. The copy thus obtained, which may be multiplied to any desired extent, is a grooved wave-line upon a strip or sheet of copper or other metal, as shown in Figs. 1, 4, 6, and 7, and for the reproduction of the recorded sounds it has the advantage over the ordinary records in tin-foil, wax, &c., that it is not sensibly attacked by the reproducing-stylus, and will stand an indefinite number of reproductions without the slightest variation in the accuracy and loudness of the reproduced sounds.

The copied record is fixed at both ends to the bars *c d*, as shown in Fig. 7, and is placed upon the elastic support f' upon the drum in the same manner as has been described with reference to the original record-strip, and as is illustrated in Figs. 1 and 4. Care must be taken that the two ends of the undulatory groove y meet exactly, as will be readily understood. This condition of the apparatus is shown in Fig. 1 with the engraven record upon the drum and the free end of the stylus entering the undulatory groove. If, now, the drum is rotated with uniform speed, the end of the stylus will be forced to follow the undulations of the groove y, and the diaphragm will be vibrated positively in both directions in strict accordance therewith, and will therefore reproduce the exact sounds which originally produced the record. This peculiarity of positive vibratory movement in both directions of the diaphragm is a feature which also distinguishes my method and my apparatus from others heretofore used.

In the phonograph and graphophone the end of the reproducing-stylus which bears upon the indented or engraved record has a vertical upward and downward movement. It is forced upwardly in a positive manner by riding over the elevated portion of the record, but its downward movement is effected solely by the elastic force of the diaphragm, which latter is always under tension. In my improved apparatus the stylus travels in a groove of even depth and is moved positively in both directions. It does not depend upon the elasticity of the diaphragm for its movement in one direction. This I consider to be an advantage, since by this method the whole movement of the diaphragm is positively controlled by the record, and is not affected or modified by the physical conditions of the diaphragm, which conditions necessarily vary from time to time and constitute some of the causes of imperfect reproduction of recorded sounds.

In practicing my method of recording and producing sounds I am not limited to the use of the identical apparatus herein shown and described. This apparatus may be varied indefinitely without seriously impairing its utility for the purposes in view. Thus it is not absolutely necessary that a diaphragm should be used for receiving the impact of sound-waves in recording and for remitting sounds in reproducing. Any sonorous body of whatever shape and material may be used in lieu of a diaphragm proper. The recording-surface need not be mounted upon a drum, but may be supported in any suitable manner upon a support of any description which is adapted to move the same under the stylus evenly and with approximately uniform speed. Nor do I confine myself to the use of lamp-black as a substratum for the phonautographic record, although I have found this substance to yield excellent results. Any other substance which adheres well to the support and may at the same time be removed from the same with a minimum force may be employed.

While I have found the process of photo-engraving to yield admirable copies of the phonautographic record, I do not mean to confine myself to this process to the exclusion of other processes for copying and multiplying the original record in solid resisting material; and it will be readily understood that the details of construction of my apparatus and the manipulations of the same may be greatly changed without departing from the fundamental idea of my invention.

I do not herein claim the apparatus shown and described, either generically or specifically, as a whole or in part, since the same forms the subject of another application for patent previously filed by me and of which this is a division.

What I do claim, and desire to secure by Letters Patent, is—

1. The method or process of recording and reproducing spoken words and other sounds, which consists in first drawing an undulatory line of even depth in a traveling layer of non-resisting material by and in accordance with sound-vibrations, then producing the record thus obtained in solid resisting material, and finally imparting vibrations to a sonorous body by and in accordance with the resisting record, substantially as described.

2. The method or process of reproducing sounds recorded phonautographically, which consists in copying the phonautographic record in solid resisting material, and then imparting vibrations to a sonorous body by and in accordance with the copy of the orignal record, substantially as described.

3. The method or process of reproducing sounds recorded phonautographically, which consists in copying the phonautographic record in solid resisting material by the process of photo-engraving, and then imparting positive to-and-fro movements to a sonorous body by and in accordance with the copy of the original record, substantially as described.

In testimony whereof I have signed my name to this specification in the presence of two subscribing witnesses.

EMILE BERLINER.

Witnesses:

JULIUS SOLGER,

JACOB G. COHEN.

Feb. 11, 1941. H. FORD 2,231,710

TRACTOR

Filed Feb. 23, 1939 2 Sheets-Sheet 1

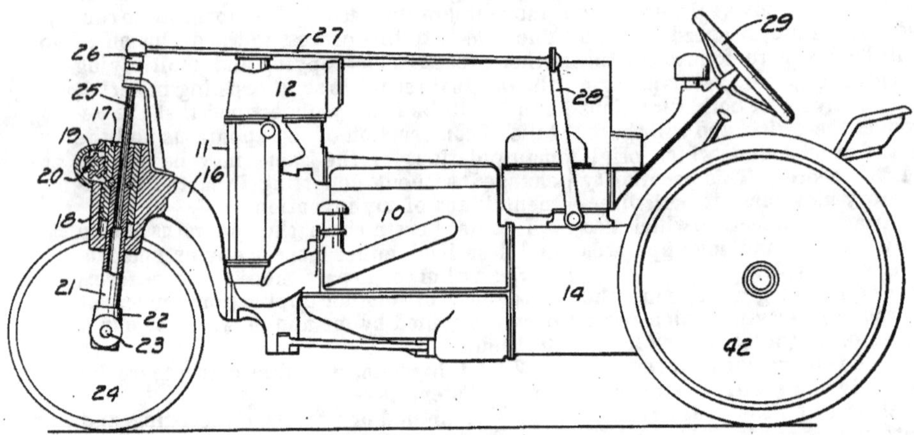

FIG. 1

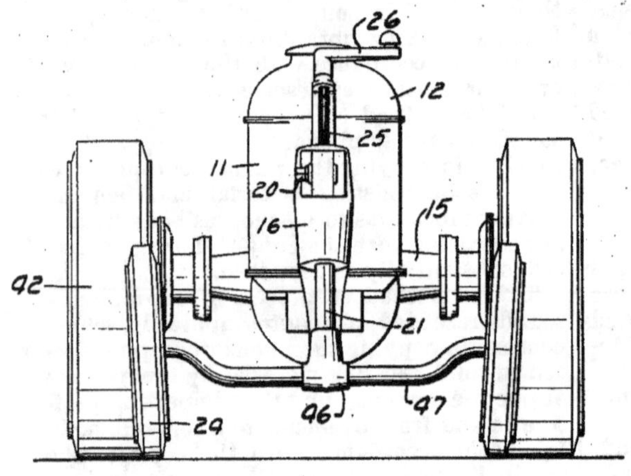

FIG. 5

WITNESS

E. Witzke

INVENTOR
Henry Ford

BY
E. L. Davis
E. C. McRae
ATTORNEY

Feb. 11, 1941. H. FORD 2,231,710

TRACTOR

Filed Feb. 23, 1939 2 Sheets-Sheet 2

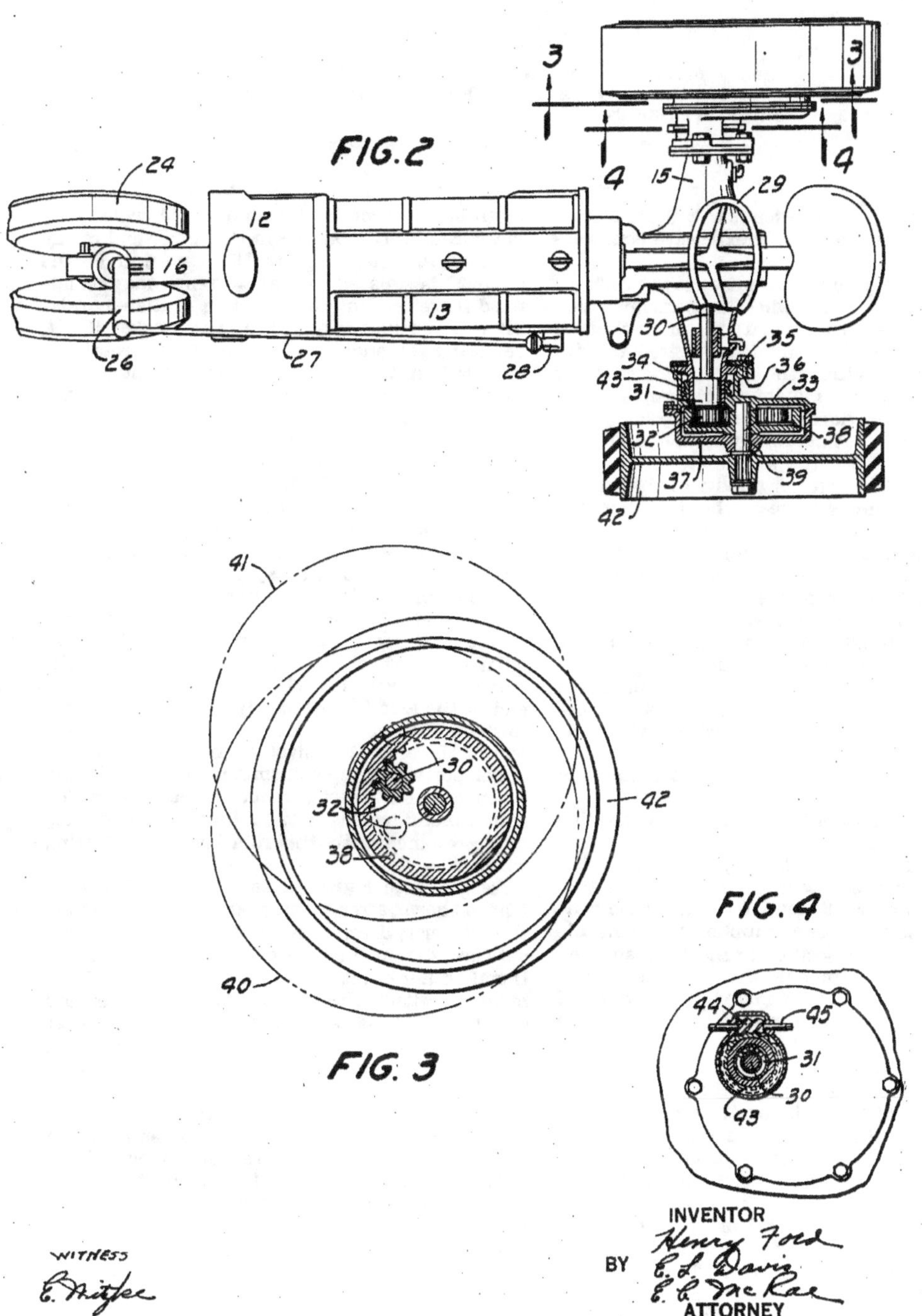

WITNESS

E. McFee

INVENTOR

Henry Ford

BY E. L. Davis

E. C. McRae

ATTORNEY

Patented Feb. 11, 1941 **2,231,710**

UNITED STATES PATENT OFFICE

2,231,710

TRACTOR

Henry Ford, Dearborn, Mich., assignor to Ford Motor Company, Dearborn, Mich., a corporation of Delaware

Application February 23, 1939, Serial No. 257,773

1 Claim. (Cl. 180—41)

The object of my invention is to provide a tractor of simple, durable and inexpensive construction.

A further object of my invention is to provide a tractor having a novel wheel suspension which differs from the conventional tractor in that all of the wheels may be raised or lowered relative to the tractor. When the wheels are in their lowermost positions, maximum clearance above the ground is provided so that the tractor may be used for culitvating relatively tall corn and similar crops. When the wheels are in their upper positions, the center of gravity of the tractor is lowered to thereby increase the draw-bar pull of the tractor.

With these and other objects in view, my invention consists in the arrangement, construction and combination of the various parts of my improved device, as described in the specification, claimed in my claim, and illustrated in the accompanying drawings, in which:

Figure 1 is a side elevation of my improved tractor, a portion of the steering mechanism being broken away to better illustrate the conconstruction.

Figure 2 is a plan view of the tractor shown in Figure 1.

Figure 3 is a sectional view, taken upon the line 3—3 of Figure 2.

Figure 4 is a sectional view, taken upon the line 4—4 of Figure 2, and

Figure 5 is a front view of an alternate construction, this unit being supplied with an axle which is detachably secured to the front steering spindle in place of the axle shown in Figure 1.

Referring to the accompanying drawings, I have used the reference numeral **10** to indicate a conventional tractor internal-combustion engine having a radiator **11** secured to the forward end thereof. A water reservoir **12** is fixed to the upper portion of the radiator **11** and a fuel tank **13** extends from the rearward portion of the reservoir **12** rearwardly in the conventional manner. An axle and transmission housing **14** is fixed to the rear end of the engine **10**, and axle tubes **15** extend outwardly from the respective sides of the rear portion of the housing **14**.

A goose neck bracket **16** is fixed to the front end of the engine **10** and is adapted to support the tractor upon the front wheels. The upper end of the bracket **16** is provided with a vertical bore therethrough in which a quill **17** is reciprocally mounted. Gear teeth **18** are machined in the form of a rack upon one side of the quill **17**, these teeth being in mesh with a pinion **19**, which pinion is rotatably mounted in the upper end of the bracket **16** upon a shaft **20**. One end of the shaft **20** is squared so that a wrench may be applied thereto to rotate the pinion **19** thereby raising or lowering the quill **17** in the bracket **16**. A vertically extending king-pin **21** is rotatably mounted in the quill **17** but is prevented from axial movement relative thereto. Consequently, when the quill **17** is raised or lowered the pin **21** is correspondingly raised and lowered.

A head **22** is detachably secured to the lower end of the pin **21**, which head is provided with a pair of wheel spindles projecting from the respective sides thereof, upon which spindles wheels **24** are rotatably mounted. The upper end of the pin **21** is provided with splined bore therein in which a splined shaft **25** is fixed. The upper end of the shaft **25** has a steering arm **26** secured thereto. Consequently, when the arm **26** is oscillated the wheels **24** will be turned to steer the tractor. A drag link **27** extends from the outer end of the arm **26** rearwardly where it is secured to the upper end of a steering arm **28**. The lower end of the arm **28** is pivotally mounted upon the upper portion of the transmission housing **14** and a conventional steering reducion gear is provided which oscillates the arm **28** forwardly and back in accordance with the rotation of a steering wheel **29**.

Referring to Figures 2 and 3 of the drawings, I have shown a reduction gearing through which the tractor is driven. Rear axle shafts **30** extend outwardly from the respective sides of a conventional differential, not shown, but which is mounted within the housing **14**. The outer end of each shaft is supported upon a bearing **31** which is in turn supported by the outer end of the adjacent axle tube **15**. Pinions **32** are fixed to the outer ends of the shafts **30**.

It will be noted from Figure 2 that a housing **33** is rotatably mounted upon the outer end of each tube **15**, which housings are held in place by a flange **34** formed on each tube against which a plate **35** is clamped by means of cap screws **36**. Each housing **33** is formed of two parts with an outer drum **37** forming an enclosure for an internal gear **38**, which gear is fixed to a stub-axle shaft **39**. Each shaft **39** is rotatably mounted in suitable bearings in the drum **37** and in the housing **33** in such position that the internal gear **38** is in mesh with the pinion **32**.

It will be noted that the axis of each shaft **39** is parallel to but is displaced rearwardly from the axis of the pinion 32. Consequently, each housing 33 may be lowered to the position shown by lines 40 in Figure 3 or it may be swung upwardly to the position shown by lines 41. The housing may, of course, be retained in any intermediate position as shown by the full lines in Figure 3. A driving wheel 42 is fixedly secured to the outer end of each shaft 39 and swings up and down as the housing 33 is oscillated around the axle tube.

Inasmuch as the housings 33 are supported upon their respective axle tubes independently of each other, it will be readily seen that they both may oscillate simultaneously or independently, as desired. When the housings are moved to their lowermost position, as shown by lines 40, then the rear portion of the tractor is elevated a considerable distance above the position that it assumes when the housings are rotated to the positon shown by lines 41. The purpose of raising the rear end of the tractor is to obtain clearance under the axle tubes when it is desired to cultivate crops of a considerable height as the crops must pass beneath the axle tubes. When ploughing or doing other work requiring the maximum draw-bar pull by the tractor such pull may be increased by lowering the center of gravity. The front end of the tractor may be conveniently raised or lowered by means of the pinion 19 in accordance with the position of the rear end so that the tractor will remain on an even level over all ranges of adjustment.

Another important feature of this construction is that when ploughing is being done it is necessary that one of the drive wheels remain at the bottom of the furrow while the other rolls upon the unploughed surface. Consequently, all other tractors have a list to one side when being used for this work. With my improved tractor, the wheel which is riding in the furrow may be lowered thereby permitting the tractor to remain level. This also produces greater tractor effort on the wheels because the wheel treads remain flat on the ground.

In order to conveniently raise and lower each housing 33, I have provided a worm wheel 43 which is fixed on the outer end of each axle tube 15. A worm gear 44 is rotatably mounted in each housing 33 upon a shaft 45 in mesh with the worm 43 so that when the shaft 45 is rotated the housings are swung up and down at a reduced rate by the worms 44.

In the cultivation of certain crops it is necessary that the steering wheels 24 be spread as they interfere with the center row of crops. In this case I have provided an auxiliary head 46 which has a solid axle shaft 47 extending crosswise through the bottom thereof. The head 22 may be removed from the pin 21 and replaced by the head 46. The wheels 24 are then mounted upon suitable wheel spindles at the outer ends of the axle shaft 47 to thereby provide a standard tread for the front end of the tractor. A conventional front axle may also be used if desired.

Among the many advantages arising from the use of my improved construction, it may be well to mention that the principal advantage results in that the center of gravity of the tractor may be lowered close to the ground when heavy work is being done but if cultivating or work which requires considerable clearance over the axle is to be done then the tractor may be raised an appreciable distance to accomplish this purpose.

Some changes may be made in the arrangement, construction and combination of the various parts of my improved device without departing from the spirit of my invention and it is my intention to cover by my claim such changes as may reasonably be included within the scope thereof.

I claim as my invention:

A tractor comprising, an engine, an axle housing secured to one end of said engine, an axle tube extending transversely from the end of said axle housing opposite said engine, an axle shaft extending outwardly through said axle tube, a pinion disposed upon the outer end of said shaft, a gear housing having a cylindrical sleeve projecting from one side thereof, the axial center of said sleeve being spaced radially from the axial center of said gear housing, said sleeve being rotatably mounted upon the outer end of said axle tube, a flange formed on said axle tube which coacts with and supports the inner end of said sleeve, means disposed upon the inner end of said sleeve which operatively engages said flange to lock said sleeve in any one of a plurality of rotatable positions around said axle tube, a worm wheel fixed upon said axle tube within said sleeve, a worm shaft rotatably mounted in said sleeve transverse to its axial center, said worm shaft having a worm fixed thereon which meshes with said worm wheel so that rotation of said worm shaft rotates said sleeve and worm around said axle tube, a wheel shaft rotatably mounted within said housing upon its axial center, a driving wheel fixed to the outer end of said wheel shaft, and an internal gear fixed to said shaft within said housing, said internal gear being in mesh with said pinion.

HENRY FORD.

April 13, 1943. G. B. WEAVER 2,316,477

SHAFT DRIVE FOR MOTORCYCLES

Filed Aug. 18, 1941 3 Sheets-Sheet 1

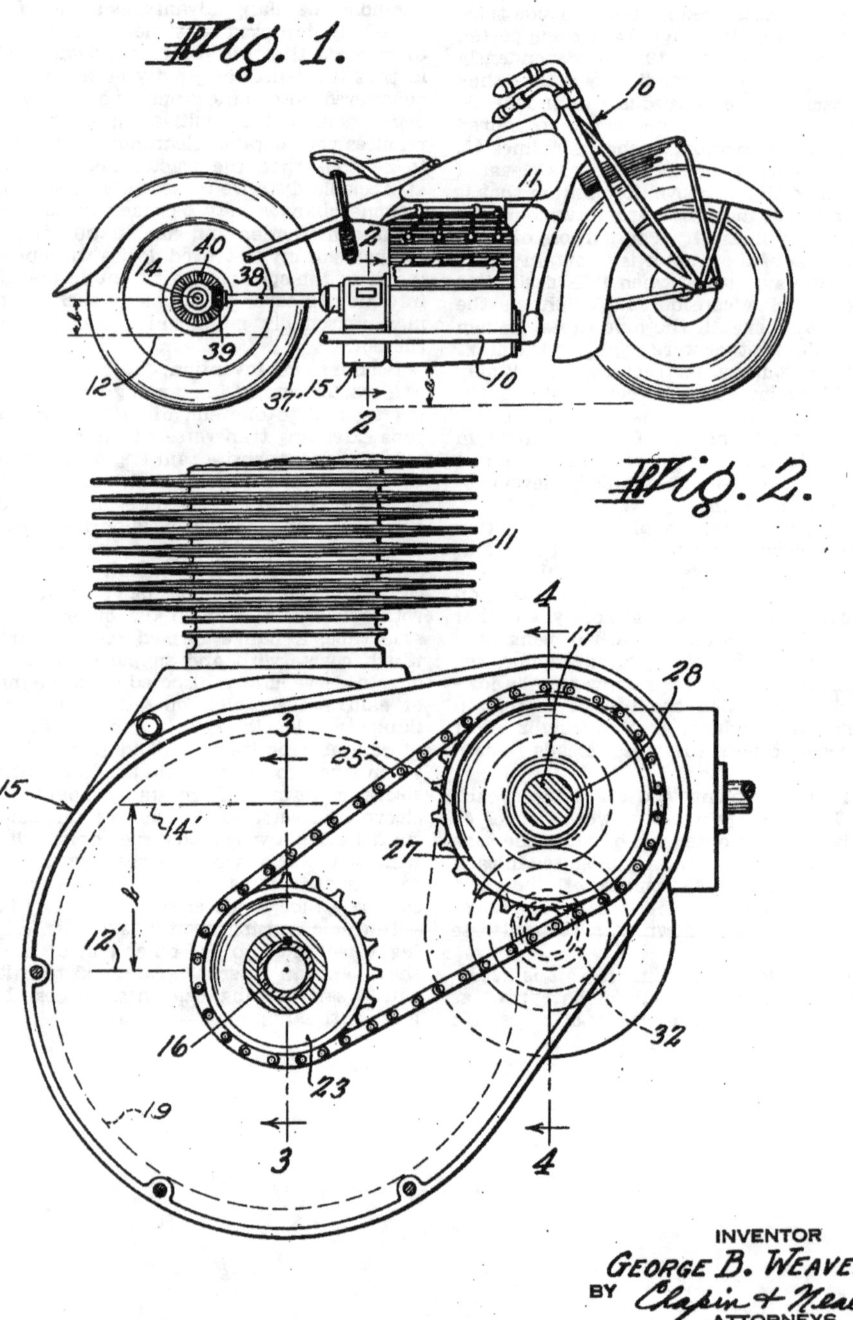

INVENTOR
GEORGE B. WEAVER
BY Chapin & Neal
ATTORNEYS

April 13, 1943. G. B. WEAVER 2,316,477

SHAFT DRIVE FOR MOTORCYCLES

Filed Aug. 18, 1941 3 Sheets-Sheet 2

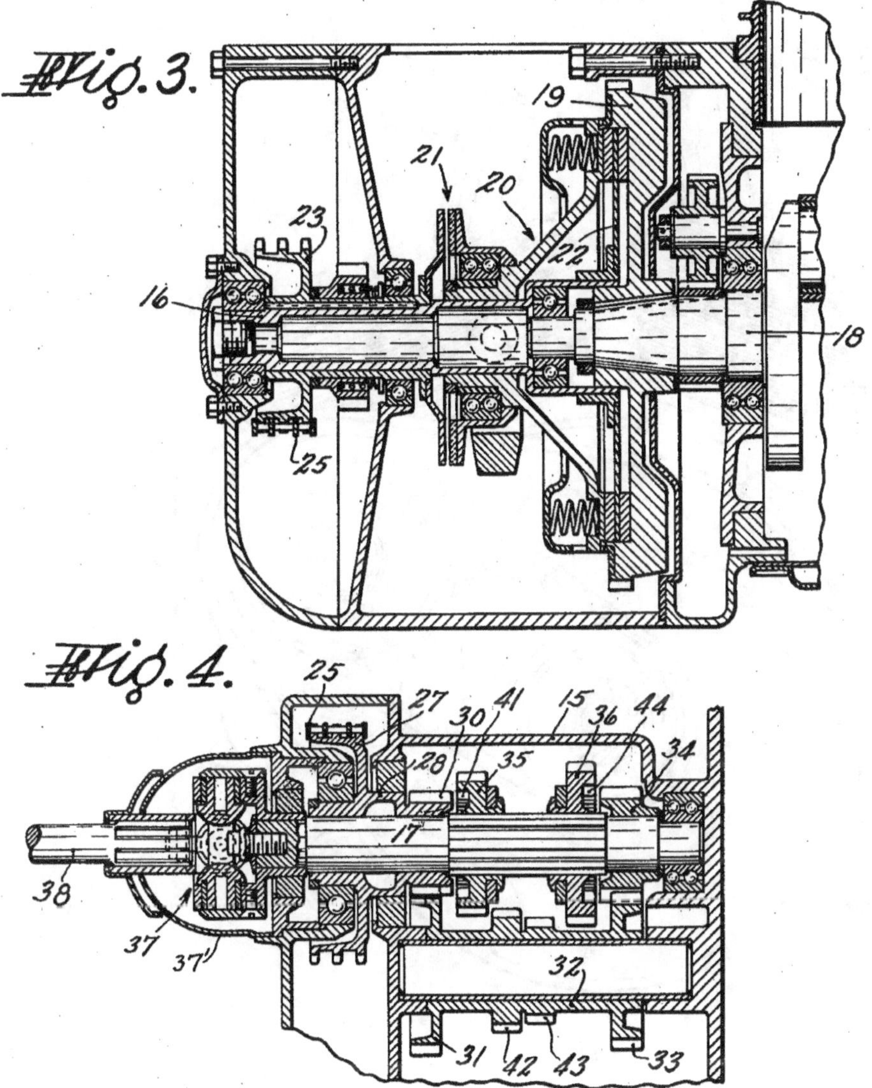

INVENTOR
GEORGE B. WEAVER
BY Chapin & Neal
ATTORNEYS

April 13, 1943. G. B. WEAVER 2,316,477

SHAFT DRIVE FOR MOTORCYCLES

Filed Aug. 18, 1941 3 Sheets-Sheet 3

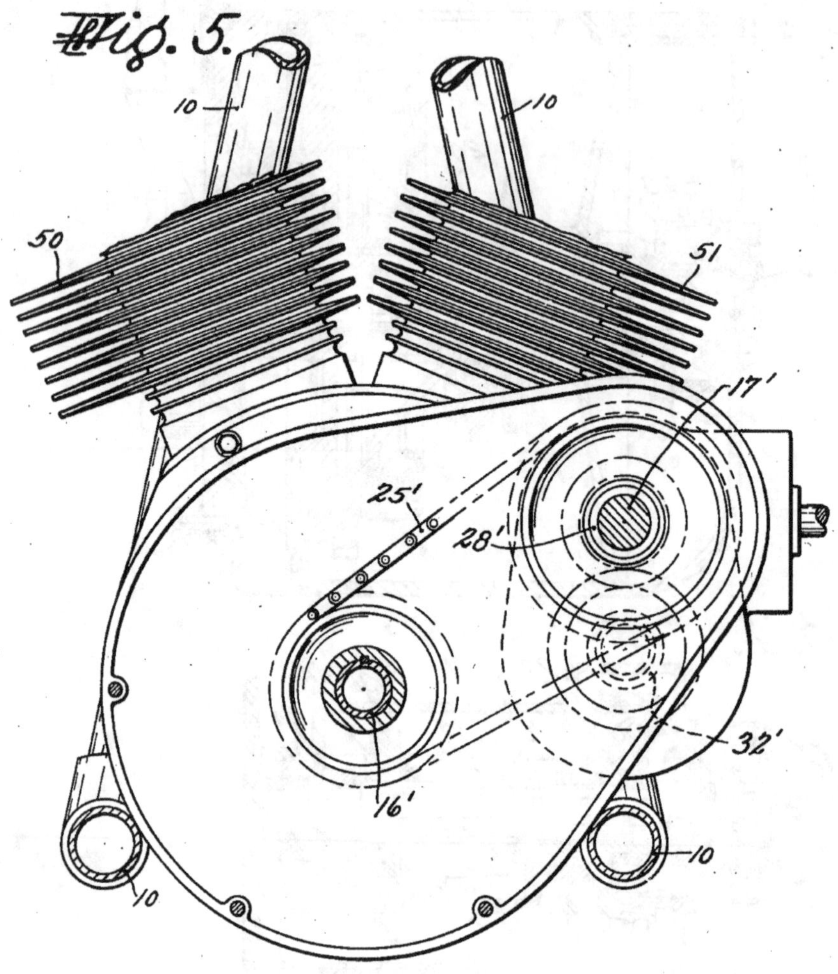

INVENTOR
GEORGE B. WEAVER
BY Chapin & Neal
ATTORNEYS

Patented Apr. 13, 1943 2,316,477

UNITED STATES PATENT OFFICE

2,316,477

SHAFT DRIVE FOR MOTORCYCLES

George B. Weaver, Springfield, Mass., assignor to Indian Motocycle Company, Springfield, Mass., a corporation of Massachusetts

Application August 18, 1941, Serial No. 407,255

11 Claims. (Cl. 180—33)

This invention relates to an improvement in motorcycles, and more particularly to a novel arrangement of the motor and the transmission and driving elements of such vehicles with respect to each other and the frame of the vehicles.

Among the objects of the invention are the more efficient use of shaft drives as distinguished from chain drives, shortened wheel bases, or the more efficient use of the space afforded by a given wheel base, and improved motor cooling.

In order to lower the center of gravity and secure desired stability, the motor is placed as low in the frame of the motor cycle as possible and provide the necessary road clearance. This practice brings the crank shaft of the motor below the level of the axle of the rear wheel of the motor cycle, to which power must be transmitted from the motor. It has proved impractical under these conditions to employ a drive or propeller shaft as the means of delivering power to the rear wheel from the motor since the inclination of the shaft (from the level of the motor crank shaft up to the level of the rear axle) imposes a destructive degree of angular movement in the universal joint between the propeller shaft and the crank shaft or "transmission" of the motor. As a result recourse has been had to chain drives.

Where shaft drives are used it has been necessary to raise the motor, as shown for example in the patent to Notman 1,420,638, issued June 27, 1922, so that the motor crank shaft is substantially on a level with the rear axle, thus sacrificing the advantages of a low center of gravity. This Notman patent also illustrates the increased wheel base which has heretofore been found necessary to accommodate such an arrangement, that is, the length of the clutch mechanism and the transmission are successively added to the length of the four cylinder motor.

A shaft drive has many advantages, well known to the art, as compared with chain drives, and my invention makes possible a much more extensive use of the shaft drive in motor cycle construction and without sacrifice of stability or the use of excessively long wheel bases. Also, as later pointed out, it permits a more efficient use of V-type motors.

In the accompanying drawings:

Fig. 1 is a side elevational view of a four cylinder motorcycle embodying my invention, parts being broken away;

Fig. 2 is a sectional view, on a larger scale, taken substantially on line 2—2 of Fig. 1, showing the manner of connecting the crank shaft assembly to the transmission assembly;

Fig. 3 is a sectional view substantially on line 3—3 of Fig. 2;

Fig. 4 is a sectional view substantially on line 4—4 of Fig. 2; and

Fig. 5 is a view similar to Fig. 2, indicating the application of my invention to a V-type motor.

Referring to the drawings, the frame of the motorcycle is generally indicated at 10. A four cylinder motor, with its cylinders in line, is shown at 11. The motor 11 is placed as low in the frame as is consistent with providing the safe minimum road clearance *a*, (Fig. 1) thereby securing a low center of gravity, and the maximum stability. This position of the motor relative to the frame brings the axis of rotation of the motor crank shaft, indicated at 12, a distance *b* below the level of the axle 14 of the rear wheel. In Fig. 2 the dotted lines 12′ and 14′ indicate the levels respectively of the motor shaft and rear axle. This is the condition which under previous practice would prohibit the use of a propeller shaft to the rear wheel. According to my invention, I meet the condition by placing the "transmission" to one side, above and parallel to the clutch mechanism, instead of in the conventional end to end relation. These elements are conveniently enclosed in a single housing indicated at 15. The novel upwardly offset parallel relation of the clutch mechanism and transmission is best shown in Fig. 2, where the driven clutch shaft, which is in line with the motor crank shaft, is shown at 16 and the transmission tail shaft is shown at 17. In this view the dotted line 19 indicates the outer circumference of the fly wheel which is located just forward of the transmission and partially overlapping the same transversely as shown.

Referring to Fig. 3, the motor crankshaft is shown at 18, the motor flywheel at 19, and the usual clutch and associated mechanism is indicated generally at 20 including clutch brake mechanism 21 and clutch shaft 16. The driven clutch member 22 is secured to the front end of the clutch shaft 16. A sprocket 23, affixed to the rear end of said clutch shaft 16 is connected by chain 25 to a sprocket 27 forming part of a quill gear or hollow shaft 28 which is rotatably mounted on the transmission tail shaft 17. The primary shaft structure of the transmission comprises the aforesaid quill gear 28 as its driving element and the tail shaft 17 (extending through quill gear 28) as its driven element. The gear teeth 30 of the quill gear 28 are in constant mesh with the gear teeth 31 formed on jack shaft 32, which constitutes the secondary shaft structure of the transmission. A gear 33 formed on the forward end of the jack shaft is constantly in mesh with a pinion 34 rotatably mounted on shaft 17. A pair of gears 35 and 36 are separately splined on shaft 17 intermediate gears 30 and 34 and are adapted to be independently moved longitudinally thereon by the usual shift lever, not shown.

While the arrangement described, viz., that of mounting the transmission in parallelism with the clutch mechanism, makes possible the shorter wheel base, it is also desirable for compactness to keep the transmission laterally in close proximity to the clutch mechanism, and avoid its extending forward beyond the fly wheel. To this end the transmission itself is folded up, so to speak, in compact form by arranging the two elements of the primary shaft, viz., the driving and driven elements one partially within the other instead of wholly in end to end relation as is usually the case. This serves the advantage of making the transmission compact enough so that it may lie wholly behind the transverse plane of the fly wheel. In order to extend forward beyond the fly wheel, either the transmission must be set at a relatively wide space from the clutch mechanism or the fly wheel must be made small, and neither of these alternatives is as desirable as the arrangement I have shown.

The tail shaft 17 is positioned at the level of the rear axle 14, and is directly connected by a suitable universal joint 37 (Fig. 4) to the propeller shaft 38 which carries at its rear end a bevel gear 39, meshing with a bevel gear 40 secured to axle 14, Fig. 1. A cover casing 37' is provided as shown for the universal joint.

Referring to Fig. 4, which shows the parts in neutral position, gear 35 is provided with an internal ring of clutch teeth 41 adapted when the gear 35 is moved to the left to engage the external ring of clutch teeth 30 and connect the quill gear 28 directly to tail shaft 17 which is connected by universal joint 37 to the motorcycle propeller shaft 38. When gear 35 is moved to the right in Fig. 4 it is brought into mesh with a gear 42 formed on jack shaft 32 forwardly of gear 31 and the drive is through gears 30, 31, shaft 32, and gears 42, 35, to shafts 17, 38. When gear 36 is shifted to the left in Fig. 4 it is brought into mesh with a gear 43 formed on the jack shaft, forwardly of gear 42, and the drive is through gears 30, 31, shaft 32 and gears 43, 36, to shafts 17, 38. Gear 36 is formed with internal clutch teeth 44 which engage external clutch teeth 34 at the forward end of the jack shaft when gear 36 is shifted to the right in Fig. 4, thereby locking gear 34 to shaft 17, whereby the drive is through gears 30—31, shaft 32, gears 33, 34, to shafts 17, 38.

It will be noted that the power train after proceeding rearwardly through the clutch to the end of the clutch shaft 16, turns at right angles through chain 25 to the upwardly offset hollow shaft or quill gear 28 and then proceeds forwardly via the jack shaft 32 to the transmission tail shaft 17 which carries it rearwardly through the aforesaid quill gear or hollow shaft 28 to the propeller shaft 38 and the rear wheel. This "overlapping" of the power train which places the clutch shaft and the four speed transmission shaft substantially side by side, rather than end to end, reduces the overall length of the power train and consequently the necessary wheelbase. It has further advantage of making possible the use of a horizontal propeller shaft and a relatively low crank shaft. The shafts 17 and 38 being normally in line, no substantial angular movement takes place in the universal joint 37.

A substantial advantage of my construction is that the basic speed ratio between the crank shaft and drive shaft can be changed by substituting sprockets of different diameter at 23 and 27 without requiring any change in the rest of the structure.

By my invention it is possible to set a V-motor with its crank shaft extending longitudinally, instead of transversely, of the frame, as shown in Fig. 5. The motor cylinders 50 and 51 extend outwardly from the frame, providing a more efficient, and equal, cooling of both cylinders. The driving connection of the V-motor shown in Fig. 5 is exactly similar to that previously described, the clutch shaft 16' being connected to the hollow shaft or quill gear 28' by a silent chain 25'. From shaft 28' power is transmitted to the forwardly extending jack shaft 32' to shaft 17' and thence to the rear wheel by means of a universal coupling and a drive shaft, with all the advantages previously described.

I claim:

1. In a motorcycle which includes a frame, front and rear wheels mounted in the frame, and a motor carried by the frame intermediate the wheels, the crank shaft of the motor being below the level of, and at right angles to the axle of the rear wheel, a power train from the crank shaft to the axle of the rear wheel which comprises a clutch mechanism and a speed change transmission mechanism arranged in sidewise parallel relationship, the forward end of the clutch mechanism being connected to the crank shaft and the rear end of the clutch mechanism being connected by a laterally directed drive means to the rear end of the transmission mechanism, said transmission mechanism extending forwardly from its connection with said laterally directed drive means, and a horizontal drive shaft including a universal joint connected to the axle of the rear wheel and connected to said transmission mechanism forwardly of said laterally directed drive member.

2. In a motorcycle which includes a frame, front and rear wheels mounted in the frame, and a motor carried by the frame intermediate the wheels, the crank shaft of the motor being below the level of, and at right angles to, the axle of the rear wheel, a power train from the crank shaft to the axle of the rear wheel which comprises a rearwardly extending clutch mechanism in line with the crank shaft, a driving means connected to the rear end of the clutch mechanism and directed laterally upwardly to the level of the axle of the rear wheel and terminating in a hollow shaft, a forwardly directed speed change transmission driven from said hollow shaft and driving a horizontal shaft extending rearwardly through said hollow shaft to a driving connection with the axle of the rear wheel, and a universal joint in said horizontal shaft intermediate its connection to said axle and the hollow shaft.

3. In a motorcycle which includes a frame, front and rear wheels mounted in the frame, and a motor carried by the frame intermediate the wheels, the crank shaft of the motor being below the level of, and at right angles to, the axle of the rear wheel, a clutch mechanism extending rearwardly from and aligned with the crank

shaft, a hollow shaft positioned substantially at the level of the axle of the rear wheel, a driving connection between the driven member of the clutch mechanism and the rear end of the hollow shaft, a forwardly extending jack shaft driven from the forward end of the hollow shaft, a transmission shaft extending forwardly and rearwardly through said hollow shaft, selective speed gearing connections between the jack shaft and the transmission shaft forwardly of the hollow shaft, and a horizontal drive shaft connected at its rear end by bevel gears to the axle of the rear wheel and connected at its forward end to the transmission shaft by a universal joint, rearwardly of the hollow shaft.

4. In a motorcycle which includes a frame and front and rear wheels mounted in the frame, a V-motor carried by the frame intermediate the wheels and having its crank shaft below the level of, and at right angles to the axle of the rear wheel, a clutch mechanism extending rearwardly from the crank shaft, a driving member connected to the rear end of the clutch mechanism, and directed laterally upwardly to the level of the axle of the rear wheel and terminating in a hollow shaft, a forwardly extending change speed transmission mechanism driven from the forward end of said hollow shaft and a shaft driven from said change speed transmission mechanism and extending rearwardly through said hollow shaft to a driving connection with the axle of the rear wheel, said last named shaft being provided with a universal joint intermediate its connection to the rear axle and the hollow shaft.

5. In a motorcycle which includes a frame, front and rear wheels mounted in the frame, a motor carried by the frame intermediate the wheels, the crank shaft of the motor extending longitudinally of the motorcycle, and being located below the level of the axle of the rear wheel, a fly wheel at the rear of said crank shaft, clutch mechanism extending rearwardly from said fly wheel, speed change transmission mechanism arranged in side by side relationship with said clutch mechanism, a driving connection between the rear end of said clutch mechanism and the rear end of said transmission mechanism, said transmission mechanism being extended forwardly of said driving connection and terminating at the rear of the transverse plane of said fly wheel, and a driving connection from said transmission mechanism to the rear wheel comprising a propeller shaft and a universal joint.

6. In a motorcycle which includes a frame, front and rear wheels mounted in the frame, a motor carried by the frame intermediate the wheels, the crank shaft of the motor extending longitudinally of the motorcycle, and being located below the level of the axle of the rear wheel, a fly wheel at the rear of said crank shaft, clutch mechanism extending rearwardly from said fly wheel, speed change transmission mechanism arranged in side by side relationship with said clutch mechanism, a driving connection between the rear end of said clutch mechanism and the rear end of said transmission mechanism, said transmission mechanism being extended forwardly of said driving connection and terminating at the rear of the transverse plane of said fly wheel, said transmission mechanism comprising a tail shaft positioned on a level with the axle of the rear wheel and a driving connection between said transmission tail shaft and the rear wheel comprising a horizontally extended propeller shaft and a universal joint.

7. In a motorcycle which includes a frame, front and rear wheels mounted in the frame, a motor carried by the frame intermediate the wheels, the crank shaft of the motor extending longitudinally of the motor cycle and being located below the level of the axle of the rear wheel, a fly wheel at the rear of said crank shaft, clutch mechanism comprising a clutch shaft in line with said crank shaft and extending rearwardly from said fly wheel, a speed change transmission comprising shafts arranged parallel to said clutch shaft, a driving connection between said clutch shaft and a shaft of said transmission, the shafts of said transmission extending forwardly of said driving connection and terminating at the rear of the tranverse plane of said fly wheel, a tail shaft of said transmission being located upwardly and laterally of said clutch shaft and positioned on a level with the axle of the rear wheel, and a propeller shaft in line with said transmission tail shaft and extending horizontally to the rear wheel for operating the same.

8. In a motorcycle which includes a frame, front and rear wheels mounted in the frame, a motor carried by the frame intermediate the wheels, the crank shaft of the motor extending longitudinally of the motorcycle and being located below the level of the axle of the rear wheel, a fly wheel at the rear of said crank shaft, clutch mechanism comprising a clutch shaft in line with said crank shaft and extending rearwardly from said fly wheel, a speed change transmission comprising primary and secondary shaft structures arranged parallel to said clutch shaft, said primary shaft structure comprising a quill gear and a transmission tail shaft extending through the hollow shaft of said quill gear, a driving connection between said clutch shaft and said quill gear, said primary and secondary shaft structures extending forwardly of said driving connection and terminating at the rear of the transverse plane of said fly wheel, said primary shaft structure being located upwardly and laterally of said clutch shaft and positioned on a level with the axle of the rear wheel, and a propeller shaft connected by a universal joint with said transmission tail shaft and extending horizontally to the rear wheel for operating the same.

9. In a motorcycle which includes a frame, front and rear wheels mounted in the frame, a motor carried by the frame intermediate the wheels, the crank shaft of the motor extending longitudinally of the motorcycle and being located below the level of the axle of the rear wheel, a fly wheel at the rear of said crank shaft, clutch mechanism comprising a clutch shaft in line with said crank shaft and extending rearwardly from said fly wheel, a speed change transmission comprising primary and secondary shaft structures arranged parallel to said clutch shaft, said primary shaft structure comprising a quill gear and a transmission tail shaft extending through the hollow shaft of said quill gear, a constantly engaged driving connection between said clutch shaft and the rear end of said quill gear, selectively engageable driving connections between the forward end of said quill gear and said tail shaft, said primary and secondary shaft structures extending forwardly of said driving connection and terminating at the rear of the transverse plane of said fly wheel, said primary shaft structure being located upwardly and laterally of said clutch shaft and positioned on a level with the axle of the rear wheel, and a propeller shaft connected by a universal joint with said trans-

mission tail shaft and extending horizontally to the rear wheel for operating the same.

10. In a motorcycle which includes a frame, front and rear wheels mounted in the frame, a motor carried by the frame intermediate the wheels, the crank shaft of the motor extending longitudinally of the motorcycle and being located below the level of the axle of the rear wheel, a fly wheel at the rear of said crank shaft, clutch mechanism comprising a clutch shaft in line with said crank shaft and extending rearwardly from said fly wheel, a speed change transmission comprising primary and secondary shaft structures arranged parallel to said clutch shaft, said primary shaft structure comprising a quill gear and a transmission tail shaft extending through the hollow shaft of said quill gear, a driving connection between said clutch shaft and said quill gear, said primary and secondary shaft structures extending forwardly of said driving connection and terminating at the rear of the transverse plane of said fly wheel, and a propeller shaft connected by a universal joint with said transmission tail shaft and extending to the rear wheel for operating the same.

11. In a motorcycle which includes a frame, front and rear wheels mounted in the frame, a motor carried by the frame intermediate the wheels, the crank shaft of the motor extending longitudinally of the motorcycle and being located below the level of the axle of the rear wheel, a fly wheel at the rear of said crank shaft, clutch mechanism comprising a clutch shaft in line with said crank shaft and extending rearwardly from said fly wheel, a speed change transmission comprising primary and secondary shaft structures arranged parallel to said clutch shaft, said primary shaft structure comprising a quill gear and a transmission tail shaft extending through the hollow shaft of said quill gear, a constantly engaged driving connection between said clutch shaft and the rear end of said quill gear, selectively engageable driving connections between the forward end of said quill gear and said tail shaft, said primary and secondary shaft structures extending forwardly of said driving connection and terminating at the rear of the transverse plane of said fly wheel, and a propeller shaft connected by a universal joint with said transmission tail shaft and extending to the rear wheel for operating the same.

GEORGE B. WEAVER.

July 16, 1946. F. WHITTLE 2,404,334

AIRCRAFT PROPULSION SYSTEM AND POWER UNIT

Filed Feb. 19, 1941 3 Sheets-Sheet 1

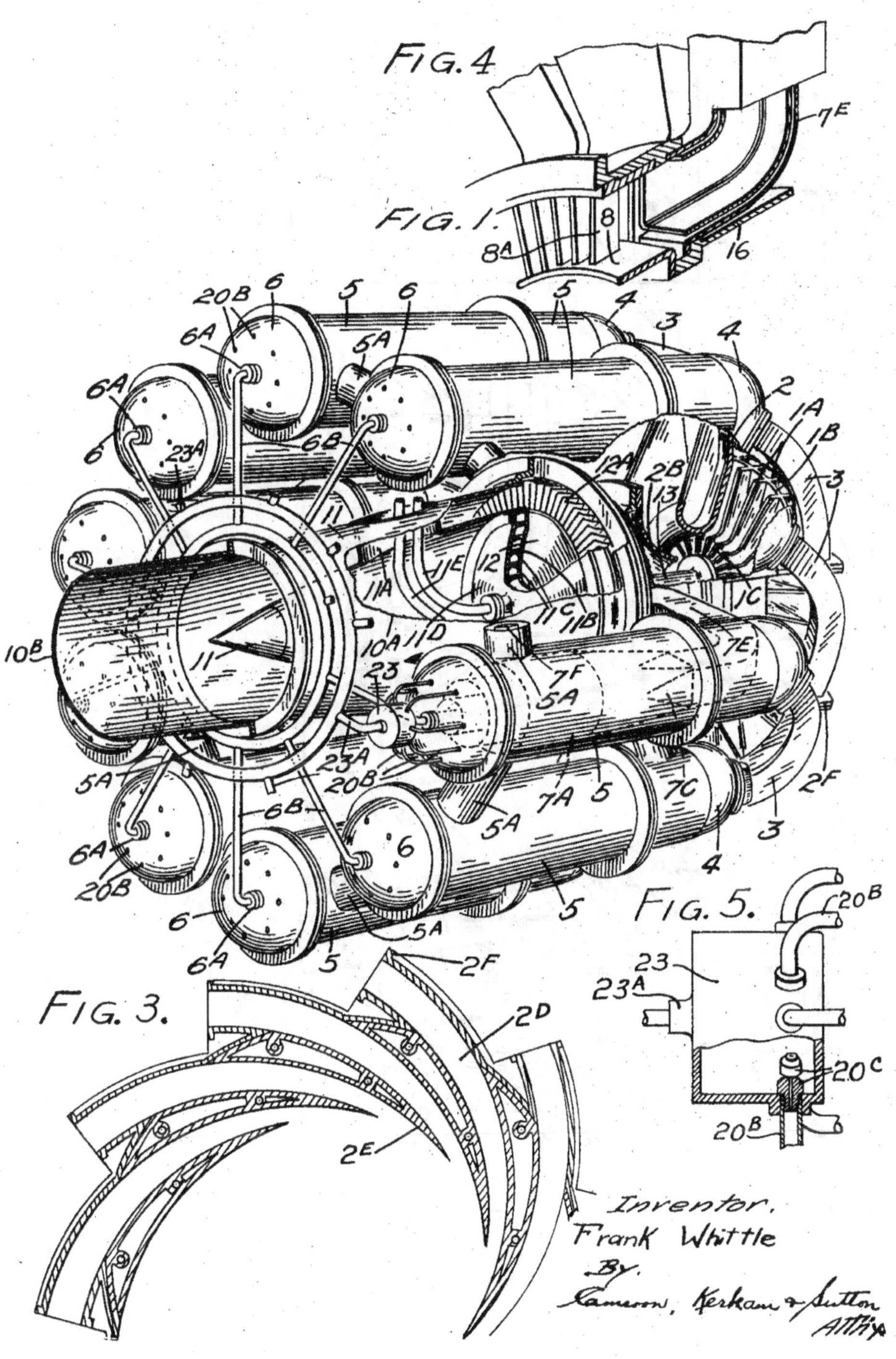

July 16, 1946. F. WHITTLE 2,404,334

AIRCRAFT PROPULSION SYSTEM AND POWER UNIT

Filed Feb. 19, 1941 3 Sheets-Sheet 2

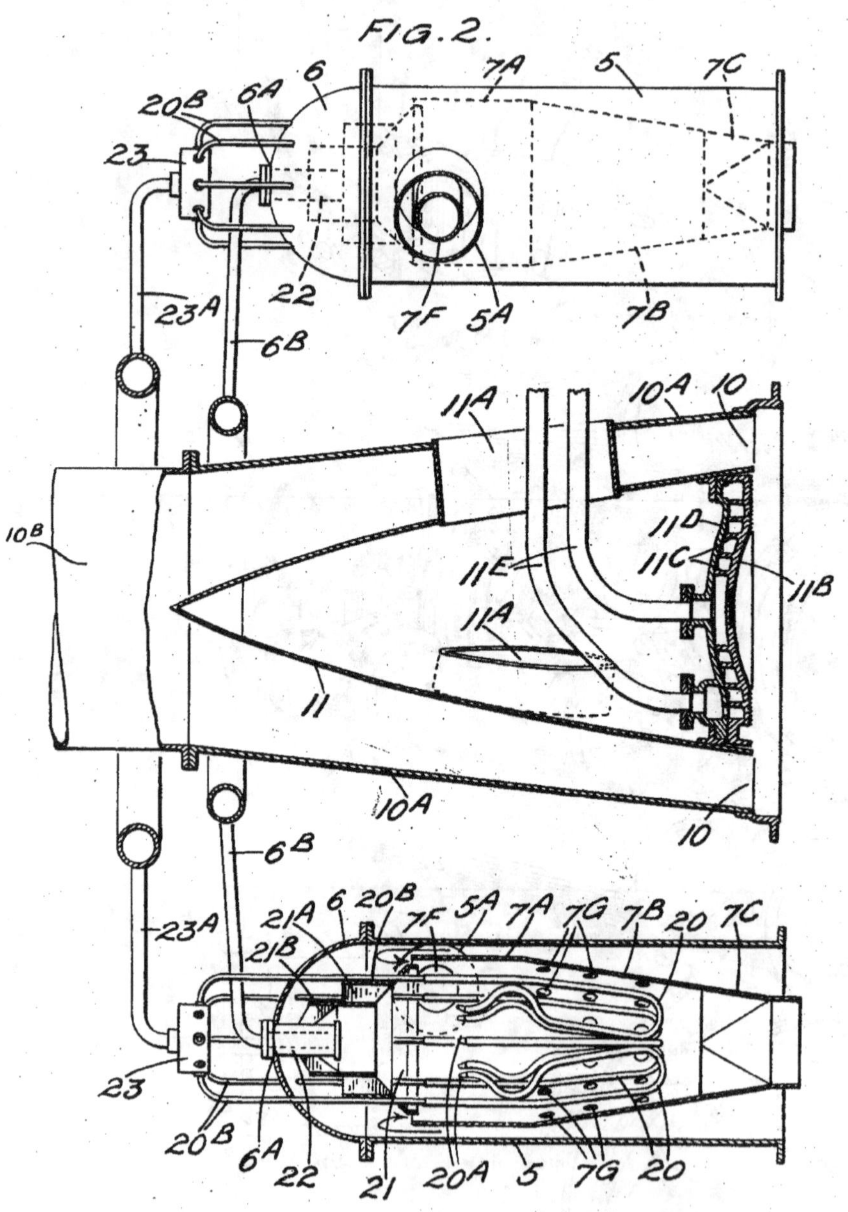

Inventor,
Frank Whittle
By,
Cameron, Kerkam & Sutton
Att'ys.

July 16, 1946. F. WHITTLE 2,404,334

AIRCRAFT PROPULSION SYSTEM AND POWER UNIT

Filed Feb. 19, 1941 3 Sheets-Sheet 3

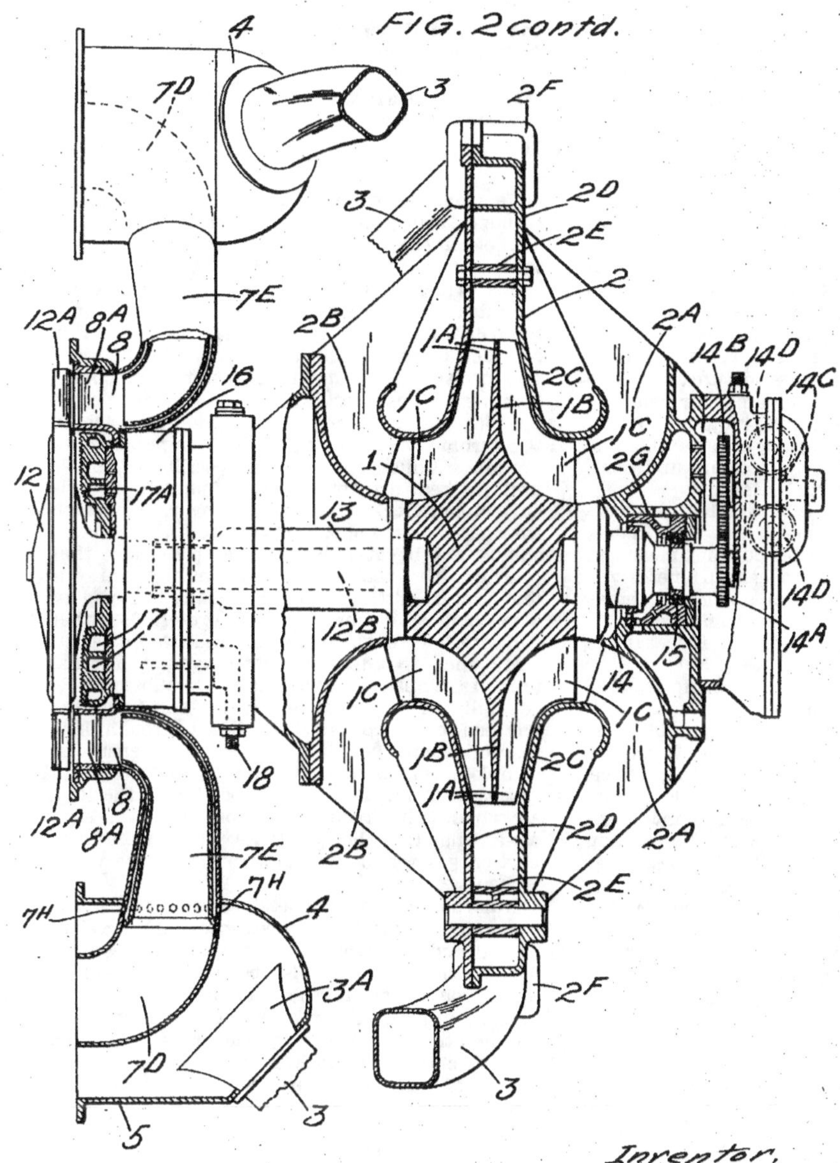

Inventor,
Frank Whittle
By,
Cameron, Kerkam & Sutton
Att'ys.

Patented July 16, 1946 **2,404,334**

UNITED STATES PATENT OFFICE

2,404,334

AIRCRAFT PROPULSION SYSTEM AND POWER UNIT

Frank Whittle, Rugby, England, assignor to Power Jets (Research & Development) Limited, London, England

Application February 19, 1941, Serial No. 379,734
In Great Britain December 9, 1939

15 Claims. (Cl. 60—35.6)

1

This invention relates to aircraft propulsion systems and power units. Whilst it is intended primarily to be applied to propulsion aircraft systems in which thrust is developed by reaction arising out of the expulsion of a stream of gas through a nozzle or jet, and this is deemed to be the application which uses the features of the invention most advantageously, yet it may be found that some or all of these features may also usefully be employed in power units for generating shaft power; or where it is required to produce a source of gas possessing considerable energy in the forms of velocity, pressure and heat, for example for driving turbines.

The kind of apparatus with which the invention is concerned, is in general that which comprises an air compressor, fuel-burning means in the compressor output, and a gas turbine driven by the combustion products and air heated thereby and mechanically driving the compressor.

Apparatus of this type is shown for example in my prior British Patent No. 347,206.

My present invention seeks to provide apparatus for a reaction propulsion system for aircraft, of the kind stated, having advantages of compactness and low weight for a given performance, which may be comparatively simple to produce, which is easy to install and which should require but little maintenance. A power unit made according to the invention, whether for the reaction propulsion of aircraft or for other purposes, has also advantages in relation to its efficiency as a prime mover, in that a highly efficient compressor is employed, bearings are few and may be simple and involve low loss, gas flows are well organised and efficiently conducted, heat is conserved, and temperature distribution is very uniform in the working parts.

The invention broadly stated, resides in an engine which comprises a centrifugal compressor with bilateral air intakes, and with a plurality of outlets arranged symmetrically about the rotor axis, which outlets lead to a like number of combustion chambers also disposed symmetrically in which fuel is burnt, the combustion chambers having ducts to the turbine nozzle ring affording continuous admission to that side of the turbine which is nearer the compressor, the turbine discharging through an axial duct around which the combustion chambers are disposed; and the compressor outlets comprise an openwork structure through which air reaches the compressor inlet which is between the compressor and turbine. The combustion chambers are preferably disposed wholly or mainly to the side of the turbine remote from the compressor, and the flow of air through them is reversed in directional sense.

2

The compressor outlets are arranged so that the air is at first allowed to flow tangentially, bends (which may have guide vanes or cascades) then directing the flow in the axial direction into the combustion chambers. Means are preferably provided to ensure substantially equal delivery of fuel into the flame tubes of the combustion chambers, as well as substantially equal quantities of air into the chambers, so that the admission to the turbine will be at a uniform temperature all around. Whilst it may not be possible to ensure uniform heat distribution (it is found to be difficult in practice) it should at any rate be such that gases reaching the tips of the turbine blades should not be at a less temperature than that reaching the roots, and substantially uniform temperature distribution peripherally is aimed at. The combustion chambers are preferably either cylindrical or slightly conical, and are domed at their ends to afford strength against internal pressure, whilst the flame tubes are substantially coaxially arranged within them, and are each adapted to receive the whole air put through their respective chambers in such a way as to ensure as complete combustion as possible, and to admit "secondary" air, i. e. that proportion of the total air which is not required for combustion, in such a way that it is thoroughly mixed with the combustion gases and the temperature of the whole is then as above stated.

The combustion chambers are preferably interconnected by short stub pipes, situated near their domed ends to balance their internal pressures. The flame tubes may also be interconnected by ducts housed within the stub pipes, to make it possible to ignite the gases in some of them from one or others, in which ignition may be started for example by a sparking plug. Alternatively, the ducts may be dispensed with, and a sparking plug provided for each flame tube.

An engine for the propulsion of an aircraft, according to the invention, is illustrated by the accompanying drawings, of which Figure 1 is an external perspective view with sufficient broken away for the identification of some internal parts, and Figure 2 is a sectional side elevation. Figure 3 shows an arrangement of diffuser vanes, Figure 4 is a detail view of part of the turbine structure, and Figure 5 is a detail view of part of the fuel feeding means.

The engine comprises, broadly, a compressor, a combustion system, a turbine, and a jet pipe for the turbine exhaust, the combustion system preferably comprising a plurality of interconnected units of the construction specifically disclosed and claimed in my co-pending application Serial No. 379,735, filed February 19, 1941.

The compressor is a centrifugal compressor with impeller 1 running in a casing 2, and bilateral air intakes. The casing has intake eyes 2A, 2B, at front and rear, a radially convergent part 2C, a parallel-walled primary diffuser part 2D, which at the periphery has tangential outlets each one of which is prolonged by an air trunk 3. The impeller 1 has radial vanes 1A on each side of an annular web 1B, the inner parts of the vanes 1A being extended axially and bent at 1C forming intake vanes to direct the inflowing air, in the eyes 2A, 2B. The vanes 1A sweep the convergent part of the casing, 2C. Between the walls of the part 2D of the casing, streamlined fixed diffuser vanes 2E are mounted (Fig. 3), and these control the path of the air and also stiffen the structure. The outlets present rectangular flanges 2F to which are bolted the trunks 3.

The trunks 3 are curved out of the plane of the compressor, and form divergent ducts further diffusing the delivered air. Their curvature brings them to the end caps 4 of combustion chambers, into which they lead with radial and axial components of direction. Within the caps 4, the trunks have bell mouths 3A.

The combustion chambers comprise the end caps 4, on the forward ends of cylindrical air casings 5, at the rear ends of which are domes 6. These parts are formed of sheet metal and are quite light, being designed to withstand such internal pressure as may be generated by the compressor and diffuser system. The axes of the chambers are parallel with that of the impeller 1 and they are disposed symmetrically and equidistantly around that axis, in a generally circular form, (as seen in Figure 1). It will be observed that the trunks 3, extending from the casing 2 to the combustion chambers, constitute an openwork structure, surrounding the rear intake eye 2B of the compressor. The necessary freedom of access of air to the rear eye of the impeller is therefore afforded.

Within each combustion chamber is a flame tube. This may have various forms in detail, the example shown being partly cylindrical at 7A, continuing as a frusto-conical part 7B, and finally changing section from circular to square at a neck 7C. The flame tubes receive air at their rear ends, fuel is burnt in them, and the combustion products proceed through the tubes, around elbows at 7D, through ducts 7E, to the turbine nozzle ring at 8. The ducts 7E are preferably double walled and the inter-wall space may carry cool air from the chambers into the turbine, the cooling air entering the inter-wall spaces through the ports 7H and thereby reducing heat losses from the ducts. The nozzle ring has fixed nozzle blades 8A, the arrangement of the ducts 7E, nozzle ring 8 and blades 8A being shown more clearly in Figure 4. Thus the ducts 7E each provide for admission of gases to a circumferential section of the nozzle ring 8, the circumferential length of these sections depending on the circumferential extent of the duct ends. With an appropriate relationship between the size of the duct ends and the number of ducts, the admission of gases may be made substantially continuous throughout the entire periphery of the ring.

Passing between the nozzle blades 8A and across an axial clearance, the gases then impinge on the blades 12A of the turbine, which is a single stage axial flow turbine, drive the turbine and are emitted therefrom through an annular passage 10. The passage 10 is formed by a frusto-conical wall 10A, coaxial with the turbine, and within which is a streamlined fairing 11 supported by (say) three streamlined hollow struts 11A. The flow from the passage 10 is conveyed through a jet pipe 10B, to the atmosphere. The fairing 11 has its forward end formed by a water cooled circular wall 11B to receive heat from the turbine wheel disc 12 and enable it to be conveyed away. The water for this is circulated through suitable passages formed by ribs 11C and a rear wall 11D, by water pipes 11E which are led through one of the struts 11A.

The turbine wheel, comprising the disc 12 and blades 12A, is mounted as an overhung wheel, integral with a shaft 12B which is fitted with splines into a hollow shaft called a quill shaft, 13, which is in turn secured to the impeller 1. On its other (forward) side, the impeller has a stub shaft 14 similarly attached to it, and this shaft drives such auxiliary mechanism as may be necessary, through appropriate trains such as pinions 14A, 14B, worm 14C, worm-wheels 14D. It may be mentioned that the turbine and impeller are aligned and interattached with the greatest care and the complete rotor assembly so formed, thoroughly balanced.

The rotor assembly is borne by two main bearings; the forward bearing is shown at 15 housed within an extension 2G of the structure of the compressor casing 2, and this bearing is adapted to take such axial loads as may arise, as well as journal loads. The rear bearing is housed within a rearward extension casing 16 of the compressor casing structure, which extension (as well as being the bearing housing) comprises an annular water jacket 17 with a rear wall 17A facing the turbine disc 12, through which jacket water is circulated similarly to that at the rear of the turbine. Bearing lubrication and cooling is preferably by force feed of air and oil provided in any suitable and convenient manner.

Returning now to the combustion arrangements it will be seen that the flame tube 7A, 7B, 7C, contains a set of vaporiser tubes 20, which are symmetrically disposed about the flame tube axis, and which terminate in jets 20A directed upstream in the air flow, so that the greater part of the tubes 20 are directly in flame. The reversal of direction of the combustion or "primary" air is indicated by arrows. The vaporiser tubes 20 are supplied by a like number of pipes 20B which extend through the dome end 6. They pass through a partial closure baffle 21 located in the otherwise open end of the flame tube part 7A. This baffle 21 has swirler vanes, for example an outer ring of vanes 21A and an inner ring 21B, which may be pitched in opposite rotational sense, to produce a high degree of swirl and local turbulence in the combustion gases. Axially through the dome 6, and within the inner vanes 21B is provided a tubular bush 22 to receive and mount a pilot jet (located at 6A) for starting purposes. There are various standard commercial atomising spray jets which may be used for this purpose, so none will be detailed.

Each combustion chamber 5 is connected to its neighbours by stub pipes 5A by which their internal pressures are balanced. Within each stub pipe 5A is a lesser pipe 7F interconnecting neighbouring flame tubes. The purpose of these ducts is to enable lighting up to be effected simply by initially procuring ignition in one flame tube, e. g.

by a sparking plug, whereafter the others light up because burning gases pass along the ducts. It appears that this action is due to the fact that, upon ignition in one tube, there is a considerable rise of pressure therein as compared with its neighbours, and therefore a flow of ignited gas to the neighbours.

The stub pipes and ducts may be dispensed with and other means adopted for lighting up, for example, the provision of a sparking plug in each flame tube.

The liquid fuel is supplied by pump, through a suitable throttle valve, and through such filters, etc. as may be required. It flows into a box 23 associated with each combustion chamber, by a manifold pipe 23A, and emerges from the box 23 by the pipes 20B. The fuel flowing from the box 23 to the pipes 20B may be caused to pass through suitable restriction orifices such as the nipples 20C (see Fig. 5), which act as weirs and prevent surging as between the vaporizer elements 20.

By this means, it is provided that the distribution of fuel is as uniform and symmetrical as possible, both around the engine as a whole, and also within each flame tube. Axially directed pilot jets of spray atomising type are also provided in the centres of the domes 6, as at 6A, and these are supplied by a fuel manifold system 6B.

The flame tubes are preferably perforated with air holes such as 7G by which secondary air flows into them, for mixture with the combustion gases. The larger proportion of the total air is regarded as secondary air, in a practical design where reasonably low turbine temperatures are to be involved. It has been found impossible to lay down the exact details of the primary and secondary air passages, as the aerodynamic, thermodynamic, and thermochemical effects are evidently very complex.

The object to be sought is, that the temperatures readable at a number of locations or stations across the outlet of a flame tube to the nozzle ring, should be as uniform as possible, and if there is inequality it should be in the sense that the temperature is higher in the gases flowing to the tips of the blades 12A, than to the roots and that the mean temperature from the individual flame tubes is as uniform as possible around the engine. This result may be achieved by choosing and if necessary altering the size, location, and number, of the holes 7G by matching the flows through the vaporiser elements 20 as nearly as possible and by equalising the flows through the pipes 23A.

It will be appreciated that the engine as a whole is practically a symmetrical arrangement about the rotor axis.

What I claim is:

1. An engine of the character described comprising a centrifugal compressor having intake eyes symmetrically disposed on opposite sides of its plane of rotation and a plurality of outlets symmetrically disposed in circular disposition about its axis of rotation, a corresponding and similarly disposed plurality of air ducts leading from said outlets towards one side of said plane, an axial flow turbine arranged coaxial with said compressor on the same side thereof as that of said ducts and adapted to directly mechanically drive the compressor, a plurality of combustion chambers arranged in circular disposition around said axis, an air duct connecting each of said outlets to one of said chambers, means for introducing fuel into each of said chambers for continuous combustion therein, a combustion product duct leading from each of said chambers to said turbine on the side thereof to which said compressor is located, the construction formed by said air ducts, combustion chambers and combustion product ducts constituting a skeleton structure leaving open acess through which air is permitted to enter the compressor intake eye situated nearer the turbine, and an exhaust conduit leading axially away from the side of the turbine opposite to the side to which said combustion products are admitted.

2. An engine according to claim 1 wherein said combustion chambers are peripherally spaced from each other and are radially spaced from the outside of said exhaust conduit.

3. An engine of the character described, comprising a centrifugal compressor having a casing, a pair of air intakes symmetrically arranged on opposite sides of its plane of rotation and a plurality of discharge outlets symmetrically disposed about and extending tangentially to the periphery of said casing, a corresponding plurality of combustion chambers symmetrically disposed with respect to the axis of rotation of said compressor, each of said chambers having an outer circular-sectioned casing and a flame tube enclosed therewithin, conduits connecting the compressor discharge outlets with the outer casings of said combustion chambers for delivering compressed air thereto, means for supplying fuel to said flame tubes, means for supplying compressed air from the interior of said combustion chamber casing to said flame tube, a turbine coaxial with said compressor and having a continuous annular nozzle chamber and an axially extending exhaust conduit, ducts symmetrically disposed with respect to the common axis of said compressor and turbine for delivering the products of combustion from said flame tubes to said turbine nozzle chamber, said combustion chambers being so constructed and arranged that the general direction of flow therethrough of the air and products of combustion is substantially parallel to said common axis, pressure equalising air conduits interconnecting the outer casings of the combustion chambers at points remote from the air delivery conduits, and gas conduits housed within said pressure equalising air conduits interconnecting the flame tubes, whereby the gases in one flame tube may be ignited by the combustion in another.

4. An engine of the character described, comprising a centrifugal compressor having a casing, a pair of air intakes symmetrically arranged on opposite sides of its plane of rotation and a plurality of discharge outlets symmetrically disposed about and extending tangentially to the periphery of said casing, a corresponding plurality of combustion chambers symmetrically disposed with respect to the axis of rotation of said compressor, each of said chambers having an outer circular-sectioned casing and a flame tube enclosed therewithin, conduits connecting the compressor discharge outlets with the outer casings of said combustion chambers for delivering compressed air thereto, means for supplying fuel to said flame tubes, means for supplying compressed air from the interior of said combustion chamber casing to said flame tube, a turbine coaxial with said compressor and having a continuous annular nozzle chamber and an axially extending exhaust conduit, ducts symmetrically disposed with respect to the common axis of said compressor and turbine for delivering the products

of combustion from said flame tubes to said turbine nozzle chamber, said combustion chambers being so constructed and arranged that the general direction of flow therethrough of the air and products of combustion is substantially parallel to said common axis, pressure equalising air conduits interconnecting the outer casings of the combustion chambers, gas conduits housed within said air conduits interconnecting the flame tubes, and ignition means associated with at least one of said flame tubes, said interconnecting gas conduits enabling ignition of the gases in those flame tubes not provided with ignition means by a flow of ignited gas from a flame tube wherein ignition has already been effected.

5. A continuous combustion gas turbine engine adapted for the jet propulsion of air craft comprising a centrifugal compressor and an axial flow turbine, said compressor having an impeller connected with said turbine for rotation therewith on a common axis and said impeller and turbine having substantially the same diameter, a casing for said compressor having a plurality of peripheral outlets, diffusion means in said casing surrounding said impeller and substantially increasing the diameter of said casing whereby the mean radius of outlet from said compressor is substantially greater than the mean radius of admission to said turbine, a plurality of combustion chambers spaced circumferentially around said common axis and each having combustion means therein, means connecting each of said chambers with one of said outlets and also with said turbine, said means and chambers conducting air and gases from said compressor to said turbine in a path which is generally axial but which comprises inward convergence, and an exhaust conduit extending axially away from said turbine on the side opposite the compressor.

6. An engine as defined in claim 5, having means interconnecting said combustion chambers to balance the pressure therein.

7. An engine as defined in claim 5, comprising flame tubes within the combustion chambers, and means interconnecting said chambers to balance the pressure therein including means interconnecting the interiors of said flame tubes.

8. An engine as defined in claim 5, said chambers being substantially cylindrical and forming a symmetrical group surrounding said common axis with the axis of each chamber in a common plane with said common axis.

9. An engine as defined in claim 5, said chambers being substantially cylindrical and forming a symmetrical group surrounding said common axis with the axis of each chamber in a common plane with said common axis, and a flame tube supported coaxially in each chamber, said flame tubes being of circular cross section and symmetrical about the chamber axes throughout their lengths.

10. An engine as defined in claim 5, said exhaust conduit being annular and continuous adjacent the turbine and leading to a propulsion exhaust jet.

11. An engine as defined in claim 5, said compressor having bilateral intakes.

12. An engine as defined in claim 5, said compressor having bilateral intakes one of which is located on the side next the turbine, said connecting means being spaced to provide access of air to said one intake.

13. An engine as defined in claim 5, said chambers extending axially beyond the turbine and comprising means for reversing the direction of axial flow of the air leaving the compressor and discharging air and gases axially toward the compressor, said connection from the chambers to the turbine converging inwardly and again reversing the direction of axial flow to admit said air and gases to that side of the turbine which is adjacent the compressor.

14. An engine according to claim 5 including common bearing means for supporting the turbine and compressor rotors located on the one hand between the two rotors and on the other hand on the side of the compressor remote from the turbine.

15. An engine according to claim 5 wherein the exhaust conduit is annular where it leaves the turbine and merges into cylindrical form at a distance more remote from the turbine.

FRANK WHITTLE.

March 22, 1960 A. L. SCHAWLOW ET AL 2,929,922

MASERS AND MASER COMMUNICATIONS SYSTEM

Filed July 30, 1958

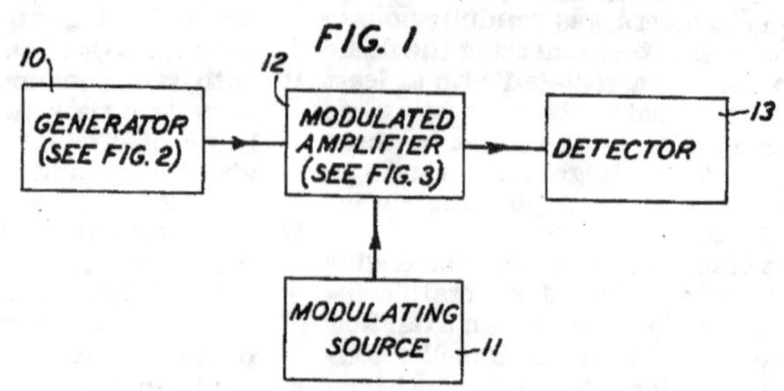

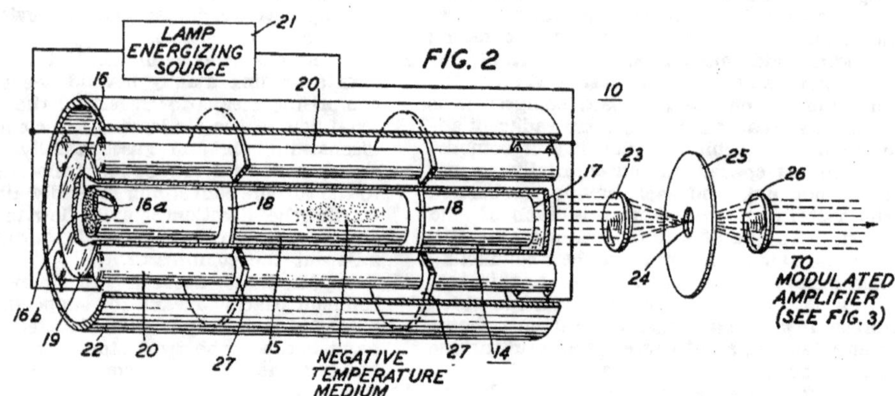

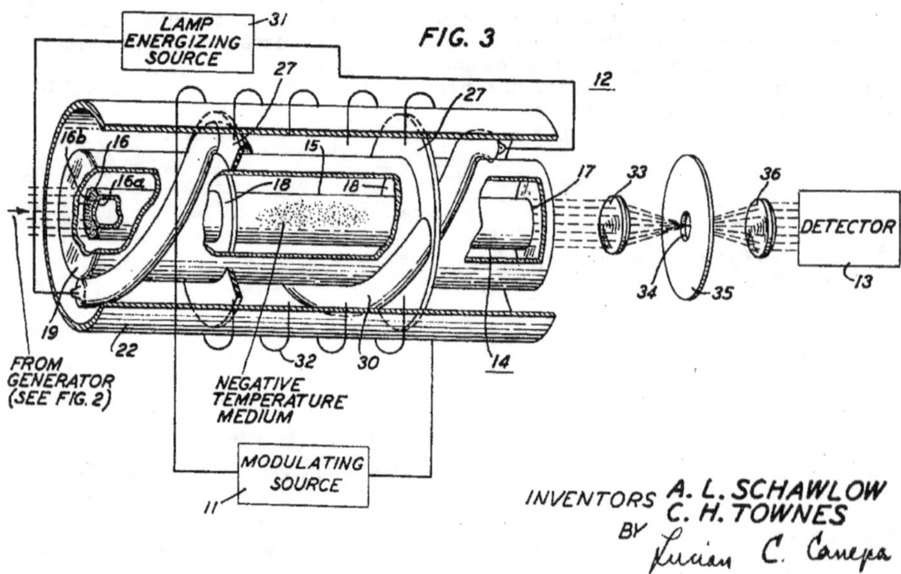

INVENTORS A. L. SCHAWLOW
C. H. TOWNES
BY Lucian C. Canepa
ATTORNEY

United States Patent Office

2,929,922

Patented Mar. 22, 1960

2,929,922

MASERS AND MASER COMMUNICATIONS SYSTEM

Arthur L. Schawlow, Madison, N.J., and Charles H. Townes, New York, N.Y., assignors to Bell Telephone Laboratories, Incorporated, New York, N.Y., a corporation of New York

Application July 30, 1958, Serial No. 752,137

11 Claims. (Cl. 250—7)

This invention relates to the generation and amplification of infrared, visible, and ultraviolet waves, and more particularly to the generation and amplification of such waves by means of devices including media in which the stimulated emission of radiation occurs; devices of this type are now generally termed "masers."

It is characteristic of a maser that it employs a medium in which there is established at least intermittently a nonequilibrium population distribution in a pair of spaced energy levels of its energy level system. In particular, the population of the higher of the selected pair of energy levels may be made larger than that of the lower. It is now usual to describe a medium which is in such a state of nonequilibrium as exhibiting a negative temperature. It is known that a competing process known as relaxation tends to return the system to equilibrium.

It is characteristic that if there be applied to a medium which is in a negative temperature state a signal of a frequency which satisfies Planck's law with respect to the two energy levels which are in nonequilibrium

$$\left(\nu=\frac{E_2-E_1}{h}\text{ where } h \text{ is Planck's constant}\right)$$

then the applied signal will stimulate the emission of radiation at the signal frequency from the medium and the signal will be amplified.

Among the more promising forms of masers known is one which employs as the negative temperature medium a material whose energy level system is characterized by at least three energy levels, with the separations of these three energy levels falling within desired operating frequency ranges. To this material, there is supplied pumping power which effects transitions from the lowest to the highest of the selected three energy levels. By power saturation of the highest energy level, whereby the populations of the highest and lowest energy levels tend to be equalized, there is established in one of these two energy levels a nonequilibrium population distribution with respect to the intermediate energy level of the selected three, whereby a negative temperature results in the material. Thereafter a signal of appropriate frequency can be amplified by being applied thereto in a manner such that the emission of radiation is stimulated therefrom.

It is to be noted that the process of relaxation from randomly overpopulated states may give rise to spontaneous emission, that is, emission caused by radiative transitions in a mode other than the desired or stimulated one.

Generators and amplifiers employing atomic and molecular processes, as do the various known varieties of masers, may in principle be extended in operation far beyond the range of frequencies which have been generated and amplified by electronic processes. As, however, the maser concept is applied to the translation of wavelengths in the infrared, visible, and ultraviolet regions of the electromagnetic wave spectrum, it is found that conventional or microwave maser techinques and structures are suitable neither for the generation of monochromatic radiation nor to provide coherent amplification.

Accordingly, an object of the present invention is a system, including a maser, for translating infrared, visible, and ultraviolet energy.

A maser designed for operation in the microwave range of the spectrum might, for example, comprise a cavity having therein an ensemble of atomic or molecular systems, the cavity being characterized by being able to support only one mode near the frequency which corresponds to the desired radiative transitions of the systems. Alternatively, such an ensemble might be located in a waveguide, which similarly would be characterized by one, or a very few, preferred modes of propagation in the frequency range of interest.

Thus, the energy emitted by a maser operating in the microwave range is typically monochromatic, due to the energy produced by stimulated emission being very much larger than the background of radiation caused by spontaneous emission. In other words, such devices are inherently monochromatic because stimulated emission produces completely coherent amplification, and spontaneous emission, which is not so coherent, is characteristically small by comparison with the stimulated emission.

On the other hand, the maintenance of a single isolated mode is a maser cavity operating at frequencies above those in the microwave range requires an impractically small cavity structure (of the order of one wavelength) and/or a high and not easily realizable density of pumping power. Hence, one is led to consider, in these higher frequency ranges, cavities which are large compared to a wavelength, and which are accordingly capable of supporting a large number of modes within the frequency ranges of interest. A disadvantage of this approach, however, is that masers including such cavities must be operated at relatively high power levels in order that the emission stimulated therefrom be at least as large as that spontaneously emitted therefrom.

Accordingly, another object of this invention is a practically realizable, efficient, low-noise maser structure which is capable of the generation of monochromatic radiation, or coherent amplification, in the infrared, visible, and ultraviolet portions of the electromagnetic spectrum.

The above and other objects of the present invention are realized in an illustrative embodiment thereof wherein a negative temperature medium is disposed between two spaced parallel reflecting plates in a configuration which is of practical size and which may be pumped by readily available power sources, and wherein a single mode corresponding to the stimulated emission can be effectively isolated.

More particularly, one specific illustrative embodiment of the present invention comprises a maser including a chamber having reflective end parallel plates and side walls. Positioned within the chamber is a negative temperature medium, which is pumped by an energy source disposed about the chamber. The side walls are transparent to the pumping energy and either transparent to or absorptive of other energy radiated thereat. Further, an optical configuration is arranged adjacent to one of the end plates of the chamber for isolating the one mode of those supported within the chamber which it is desired to selectively utilize.

The principles of the present invention may illustratively be embodied in a communications system which comprises a maser device capable of generating monochromatic radiation, a second maser device capable of modulating and coherently amplifying the output of the maser generator, and a device for detecting the output of the second maser. Alternatively, such a system may include a maser generator whose output is modulated by a nonamplifying device, or a system in which the maser generator itself is modulated.

Thus, a feature of the present invention is a system for communicating information by means of energy having wavelengths in the infrared, visible, or ultraviolet portions of the electromagnetic spectrum, comprising a monochromatic generator, a modulatable coherent amplifier, and a detector.

Another feature of this invention is a maser generator including a chamber comprising reflective parallel end members and side members, a negative temperature medium within said chamber, a pumping power source disposed about the side members, the side members being transparent to the pumping energy and either transparent to or absorptive of other energy radiated thereat, and a configuration arranged adjacent to one of the end members for abstracting from the chamber a selected one of the modes supported therein, whereby there is provided efficient, low-noise, monochromatic generation of infrared, visible, or ultraviolet waves.

A further feature of the present invention is a maser amplifier including a chamber comprising reflective parallel end members and side members, a negative temperature medium within said chamber, a pumping power source disposed about the side members, the side members being transparent to the pumping energy and nonreflective of other energy radiated thereat, and a configuration arranged adjacent to one of the end members for abstracting from the chamber an amplified replica of a wave fed through the other end member thereof, whereby there is provided an efficient, low-noise, coherent amplifier of infrared, visible, or ultraviolet waves.

A still further feature of this invention is an arrangement for modulating the signal output of a maser of the type herein-described comprising a structure for establishing a magnetic field parallel to the longitudinal axis of the chamber thereof, and an information source capable of varying the magnetic field in correspondence with the output of the source.

The principles of the present invention will be better understood from the following more detailed discussion taken in conjunction with the accompanying drawing, in which:

Fig. 1 is a block diagram of a communications system illustratively embodying aspects of the principles of the present invention;

Fig. 2 is a perspective view of a generator made in accordance with the principles of this invention; and

Fig. 3 is a perspective view of an amplifier embodying the principles of the present invention. Also, Fig. 3 depicts a modulating source and a detector, arranged in typical relationship to the amplifier.

Referring now to Fig. 1, there is shown a communications system in which the principles of the present invention are illustratively embodied. The system includes a generator or oscillator **10**, a modulating source **11**, a modulated amplifier **12**, and a detector **13**.

The generator **10**, which is shown in detail in Fig. 2, includes a chamber **14**, which typically may be about one centimeter in diameter and ten centimeters in length. The chamber **14** comprises a hollow cylinder **15** having its ends capped by two flat parallel assemblies **16** and **17**. Disposed within the chamber **14** is a negative temperature material whose radiative energy level separations correspond to frequencies in the ranges of interest, namely, the infrared, visible, and ultraviolet ranges.

Various materials are suitable for use as the active or negative temperature medium of maser devices of the general type described herein. For example, vapors of the alkali metals; namely, lithium, sodium, potassium, rubidium and cesium, and some solid rare earth salts, for example, anhydrous chlorides of europium and samarium, may be so used.

In particular, potassium maintained at a temperature of about 435 degrees Kelvin, at which temperature it exhibits a vapor pressure of about 0.001 millimeter, may advantageously be included in a specific illustrative embodiment of the principles of the present invention as the active medium thereof.

Each of the flat parallel assemblies **16** and **17** of the device shown in Fig. 2 advantageously includes as a component part thereof a material which reflects most of the energy incident thereupon. Thus, for example, an assembly comprising sapphire **16*a***, which material is characterized by good chemical inertness and excellent transmission properties, particularly for infrared wavelengths, and having a coating of gold **16*b***, typically about 500 angstrom units thick, on the outer surface of the sapphire member, may be included in specific embodiments of this invention. Such an assembly exhibits 97 percent reflectivity, 2 percent absorptivity, and 1 percent transmitivity to wavelengths in the infrared range.

The inner and outer parallel faces of each sapphire plate reflect a small portion of the radiation directed thereat. Therefore, the thickness of the sapphire plates should advantageously be chosen such that the reflections from the two faces of each plate add in phase.

It is noted that the phase angle between the reflections from the two faces or surfaces of each sapphire member depends on the thickness and refractive index thereof. Since sapphire is crystalline and the refractive index is different for ordinary and extraordinary rays, the thickness may be chosen to give constructive interference for one polarization and destructive interference for the polarization perpendicular thereto, in that manner discriminating between modes traveling in the same direction but having different polarizations.

The cylinder **15** of the chamber **14** is advantageously of a material which is transparent to the pumping energy and either transparent to or absorptive of other radiation impinging thereupon, thereby both to allow the negative temperature medium within the cylinder **15** to be pumped and to eliminate from the chamber radiation occurring in all modes except those corresponding to waves which travel back and forth between the reflective assemblies **16** and **17**. These reflected modes are coupled much more strongly to the excited atomic systems of the negative temperature medium than any other modes and hence would be strongly favored for maser oscillations.

In those specific embodiments of the present invention in which the negative temperature medium within the chamber **14** is at a pressure other than atmospheric, as in the case of the potassium vapor, for example, it is advantageous to support the chamber **14**, by means of spacer elements **18**, within a protective shell **19**, typically of glass, within which shell a pressure approximately equal to that within the chamber **14** is maintained. In this manner, the resultant forces acting on the opposing faces of the end assemblies **16** and **17** are made so small as not to be capable of distorting the assemblies and thereby disturbing their parallelism.

Arranged around the protective shell **19** are a plurality of pumping sources **20** which, in a maser generator including potassium vapor as the active medium thereof, may advantageously comprise an assembly of potassium lamps, which lamps **20** are energized by a source **21**.

The maser generator shown in Fig. 2 further includes a housing **22** in which the protective shell **19** is supported by spacer elements **27**. The inner surface of the housing **22** is of a material which is capable of reflecting a major part of the energy radiated thereupon from the pumping power sources **20**, thereby to aid in directing a substantial portion of the energy emitted by the sources **20** toward the chamber **14** and into the negative temperature medium therein.

The process of oscillation within a maser generator made in accordance with the principles of the present invention depends on the selective regeneration within the chamber 14 of a component of the energy spontaneously emitted by the negative temperature medium therein.

Mode selection in the maser generator shown in Fig. 2 is based on the phenomenon that, when energy is radiated from a chamber of the type herein-described through an end plate member which is large compared to the wavelength of the radiation, each mode radiates in a characteristic direction. Thus, if the emitted radiation is focused by a lens, each point in the focal plane thereof will correspond to a mode of a particular direction, affording thereby a separation of modes. And, if radiation falling on a very limited area in the focal plane is detected, that radiation will represent spontaneous and stimulated emission from a selected and limited number of modes, the large background of spontaneous emission produced in other modes being thereby effectively isolated.

The principles of this phenomenon are utilized in the maser generator shown in Fig. 2. Radiation in the desired mode is transmitted through the end assembly 17 and focused by a double-convex lens 23 arranged such that the desired energy is directed through an aperture 24 in an absorptive sheet 25 which lies in the focal plane of the lens 23. A second double-convex lens 26 is employed to reconvert the selected energy to the form of a plane wave, in which form the desired energy radiates to the modulated amplifier 12.

The maser amplifier shown in Fig. 3 is similar in structure to the generator described above. The amplifier includes a chamber 14 comprising a hollow cylinder 15 supported within a protective shell 19 by supporting members 18 and within which cylinder 15 there is disposed a suitable negative temperature medium. The shell 19 is supported within a reflective housing 22 by spacer or supporting members 27.

Arranged about the protective shell 19 of the amplifier 12 shown in Fig. 3 is a pumping power assembly 30 which may advantageously comprise, in a specific illustrative embodiment of the present invention wherein the negative temperature medium is potassium vapor, a potassium lamp formed in the shape of a spiral, which spiral lamp is energized by a source 31.

Energy which is directed from the generator 10 through the left-hand end of the cylinder 15 of the amplifying device of Fig. 3 may be modulated by an assembly including a coil 32 for establishing a magnetic field parallel to the longitudinal axis of the cylinder 15 and a source 11 for varying the strength of the longitudinal magnetic field, whereby broadening or splitting of the spectral lines emitted by the device 12 in correspondence with the variation of the magnetic field results, which phenomenon is generaly termed the Zeeman effect.

The device 12 shown in Fig. 3 radiates through the right-hand end of the cylinder 15 an amplified counterpart of the energy directed at the device 12 by the generator 10. The radiated energy is directed by two lenses 33 and 36 through an aperture 34 in an absorptive member 35 and to a detector 13. The detector 13 may, for example, include a photomultiplier tube.

It is noted that the admission of a signal into the region between the two end parallel plates of the amplifying device 12 is similar to the process involved in a microwave cavity. More particularly, the partially reflecting surfaces of the end plates are analogous to coupling holes; and, if a monochromatic plane wave strikes the outside of one of the partially reflecting surfaces, energy will build up in the region between the plates, and the relations between input wave, energy in the "cavity," and output wave correspond to those for a microwave impinging on an appropriate cavity with input and output coupling holes.

Thus, it is seen that the principles of the present invention may illustratively be embodied in monochromatic maser generators of infrared, visible, or ultraviolet wavelengths. It is feasible to tune such generators by varying the pressure or temperature of the negative temperature media thereof. Alternatively, tuning of such devices may be based on the Stark effect (i.e., observed changes in the spectrum of a system when the system is subjected to an electric field) or on the Zeeman effect.

Further, it has been shown that embodiments of this invention include coherent maser amplifiers of infrared, visible, or ultraviolet wavelengths. It is to be noted that these devices are capable of amplifying energy of these wavelengths with no significant change in the wavefront or phase thereof.

It is to be noted that maser devices embodying the principles of the present invention may advantageously be utilized in various spectroscopy and measurement applications, as well as in the communications field.

It is to be understood that the various specific embodiments disclosed are merely illustrative of the general principles of the invention. Thus, although the amplifying chamber 14 has been shown and described as including a hollow cylinder, it is of course clear that any other transparent structure (more specifically, transparent to the pumping energy and transparent to or absorptive of other radiation) suitable for retaining the negative temperature medium and including reflective end assemblies may be easily substituted therefor.

What is claimed is:

1. A communications system for operation in the infrared, visible, or ultraviolet regions of the electromagnetic wave spectrum comprising a monochromatic maser generator, a coherent modulated maser amplifier, a modulating source, and a detector; said generator comprising a chamber having end reflective parallel members and transparent side members, a negative temperature medium disposed within said chamber, and means arranged about said chamber for pumping said medium; said amplifier comprising a chamber having end reflective parallel members and transparent side members, a negative temperature medium disposed within said chamber, means arranged about said chamber for pumping said medium, and coupling means for abstracting from one end of said chamber an amplified counterpart of the energy transmitted into the other end thereof and for directing said amplified counterpart at said detector.

2. A communications system for operation in the infrared, visible or ultraviolet regions of the electromagnetic wave spectrum comprising a monochromatic maser generator, a coherent maser amplifier, said generator and amplifier including means for modulating the output of said generator in accordance with signal information, and a detector; said generator comprising a chamber having a length which is substantially greater than its transverse dimension and having partially reflective parallel end members and nonreflective side members, a negative temperature medium disposed within said chamber and characterized by at least three distinct energy levels, two of which have a separation in the frequency range of interest, means for pumping said medium so that a population inversion is produced therein between said two separated energy levels, and means for abstracting from said chamber and directing at the amplifier input the energy of a particular mode of electromagnetic vibration; said amplifier comprising a chamber having a length which is substantially greater than its transverse dimension and having partially reflective parallel end members and nonreflective side members, a negative temperature medium disposed within said chamber and characterized by at least three distinct energy levels, two of which have a separation in the frequency range of interest, means for pumping said medium so that a population inversion is produced therein between said two separated energy levels, means for abstracting from said chamber the energy of a particular mode of electromagnetic vibration representing an amplified and modulated replica of the generator output, and means for directing said replica at said detector.

3. A communications system for operation in the in-

frared, visible, or ultraviolet regions of the electromagnetic wave spectrum comprising a monochromatic maser generator, a coherent modulated maser amplifier, a modulating source, and a detector; said amplifier including means defining an amplifying chamber, and means for establishing a magnetic field parallel to the longitudinal axis of said chamber, said modulating source being coupled to said magnetic means, the radiative output of said generator being directed at said amplifier, and the radiative output of said amplifier, constituting an amplified and modulated counterpart of the energy radiated thereinto, being directed at said detector.

4. A maser generator comprising a chamber having end reflective parallel members and side members, a negative temperature medium disposed within said chamber, and means arranged about said chamber for pumping said medium, said side members being transparent to the pumping energy and transparent to or absorptive of other energy radiated thereat.

5. A maser generator for operation in the infrared, visible or ultraviolet regions of the electromagnetic wave spectrum comprising a chamber having a length which is substantially greater than its transverse dimension and having partially reflective parallel end members and non-reflective side members, a negative temperature medium disposed within said chamber and characterized by at least three distinct energy levels, two of which have a separation in the frequency range of interest, means for pumping said medium so that a population inversion is produced therein between said two separated energy levels, and means for abstracting from said chamber and directing at an amplifier input the energy of a particular mode of electromagnetic vibration.

6. A maser generator as in claim 5 wherein said mode selecting means includes an absorptive member having an opening therethrough, said absorptive member being positioned adjacent to one end of said chamber, and means for directing a selected portion of the energy radiated by said generator through said opening.

7. A maser generator as in claim 5 wherein said negative temperature medium comprises potassium, and said pumping means comprises an assembly of potassium lamps.

8. A maser amplifier comprising a chamber having end reflective parallel members and said members, a negative temperature medium disposed within said chamber, means arranged about said chamber for pumping said medium, said side members being transparent to the pumping energy and non-reflective of other energy radiated thereat, and coupling means for abstracting from one end of said chamber an amplified counterpart of the energy directed into the other end thereof.

9. A maser amplifier for operation in the infrared, visible or ultraviolet regions of the electromagnetic wave spectrum comprising a chamber having a length which is substantially greater than its transverse dimension and having partially reflective parallel end members and non-reflective side members, a negative temperature medium disposed within said chamber and characterized by at least three distinct energy levels, two of which have a separation in the frequency range of interest, means for pumping said medium so that a population inversion is produced therein between said two separated energy levels, and means for abstracting from said chamber and directing at a detector the energy of a particular mode of electromagnetic vibration.

10. A maser amplifier as in claim 9 wherein said negative temperature medium comprises potassium, and said pumping means comprises an assembly of potassium lamps.

11. A modulated maser amplifier comprising a chamber having end reflective parallel members, a negative temperature medium disposed within said chamber, means arranged about said chamber for pumping said medium, means coupled to and under the control of a modulating source for establishing a magnetic field parallel to the longitudinal axis of said chamber, and means for abstracting from one end of said chamber an amplified counterpart of the energy directed into the other end thereof, which counterpart is modulatable in accordance with the output of said source.

References Cited in the file of this patent

UNITED STATES PATENTS

2,836,722 Dicke et al. .... May 27, 1958

R. J. Gatling.

Machine Gun.

No. 36,836. Patented Nov. 4, 1862.

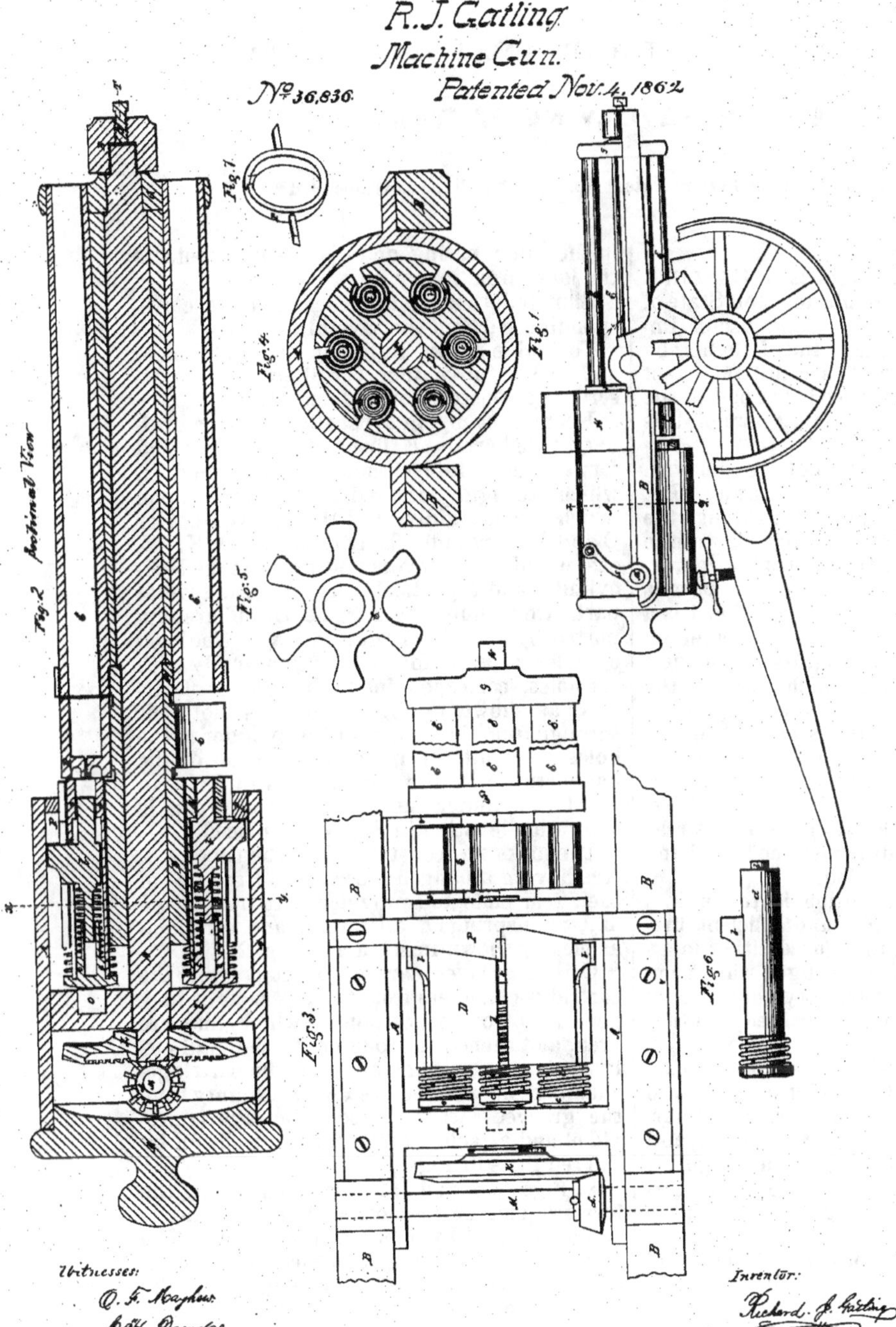

Witnesses:

O. F. Mayhew

C. H. Davidge.

Inventor:

Richard J. Gatling

UNITED STATES PATENT OFFICE.

RICHARD J. GATLING, OF INDIANAPOLIS, INDIANA.

IMPROVEMENT IN REVOLVING BATTERY-GUNS.

Specification forming part of Letters Patent No. **36,836**, dated November 4, 1862.

To all whom it may concern:

Be it known that I, RICHARD J. GATLING, of Indianapolis, county of Marion, and State of Indiana, have invented new and useful Improvements in Fire-Arms; and I do hereby declare that the following is a full and exact description thereof, reference being had to the accompanying drawings, making part of this specification, in which—

Figure 1 is a side elevation of the gun with the upper portion of the wheels cut away. Fig. 2 is a vertical longitudinal section through the center of the gun. Fig. 3 is a top view of the gun with the top half of the external casing, A, left off and the middle portion of the barrels cut away to shorten the drawing. Fig. 4 is a transverse section through lock-cylinder on line *x y* in Figs. 1 and 2. Fig. 5 is an end view of the grooved carrier C which receives the cartridges or cartridge-chambers. Fig. 6 is a side view of one of the tubes containing the mainspring and hammer of one of the locks. Fig. 7 is a perspective view of the ring P which surrounds the forward end of the lock-cylinder D, having inclined planes on its rear edge for cocking and drawing back the hammers to their proper position.

The object of this invention is to obtain a simple, compact, durable, and efficient fire-arm for war purposes, to be used either in attack or defence, one that is light when compared with ordinary field-artillery, that is easily transported, that may be rapidly fired, and that can be operated by few men.

The invention consists in a singularly-constructed revolving lock cylinder or breech, in combination with a grooved carrier and barrels all rigidly fixed upon the same shaft, and all of which revolve together when the gun is in operation, the locks and grooves in the carrier and the barrels all being parallel with the axis of revolution.

The invention also consists in the novel means employed in cocking and firing the gun without the use of a trigger by means of the inclined plane on the rear edge of the ring P, which surrounds the forward end of the lock-cylinder, and also in the novel use of the inner tubes (which contain the locks) to press the cartridge-chambers firmly against the rear ends of the barrels while being discharged, and in the outer casing and disk, which protects the locks from injury.

Similar letters of reference indicate corresponding parts in the several figures.

To enable others skilled in the art to make and use my invention, I will proceed to describe its construction and operation.

I construct my gun usually with six ordinary rifle-barrels, E, fixed at their rear and forward ends into circular plates F and G, which are rigidly secured to a shaft, N, upon which is also rigidly fixed the grooved carrier C and lock-cylinder D and cog-wheel K. A case or shield, A, covers and protects the lock-cylinder and cog-wheel. All of these several parts are mounted on a frame, B, and are supported by an ordinary gun-carriage. The lock-cylinder D is perforated longitudinally with six holes, (corresponding to the number of barrels,) as shown in Fig. 4, and has slots cut through from the surface of the cylinder to the holes to admit the projecting portion of the hammers *b*. In the perforations or holes in the lock-cylinder the locks (one of which is shown in elevation in Fig. 6) are placed.

The locks are constructed of the tubes *a a*, &c., having a flanged breech-pin, *c*, secured in their rear ends and provided with hammers *b* and mainsprings *d*, all formed and arranged as clearly shown in section in Fig. 2.

C is a grooved carrier for conveying the cartridge-chambers from the reservoir or hopper H up to the position in which they are fired, and thence on around until they fall out by their own weight; but that the cartridge-chambers may be removed with certainty from the grooved carrier C a comb or rake is provided and attached to the frame, as shown by the red lines in Figs. 2 and 3.

P, Figs. 2, 3, and 7, is a ring encircling the forward end of the lock-cylinder D, and is rigidly secured by lugs to the frame B. The rear edge of this ring is formed into two inclined planes, as clearly shown in Fig. 3, the greater inclined plane serving to push back or cock the hammers *b* as they are successively revolved, while the lesser inclined plane serves to push the hammers back into their proper places within the tubes *a* after they have struck the percussion-cap, so as to allow the cartridge-chambers to drop from the carrier.

The disk I forms a division in the case A, the forward portion of the case forming a shield or covering for the locks, while the rear division contains and protects the cog-wheel K and L. In the forward face of the disk I a small steel plug, O, is inserted, having its forward face rounded or swelled out slightly beyond the face of the disk. This swell is for the purpose of pressing the tubes *a* forward against the cartridge-chambers R, and thus pressing the cartridge-chambers firmly against the rear end of the barrel at the time of each and every discharge, thereby preventing the escape of gas from the ignited powder. The forward motion of the tubes *a*, caused by the swell O on disk I, also assists in compressing the mainsprings *d*, thereby increasing the force of the blow from the lock-hammers *b* upon the percussion-caps on the nipples of the cartridge-chambers.

The rounded heads of the breech-pin *c* bear against the forward face of the disk I, being kept in their position by the coiled springs *e e*, &c., which surround the rear ends of the tubes *a a*, &c., the springs *e* bearing against the rear end of the lock-cylinder and against the flange of the breech-pin *c*. By this arrangement the forward ends of the locks are kept flush with the forward face of the lock-cylinder until they are revolved opposite the swell *o*, when they are pressed forward, as before described.

The shaft N, upon which the lock-cylinder D, carrier C, barrels E, and cog-wheel K are rigidly secured, has a bearing near its rear end in disk I and a bearing at its forward end in a box on the frame B. A crank-shaft, M, runs through the rear part of case A and has fixed upon it the small cog-wheel or pinion L and crank S.

An adjusting-screw, T, is placed in the box opposite the forward end of shaft N, for regulating the pressure upon the cartridge-chambers R. The cartridge-chambers R, (any desired number of which may be used,) being loaded, are placed in the hopper or reservoir, with their nipple or cap ends toward the hammers, over the grooved carriers C, when, by rotating the crank S, which carries with it the shaft M, and pinion L, which meshes into the large cog-wheel K, thereby revolving the shaft N, lock-cylinder D, carrier C, and barrels E, the cartridges drop or rather roll into the grooves of carrier C and are carried by it up to the position in which they are discharged. The hammers, cartridge-chambers, and barrels all being on a line parallel to the axis of revolution, it is impossible for the cartridges to be out of place when discharged.

The hammers *b* are pushed back by the large inclined plane on the rear edge of the ring P, and when they have passed the highest point of the inclined plane they are driven forward against the percussion-cap on the nipple of the cartridge-chamber by the coiled mainspring *e* with sufficient force to explode the cap and discharge the cartridge, after which the cartridge-holder is carried on around until it drops out of the carrier by its own weight, when it is ready to be taken up and reloaded.

I do not claim the use of the grooved or fluted revolving carrier C, separately considered, and when the same is made to revolve separately and independently of the barrels and breech, the same being an old device; neither do I claim the direct combination thereof with an automatic revolving gear or with a device for pressing the cartridge-chamber against the barrel when used alone for that purpose; but

What I do claim as new and as my invention, and desire to secure by Letters Patent, is—

1. The combination of the lock-cylinder or breech D with the grooved carrier C, circular plate F, and barrels E E, &c., the lock-cylinder or breech, carrier, and circular plate being firmly fastened upon the main shaft N, and the locks, grooves in the carrier, and barrels being arranged on a line parallel with the axis of revolution, the whole revolving together when the gun is in operation, substantially as described.

2. In the construction of revolving fire-arms, the use of as many locks as there are barrels, said locks revolving simultaneously with the breech and barrels, and being arranged and operated substantially as set forth.

3. The stationary ring P, provided with inclined planes on its rear edge, in combination with lock-cylinder D and locks, when constructed and operated for the purposes substantially as set forth.

4. The tubes *a a*, &c., furnished with the flanged breech-pins *c c*, &c., and springs *e e*, &c., and which contain the lock-hammers *b b*, &c., and mainsprings *d d*, &c., in combination with the revolving breech D, disk I, and swell *o*, when constructed, arranged, and operated for the purposes substantially as set forth.

5. The disk I, in combination with the external breech-piece or casing, A, which forms a shield or covering for the lock-cylinder and which protects the locks and cog-wheels from injury.

RICHARD J. GATLING.

Witnesses:

A. F. MAYHEW,

W. O. ROCKWOOD.

E. BERLINER.
Microphone.

No. 224,573. Patented Feb. 17, 1880.

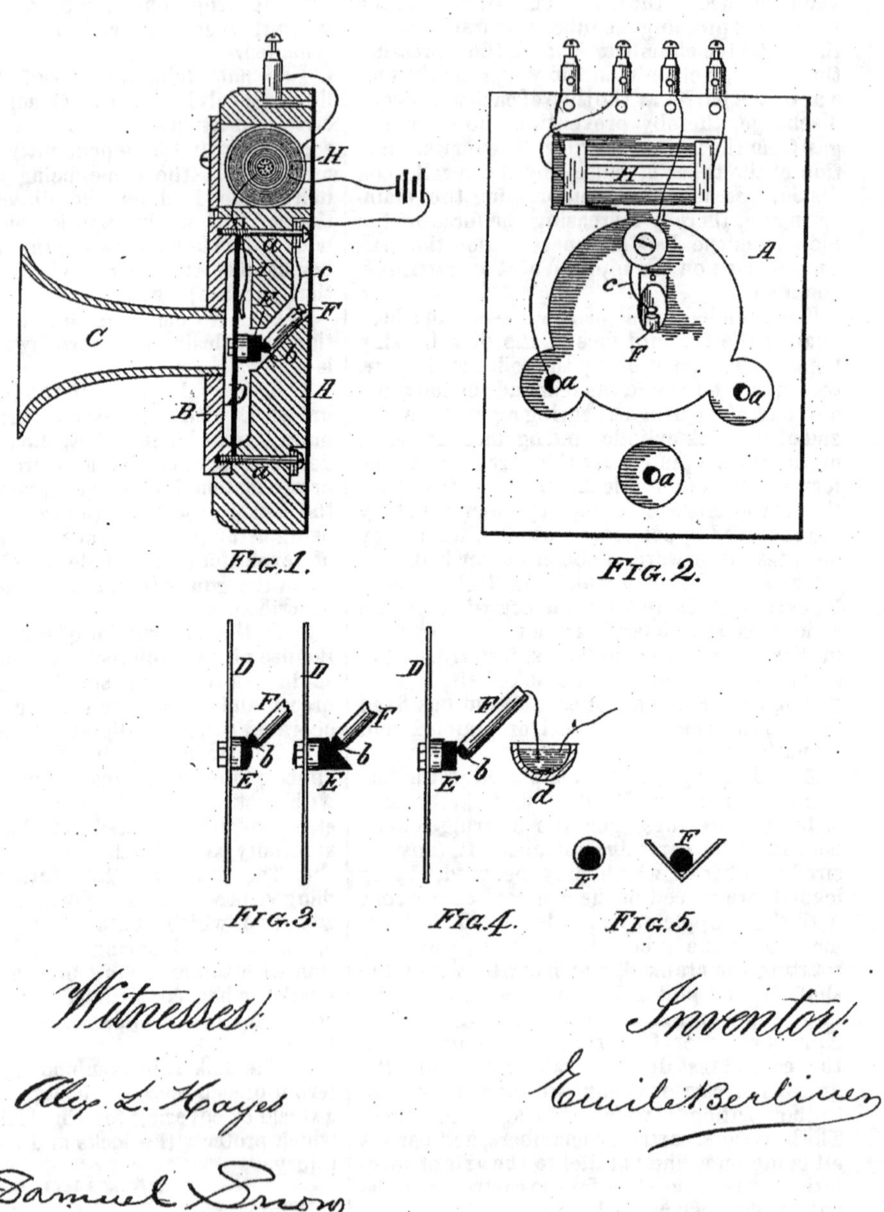

FIG. 1. FIG. 2. FIG. 3. FIG. 4. FIG. 5.

Witnesses:

Alex. L. Hayes

Samuel Snow

Inventor:

Emile Berliner

N. PETERS, PHOTO-LITHOGRAPHER, WASHINGTON, D. C.

UNITED STATES PATENT OFFICE.

EMILE BERLINER, OF BOSTON, MASSACHUSETTS.

MICROPHONE.

SPECIFICATION forming part of Letters Patent No. 224,573, dated February 17, 1880.

Application filed September 5, 1879.

To all whom it may concern:

Be it known that I, EMILE BERLINER, of Boston, in the county of Suffolk and State of Massachusetts, have invented a new and useful Improvement in Microphones or Contact-Telephones, of which the following is a specification, reference being had to the accompanying drawings.

This invention is an improvement upon the microphone for which I filed an application for a patent of the United States August 11, 1879; and it consists in dispensing with the clamping device for fixing the carbon pin in position, in maintaining it constantly in contact with the diaphragm by the action of gravity, and in connecting the carbon pin to the battery by a device that will not interfere with said action.

In the accompanying drawings, Figure 1 is a sectional view of a microphone-transmitter embodying my improvement. Fig. 2 is a rear view of the same. Fig. 3 shows modifications in the form of the carbon button on the diaphragm. Fig. 4 shows the manner of connecting the carbon pin to the battery by means of a cup of mercury, and Fig. 5 shows modifications in the form of the opening in the block.

If the vibrating diaphragm of the microphone is placed horizontally, the carbon pin may be suspended above it vertically, and in contact with the diaphragm, by a flexible conductor which will not interfere with the action of gravity; but when the vibrating diaphragm is placed perpendicularly, as is generally the case, the carbon pin may be made to bear upon the diaphragm by the action of gravity by being placed in an inclined opening or perforation in the block behind the diaphragm, to which block the diaphragm is attached. This arrangement is shown in Figs. 1 and 2.

In the several figures the same letters refer to the same parts.

A is a block, of wood, vulcanite, or other similar material, which has a recess upon its face, opposite to which recess is fastened the diaphragm D by the screws *a a a*.

B is a block, of metal, on the other side of the diaphragm, into which block the screws pass, so as to clamp the diaphragm between the blocks A and B, and C is a mouth-piece screwed into an opening in the block B.

To the back of the diaphragm, at its center, is suitably secured a block, E, of hardened carbon, and opposite to this carbon block is an inclined opening or perforation in the wooden block A, in which rests loosely the carbon pin F, which, sliding in the inclined opening, is maintained in contact with the carbon block E. This pin F may be made entirely of hardened carbon, or it may be made partly of metal and be provided with a tip of hardened carbon.

In order to obtain as small a surface of contact as possible, it is desirable that the end of the carbon pin in contact with the carbon block should be convex.

I is a metal spring attached to the block A, one end of which spring rests upon the diaphragm and has a covering of soft rubber. The function of this damper is well known, and it will not be required when a diaphragm is used which has but little vibration—as, for instance, one of carbon or hard wood.

H is the induction-coil, which is generally used with a microphone-transmitter, and is constructed and connected in the well-known manner.

The connection of the carbon pin with the battery must be such that it will not interfere with the free movement of the pin in the inclined opening. This result may be accomplished by the use of a soft and pliable wire or any flexible conductor, as shown at C in Figs. 1 and 2, or by the use of a mercury-cup, as shown in Fig. 4.

The face of the carbon block E may be parallel with the diaphragm, as shown in Fig. 1, or it may be convex or form an angle with the diaphragm, as shown in Fig. 3.

The opening in the block A may be elliptical in section, as shown in Fig. 2, or circular or angular in section, as shown in Fig. 5.

What I claim as my invention, and desire to secure by Letters Patent of the United States, is—

1. In combination with a vibrating surface forming one electrode of an electric current, an opposite electrode maintained in contact with the vibrating surface by the action of gravity, and connected to the battery by a conductor which will not interfere with said action, substantially as and for the purpose set forth.

2. In a contact-telephone or microphone

having no frictional or rubbing electric contact affected by sound-waves, the combination, with a vibrating surface forming one electrode of an electric current, of a pin forming the opposite electrode, sliding freely on a suitable support inclined at an angle toward the vibrating surface, and connected to the battery in such a manner that its free movement on its support will not be interfered with, substantially as and for the purpose set forth.

3. In combination with a vibrating surface forming one electrode of an electric current, the block A, provided with an opening or perforation inclined downward to the vibrating surface, a pin, F, having a carbon extremity and sliding freely in the inclined opening, so as to make contact at its carbon extremity with the vibrating surface, and a connection between said pin and the battery which will not interfere with the free movement of the pin, substantially as and for the purpose set forth.

4. The combination of the diaphragm D, the carbon block E, the damper I, the block A, the pin F, sliding freely in an inclined opening in the block A, and the flexible conductor *c*, substantially as and for the purpose set forth.

5. A contact-telephone or microphone having no frictional or rubbing electric contact affected by sound-waves, in which a variation of electric contact is effected by a variation in the pressure exerted by a body moving or sliding freely by the action of gravitation in an angle inclined toward said contact, substantially as and for the purpose set forth.

In witness whereof I have hereunto set my hand.

EMILE BERLINER.

Witnesses:
ALEX. L. HAYES,
SAMUEL SNOW.

Sheet 1. 3 Sheets

S. F. B. Morse.

Telegraph Signs.

No. 1,647. Patented Jun. 20, 1840.

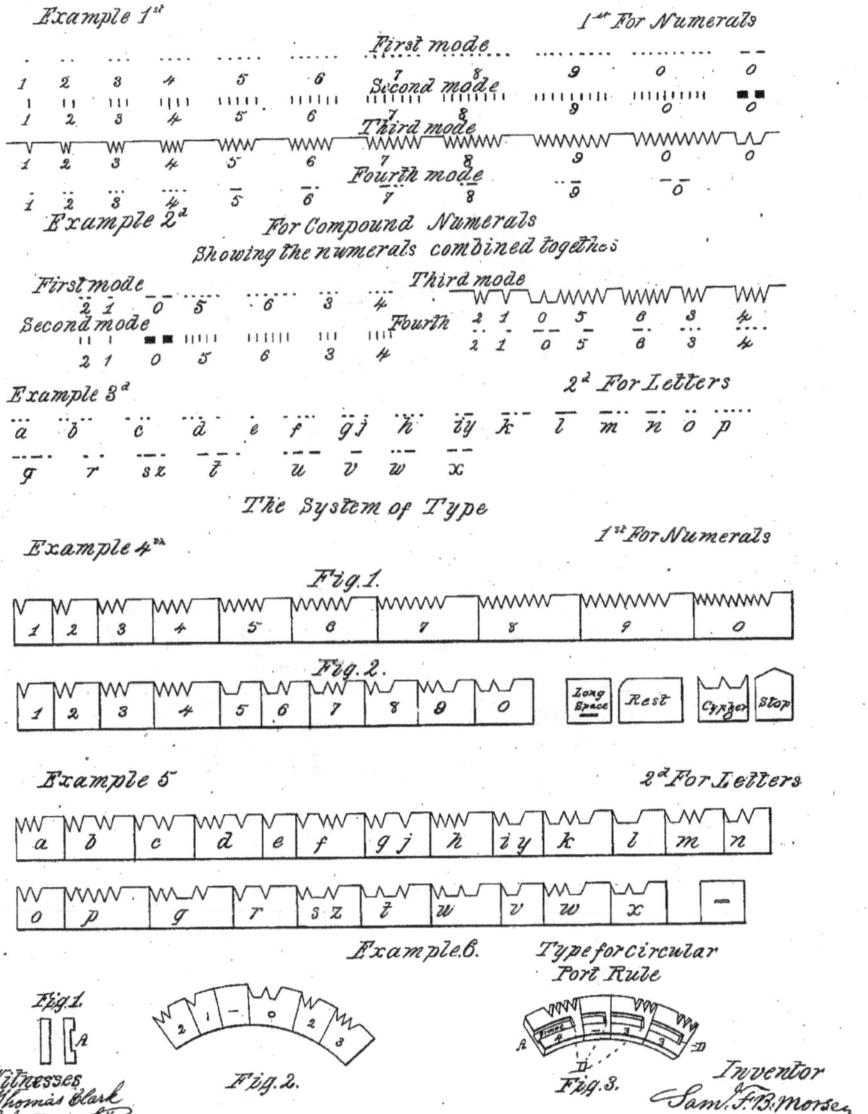

Example 1st — 1st For Numerals

First mode

Second mode

Third mode

Fourth mode

1 2 3 4 5 6 7 8 9 0

Example 2d — For Compound Numerals

Showing the numerals combined together

First mode — Second mode — Third mode — Fourth

2 1 0 5 6 3 4

Example 3d — 2d For Letters

a b c d e f g j h i y k l m n o p

q r s z t u v w x

The System of Type

Example 4th — 1st For Numerals

Fig. 1.

1 2 3 4 5 6 7 8 9 0

Fig. 2.

1 2 3 4 5 6 7 8 9 0 Long Space Rest Cypher Stop

Example 5 — 2d For Letters

a b c d e f g j h i y k l m n

o p q r s z t u v w x —

Example 6. — Type for circular Port Rule

Fig. 1. A

Fig. 2.

Fig. 3.

Witnesses
J. Thomas Clark
Alexr. Jackson

Inventor
Saml. F. B. Morse

S. F. B. Morse.

Telegraph Signs.

N.º 1,647. Patented Jun. 20, 1840.

Example. 7.

Type Rule.

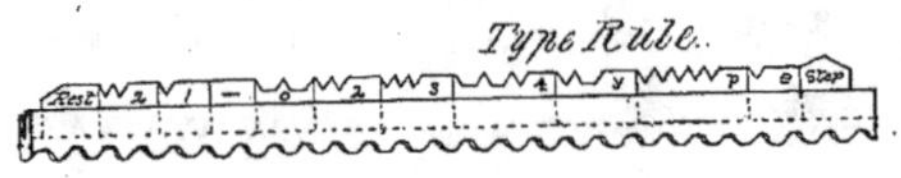

Example. 8.

Straight Port Rule

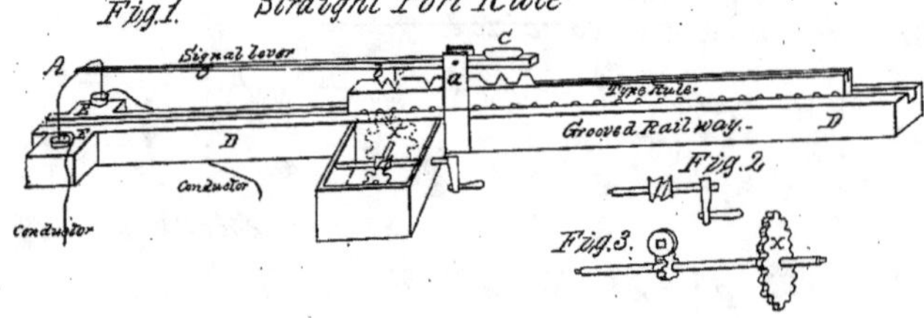

Example. 9.

Circular Port Rule.

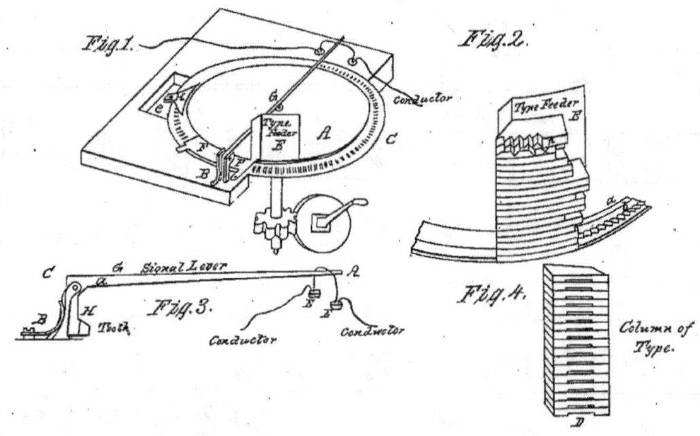

Witnesses.
J. Thomas Clark
Alex. Jackson

Inventor
Sam.l F. B. Morse

Sheet 3. 3 Sheets

S. F. B. Morse.

Telegraph Signs.

No 1,647. Patented Jun. 20, 1840.

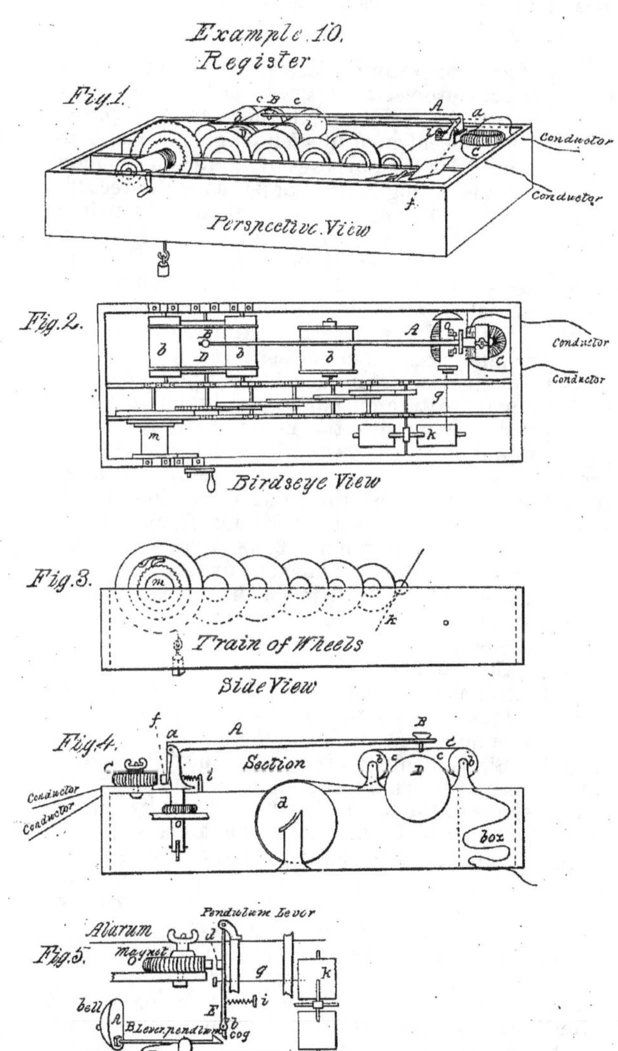

Witnesses.
J. Thomas Clark
" "
Alexr. Jackson

Inventor
Saml. F. B. Morse.

UNITED STATES PATENT OFFICE.

SAMUEL F. B. MORSE, OF NEW YORK, N. Y.

IMPROVEMENT IN THE MODE OF COMMUNICATING INFORMATION BY SIGNALS BY THE APPLICATION OF ELECTRO-MAGNETISM.

Specification forming part of Letters Patent No. 1,647, dated June 20, 1840.

To all whom it may concern:

Be it known that I, the undersigned, SAMUEL F. B. MORSE, of the city, county, and State of New York, have invented a new and useful machine and system of signs for transmitting intelligence between distant points by the means of a new application and effect of electro-magnetism in producing sounds and signs, or either, and also for recording permanently by the same means, and application, and effect of electro-magnetism, any signs thus produced and representing intelligence, transmitted as before named between distant points; and I denominate said invention the "American Electro-Magnetic Telegraph," of which the following is a full and exact description, to wit:

It consists of the following parts—first, of a circuit of electric or galvanic conductors from any generator of electricity or galvanism and of electro-magnets at any one or more points in said circuit; second, a system of signs by which numerals, and words represented by numerals, and thereby sentences of words, as well as of numerals, and letters of any extent and combination of each, are communicated to any one or more points in the before-described circuit; third, a set of type adapted to regulate the communication of the above mentioned signs, also cases for convenient keeping of the type and rules in which to set and use the type; fourth, an apparatus called the "straight port-rule," and another called the "circular port-rule," each of which regulates the movement of the type when in use, and also that of the signal-lever; fifth, a signal-lever which breaks and connects the circuit of conductors; sixth, a register which records permanently the signs communicated at any desired points in the circuit; seventh, a dictionary or vocabulary of words to which are prefixed numerals for the uses hereinafter described; eighth, modes of laying the circuit of conductors.

The circuit of conductors may be made of any metal—such as copper, or iron wire, or strips of copper or iron, or of cord or twine, or other substances—gilt, silvered, or covered with any thin metal leaf properly insulated and in the ground, or through or beneath the water, or through the air. By causing an electric or galvanic current to pass through the circuit of conductors, laid as aforesaid, by means of any generator of electricity or galvanism, to one or more electro-magnets placed at any point or points in said circuit, the magnetic power thus concentrated in such magnet or magnets is used for the purposes of producing sounds and visible signs, and for permanently recording the latter at any and each of said points at the pleasure of the operator and in the manner hereinafter described—that is to say, by using the system of signs which is formed of the following parts and variations, viz:

Signs of numerals consist, first, of ten dots or punctures, made in measured distances of equal extent from each other, upon paper or any substitute for paper, and in number corresponding with the numeral desired to be represented. Thus one dot or puncture for the numeral 1, two dots or punctures for the numeral 2, three of the same for 3, four for 4, five for 5, six for 6, seven for 7, eight for 8, nine for 9, and ten for 0, as particularly represented on the annexed drawing marked Example 1, Mode 1, in which is also included a second character, to represent a cipher, if prefered.

Signs of numerals consist, secondly, of marks made as in the case of dots, and particularly represented on the annexed drawing marked Example 1, Mode 2.

Signs of numerals consist, thirdly, of characters drawn at measured distances in the shape of the teeth of a common saw by the use of a pencil or any instrument for marking. The points corresponding to the teeth of a saw are in number to correspond with the numeral desired to be represented, as in the case of dots or marks in the other modes described, and as particularly represented in the annexed drawing marked Example 1, Mode 3.

Signs of numerals consist, fourthly, of dots and lines separately and conjunctively used as follows, the numerals 1, 2, 3, and 4 being represented by dots, as in Mode 1, first given above: The numeral 5 is repesented by a line equal in length to the space between the two dots of any other numeral; 6 is represented by the addition of a dot to the line representing 5; 7 is represented by the addition of two dots to said line; 8 is represented by prefixing a dot to said line; 9 is represented by two dots prefixed to said line; and 0 is represented by two lines, each of the length of said line that represents the number 5; said signs are particularly set forth in the annexed drawings, marked Example 1, Mode 4.

Either of said modes are to be used as may be preferred or desired and in the method hereinafter described.

The sign of a distinct numeral, or of a compound numeral when used in a sentence of words or of numerals, consists of a distance or space of separation between the characters of greater extent than the distance used in separating the characters that compose any such distinct or compound numeral. An illustration of this sign is particularly exhibited in the annexed drawing marked Example 2.

Signs of letters consist in variations of the dots, marks, and dots and lines, and spaces of separation of the same formation as compose the signs of numerals, varied and combined differently to represent the letters of the alphabet in the manner particularly illustrated and represented in the annexed drawing marked Example 3.

The sign of a distinct letter, or of distinct words, when used in a sentence, is the same as that used in regard to numerals and described above.

Signs of words, and even of set phrases or sentences, may be adopted for use and communication in like manner under various forms, as convenience may suggest.

The type for producing the signs of numerals consist, first, of fourteen pieces or plates of thin metal, such as type-metal, brass, iron, or like substances, with teeth or indentations upon one side or edge of ten of said type, corresponding in number to the dots or punctures or marks requisite to constitute the numerals respectively heretofore described in the system of signs, and having also a space left upon the side or edge of each type, at one end thereof, without teeth or indentations, corresponding in length with the distance or separation desired between each sign of a numeral. Another of said type has two indentations, forming thereby three teeth only, and without any space at either end, to correspond with the size of a cypher, as heretofore described by reference to Example 1, Modes 1, 2, 3, of drawings in said system of signs. One other of said type is without any indentation on its side or edge, and being in length to correspond with the distance or separation desired between distinct or compound numerals, and with the sign heretofore described for that purpose. One of the remaining two of said type is formed with one corner of it beveled, (system of type, Example 4, Fig. 1,) and is called a "rest," and the other is in a pointed form and called a "stop."

Each of said type is particularly delineated on the annexed drawing marked Example 4, Fig. 1, and numbered or labeled in accordance with the purpose for which they are designed respectively, and are used, in like manner, for producing each of the several signs of numerals heretofore described in the system of signs.

The type for producing the signs of numerals consist, secondly, of five pieces or plates of metal first described above, four of which are the same as are numbered 1, 2, 3, and 4 in the annexed drawing marked Example 4, Fig. 1, and the fifth one being the same as is denominated in the same example "the long space," and heretofore alluded to; also, of six other pieces or plates of said metal, varied in indentations and teeth and spaces, as represented on the annexed drawings marked Example 4, Fig. 2, to produce signs of the denominations described in the fourth mode of the before-mentioned system of signs, Example 1.

The type for producing the signs of letters are of the same denomination with those used in producing signs of numerals, and only varied in form, from one to twenty-three, as exhibited in the annexed drawing marked Example 5.

The type for producing both signs of numerals and signs of letters are adapted for use to either a straight rule, called the "straight port-rule," and are in that case made straight lengthwise, as described in the drawings annexed and heretofore referred to in Example 5, or to a circular port-rule, in which case they are lengthwise circular or formed into sections of a circle, as represented in the drawings annexed marked Example 6, Figs. 2 and 3, and as will be further understood by the descriptions hereinafter contained of the straight and circular port-rules. On the under side of the type for the circular port-rule (which type are of greater thickness than those for the straight port-rule) is a groove (system of type, Example 6, A in Figs. 1 and 3) about midway of their width, and in depth about half the thickness aforesaid, and extending from the space ends, as B, Example 6, Fig. 3—that is, the ends without indentation—of said type, along the length, and conforming to the curve thereof, to a point, D D, equal in distance from the opposite ends to half the width of the pointed teeth cut upon their edges. For a delineation of these type reference is made to sections thereof in Figs. 1 and 3 upon the annexed drawings marked Example 6.

The type-cases are wood, or of any other material, with small compartments of the exact length of the type, for greater convenience in distributing, and resembling those in common use among printers.

The type-rules are of wood or metal, or other material that may be preferred, and about three feet in length, with a groove, into which the type, when used, are placed. On the under side of each type-rule are cogs, by which they are adapted to a pinion-wheel having corresponding cogs and forming part of a port-rule. The type-rule in use is moved onward as motion is given to the said wheel. A delineation of the type-rule is contained in the annexed drawing marked Example 7.

The straight port-rule consists of a pinion-wheel, before mentioned, turned by a hand-crank attached to a horizontal screw that plays into the cogs of the pinion-wheel as the latter do into the cogs of the type-rule, or by any other power in any of the well-known methods of mechanism. It is connected with a railway or groove, in and by which the type-rule, from the motion imparted to it by said wheel, is conveyed in a direct line beneath a lever that breaks and connects the galvanic circuit in the manner hereinafter mentioned. A delineation of said wheel, crank, and screw is contained in the drawings hereunto annexed marked Example 8, Figs. 1, 2, 3.

The circular port-rule is a substitute, when preferred, for both the type-rule and the straight port-rule, and consists of a horizontal or inclined wheel, Example 9, Fig. 1, A, of any convenient diameter, of wood or metal, having its axis connected on the under side of the wheel, with a pinion-wheel, K, and as in the case of the straight port-rule. It is moved by the motion of the pinion-wheel, as is the type-rule in the former description. On the entire circumference of said horizontal or inclined wheel, and upon its upper surface, is a shoulder or cavity, *a*, Figs. 1, 2, corresponding in depth with the thickness of the type used, and in width, *b*, equal to that of the type, exclusive of their teeth or indentations. Near the outer edge of the surface of said shoulder or cavity are cogs *c*, throughout the circumference of the wheel, projecting upward at a distance from each other equal to one-half of the width of the teeth or indentations of the type, and otherwise corresponding in size to the width and depth of the groove D D, Fig. 4, in the under side of the circular type before described and illustrated by reference to Example 6, Figs. 1 and 3. Directly over said shoulder or cavity and cogs, and at one or more points on the circumference of said wheel, is extended from a fixture outside of the orbit of the wheel a stationary type-feeder, E, Fig. 1, formed of one end, *e*, and one side, E, perpendicular, of tin or brass plate or other substance, and of interior size and shape to receive any number of the type which are therein deposited with their indentations projecting outward, as in Fig. 2, and their grooves downward, as in Fig. 4. Said type-feeder is so suspended from its fixture F F over the shoulder or cavity of the wheel A, before described, as to admit of the passage under it of said wheel in its circuit as near the bottom of the feeder as practicable, without coming in contact therewith. The type deposited in the feeder as before mentioned form a perpendicular column, as in Fig. 2, the lower type of which rests upon the surface of the before-named shoulder of the wheel *b*, Fig. 2, and the cog of the wheel, projecting upward, enters the groove D D, Fig. 4, of the type hereinbefore described.

The operation of said circular port-rule in regulating the movement of the type in sue is as follows: When the wheel A is set in motion the type resting immediately upon the shoulder of the wheel, in the manner mentioned above, as in Fig. 2, is carried forward on the curvature of the wheel from beneath the column of type resting upon it in the stationary type-feeder by means of one of the before-named cogs coming in contact with that point D, Fig. 3, Example 6, in the groove of the type, hereinbefore described as forming the termination of said groove, and which is particularly delineated at the points D D in the annexed drawings, marked Example 6, Fig. 3. As by said process the lower type in the column that is held by the stationary feeder is carried forward and and removed, the next type settles immediately upon the shoulder of the wheel, and, after the manner of the removed type, is brought in contact with another cog of said shoulder within the groove of the type, and thence carried forward from beneath the incumbent column, as was its predecessor. Then follows consecutively in the same method each type deposited within the feeder so long as the wheel is kept in motion. The deposit of the type in the stationary feeder is regulated by the order in which the letters or numerals or words they represent are designed to be communicated at any distant point or points. After the type are respectively carried forward on the curvature of the wheel in the manner stated above, beyond the point where they are acted upon by the signal-lever, as is hereinafter described, they are lifted, each in its turn, from the shoulder of the wheel A and cast off into a box or pocket, G, below the wheel by means of a slender shaft or spindle, H, made of any metal, and resembling in form a common plowshare, extending downward from a fixture, *o*, placed outside of the wheel, into a groove, K, within the before-named shoulder of said wheel A, and on the inner side of the cogs *c*, already described. By means of said groove the downward point of said shaft or spindle H is brought within the curvature and below the surface of said shoulder *b*, Fig. 2, and consequently under the approaching end of the type, so that each type successively, as it is carried forward on said curvature, in the manner before described, is lifted from the shoulder and forced upward on the inclined shaft or spindle by the type in contact with it at the other end until turned off into the before-named box or pocket G below, ready for a redistribution.

For a more particular delineation of the several parts of said circular port-rule reference is made to the annexed drawings marked Example 9, Figs. 1 and 2.

The signal-lever, Example 9, Fig. 3, consists, first, for use with the straight port-rule, Example 8, Fig. 1, A, of a strip of wood of any length from six to twenty-four inches, resting upon a pivot, *a*, or in a notched pillar formed into a fulcrum by a metal pin, *a*, passing through it and the lever. At one end of the lever a metallic wire, bent to a semicircular or half-square form, as at A, or resembling the prongs of a fork distended, is attached by its center, as described in the annexed drawings, Example 8, at the point marked A. Between said end of the lever and the fulcrum *a*, and near the latter, on the under side of the lever

A, is inserted a metallic tooth or cog, b, curved on the side nearest to the fulcrum, and in other respects corresponding to the teeth or indentations upon the type already described. On the opposite extremity of the lever is a small weight, C, to balance or offset, in part, when needed, the weight of the lever on the opposite side of the fulcrum. The lever thus formed is stationed directly over the railway or groove D D, heretofore described as forming a connected part of the straight port-rule. The movement of the type-rule brings the tooth of each type therein set in contact with the tooth or cog of the lever, and thereby forces the lever upward until the points of the two teeth in contact have passed each other, when the lever again descends as the teeth of the type proceeds onward from the tooth of the lever. This operation is repeated as frequently as the teeth of the type are brought in contact with the tooth of the lever. By thus forcing the said lever upward and downward the ends of the semicircular or pronged wire are made alternately to rise from and fall into two small cups or vessels of mercury, E E, in each of which is an end or termination of the metallic circuit-conductors, first described above. This termination of the metallic circuit in the two cups or vessels breaks and limits the current of electricity or galvanism through the circuit; but a connection of the circuit is effected or restored by the falling of the two ends of the pronged wire A attached to said lever into the two cups, connecting the one cup with the other in that way. By the rising of the lever, and consequently the wire upon its end, from its connection with said cups, said circuit is in like manner again broken, and the current of electricity or galvanism destroyed. To effect at pleasure these two purposes of breaking and connecting said circuit is the design of said motion that is imparted in the before-mentioned manner to said lever, and to regulate this motion, and reduce it to the system of intelligible signs before described, is the design and use of the variations in the form of the type, also before described. A plate of copper, silver, or other conductor connected with the broken parts of said circuit of conductors, and receiving the contact of the wire attached to said lever, may be substituted, if preferred, for said cups of mercury. For a particular delineation of the several parts of said lever, reference is made to the annexed drawing marked Example 8.

The signal-lever consists, secondly, for use with the circular port-rule, Example 9, Fig. 3, of a strip of wood, G, with a metallic wire, A, at one end, of the form and for the purposes of the lever already described above. It turns on a pivot or fulcrum, a, placed either near the middle or in the end of the lever. At the end of the lever, at C, opposite to the metallic wire A, an elbow, c, is formed on a right angle with the main lever, and extending downward from the level with the pivot or fulcrum sufficiently for a metallic tooth, H, in the end thereof, corresponding with the teeth or indentations of the type, already described, to press against the type projecting from the shoulder or cavity of the wheel A, Fig. 1, that forms the circular port-rule, before described. Said wheel is placed beneath the said lever, as seen at G, Fig. 1, in a position to be reached by the extremity or tooth H of the arm of the lever just mentioned. The tooth H in the arm of the lever is kept in constant contact with the type of the circular port-rule by the pressure of a spring, B, upon it, as described in the annexed drawings marked Example 9, at B. Figs. 1 and 3 in the same example exhibit sections of the said lever. The action thus produced by the contact of the teeth of the type in the port-rule, when said wheel is in motion, with the tooth in the arm of the lever, lifts up and drops down the opposite extremity, A, of said lever, having the metallic wire upon it, as the tooth of said lever passes into or out of the indentations of the type, and in the same manner and to the same effect as the first-described lever rises and falls, and accordingly breaks and closes the circuit of conductors, as in the former instance. In the use of this circular port-rule and its appropriate lever, Fig. 3, type may be used having the points of their teeth and their indentations shaped as counterparts or reverses to those delineated in the annexed drawings heretofore referred to and marked Examples 4, 5, and 6, and thereby the forms of the recorded signs will be changed in a corresponding manner.

The register consists, first, of a lever of the shape of the lever connected with the circular port-rule above described, and is delineated in the annexed drawings marked Example 10, Figs. 1, 2, and 4, at A. Said lever A operates upon a fulcrum, a, that passes through the end that forms the elbow a, upon the lower extremity of which, and facing an electro-magnet, is attached the armature of a magnet, f. In the other extreme of the lever, at, B is inserted one or more pencils, fountain-pens, printing-wheels, or other marking-instruments, as may be seen in the Fig. 4 of the example last mentioned, at letter B. The magnet is at letter C in the same figure.

Secondly, of a cylinder or barrel of metal or wood, and covered with cloth or yielding coating, to turn upon an axis and occupying a position directly beneath the pencil, fountain-pen, printing-wheel, or other marking-instrument to be used, as exhibited in the last-mentioned example of drawings, Fig. 4, D. Two rollers, marked b b in said figure of drawings, are connected with said cylinder, on the upper-side curvatures thereof, and being connected with each other by two narrow bands of tape passing over and beneath each, near the ends thereof, and over the intervening surface of the cylinder, in a manner to cause a friction of the bands of tape upon the latter when in motion, as delineated in the last-named example, Fig. 4, at points marked c c c. The distance between said bands of tape on the rollers is such as to admit of the pencil, or other marking instrument in the lever, to drop upon the intervening space of the cylinder. Near by said cylinder is a spool to turn on an axis, and marked *d* in the said figure, to receive any desired length of paper or other substance formed into slips or a continuous ribbon, and for the purpose of receiving a record of the signs of intelligence communicated. When the register is in motion one end of the paper on said spool being inserted between the under surfaces of said two rollers, under the strips of tape that connect them and the cylinder, it is drawn by the friction or pressure thus caused upon it forward from said spool gradually, and passed over said cylinder, and is thence deposited in a box on the opposite side, or is cut off at any desired length as it passes from the cylinder and rollers.

Thirdly, of an alarm-bell, A, Example 10, Fig. 5, which is struck by means of a lever-hammer, B, that is acted upon by a movable cog, *b*, placed upon an axis or pin, *b*, that confines it in the lower extremity of a pendulum-lever, (marked E in Fig. 5 of Example 10,) having an armature of a magnet attached to it at *d*, and acted upon by an electro-magnet, *o*, placed near it and the before-named magnet, and in the same circuit of conductors with the latter. Said cog *b* moves in a quarter-circle only, as the motion of said arm of the lever passes backward and forward in the act of recording, as hereinafter described. When forced into a horizontal position in said quarter-circle it ceases to act upon the hammer; but when moved from a perpendicular position it presses upon the projection in the end of the hammer, causing the opposite end of the hammer to be raised, from which elevation it again falls upon a stationary bell, A, as soon as said cog reaches a horizontal position, and ceases, as before mentioned, to press upon the hammer. Thus a notice, by sound or an alarm, is given at the point to which intelligence is to be communicated as soon as the register begins to act, and such sound may be continued or not, at pleasure, for the purpose mentioned or for any other uses, as the hammer shall be suspended or not from contact with the bell, or with any number of bells that may be employed. Fig. 5 of said example, marked 10 in the annexed drawings, represents sections of said hammer and bell.

Said several parts of the register are set in motion by the communication to or action upon the before-named armature of a magnet, attached to the lever of the register, of the electric or galvanic current in the circuit of conductors, and from an electro-magnet in said circuit, as before described, stationed near the said armature. As said armature is drawn or attracted from its stationary and horizontal position toward the said magnet when the latter is charged from the circuit of conductors, said lever is turned upon its fulcrum, and the opposite end thereof necessarily descends and brings the pen, or marking-instrument which it contains in contact with the paper or other substance on the revolving cylinder directly beneath it. As said armature ceases to be thus drawn or attracted by said magnet, as is the case as soon as said magnet ceases to be charged from the circuit of conductors, or as the current in said circuit is broken in the manner hereinbefore described, the said armature is forced back by its own specific gravity, or by a spring or weight, as may be needed, to its former position, and the pen or marking-instrument in the opposite end of the lever is again raised from its contact with the paper or other substance on the before-named revolving cylinder. This same action is communicated simultaneously from the same circuit of conductors to as many registers as there are corresponding magnets provided within any circuit and at any desired distances from each other.

The cylinder and its two associate rollers are set in motion simultaneously with the first motion of the lever by the withdrawal of a small wire or spindle, *g*, Example 10, Figs. 2 and 5, from beneath one branch of a fly-wheel, *k*, that forms a part of the clock machinery hereinafter named. Said wire *g* is withdrawn by the action upon said wire of a small electro-magnet, *o*, Figs. 2 and 5, stationed in the circuit and near the large magnet before named, as delineated in Fig. 5 of Example 10. Said cylinder and rollers are subsequently kept in motion by a train of wheels similar to common clock-wheels, as in Figs. 2 and 3, acted upon by a weight, raised as occasion may require by a hand-crank, and their motion is regulated by the same wheels to correspond with the action of the registering-pen or marking-instrument. Said train is represented in Figs. 1, 2, and 3 of said Example 10.

The electro-magnet thus used is made in any of the usual modes, such as winding insulated copper wire, or strips of copper, or tin-foil, or other metal around a bar of soft iron, either straight or bent into a circular form, and having the two extremities of the coils connected with the circuit of conductors, so that the coils around the magnet make part of the circuit.

To extend more effectually the length of any desired circuit of conductors, and to perpetuate the power of the electric or galvanic current equally throughout the same, I adopt the following mode, and also for connecting and using any desired number of additional and intervening batteries or generators of said current, and for connecting progressively any number of consecutive circuits, viz: Place at any point in a circuit an electro-magnet of the denomination already described, with an armature upon a lever of the form and structure, and in the position of that used at the register to hold and operate the marking-instrument, with only a substitution therein for such marking-instrument of a forked wire, A, Example 9, Fig. 3, like that upon the end of the signal-lever heretofore described. Directly beneath the latter wire place two cups of mercury, E E, or two metallic plates joined to terminations of a circuit leading from the fresh or additional battery or generator of said circuit in the same manner as they are to be provided in the first circuit of conductors at the points where the cups of mercury are hereinbefore described. As the current in the first circuit acts upon the magnet thus provided the armature thereof and lever are thereby moved to dip the forked wire A into the cups of the second circuit, as in the circuit first described. This operation instantly connects the break in said second circuit, and thus produces an additional and original power or current of electricity or galvanism from the battery of said second circuit to the magnet or magnets placed at any one or more points in such circuit, to be broken at pleasure, as in the first circuit; and from thence by the same operation the same results may again be repeated, extending and breaking at pleasure such current through yet another and another circuit, *ad infinitum*, and with as many intervening registers for simultaneous action as may be desired, and at any distances from each other.

The dictionary or vocabulary consists of words alphabetically arranged and regularly numbered, beginning with the letters of the alphabet, so that each word in the language has its telegraphic number, and is designated at pleasure, through the signs of numerals.

The modes which I propose of insulating the wires or other metal for conductors, and of laying the circuits, are various. The wires may be insulated by winding each wire with silk, cotton, flax, or hemp, and then dipping them into a solution of caoutchouc, or into a solution of shellac, or into pitch or resin and caoutchouc. They may be laid through the air, inclosed above the ground, in the ground, or in the water. When through the air they may be insulated by a covering that shall protect them from the weather, such as cotton, flax, or hemp, and dipped into any solution which is a non-conductor, and elevated upon pillars. When inclosed above the ground they may be laid in tubes of iron or lead, and these again may be inclosed in wood, if desirable. When laid in the ground they may be inclosed in iron, leaden, wooden, or earthen tubes, and buried beneath the surface. Across rivers the circuit may be carried beneath the bridges, or, where there are no bridges, inclosed in lead or iron, and sunk at the bottom, or stretched across, where the banks are high, upon pillars elevated on each side of the river.

What I claim as my invention, and desire to secure by Letters Patent, is as follows:

1. The formation and arrangement of the several parts of mechanism constituting the type-rule, the straight port-rule, the circular port-rule, the two signal-levers, and the register-lever, and alarm-lever, with its hammer, as combining respectively with each of said levers one or more armatures of an electro-magnet, and as said parts are severally described in the foregoing specification.

2. The combination of the mechanism constituting the recording-cylinder, and the accompanying rollers and train-wheels, with the formation and arrangement of the several parts of mechanism, the formation and arrangement of which are claimed as above, and as described in the foregoing specification.

3. The use, system, formation, and arrangement of type, and of signs, for transmitting intelligence between distant points by the application of electro-magnetism and metallic conductors combined with mechanism described in the foregoing specification.

4. The mode and process of breaking and connecting by mechanism currents of electricity or galvanism in any circuit of metallic conductors, as described in the foregoing specification.

5. The mode and process of propelling and connecting currents of electricity or galvanism in and through any desired number of circuits of metallic conductors from any known generator of electricity or galvanism, as described in the foregoing specification.

6. The application of electro-magnets by means of one or more circuits of metallic conductors from any known generator of electricity or galvanism to the several levers in the machinery described in the foregoing specification, for the purpose of imparting motion to said levers and operating said machinery, and for transmitting by signs and sounds intelligence between distant points and simultaneously to different points.

7. The mode and process of recording or marking permanently signs of intelligence transmitted between distant points, and simultaneously to different points, by the application and use of electro-magnetism or galvanism as described in the foregoing specification.

8. The combination and arrangement of electro-magnets in one or more circuits of metallic conductors with armatures of magnets for transmitting intelligence by signs and sounds, or either, between distant points and to different points simultaneously.

9. The combination and mutual adaptation of the several parts of the mechanism and system of type and of signs with and to the dictionary or vocabulary of words, as described in the foregoing specification.

In testimony whereof I, the said SAMUEL F. B. MORSE, hereto subscribe my name in the presence of the witnesses whose names are hereto subscribed, on the 7th day of April, A. D. 1838.

SAML. F. B. MORSE.

Witnesses:
B. B. FRENCH,
CHARLES MONROE.

Sheet 1
3 Sheets

E. Howe, Jr.

Sewing Machine.

No 4750 Patented Sep. 10, 1846.

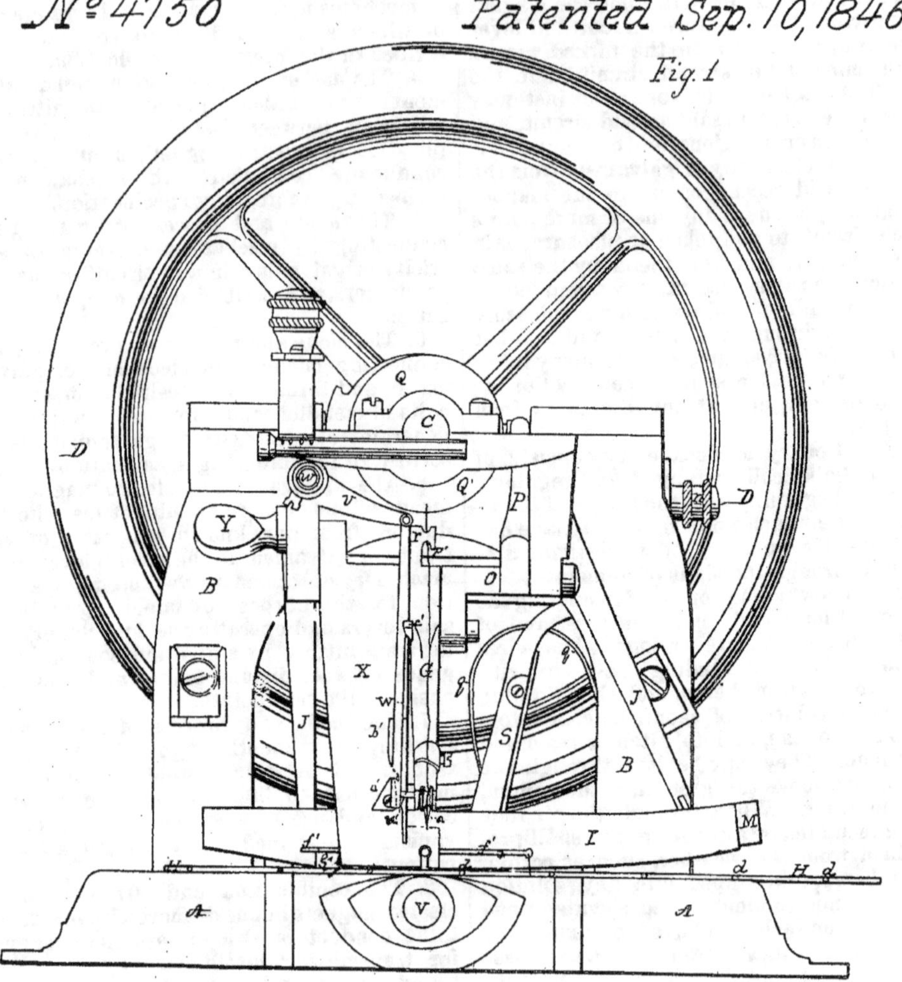

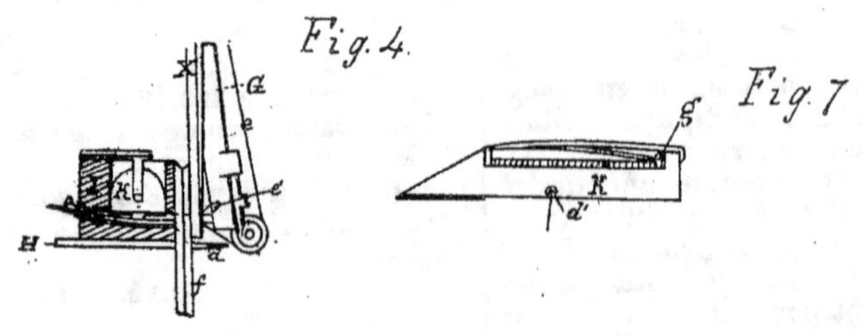

Sheet 2
3 Sheets

E. Howe, Jr.

Sewing Machine.

No. 4750 Patented Sep. 10, 1846.

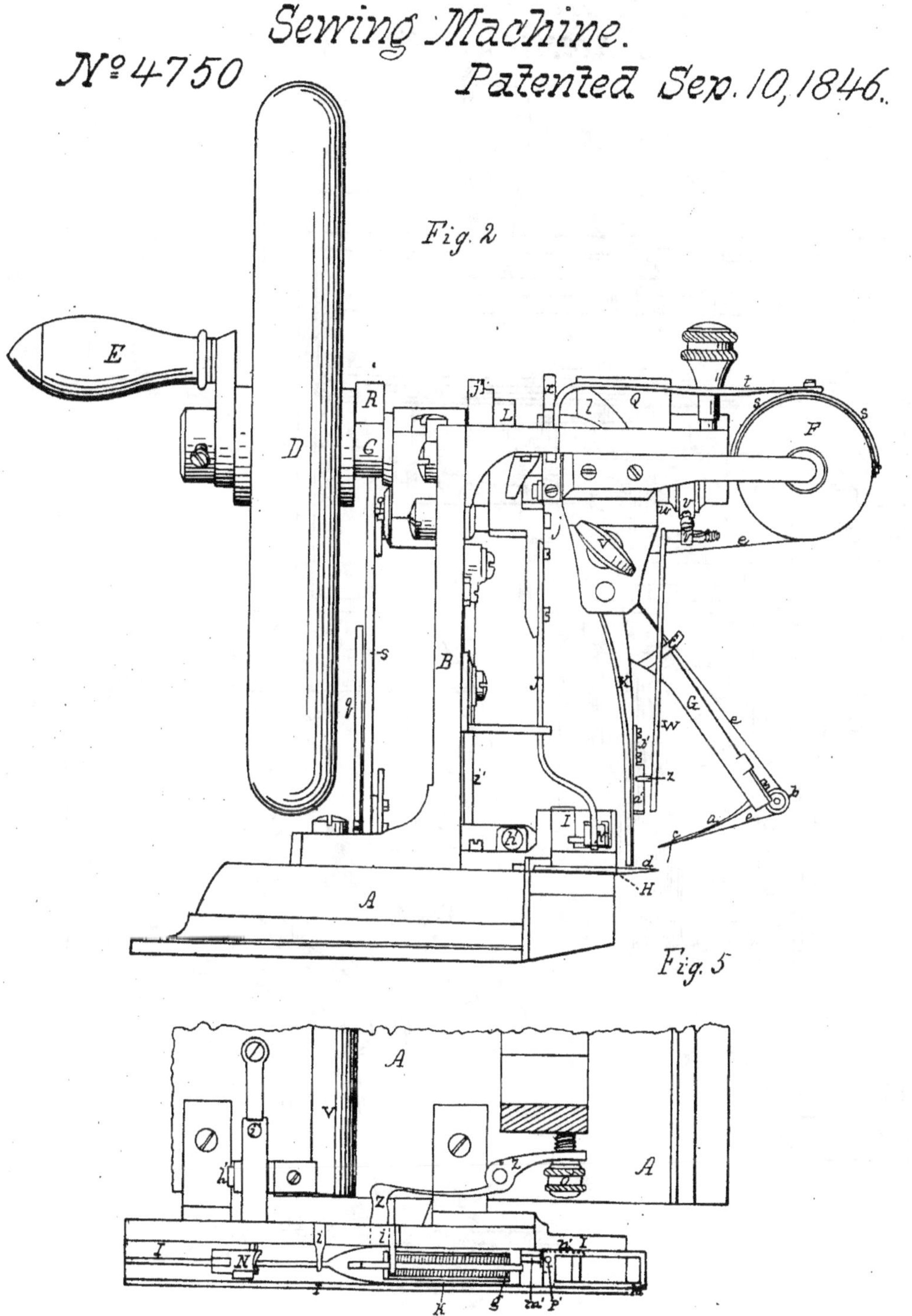

Sheet 3
3 Sheets

E. Howe, Jr.

Sewing Machine.

Patented Sep. 10, 1846.

No 4750

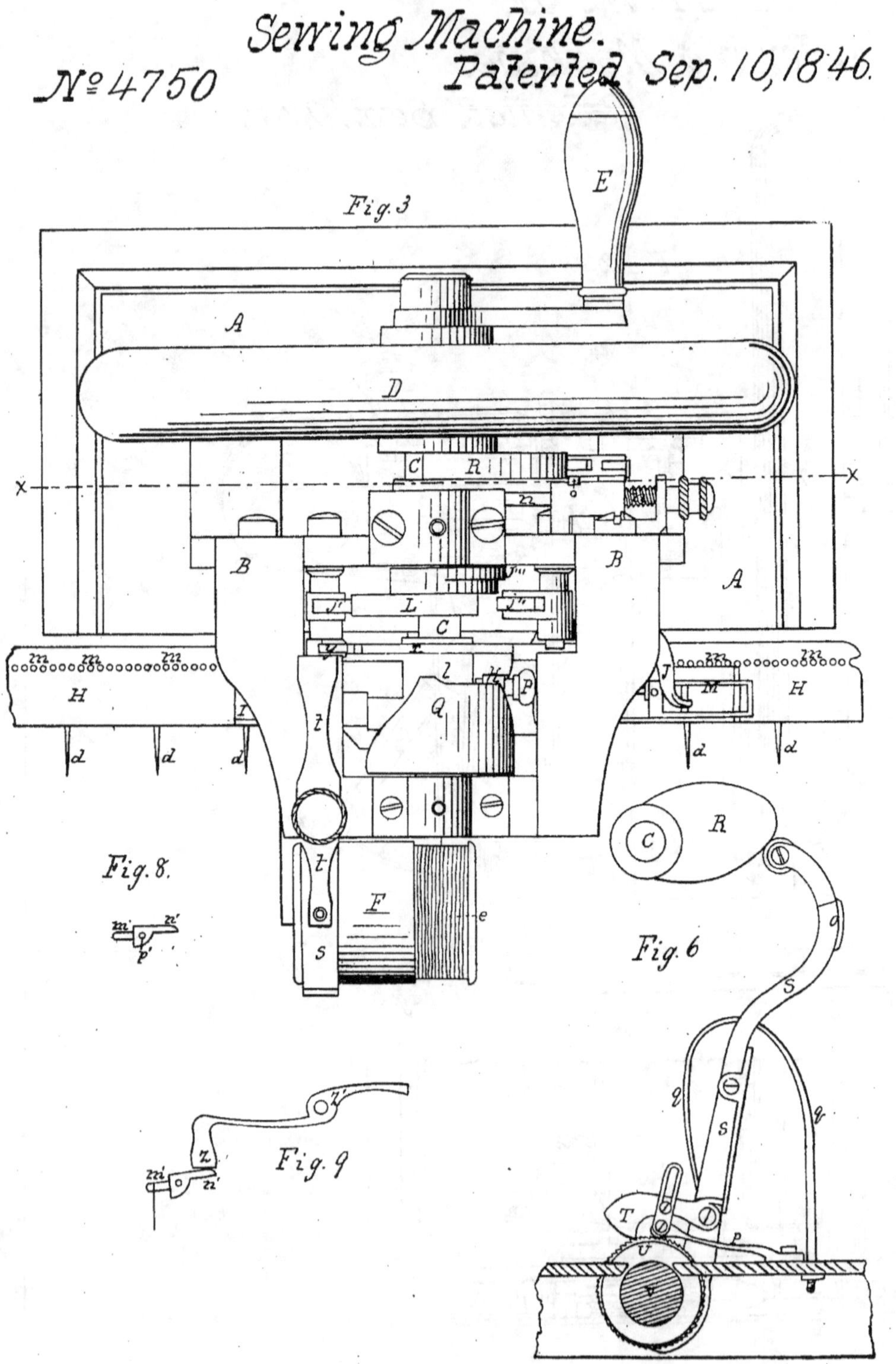

UNITED STATES PATENT OFFICE.

ELIAS HOWE, JR., OF CAMBRIDGE, MASSACHUSETTS.

IMPROVEMENT IN SEWING-MACHINES.

Specification forming part of Letters Patent No. 4,750, dated September 10, 1846.

To all whom it may concern:

Be it known that I, ELIAS HOWE, Jr., of Cambridge, in the county of Middlesex and State of Massachusetts, have invented a new and useful machine for sewing seams in cloth or other articles requiring to be sewed; and I do hereby declare that the following is a full and exact description thereof.

In sewing a seam with my machine two threads are employed, one of which threads is carried through the cloth by means of a curved needle, the pointed end of which is to pass through said cloth. The needle used has the eye that is to receive the thread within a small distance—say, an eighth of an inch—of its inner or pointed end. The other or outer end of the needle is held by an arm that vibrates on a pivot or joint pin, and the curvature of the needle is such as to correspond with the length of the arm as its radius. When the thread is carried through the cloth, which may be done to the distance of about three-fourths of an inch, the thread will be stretched above the curved needle, something in the manner of a bow-string, leaving a small open space between the two. A small shuttle carrying a bobbin filled with silk or thread is then made to pass entirely through this open space between the needle and the thread which it carries, and when the shuttle is returned, which is done by means of a picker-staff or shuttle-driver, the thread which was carried in by the needle is surrounded by that received from the shuttle, and as the needle is drawn out it forces that which was received from the shuttle into the body of the cloth, and as this operation is repeated a seam is formed which has on each side of the cloth the same appearance as that given by stitching, with this peculiarity, that the thread shown on one side of the cloth is exclusively that which was given out by the needle, and the thread seen on the other side is exclusively that which was given out by shuttle. It will therefore be seen that a stitch is made at every back-and-forth movement of the shuttle. The two thicknesses of cloth that are to be sewed are held upon pointed wires which project out from a metallic plate, like the teeth of a comb, but at a considerable distance from each other—say three-fourths of an inch, more or less—these pointed wires sustaining the cloth and answering the purpose of ordinary basting. The metallic plate from which these wires project has numerous holes through it, which answer the purpose of rack-teeth in enabling the plate to be moved forward by means of a pinion as the stitches are taken. The distance to which said plate is moved, and consequently the length of the stitches, may be regulated at pleasure.

In the accompanying drawings, Figure 1 is a front elevation of the machine; Fig. 2, an end elevation thereof, and Fig. 3 a top view. The other figures represent sections and parts in detail, which will be presently explained.

A A is the bed or base of the machine, and B B standards rising therefrom, which sustain the main shaft and other parts of the apparatus.

C C is the main shaft, which carries the cams that operate the needle, the shuttle-drivers, and other parts of the machine. D is a fly-wheel, and E a winch, on said shaft.

F is a bobbin on which the silk is wound that is to supply the needle.

G is the needle-arm, that carries the curved needle *a*. This is seen most distinctly in the end elevation, Fig. 2. The thread from the bobbin F passes round a small friction-roller, *b*, or round a smooth groove in the situation of said roller, then up through the eye of the needle at *c*, which eye is situated near to the needle-point. The cloth is stuck on the points *d d*, that project from the metallic plate H, which I will call the "baster-plate." This plate is shown most distinctly in the top view, Fig. 3. When the thread *e* is carried through the cloth by the needle *a*, the upper portion of said thread will be above the needle and will allow the point of the shuttle (to be presently described) to pass between them. To enable it to enter readily, the needle, after entering the cloth, is immediately drawn back to a short distance, which opens the loop slightly. The cam which operates the needle-arm being so formed as to cause such drawing back, the shuttle will, in order to give itself the necessary room, draw a portion of the thread which had been given out by the needle through the cloth, said thread having been left in a loop or slack state for that purpose.

Fig. 4 represents a part of the same portion of the machine that is shown in Fig. 2, but with the needle-arm down and with the needle passed through the cloth. *f* is the cloth, (seen in section, but not shown in any of the other figures.) e' is the loop or slack thread formed on the outside of the cloth, and which is to be drawn through it by the passing of the shuttle.

I in the respective figures is the shuttle box or trough, within which the shuttle is moved back and forth by means of the picker-staves or shuttle-drivers J J. In Fig. 5 I have given a top view of this box with the shuttle K within it. This shuttle is in its general construction similar to the larger shuttle used in weaving, and its spool g is capable of containing an ordinary skein of silk. The shuttle-box I is represented as made convex on its under side, by which it is adapted to admit a baster-plate that may be in a curved form, although for most purposes a straight baster-plate may be used. The pieces marked i i are light springs above the shuttle, which bear slightly upon it and serve to steady its motion. The shuttle-drivers work on joint-pins, as shown at j, Fig. 2, there being a corresponding fixture for the drivers on the other side.

L, Fig. 3, is the cam that operates the shuttle-drivers, on the upper ends of which drivers there may be friction-rollers j' j'. The cam L acts upon the shuttle-drivers alternately.

M, Fig. 5, is a sliding box fitted into the shuttle-box and moved back and forth in the rear of the shuttle by one of the drivers, and N is a corresponding sliding piece moved by the other driver and adapted to the fore or pointed end of the shuttle. The needle-arm is attached to the rock-shaft O, Fig. 1, which vibrates on a center pin or pivots, and from this shaft rises an arm, P, that carries a pin and friction-roller, k, which enters a space, l, in the cam Q, which space operates as a zigzag groove, and is of course so formed as to give the proper vibration to the needle-arm. There is a groove or narrow channel made across the bottom of the shuttle-box to receive the needle, in order that its upper part may be even with said bottom and allow the shuttle to pass freely over it.

The baster-plate H, Fig. 3, which receives the cloth to be sewed, is furnished with a row of small holes, m m, drilled at a regular distance from each other, serving the purpose of rack-teeth, and into these round pinion-teeth enter for the purpose of carrying the plate forward to a proper distance at every stitch.

Fig. 6 shows the principal portion of the feeding apparatus as it would appear were a vertical section made through the machine in the line x x of Fig. 3. R is a cam on the cam-shaft C, that vibrates an arm, S, carrying a feeding-claw, T, that takes into a ratchet-wheel, U, on the shaft V, which shaft crosses the bed A of the machine, its fore end being seen at V, Fig. 1. This shaft has on it near its fore end the pinion that carries the pins or teeth that take into the holes m in the baster and cause it to advance between every stitch. The length of the stitch may be regulated by regulating the play of the arm S, and this is effected by the regulating-screw n, Fig. 3, that moves a pin back and forth that serves as a stop to said arm. The pin is represented by the dot o, Fig. 6, and is seen at o, Figs. 2 and 3. p is a spring that retains the ratchet-wheel in place as the claw is taking a new hold. q is a spring for holding the arm S against the cam.

In sewing with this machine, the thread from the bobbin F is passed over a notch, r, Fig. 1, at the upper end of the needle-arm, and is returned through the notch r'. It then passes down in front of said arm, then around the roller b, and through the needle-eye. To regulate the giving out of the thread from the bobbin, friction is made on it by the semicircular clasp s, that is made to press on it by a spring, t, regulated by a tempering-screw. Before the needle passes through the cloth the thread, which extends from the needle-eye to said cloth, is raised or drawn up by a lifting-pin, so as to form the loop or slack, which is subsequently to be drawn in by the passing of the shuttle between the thread and the needle.

W, Figs. 1 and 2, is a lifting-rod, from the side of which projects the lifting-pin u. The lifting-rod is attached at its upper end to a crank-arm, v, which works on a shaft, w, and this shaft is made to vibrate by means of the cam x on the cam-shaft. This cam operates on a friction-roller, y, on a short arm on the inner end of the shaft w. The lifting-rod stands in front of a plate, X, Figs. 1 and 2, which is attached at its upper end to the frame of the machine, and between the lower end of this plate and the shuttle-box the cloth is to pass. The plate X is furnished with a hinge-joint at its upper end, in order that its distance from the shuttle-box may be regulated to suit cloth of different thicknesses.

Y, Fig. 1, is a set-screw, by which it is held in place. From the back part of the lifting-rod proceeds a guide-pin, z, that moves the lifting-rod laterally, so as to govern the action of the lifting-pin u. This guide-pin works against guide-pieces a' b', affixed on the front of the plate X. The dotted lines show the groove formed by the pieces a' b', along which the guide-pin is to pass. The lifting-rod is carried toward the piece b' by means of a spiral spring around its shaft, or in any other convenient mode. In the position in which the apparatus is shown in Fig. 1 the lifting-pin is partially raised, and will have lifted the thread. In raising it the guide-pin passes through the groove between a' b', (shown by dotted lines,) and when at the upper end of this groove the needle-arm acts and carries the needle through the cloth. On the side of the needle-arm there is a projecting piece, c', the inclined edge of which, coming in contact with the lifting-rod, pushes it laterally over the angular point of the piece a', and the crank-arm v descending at this moment, the lifting-pin is withdrawn from the thread, which is thereby left slack to a sufficient extent for the purpose designated.

The shuttle (shown separately in Fig. 7) has a hole, d', through its side for the thread to pass from the spool; and a slot, f' f', is made through the side of the shuttle-box to allow of the play of the shuttle-thread back and forth. At the time when the shuttle has completed its passage between the needle and its thread, the needle is to be withdrawn from the cloth; and when this is taking place, it is necessary that the shuttle-thread should be held firmly, or the withdrawing of the needle, instead of drawing the shuttle-thread firmly into the body of the cloth and making a perfect seam, would draw a portion of it from the spool and cause it to pass entirely through said cloth.

In Fig. 1, g' is the outer end of a lever which is made to rise at the proper moment, and to clip the thread between it and the upper edge of the slot f'. This lever is seen in Fig. 2, its fulcrum being at h'. The rod i' serves to depress the inner end of said lever and to raise its outer end, the cam j'' on the cam-shaft performing this office.

The sliding box M does not bear directly against the rear end of the shuttle-box, but has a pin, m', projecting from its fore end, which pin acts against the shuttle. The pin m' constitutes a part of a small lever shown separately in Fig. 8. The part n' of this lever is received within a suitable slot in the sliding box M, and it turns on a fulcrum-pin, p'. When the shuttle has passed through the loop formed by the needle-thread, it is received upon the pin m', and as the needle is retracted the thread will be drawn taut upon said pin. At this time the head of an adjustable spring-piece, z z', bears against the end n' of the small lever, and the force of its pressure has to be overcome before the thread escapes from the pin, which it does by drawing over against the power of the spring. As the loop then escapes, it will draw up the filling-thread from the shuttle firmly against the cloth and embed it within it. The head of the spring Z passes through a mortise in the shuttle-box, as shown by the dotted lines. o' is an adjusting-screw by which the force of the spring Z may be regulated.

Having thus fully described the manner in which I construct my machine for sewing seams, and shown the operation thereof, what I claim therein as new, and desire to secure by Letters Patent, is—

1. The forming of the seam by carrying a thread through the cloth by means of a curved needle on the end of a vibrating arm, and the passing of a shuttle furnished with its bobbin, in the manner set forth, between the needle and the thread which it carries under a combination and arrangement of parts substantially the same with that described.

2. The lifting of the thread that passes through the needle-eye by means of the lifting-rod W, for the purpose of forming a loop of loose thread that is to be subsequently drawn in by the passage of the shuttle, as herein fully described, said lifting-rod being furnished with a lifting pin, u, and governed in its motions by the guide-pieces and other devices, arranged and operating substantially as described.

3. The holding of the thread that is given out by the shuttle, so as to prevent its unwinding from the shuttle-bobbin after the shuttle has passed through the loop, said thread being held by means of the lever or clipping-piece g', as herein made known, or in any other manner that is substantially the same in its operation and result.

4. The manner of arranging and combining the small lever m' n' with the sliding box M, in combination with the spring-piece Z, for the purpose of tightening the stitch as the needle is retracted, as described.

5. The holding of the cloth to be sewed by the use of a baster-plate furnished with points for that purpose, and with holes enabling it to operate as a rack in the manner set forth, thereby carrying the cloth forward and dispensing altogether with the necessity of basting the parts together.

ELIAS HOWE, JR.

Witnesses:
THOS. P. JONES,
GEORGE FISHER.

N. A. OTTO.

GAS-MOTOR ENGINES.

No. 194,047. Patented Aug. 14, 1877.

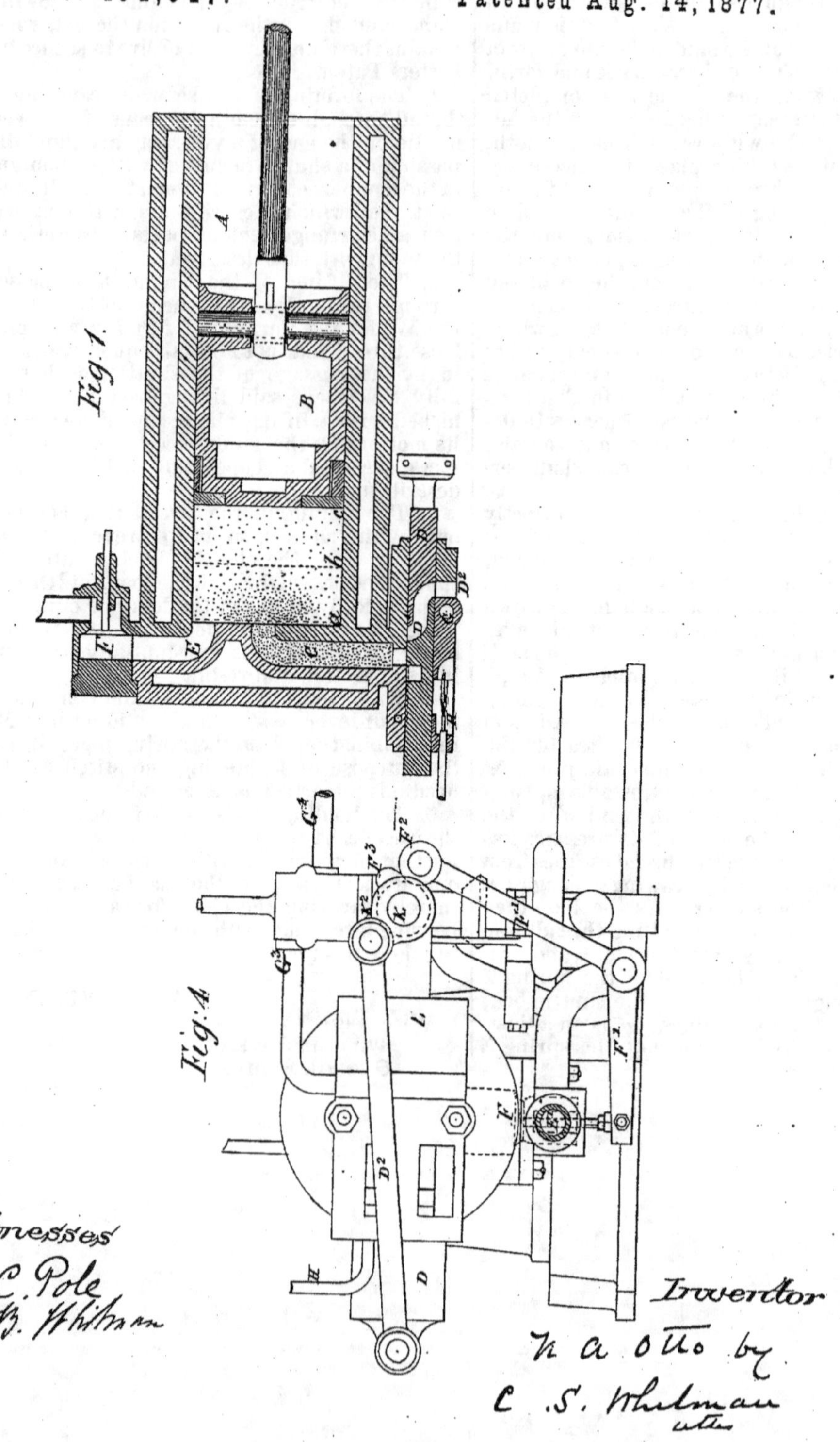

N. PETERS, PHOTO-LITHOGRAPHER, WASHINGTON, D. C.

4 Sheets—Sheet 2.

N. A. OTTO.

GAS-MOTOR ENGINES.

No. 194,047. Patented Aug. 14, 1877.

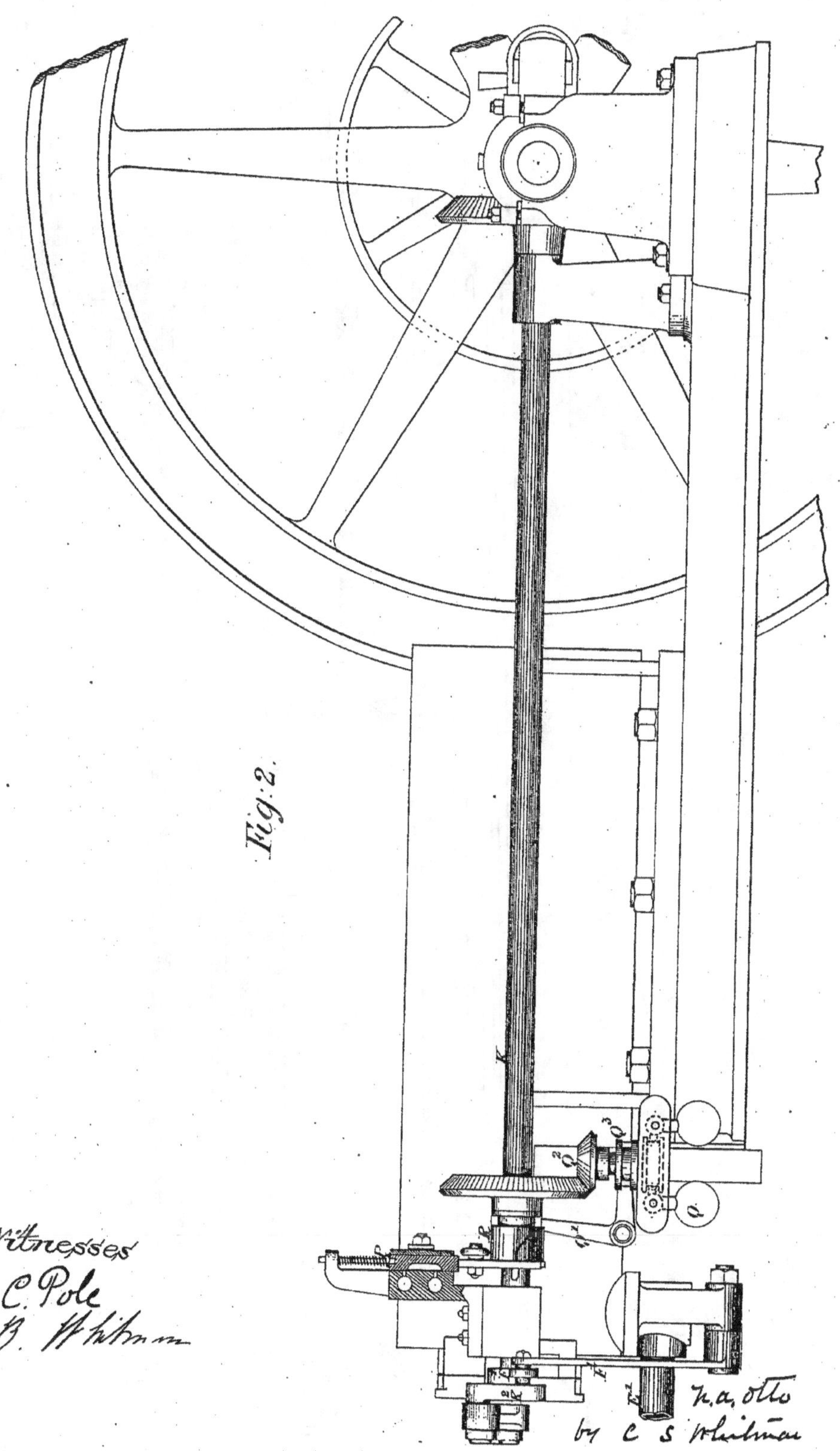

Witnesses

B. C. Pole

N. B. Whitman

N. A. Otto

by C S Whitman

N. PETERS, PHOTO-LITHOGRAPHER, WASHINGTON, D. C.

4 Sheets—Sheet 3.

N. A. OTTO.

GAS-MOTOR ENGINES.

No. 194,047. Patented Aug. 14, 1877.

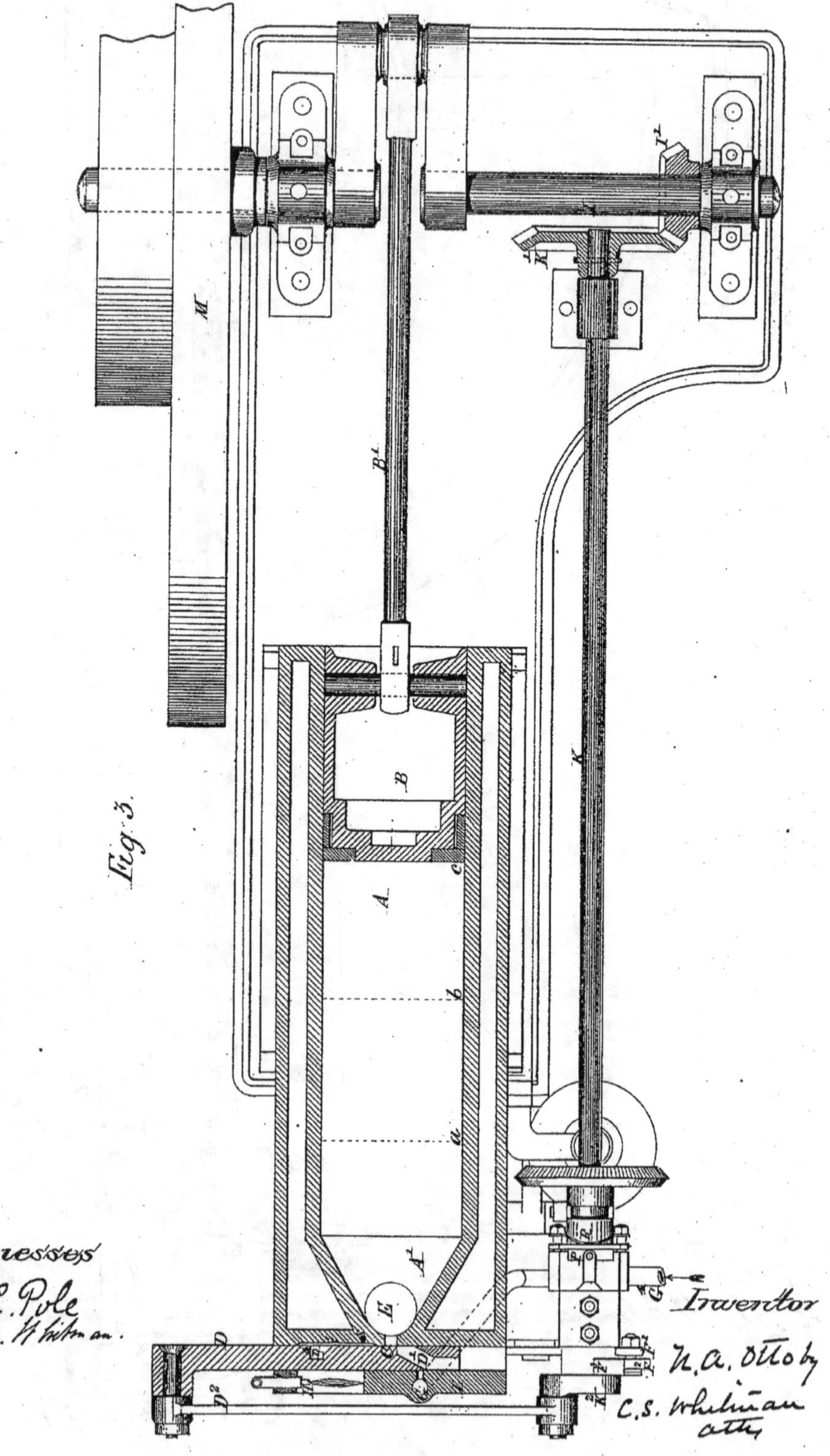

N. PETERS, PHOTO-LITHOGRAPHER, WASHINGTON, D. C.

4 Sheets—Sheet 4.

N. A. OTTO.

GAS-MOTOR ENGINES.

No. 194,047. Patented Aug. 14, 1877.

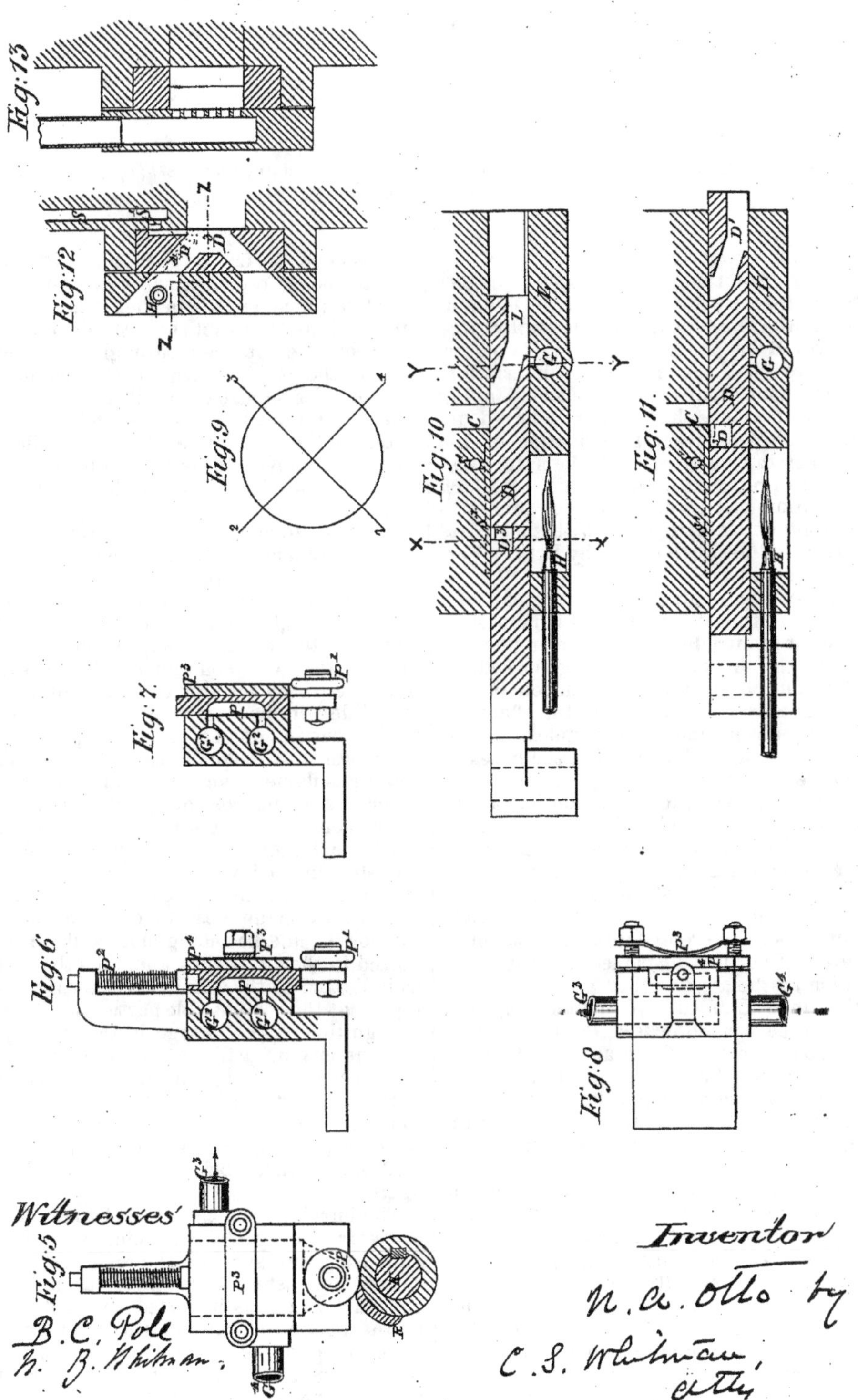

Witnesses

B. C. Pole

N. B. Whitman.

Inventor

N. A. Otto by

C. S. Whitman, atty.

N. PETERS, PHOTO-LITHOGRAPHER, WASHINGTON, D. C.

UNITED STATES PATENT OFFICE.

NICOLAUS A. OTTO, OF DEUTZ, GERMANY.

IMPROVEMENT IN GAS-MOTOR ENGINES.

Specification forming part of Letters Patent No. **194,047**, dated August 14, 1877; application filed July 13, 1876.

To all whom it may concern:

Be it known that I, NICOLAUS AUGUST OTTO, of the Gas-Motoren Fabrik-Deutz, at Deutz, in the German Empire, have invented an Improved Gas-Motor Engine; and do hereby declare that the following description, taken in connection with the accompanying sheets of drawings, hereinafter referred to, forms a full and exact specification of the same, wherein I have set forth the nature and principles of my said improvement, by which my invention may be distinguished from others of a similar class, together with such parts as I claim and desire to secure by Letters Patent—that is to say:

In gas-motor engines as at present constructed, an explosive mixture of combustible gas and air is introduced into the engine-cylinder, where it is ignited, resulting in a sudden development of heat and expansion of the gases, a great portion of the useful effect being lost by absorption of heat, unless special provisions are made for allowing the gases to expand very rapidly.

According to my present invention an intimate mixture of combustible gas or vapor and air is introduced into the cylinder, together with a separate charge of air or other gas, that may or may not support combustion, in such a manner and in such proportions that the particles of the combustible gaseous mixture are more or less dispersed in an isolated condition in the air or other gas, so that on ignition, instead of an explosion ensuing, the flame will be communicated gradually from one combustible particle to another, thereby effecting a gradual development of heat and a corresponding gradual expansion of the gases, which will enable the motive power so produced to be utilized in the most effective manner.

In order more clearly to describe my invention, I will refer to the accompanying drawings, in which Figure 1 shows a longitudinal section of an engine-cylinder, A, having a piston, B, connected to a fly-wheel shaft, and a port or passage, C, for the admission of combustible gaseous mixture and air, controlled by the slide D, and having also a passage, E, for the emission of the products of combustion, closed by a valve, F. Assuming the piston to be at the end of its instroke, its bottom surface being represented by the dotted line *a*, while the slide D is in such a position that its passage D^1 establishes a communication between the outer air through the aperture D^2 and the port C, then, on the piston commencing its outstroke, it will draw in atmospheric air until it arrives at the point indicated by the dotted line *b*, when the slide will have been moved so as to cut off the air-supply and establish a communication between the passage G in the slide-cover, for an intimate mixture of coal-gas or petroleum vapor and air, (in such proportions that the mixture will burn of itself, but, owing to the presence of the first admitted air, will not explode,) and the port C through the passage D^1. On the continued motion of the piston, combustible gaseous mixture will consequently be drawn in until the piston has arrived at a point, *c*, when the slide will have moved into the position shown, cutting off the gas-supply, and about to establish a communication between the small gas-flame H and the charge in the cylinder, for the purpose of igniting the latter.

The combustible gaseous mixture, in entering the cylinder behind the charge of air previously admitted, will, to a certain extent, mix with the latter, the particles of the combustible mixture being close together in and near the port C, and becoming more and more dispersed in the air as they approach the piston, as indicated by the dots in the drawing, which represent the combustible particles. Thus, on the ignition of the charge in the port C, the gaseous mixture will at first burn with comparative rapidity, the flame spreading from particle to particle; but as the ignition extends toward the front end of the charge, it will proceed more and more slowly, owing to the combustible particles being farther and farther apart.

The burning particles impart their heat to the surrounding air, producing a gradually increasing pressure in the cylinder, which causes the piston to complete its outstroke. Motion being thus imparted to the fly-wheel by the piston-rod, its momentum causes the piston to perform its return stroke, whereby the products of combustion are expelled through the valve F, and, the fly-wheel also causing the piston to commence its next outstroke, a fresh charge of air and combustible mixture is drawn in, as before described.

In order to vary the power of the engine, the charge of combustible mixture (represented by the space *a* to *b*) may be varied, as may also the proportions of air and coal-gas or vapor of which it is composed, and such variation may be controlled by connecting the valve-gear with any suitable construction of governor, as will be presently described.

From the foregoing general description it will be seen that as in the improved mode of operating there is no sudden explosion of the gaseous charge, but a gradual development of heat and expansion of the gases, there will be no such losses of effect as result in gas-engines of present construction through shocks produced by the sudden development of motive power, and by the absorption of heat consequent upon the inability of the gases to expand with sufficient rapidity.

The above-described beneficial effect of the improved mode of working will be further increased by the fact that the charge of air interposed between the combustible mixture and the piston will operate as a cushion or buffer in still further reducing the suddenness of the expansive force generated as it transmits it to the piston.

Engines operating according to my invention may either be single-acting—the return stroke being effected by the momentum of the fly-wheel—or they may be double-acting, a gaseous charge being introduced at each end of the cylinder. They may also operate with the gases either at atmospheric pressure or compressed to any desired degree. In the latter case the engine may be arranged in a similar manner to that above described, the gases being compressed by any suitable known means before being introduced into the engine; but, by preference, I dispense with any such additional compressing mechanism by arranging the engine to operate in the manner I will now proceed to describe with reference to Figs. 2 to 13 of the drawings, of which—

Fig. 2 shows a side elevation; Fig. 3, a sectional plan; Fig. 4, a back-end view, and Figs. 5 to 13 details of the valve-gear.

The engine is here represented as being single-acting, the cylinder A being open to the atmosphere at its front end. At its closed back end it has a space, A′, beyond the stroke of the piston B, which space is, by preference, made conical at the end, as shown, tapering to the inlet-port C for combustible gas and air, and also communicating by the passage E with the escape-valve F, Fig. 3, for the products of combustion.

The piston B is connected by the rod B′ to the crank-shaft I, on which is a bevel-pinion, I′, in gear with a bevel-wheel, K^1, on a shaft, K. On the other end of this shaft is a crank, K^2, connected by a link, D^2, to the slide D, governing the admission of gas and air to the cylinder. The gearing I′ K^1 is so proportioned that the crank K^2 makes one revolution, and, consequently, the slide one to-and-fro motion, while the piston makes two double strokes.

The mode of operating with this engine is as follows: Assuming the piston to be at the end of its instroke (represented by the dotted line *a*, Fig. 3,) and about to be moved through its outstroke by the momentum of the fly-wheel M, then, the slide D (the construction of which will be presently explained) being in position to admit atmospheric air through the passage D^1 and port C, air will be drawn into the cylinder until the piston has reached the point represented by the dotted line *b*, when, the slide having established a communication with the combustible-gas supply and the cylinder, combustible gas intimately mixed with air will be drawn in until the piston has arrived at the end of its outstroke, the position shown at Fig. 3. As before explained with reference to Fig. 1, the combustible gaseous mixture, in entering, will mix to a certain extent with the air previously introduced, the particles of gaseous mixture being close together at the back end of the cylinder, and more and more separated from each other toward the front end. The slide having moved so as to close the inlet-port C, the piston is caused, by the momentum of the fly-wheel, to perform its instroke, whereby the charge of gaseous mixture and air that filled the cylinder at atmospheric pressure will be compressed into the space from the line *a* to the back end of the cylinder, the particles of gaseous mixture remaining in much the same unequally-distributed condition in the air as they did before compression. The slide now establishes a communication between the gas-flame H and the interior of the cylinder, so as to ignite the charge, resulting in a gradual development of heat and expansion of the gases, as before explained, whereby the piston will be caused to perform its outstroke, imparting fresh momentum to the fly-wheel. This momentum will again cause the piston to perform its instroke, whereby the products of combustion will be expelled through the valve F, which has been opened by the lever N, acted on by a cam, O, on the shaft K. As the piston only moves back to the line *a*, it will be seen that a certain portion of the products of combustion will remain in the cylinder, and will consequently mix to a certain extent with the air drawn in behind them at the next outstroke; but as the mixture of combustible gas and air afterward introduced will burn independently of the air or other gas surrounding its particles, it will be seen that the presence of such products of combustion in the charge will be of no consequence.

As before stated, the power of the engine may be regulated by regulating the quantity of combustible gas introduced at each charge. This is effected by the gas-slide P, controlled by the governor Q, operating on the sliding cam R as follows: Fig. 5 shows an enlarged front view of the gas-slide; Figs. 6 and 7, vertical sections, and Fig. 8 a plan, of the same. In the casing of the slide are formed two passages, G^1 and G^2, the former communicating with a pipe, G^3, leading to the gas-passage G in the slide D, and the other with the gas-supply pipe G^4. These passages have small side openings, as shown, which, when the slide is in the position shown in Fig. 7, both communicate with the cavity of the slide P, so that gas can pass from G^2 into G^1, and thence into the passage G of the slide D. When the slide is moved into the position shown in Fig. 6, this communication, and consequently the gas-supply, is cut off. The slide P rests with a small roller, P^1, upon a cam, R, which revolves with, but can slide somewhat upon, the shaft K, the raising of the slide being effected by the cam, while its downward motion is effected by the spring P^2. According as the cam is shifted relatively to the roller P^1 by the action of the governor Q and lever Q^1, (which has a fork taking into a collar on the cam, as shown,) the slide is made to establish the communication between G^1 and G^2 for a longer or a shorter period, thus allowing a greater or less quantity of the combustible gas for one charge to pass into the cylinder A independently of the action of the slide D. The gas-slide P is held against the face of the casing by a spring, P^3, pressing against a cover, P^4, on the back of the slide.

The construction and mode of operating of the engine-slide D will be understood on reference to Figs. 10 to 13, of which Figs. 10 and 11 represent two longitudinal sections of the slide and casing on line Z Z, Fig. 12, with the slide in two different positions, and Figs. 12 and 13 show transverse sections, respectively on lines X X and Y Y, Fig. 10.

From the previous description of the action of the engine, it will be seen that there are four strokes of the piston required for one complete operation—namely, an outstroke for drawing in the charge of combustible mixture and air, an instroke for compressing the gases, a second outstroke when the piston is propelled on the ignition of the gases, and a second instroke for expelling the products of combustion. The slide D consequently has to perform one to-and-fro motion while the piston is performing the above-mentioned four operations, for which purpose, as before stated, the slide-crank K^2 makes one revolution while the engine-shaft makes two. The circle at Fig. 9 represents a diagram of the path of the crank K^2, in which the part from 1 to 2 represents the motion of the slide during the time of drawing in the gaseous charge, the part from 2 to 3 the motion during the compression of the charge, 3 to 4 the motion during the working stroke, and 4 to 1 the motion during the expulsion of the products of combustion. Figs. 10 and 11 each show two positions of the slide, Fig. 10 showing, first, its position at the point 1 of the crank-path when the air-passage D^1 is just about to communicate with the port C, and, secondly, its position at point 2, the gas and air supply having just been cut off. It will be seen that in the first position the gas-passage G is also about to open; but the before-described action of the gas-slide P will prevent the admission of combustible gas until the requisite charge of air is introduced. Fig. 11 shows, first, the position of the slide at the point 3 when the flame of the gas jet H is about to be communicated to the gaseous charge by a small quantity of inflamed gas in the passage D^3, and, secondly, its position at the point 4 when the escape-valve F is about to be opened.

For effecting the ignition of the charge, a small quantity of combustible gas is made to pass down a pipe, S, into a recess, S′, in the end of the cylinder, whence it issues through a small channel, D^4, in the slide into the passage D^3. Here it is ignited by the jet H, and the flame is, by the motion of the slide, conveyed to the port C, the slide-cover L being made to close the outer opening of D^3 before its inner opening communicates with C, as shown at Fig. 11.

The gas-passage G communicates with the air-passage D^1 through a number of small openings, as shown at Fig. 13, so that the gas, in issuing in small divided jets into D^3, becomes intimately mixed with the air therein in the requisite proportions for producing the combustible mixture before described.

The opening of the escape-valve F at the commencement of the second instroke of the piston (point 4 at Fig. 9) is effected by the bell-crank lever F^1, connected at one end to the stem of the valve, and having at the other end a roller, F^2, which is acted upon by the cam F^3 on the shaft K. E′ is the pipe for conducting away the products of combustion.

The governor Q is driven by bevel-gearing from the shaft K, its arms being made to move a sliding collar, Q^3, up or down, thus imparting motion through the lever Q^1 to the cam R, as before described.

The cylinder A is, by preference, provided with a jacket, as shown.

As before stated, the engine may be arranged double-acting by providing the requisite valve-gear for each end of the cylinder. It may also be arranged in a vertical or inclined position, instead of horizontal; and if single-acting, or if great regularity of motion be required, two or more engines may be connected to one and the same crank-shaft.

Having thus described the nature of my invention, and in what manner the same is to be performed, I wish it to be understood that I do not claim generally the separate introduction of combustible gas and air into the cylinder of a gas-engine, as I am aware that is to a certain extent described in the English Patents No. 1,655 of 1857, and 335 of 1860; but,

I claim—

1. A gas-motor engine wherein an intimate mixture of combustible gas or vapor and air is introduced into the cylinder, separate from a charge of air or other incombustible gas, in such manner and in such proportions that the particles of combustible mixture will be close together at the point of ignition, but will be more and more dispersed in the charge of air forward of that point, whereby the development of heat and the expansion or increase of pressure produced by the combustion are rendered gradual, substantially as herein described.

2. A gas-motor engine wherein an intimate mixture of combustible gas or vapor and air is introduced into the cylinder separate from and subsequent to a charge of air, such introduction being effected through an aperture or apertures in the end surface of the cylinder, in order to cause the charge of air to move forward in the cylinder as the combustible mixture is introduced, substantially as and for the purposes set forth.

3. A gas-motor engine wherein, by one outstroke of the piston, separate charges of combustible gaseous mixture and of air are drawn into the cylinder, which charges are compressed by the instroke and then ignited, so as to propel the piston, which, by its return stroke, expels the products of combustion, substantially as herein described with reference to Figs. 2 to 13 of the drawings.

4. In gas-motor engines wherein charges of combustible gas and air are introduced separately into the cylinder, regulating the power of the engine by controlling the gas-supply by means of a valve operated by a governor, substantially as herein described.

5. In gas-motor engines, the shaft K, driven from the engine-shaft, with crank K^2, imparting motion to the slide D, cam R, for regulating the gas-supply, and cam F^3, for opening the escape-valve F, substantially as herein described.

6. In gas-motor engines, the combination of the cylinder A, piston B, engine-shaft I, counter-shaft K, crank K^2, slide D, gas-slide P, cam R, escape-valve F, lever F^1, and cam F^3, all arranged and operating substantially as and for the purposes herein described.

In testimony whereof I have signed my name to this specification in the presence of two subscribing witnesses this 1st day of June, 1876.

NICOLAUS AUGUST OTTO.

Witnesses:

FRIEDRICH ALBERT SPIECKER,

GUSTAV KLEINJING.

(No Model.) 3 Sheets—Sheet 1.

G. MARCONI.

TRANSMITTING ELECTRICAL SIGNALS.

No. 586,193. Patented July 13, 1897.

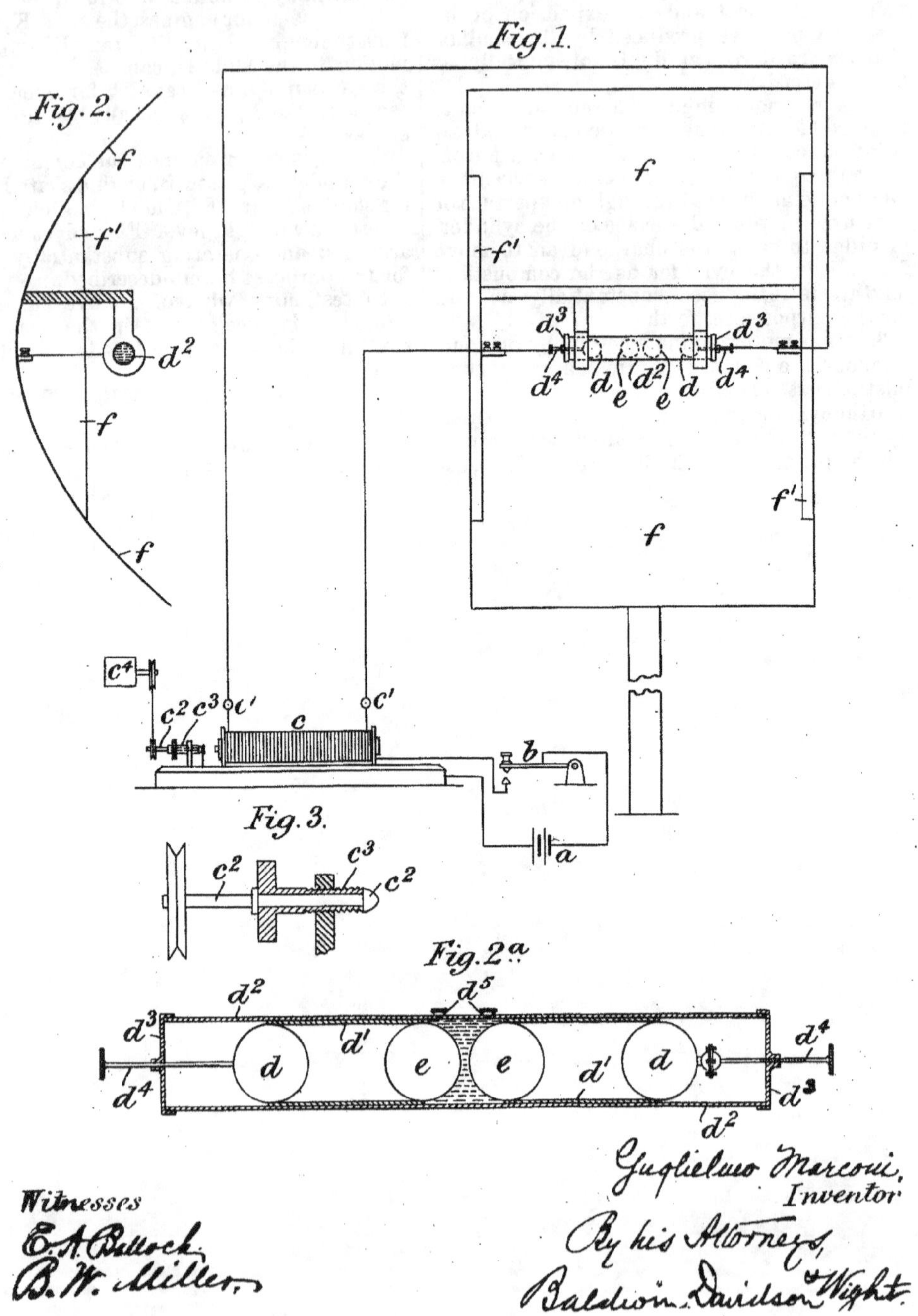

(No Model.) 3 Sheets—Sheet 2.

G. MARCONI.

TRANSMITTING ELECTRICAL SIGNALS.

No. 586,193. Patented July 13, 1897.

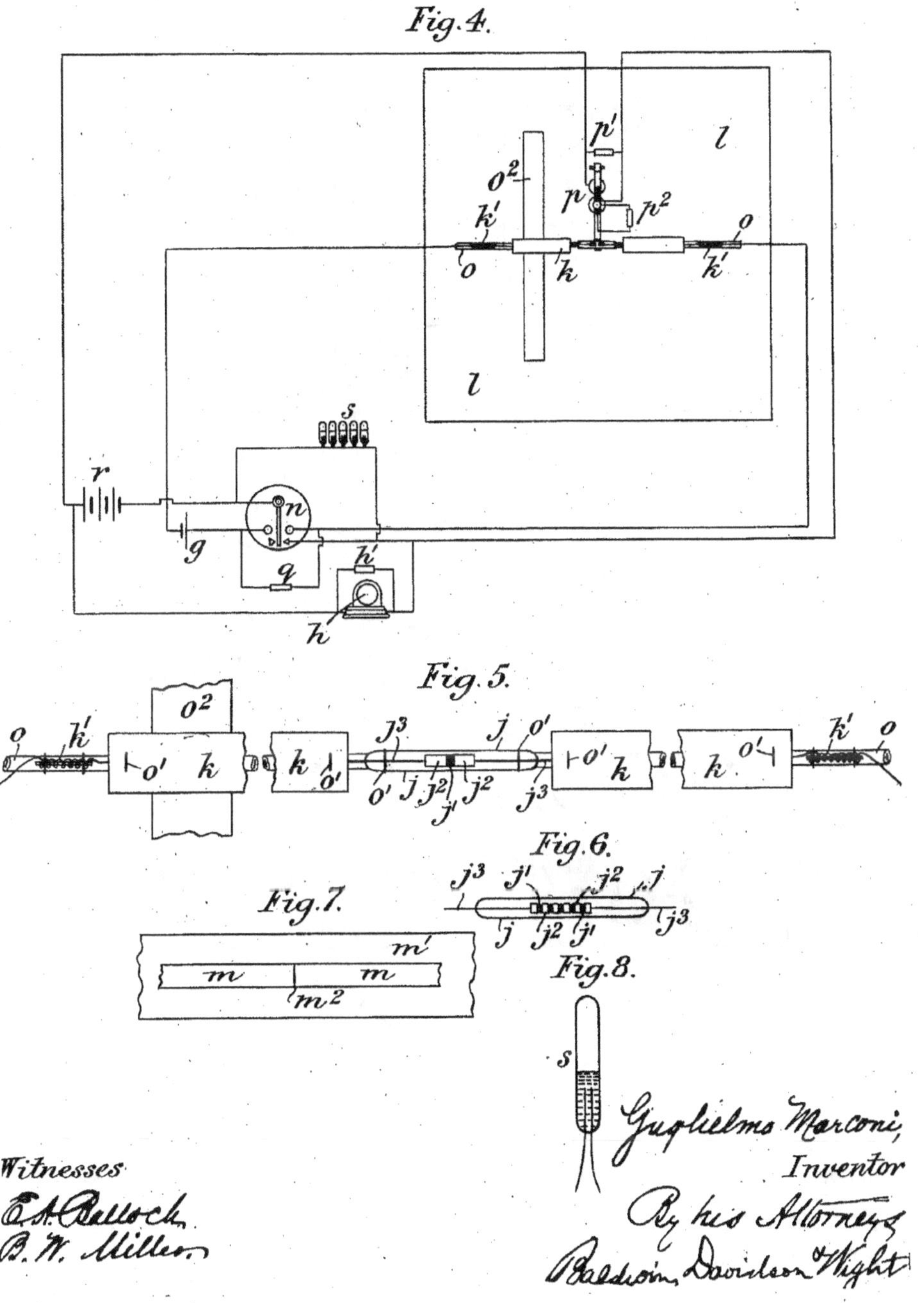

Witnesses
E. A. Ballock
B. W. Miller

Guglielmo Marconi,
Inventor
By his Attorneys
Baldwin, Davidson & Wight

(No Model.) 3 Sheets—Sheet 3.

G. MARCONI.

TRANSMITTING ELECTRICAL SIGNALS.

No. 586,193. Patented July 13, 1897.

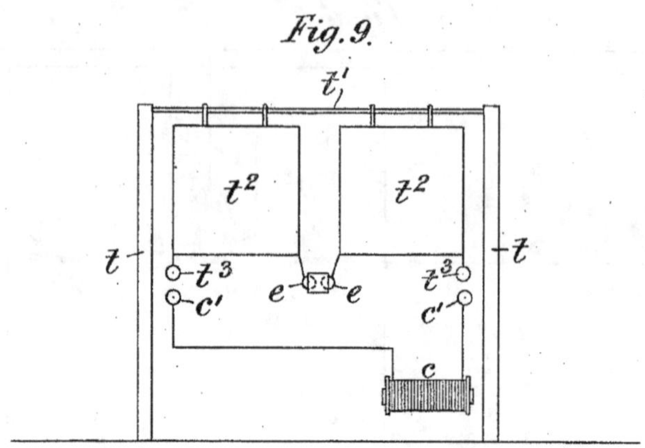

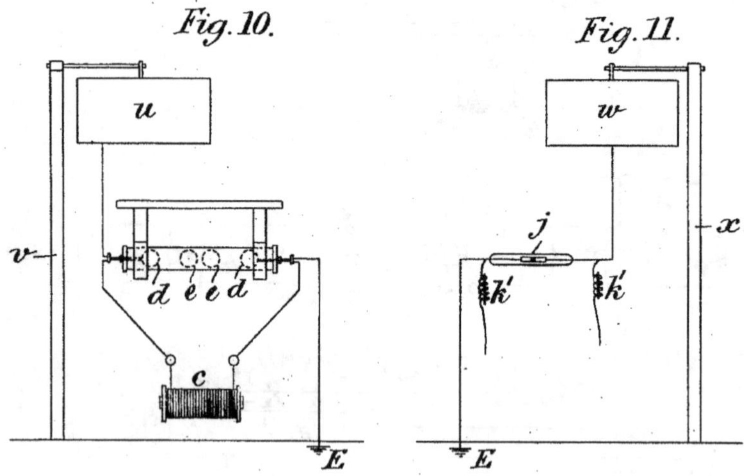

Witnesses

E. A. Bullock

B. W. Miller

Guglielmo Marconi

Inventor

By his Attorneys

Baldwin, Davidson & Wight

UNITED STATES PATENT OFFICE.

GUGLIELMO MARCONI, OF LONDON, ENGLAND.

TRANSMITTING ELECTRICAL SIGNALS.

SPECIFICATION forming part of Letters Patent No. 586,193, dated July 13, 1897.

Application filed December 7, 1896. Serial No. 614,838. (No model.)

To all whom it may concern:

Be it known that I, GUGLIELMO MARCONI, student, a subject of the King of Italy, residing at 21 Burlington Road, London, in the county of Middlesex, England, have invented certain new and useful Improvements in Transmitting Electrical Impulses and Signals and in Apparatus Therefor, of which the following is a specification.

According to this invention electrical signals, actions, or manifestations are transmitted through the air, earth, or water by means of oscillations of high frequency, such as have been called the "Hertz rays" or "Hertz oscillations." Usually all line-wires are dispensed with. At the transmitting-station I employ a Ruhmkorff coil, having in its primary circuit a Morse key or other signaling instrument and at its poles appliances for producing the desired oscillations. The Ruhmkorff coil may, however, be replaced by any other source of high-tension electricity. When working with large amounts of energy, it is, however, better to keep the coil or transformer constantly working for the time during which one is transmitting, and instead of interrupting the current of the primary interrupting the discharge of the secondary. In this case the contacts of the key should be immersed in oil, as otherwise, owing to the length of the spark, the current will continue to pass after the contacts have been separated. At the receiving-station there is a local-battery circuit, containing any ordinary receiving instrument and an appliance for closing the circuit, the latter being actuated by the oscillations from the transmitting-station. When transmitting through the air and it is desired that the signal should only be sent in one direction, I place the oscillation-producer at the transmitting-station in the focus or focal line of a reflector directed to the receiving-station, and I place the circuit-closer at the receiving-station in a similar reflector directed toward the transmitting-station. When transmitting signals through the earth, I connect one end of the oscillation-producer and one end of the circuit-closer to earth and the other ends to similar plates, preferably electrically tuned with each other in the air and insulated from earth.

Figure 1 is a diagrammatic front elevation of the instruments at the transmitting-station when signaling through the air, and Fig. 2 is a vertical section of the transmitter. Fig. 2^a is a longitudinal section of the oscillator to a larger scale. Fig. 3 shows a detail on a larger scale. Fig. 4 is a diagrammatic front elevation of the instruments at the receiving-station. Fig. 5 is a full-sized view of the receiver. Fig. 6 shows a modification of the tube *j*. Fig. 7 shows the detector. Fig. 8 is a full-sized view of the liquid-resistance. Figs. 9 and 10 show modifications of the arrangements at the transmitting-station. Fig. 11 shows a modification of the arrangements at the receiving-station.

Referring now to Fig. 1, *a* is a battery, and *b* an ordinary Morse key closing the circuit through the primary of a Ruhmkorff coil *c*. The terminals c' of the secondary circuit of the coil are connected to two metallic balls *d d*, fixed by heat or otherwise at the ends of tubes d' d', Fig. 2^a, of insulating material, such as ebonite or vulcanite. *e e* are similar balls fixed in the other ends of the tubes d'. The tubes d' fit tightly in a similar tube d^2, having covers d^3, through which pass rods d^4, connecting the balls *d* to the conductors. One (or both) of the rods d^4 is connected to the ball *d* by a ball-and-socket joint and has a screw-thread upon it working in a nut in the cover d^3. By turning the rod therefore the distance of the balls *e* apart can be adjusted. d^5 are holes in the tube d^2, through which vaseline, oil, or like material is introduced into the space between the balls *e*

The balls *d* and *e* are preferably of solid brass or copper, and the distance they should be apart depends on the quantity and electromotive force of the electricity employed, the effect increasing with the distance so long as the discharge passes freely. With a coil giving an ordinary eight-inch spark the distance between *e* and *e* should be from one twenty-fifth to one-thirtieth of an inch and the distance between *d* and *e* about one and a half inches. *f* is a cylindrical parabolic reflector made by bending a metallic sheet, preferably of brass or copper, to form and fixing it to metallic or wooden ribs f'. Other conditions being equal the larger the balls the greater is the distance at which it is possible to communicate. I have generally used balls of solid brass of four inches diameter, giving oscillations of ten inches length of wave.

The reflectors applied to the receiver and transmitter ought to be preferably in length and opening the double at least of the length of wave emitted from the oscillator.

If a very powerful source of electricity giving a very long spark be employed, it is preferable to divide the spark gap between the central balls of the oscillator into several smaller gaps in series. This may be done by introducing between the big balls smaller ones, (of about half an inch diameter,) held in position by ebonite frames.

I find that the regularity and power of the discharge of an ordinary Ruhmkorff coil with a trembler-break on its primary is greatly improved by causing one of the contacts of the vibrating break to revolve rapidly. I do this by having a revoluble central core c^2, Fig. 3, in the ordinary screw c^3, which is in communication with the platinum contacts. I cause the said central core with one of the platinum contacts attached to it to revolve by connecting it to a small electric motor c^4. This motor can be worked by the same circuit that works the coil, or, if necessary, by a separate circuit. The connections are not shown in the drawings. By this means the platinums are kept smooth and any tendency to stick is removed. They last also much longer. At the receiving-station is a battery whose circuit includes an ordinary telegraphic instrument (or it may be a relay or other apparatus which it is desired to work from a distance) and a circuit-closer.

In Fig. 4, g is the battery, and h a telegraphic instrument on the derived circuit of a relay n.

The appliance I employ as a circuit-closer is shown full size at Fig. 5 and consists of a glass tube j, containing metallic powder or grains of metal j', each end of the column of powder being connected to a metallic plate k of suitable length to cause the system to resonate electrically in unison with the electrical oscillations transmitted. The glass tube may be replaced in some cases by one of gutta-percha or like material. Two short pieces of thick silver wire j^2 of the same diameter as the internal diameter of the tube j, so as to fit tightly in it, are joined to two pieces of platinum wire j^3. The tube is closed and sealed onto the platinum wires j^3 at both ends.

Many metals can be employed for producing the powder or filings j'; but I prefer to use a mixture of two or more different metals. I find hard nickel to be the best metal, and I prefer to add to the nickel filings about ten per cent. of hard-silver filings, which increase greatly the sensitiveness of the tube to electric oscillations. By increasing the proportion of silver powder or grains the sensitiveness of the tube also increases; but it is better for ordinary work not to have a tube of too great sensitiveness, as it might be influenced by atmospheric or other electricity. The sensitiveness can also be increased by adding a very small amount of mercury to the filings and mixing up until the mercury is absorbed.

The mercury must not be in such a quantity as to clot or cake the filings. An almost imperceptible globule is sufficient for a tube. Instead of mixing the mercury with the powder one can obtain the same effects by slightly amalgamating the inner surfaces of the plugs which are to be in contact with the filings. Very little mercury must be used, just sufficient to brighten the surface of the metallic plugs without showing any free globules. The size of the tube and the distance between the two metallic stops may vary under certain limits. The greater the space allowed for the powder the larger and coarser ought to be the filings or grains.

I prefer to make my sensitive tubes of the following size: The tube j is one and one-half inches long and one-tenth or one-twelfth of an inch internal diameter. The length of the stops j^2 is about one-fifth of an inch, and the distance between the stops is about one-thirtieth of an inch. I find that the smaller the space between the stops in the tube the more sensitive it proves, but the space cannot under ordinary circumstances be excessively shortened without injuring the fidelity of the transmission.

The metallic powders ought not to be fine, but rather as coarse as can be produced by a large and rough file.

All the very fine powder ought to be removed by blowing or sifting.

The powder ought not to be compressed between the stops, but rather loose and in such a condition that when the tube is tapped the powder may be seen to move.

The tube must be sealed, but a vacuum inside it is not essential, except the slight vacuum which results from having heated it while sealing it. Care must also be taken not to heat the tube too much in the center when sealing it, as it would oxidize the surfaces of the silver stops and also the powder, which would diminish its sensitiveness. I use in sealing the tubes a hydrogen and air flame. A vacuum is, however, desirable, and I have used one of about one one-thousandth of an atmosphere, obtained by a mercury-pump. It is also necessary for the powder or grains to be dry and free from grease or dirt, and the files used in producing the same ought to be frequently washed and dried and used when warm.

If the tube has been well made, it should be sensitive to the induction of an ordinary electric bell when the same is working at one to two yards or more from the tube.

In order to keep the sensitive tube j in good working order, it is desirable, but not absolutely necessary, not to allow more than one milliampere to flow through it when active. If a stronger current is necessary, several tubes may be put in derivation between the tuned plates, but this arrangement is not quite as satisfactory as the single tube. It is necessary when using tubes of the type I have described not to insert in the circuit more than one cell of the Leclanché type, as a higher electromotive force than 1.5 volts is apt to pass a current through the tube even when no oscillations are transmitted. I can, however, construct tubes capable of working with a much higher electromotive force. Fig. 6 shows one of these tubes. In this tube instead of one space or gap filled with filings there are several spaces separated by sections of tight-fitting silver wire. A tube thus constructed, observing also the rules of construction of my tubes in general, will work satisfactorily if the electromotive force of the battery in circuit with the tube is equal to 1.2 volts multiplied by the number of gaps. With this tube also it is well not to allow a current of more than one milliampere to pass.

The tube j may be replaced by other forms of imperfect electrical contacts, but this is not desirable.

The plates k are of copper or aluminium or other metal, about half an inch or more broad, about one-fiftieth of an inch thick, and preferably of such a length as to be electrically tuned with the electric oscillations transmitted. The means I adopt for fixing the length of the plates is as follows: I stick a rectangular strip of tin-foil m (see Fig. 7) about twenty inches long (the length depends on the supposed length of wave that one is measuring) by means of a weak solution of gum onto a glass plate m'. Then by means of a very sharp penknife or point I cut across the middle of the tin-foil, leaving a mark of division m^2. If this detector is held in the proximity (four or five yards) and parallel with the axis of the oscillator in action, it will show little sparks at m^2. If the length of the pieces of tin-foil approximates to the length of wave emitted from the oscillator, the spark will take place between them at a certain distance from the transmitter, which is a maximum when they are of suitable length. By shortening or lengthening the strips, therefore, it is easy to find the length most appropriate to the length of wave emitted by the oscillator. It is desirable to try this detector in the focus or focal line of the reflector. The length so found is the proper length for the plates k, or rather these should be about half an inch shorter on account of the length of the sensitive tube j, connected between them.

l is a cylindrical parabolic reflector similar to that used at the transmitting-station.

The plates k may be in the form of tubes or even wires.

It is slightly advantageous for the focal distance of the reflector to be equal to one-fourth or three-fourths of the wave length of the oscillation transmitted.

When no oscillations are sent from the transmitting-station, the tube j does not conduct the current, and the local-battery circuit is broken, but when the powder or tube is influenced by the electrical oscillations from the transmitter it conducts and closes the circuit. I find, however, that when once started the powder in the tube continues to conduct even when the oscillations from the transmitter have ceased, but if it be shaken or tapped the circuit is broken. A tube well prepared will instantly interrupt the current passing through it at the slightest tap, provided it is inserted in a circuit in which there is little self-induction and small electromotive force, such as a single cell, and where the effects of self-induction have been removed by one of the methods which I will presently describe.

The two plates k communicate with the local circuit through two very small coils k', which I will call "choking-coils," formed by winding a few inches of very thin and insulated copper wire around a bit of iron wire about an inch and a half long. The object of these choking-coils is to prevent the high-frequency oscillation induced across these plates by the transmitter from dissipating itself by running along the local-battery wires which might weaken its effect on the sensitive tube j. These choking-coils may, however, be sometimes replaced by simple thin wires. They may also be connected directly to the tube j. The local circuit in which the sensitive tube j is inserted contains a sensitive relay n, preferably wound to a resistance of about twelve hundred ohms. This resistance need not be necessarily that of the relay, but may be the sum of the resistance of the relay and another additional resistance. The relay ought to be one possessing small self-induction.

The plates k, tube j, and coils k' are fastened by means of wire stitches o' to a thin glass tube o, preferably not longer than twelve inches, firmly fixed at one end to a strong piece of timber o^2. This may be done by means of wood or ebonite grasping-screws.

I do the tapping automatically by the current started by the tube, employing a trembler p on the circuit of the relay n similar in construction to that of an electric bell, but having a shorter arm. The vibrator must be carefully adjusted. Preferably the blows should be directed slightly upward to prevent the filings from getting caked. In place of tapping the tube the powder can be disturbed by slightly moving outward and inward one or both of the stops j^2, the trembler p being replaced by a small electromagnet (or magnets) whose armature is connected to the stop.

I ordinarily work the telegraphic receiver h (or other instruments) by a derivation, as shown, from the circuit which works the trembler p. They can also, however, be worked in series with the trembler. When working ordinary sounders or Morse apparatus, a special adjustment of the same is sometimes needed to enable one to obtain dots and dashes. Sometimes it is necessary to work the telegraphic instruments or relays from the back-stops of the first relay, as is done in some systems of multiple telegraphy. Such adjustments are known to telegraphic experts.

By means of a tube with multiple gaps it is possible to work the trembler and also the signaling or other apparatus direct on the circuit which contains the tube, but I prefer when possible to work with the single-gap tube and the relay, as shown. With a sensitive and well-constructed trembler it is also possible to work the trembler with the single-gap tube in series with it without the relay.

In derivation on the terminals of the relay n is placed an ordinary platinoid resistance double-wound (or wound on the "bight," as it is sometimes termed) coil q of about four times the resistance of the relay, which prevents the self-induction of the winding of the relay from affecting the sensitive tube.

The circuit actuated by the relay contains an ordinary battery r of about twelve cells and the trembler p, the resistance of the winding of which should be about one thousand ohms, and the nucleus ought preferably to be of soft iron, hollow and split lengthwise, like most electromagnets used in telegraph instruments. In series or derivation from this circuit is inserted the telegraphic or other apparatus h which one may desire to work. It is desirable that this instrument or apparatus, if on a derivation, should have a resistance equal to the resistance of the trembler p. A platinoid resistance h' of about five times the resistance of the instrument is inserted in derivation across the terminals of the instrument and connected as close to the same as possible. In derivation across the terminals of the trembler p is placed another platinoid resistance p', also of about five times the resistance of the trembler. A similar resistance p^2 is inserted in a circuit connecting the vibrating contacts of the trembler. In derivation across the terminals of the relay-circuit it is well to have a liquid resistance s, which is constituted of a series of tubes, one of which is shown full size in Fig. 8, filled with water acidulated with sulfuric acid. The number of these tubes in series across the said terminals ought to be about ten for a circuit of fifteen volts, so as to prevent, in consequence of their counter electromotive force, the current of the local battery from passing through them, but allowing the high-tension jerk of current generated at the opening of the circuit in the relay to pass smoothly across them without producing perturbing sparks at the movable contact of the relay. It is also necessary to insert a platinoid resistance in derivation on any apparatus one may be working on the local circuits. These resistances ought also to be inserted in derivation on the terminals of any resistance which may be apt to give self-induction.

I have hitherto only mentioned the use of cylindrical reflectors, but it is also possible to use ordinary concave reflectors, preferably parabolic, such as are used for projectors.

It is not essential to have a reflector at the transmitters and receivers, but in their absence the distance at which one can communicate is much smaller.

I find it convenient when transmitting across long distances to make use of the transmitter shown in Fig. 9.

t t are two poles connected by a rope t', to which are suspended by means of insulating suspenders two metallic plates t^2 t^2, preferably in the form of cylinders closed at the top, connected to the spheres e (in oil or other dielectric, as before) and to the other balls t^3 in proximity to the spheres c', in communication with the coil or transformer c. The balls t^3 are not absolutely necessary, as the plates t^2 may be made to communicate with the coil or transformer by means of thin insulated wires. The receiver I adopt with this transmitter is similar to it, except that the spheres e are replaced by the sensitive tube j and plates k, while the spheres t^3 are replaced by the choking-coils k', in communication with the local circuit. It may be observed that, other conditions being equal, the larger the plates at the transmitter and receiver and the higher they are from earth and to a certain extent the farther apart they are the greater is the distance at which correspondence is possible.

When transmitting through the earth or water, I use a transmitter as shown in Fig. 10. I connect one of the spheres d to earth E, preferably by a thick wire, and the other to a plate or conductor u, suspended on a pole v and insulated from earth; or the spheres d may be omitted and one of the spheres e be connected to earth and the other to the plate or conductor u. At the receiving-station, Fig. 11, I connect one terminal of the sensitive tube j to earth E, also by a thick wire, and the other to a plate or conductor w, preferably similar to u. The plate w may be suspended on a pole x and must be insulated from earth. The larger the plates of the receiver and transmitter and the higher from the earth the plates are suspended the greater is the distance at which it is possible to communicate. When using the last-described apparatus, it is not necessary to have the two instruments in view of each other, as it is of no consequence if they are separated by mountains or other obstacles. At the receiver it is possible to pick up the oscillations from the earth or water without having the plate w. This may be done by connecting the terminals of the sensitive tube j to two earths, preferably at a certain distance from each other and in a line with the direction from which the oscillations are coming. These connections must not be entirely conductive, but must contain a condenser of suitable capacity—say one square yard of surface. Balloons can also be used instead of plates on poles, provided they carry up a plate or are themselves made conductive by being covered with tin-foil. As the height to which they may be sent is great, the distance at which communication is possible becomes greatly multiplied. Kites may also be successfully employed if made conductive by means of tin-foil.

The apparatus above described is so sensitive that it is essential either that the transmitters and receivers at each station should be at a considerable distance from each other or that they should be screened from each other by stout metal plates. It is sufficient to have all the telegraphic apparatus in a metal box and any exposed part of the circuit of the receiver inclosed in metallic tubes which are in electrical communication with the box. (Of course the part of the apparatus which has to receive the radiation from the distant station must not be inclosed, but possibly screened from the local transmitter by means of metallic sheets.) When working through the earth or water, the local receiver must be switched out of circuit when the transmitter is at work, and this may also be done when working through air.

What I claim is—

1. In a receiver for electrical oscillations the combination of an imperfect electrical contact, a circuit through the contact and means actuated by the circuit for shaking the contact.

2. In a receiver for electrical oscillations the combination of an imperfect electrical contact, metallic plates connected to it, a circuit through the contact and means actuated by the circuit for shaking the contact.

3. In a receiver for electrical oscillations the combination of an imperfect electrical contact, metallic plates connected to the contact, choking-coils connected to the contact, a circuit through the coils, and contact and means actuated by the circuit for shaking the contact.

4. In a receiver for electrical oscillations the combination of a tube containing metallic powder, a circuit through the powder and means actuated by the circuit for shaking the powder.

5. In a receiver for electrical oscillations the combination of a tube containing metallic powder, metallic plates connected to the powder, a circuit through the powder and means actuated by the circuit for shaking the powder.

6. In a receiver for electrical oscillations the combination of a tube containing metallic powder, metallic plates connected to the powder, choking-coils connected to the powder, a circuit through the coils and powder and means actuated by the circuit for shaking the powder.

7. In a receiver for electrical oscillations the combination of a tube containing a mixture of metallic powders, a circuit through the powder, and means actuated by the circuit for shaking the powder.

8. In a receiver for electrical oscillations the combination of a tube containing a mixture of metallic powders, metallic plates connected to the powder, a circuit through the powder and means actuated by the circuit for shaking the powder.

9. In a receiver for electrical oscillations the combination of a tube containing a mixture of metallic powders, metallic plates connected to the powder, choking-coils connected to the powder, a circuit through the coils, and powder and means actuated by the circuit for shaking the powder.

10. In a receiver for electrical oscillations the combination of a tube containing a mixture of metallic powder and mercury, a circuit through the powder and means actuated by the circuit for shaking the powder.

11. In a receiver for electrical oscillations the combination of a tube containing a mixture of metallic powder and mercury, metallic plates connected to the powder, a circuit through the powder and means actuated by the circuit for shaking the powder.

12. In a receiver for electrical oscillations the combination of a tube containing a mixture of metallic powder and mercury, metallic plates connected to the powder, choking-coils connected to the powder, a circuit through the coils and powder and means actuated by the circuit for shaking the powder.

13. In a receiver for electrical oscillations the combination of a tube, metallic plugs in the tube, metallic powder between the plugs, a circuit through the plugs and powder and means actuated by the circuit for shaking the powder.

14. In a receiver for electrical oscillations the combination of a tube, metallic plugs in the tube, metallic powder between the plugs, metallic plates connected to them, a circuit through the plugs and powder and means actuated by the circuit for shaking the powder.

15. In a receiver for electrical oscillations the combination of a tube, metallic plugs in the tube, metallic powder between the plugs, metallic plates connected to the plugs, choking-coils connected to the plugs, a circuit through the coils and plugs and means actuated by the circuit for shaking the powder.

16. In a receiver for electrical oscillations the combination of a tube, metallic plugs in the tube, a mixture of metallic powders between the plugs, a circuit through the plugs and powder and means actuated by the circuit for shaking the powder.

17. In a receiver for electrical oscillations the combination of a tube, metallic plugs in the tube, a mixture of metallic powders between the plugs, metallic plates connected to the plugs, a circuit through the plugs and powder and means actuated by the circuit for shaking the powder.

18. In a receiver for electrical oscillations the combination of a tube, metallic plugs in the tube, a mixture of metallic powders between the plugs, metallic plates connected to the plugs, choking-coils connected to the

plugs, a circuit through the coils plugs and powder and means actuated by the circuit for shaking the powder.

19. In a receiver for electrical oscillations the combination of a tube, metallic plugs in the tube, a mixture of metallic powder and mercury between the plugs, a circuit through the plugs and powder and means actuated by the circuit for shaking the powder.

20. In a receiver for electrical oscillations the combination of a tube, metallic plugs in the tube, a mixture of metallic powder and mercury between the plugs, metallic plates connected to the plugs, a circuit through the plugs and powder and means actuated by the circuit for shaking the powder.

21. In a receiver for electrical oscillations the combination of a tube, metallic plugs in the tube, a mixture of metallic powder and mercury between the plugs, metallic plates connected to the plugs, choking-coils connected to the plugs, a circuit through the coils plugs and powder and means actuated by the circuit for shaking the powder.

22. In a receiver for electrical oscillations the combination of an imperfect electrical contact, a circuit through the contact, a relay actuated by the circuit and means actuated by the relay for shaking the contact.

23. In a receiver for electrical oscillations the combination of an imperfect electrical contact, metallic plates connected to it, a circuit through the contact, a relay actuated by the circuit and means actuated by the relay for shaking the contact.

24. In a receiver for electrical oscillations the combination of an imperfect electrical contact, metallic plates connected to the contact, choking-coils connected to the contact, a circuit through the coils and contact, a relay actuated by the circuit and means actuated by the relay for shaking the contact.

25. In a receiver for electrical oscillations the combination of a tube containing metallic powder, a circuit through the powder, a relay actuated by the circuit and means actuated by the relay for shaking the powder.

26. In a receiver for electrical oscillations the combination of a tube containing metallic powder, metallic plates connected to the powder, a circuit through the powder, a relay actuated by the circuit and means actuated by the relay for shaking the powder.

27. In a receiver for electrical oscillations the combination of a tube containing metallic powder, metallic plates connected to the powder, choking-coils connected to the powder a circuit through the coils and powder a relay actuated by the circuit and means actuated by the relay for shaking the powder.

28. In a receiver for electrical oscillations the combination of a tube containing a mixture of metallic powders, a circuit through the powder, a relay actuated by the circuit and means actuated by the relay for shaking the powder.

29. In a receiver for electrical oscillations the combination of a tube containing a mixture of metallic powders, metallic plates connected to the powder, a circuit through the powder, a relay actuated by the circuit and means actuated by the relay for shaking the powder.

30. In a receiver for electrical oscillations the combination of a tube containing a mixture of metallic powders, metallic plates connected to the powder, choking-coils connected to the powder, a circuit through the coils and powder, a relay actuated by the circuit and means actuated by the relay for shaking the powder.

31. In a receiver for electrical oscillations the combination of a tube containing a mixture of metallic powder and mercury, a circuit through the powder, a relay actuated by the circuit and means actuated by the relay for shaking the powder.

32. In a receiver for electrical oscillations the combination of a tube containing a mixture of metallic powder and mercury, metallic plates connected to the powder, a circuit through the powder, a relay actuated by the circuit and means actuated by the relay for shaking the powder.

33. In a receiver for electrical oscillations the combination of a tube containing a mixture of metallic powder and mercury, metallic plates connected to the powder, choking-coils connected to the powder, a circuit through the coils and powder, a relay actuated by the circuit and means actuated by the relay for shaking the powder.

34. In a receiver for electrical oscillations the combination of a tube, metallic plugs in the tube, metallic powder between the plugs, a circuit through the plugs and powder, a relay actuated by the circuit and means actuated by the relay for shaking the powder.

35. In a receiver for electrical oscillations the combination of a tube, metallic plugs in the tube, metallic powder between the plugs, metallic plates connected to the plugs, a circuit through the plugs and powder, a relay actuated by the circuit and means actuated by the relay for shaking the powder.

36. In a receiver for electrical oscillations the combination of a tube, metallic plugs in the tube, metallic powder between the plugs, metallic plates connected to the plugs, choking-coils connected to the plugs, a circuit through the coils, plugs and powder, a relay actuated by the circuit, and means actuated by the relay for shaking the powder.

37. In a receiver for electrical oscillations the combination of a tube, metallic plugs in the tube, a mixture of metallic powders between the plugs, a circuit through the plugs and powder, a relay actuated by the circuit, and means actuated by the relay for shaking the powder.

38. In a receiver for electrical oscillations the combination of a tube, metallic plugs in the tube, a mixture of metallic powders between the plugs, metallic plates connected to

the plugs, a circuit through the plugs and powder, a relay actuated by the circuit, and means actuated by the relay for shaking the powder.

39. In a receiver for electrical oscillations the combination of a tube, metallic plugs in the tube, a mixture of metallic powders between the plugs, metallic plates connected to the plugs, choking-coils connected to the plugs, a circuit through the coils, plugs and powder, a relay actuated by the circuit and means actuated by the relay for shaking the powder.

40. In a receiver for electrical oscillations the combination of a tube, metallic plugs in the tube, a mixture of metallic powder and mercury between the plugs, a circuit through the plugs and powder, a relay actuated by the circuit and means actuated by the relay for shaking the powder.

41. In a receiver for electrical oscillations the combination of a tube, metallic plugs in the tube, a mixture of metallic powder and mercury between the plugs, metallic plates connected to the plugs, a circuit through the plugs and powder, a relay actuated by the circuit and means actuated by the relay for shaking the powder.

42. In a receiver for electrical oscillations the combination of a tube, metallic plugs in the tube, a mixture of metallic powder and mercury between the plugs, metallic plates connected to the plugs, choking-coils connected to the plugs, a circuit through the coils, plugs and powder, a relay actuated by the circuit and means actuated by the relay for shaking the powder.

43. The combination of a spark-producer at the transmitting-station, an earth connection to one end of the spark-producer, an insulated conductor connected to the other end, an imperfect electrical contact at the receiving-station, an earth connection to one end of the contact an insulated conductor connected to the other end and a circuit through the contact.

44. The combination of a spark-producer at the transmitting-station, an earth connection to one end of the spark-producer, an insulated conductor connected to the other end, an imperfect electrical contact at the receiving-station, an earth connection to one end of the contact an insulated conductor connected to the other end, a circuit through the contact and means actuated by the circuit for shaking the contact.

45. The combination of a spark-producer at the transmitting-station, an earth connection to one end of the spark-producer, an insulated conductor connected to the other end, an imperfect electrical contact at the receiving-station, choking-coils connected to each end of the contact, an earth connection to one end of the imperfect contact an insulated conductor connected to the other end and a circuit through the coils and contact.

46. The combination of a spark-producer at the transmitting-station, an earth connection to one end of the spark-producer, an insulated conductor connected to the other end, an imperfect electrical contact at the receiving-station, choking-coils connected to each end of the contact, an earth connection to one end of the imperfect contact, an insulated conductor connected to the other end, a circuit through the coils and contact and means actuated by the circuit for shaking the contact.

47. The combination of a spark-producer at the transmitting-station, an earth connection to one end of the spark-producer, an insulated conductor connected to the other end, a tube containing metallic powder at the receiving-station, an earth connection to one end of the powder, an insulated conductor connected to the other end and a circuit through the powder.

48. The combination of a spark-producer at the transmitting-station, an earth connection to one end of the spark-producer, an insulated conductor connected to the other end, a tube containing metallic powder at the receiving-station, an earth connection to one end of the powder an insulated conductor connected to the other end, a circuit through the powder and means actuated by the circuit for shaking the powder.

49. The combination of a spark-producer at the transmitting-station, an earth connection to one end of the spark-producer, an insulated conductor connected to the other end, a tube containing metallic powder at the receiving-station, choking-coils connected to each end of the powder, an earth connection to one end of the powder, an insulated conductor connected to the other end and a circuit through the coils and powder.

50. The combination of a spark-producer at the transmitting-station, an earth connection to one end of the spark-producer, an insulated conductor connected to the other end, a tube containing metallic powder at the receiving-station, choking-coils connected to each end of the powder, an earth connection to one end of the powder, an insulated conductor connected to the other end, a circuit through the coils and powder and means actuated by the circuit for shaking the powder.

51. The combination of a spark-producer at the transmitting-station, an earth connection to one end of the spark-producer, an insulated conductor connected to the other end, a tube containing metallic powder at the receiving-station, choking-coils and earth connection through condensers connected to each end of the powder, a circuit through the coils and powder and means actuated by the circuit for shaking the powder.

52. In a receiver for electrical oscillations, the combination of an imperfect electrical contact, a circuit through the contact, an electric trembler shaking the contact, and means for preventing the self-induction of the trembler from affecting the contact.

53. A receiver for electrical oscillatory impulses having a medium whose electrical resistance is altered by the received electrical oscillations, a trembler or shaker for acting upon the variable-resistance medium to restore it to its normal condition of electrical resistance, and means for controlling such trembler to cause it to act upon the variable-resistance medium to restore it to its normal condition after each reception of such oscillatory impulses.

54. A receiver for electrical oscillatory impulses having a medium whose electrical resistance is altered by the received electrical oscillations, a trembler or shaker for acting upon the variable-resistance medium to restore it to its normal condition of electrical resistance, means controlling such trembler to cause it to act upon the variable-resistance medium to restore it to its normal condition after each reception of such oscillatory impulses, and means for rendering manifest said electrical oscillatory impulses consecutively received, whereby defined signals may be given out by the receiver.

55. The combination of a transmitter capable of producing at will of the operator electric oscillatory impulses or rays, and a receiver responsive thereto having a variable-resistance medium whose resistance is altered by such received oscillatory impulses, means controlled by the received oscillations for restoring such medium to its normal condition after each reception of such oscillations, and means for rendering manifest the received oscillations, whereby signals sent from the transmitter may be received upon the receiver.

56. The combination of a transmitter capable of producing electrical oscillations or rays at the will of the operator, and a receiver located at a distance and having a conductor tuned to respond to such oscillations, a variable-resistance medium, in circuit with the conductor, whose resistance is altered by the received oscillations, means controlled by the received oscillations for restoring the resistance medium to its normal condition after each reception of such oscillations, and means for rendering the received oscillations manifest.

GUGLIELMO MARCONI.

Witnesses:

WILFRED CORPMAEL,

FRED C. HARIES.

Dec. 17, 1957 R. J. TAYLOR 2,816,721

ROCKET POWERED AERIAL VEHICLE

Filed Sept. 15, 1953

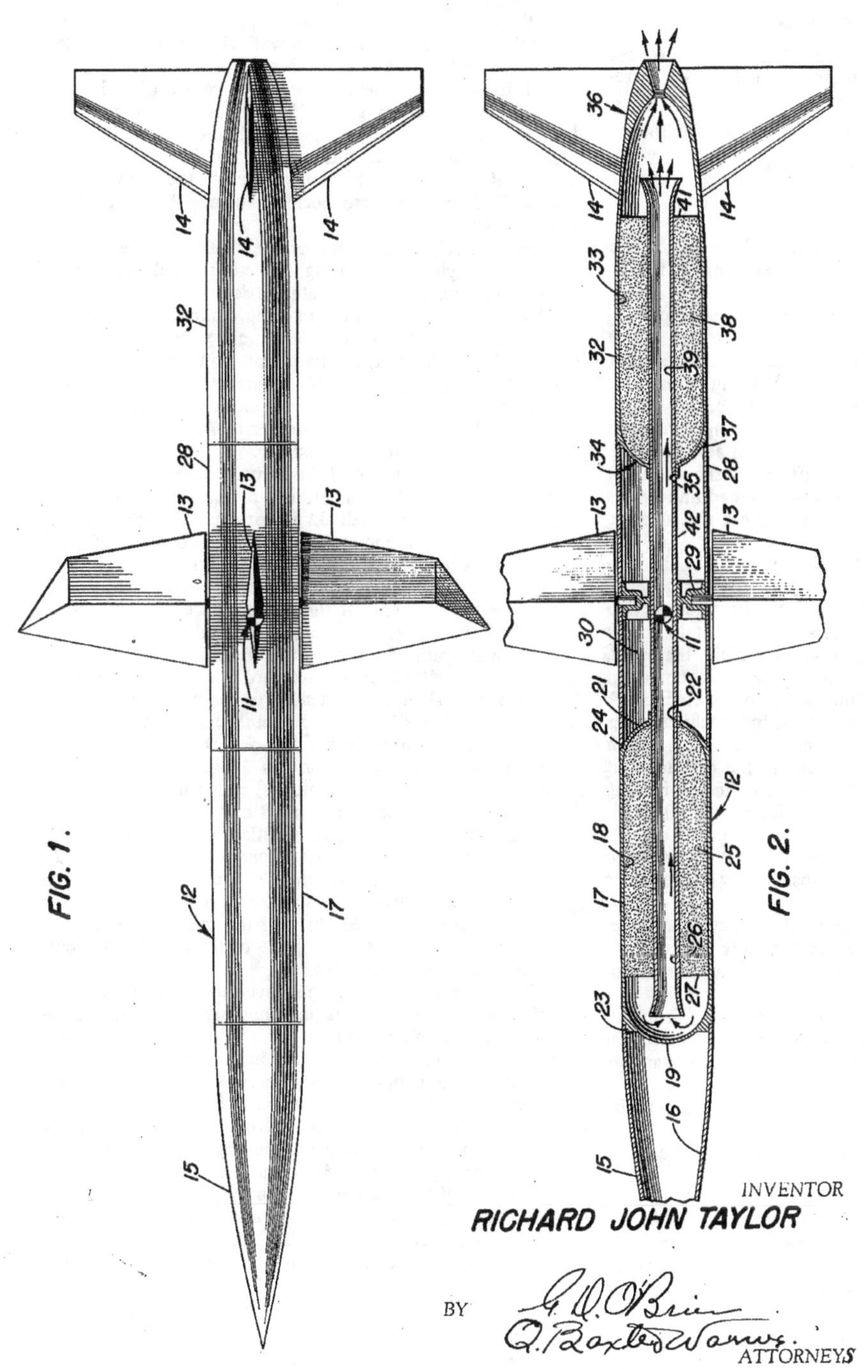

INVENTOR
RICHARD JOHN TAYLOR

BY G. W. O'Brien
Q. Baxter Warner
ATTORNEYS

United States Patent Office

2,816,721

Patented Dec. 17, 1957

1

2,816,721

ROCKET POWERED AERIAL VEHICLE

Richard John Taylor, Buffalo, N. Y., assignor to the United States of America as represented by the Secretary of the Navy

Application September 15, 1953, Serial No. 380,380

4 Claims. (Cl. 244—74)

This invention relates generally to aerial vehicles and more particularly to an improved rocket engine for an aerial vehicle.

The conventional rocket engine employs a solid propellant fuel in the form of a grain or several grains which may be designed for end burning. A grain designed for end burning is characterized by even burning to produce a constant thrust throughout the life of the grain. However, the ordinary end burning grain in being consumed causes a shift in the center of gravity of an aerial vehicle propelled by the grain. Such a shift upsets the optimum conditions for which the vehicle was designed and brings into play aerodynamic forces which affect the steering control of the vehicle.

It is customary in the construction of aerial vehicles to locate wing surfaces at the center of gravity. In general, the center of gravity is positioned in the heaviest portion of the vehicle which in most cases is that containing the propulsion system. A rocket powered vehicle, therefore, is generally constructed with wing surfaces mounted on that area of the vehicle constituting the rocket engine. A serious problem is presented, however, in attaching wings to a rocket engine because of the corrosive effects of hot combustion gases normally produced in the operation of the engine. This problem is further complicated if wing control apparatus is to be employed for varying the attitude of the wings.

It is, therefore, the principal object of this invention to provide a rocket engine having a substantially stable center of gravity.

A further object of this invention is to provide a rocket engine in which combustion gases normally produced in operation are handled in such a manner that their corrosive effect is restricted.

Further objects and attendant advantages of this invention will become evident from the following detailed description taken in conjunction with the accompanying drawings, in which:

Fig. 1 is a side elevation of a rocket powered aerial vehicle; and

Fig. 2 is an axial section of the main body of the aerial vehicle shown in Fig. 1, particularly illustrating the rocket engine constituting the present invention.

Referring now to Fig 1, the aerial vehicle shown therein has a center of gravity **11** and includes a hollow main body **12**, wings **13** mounted on said main body at the center of gravity **11** and tail fins **14** mounted on the rear end portion of the main body.

As best shown in Fig. 2, the body **12** comprises four sections. The forwardmost section includes a casing **15**, in the shape of an ogive forming the nose of the body **12**. A compartment **16** is defined by the casing **15** and is adapted to accommodate a warhead and/or guidance instrumentation.

The section next to the rear of the casing **15** is a cylindrical shell **17** constituting a forward combustion chamber **18**. The shell **17** is constructed with a hemispherical for-

2

ward end wall **19** and a hemispherical rear end wall **21** having an opening **22**. The forward end wall **19** is formed with an external annular shoulder **23** for receiving the rear end of the casing **15** and for connecting said casing to the shell **17**, by way of welding, riveting or by other suitable means. The rear end wall **21** is similarly formed with an external annular recess **24** for connecting the shell to the section next rearward, to be described hereinafter. A propellant grain **25**, cylindrically shaped with an axial bore **26**, is disposed in the chamber **18**. As shown in Fig. 2, the grain **25** fills the rear and intermediate portions of the shell **17** but ends short of the forward end wall **19** in a transverse burning surface **27** confronting said forward end wall.

Immediately to the rear of the shell **17** there is positioned a cylindrical casing **28** constructed with wing sockets **29** on its intermediate portion and defining a cell **30** for the accommodation of wing control and other apparatus. The wing sockets **29** mount wings on the body **12** and may be adapted to permit wing movements induced by the wing control apparatus contained in the cell **30**. The forward end of the casing **28** is received by the annular shoulder **24** in the shell **17** for attachment to said shell in any suitable manner.

The rearmost section of the body **12** comprises a second cylindrical shell **32** constituting a rearward combustion chamber 33. Thes hell **32** is constructed with a hemispherical forward end wall **34** having an opening **35**, and with an exhaust nozzle **36** at its rear end. The forward end wall **34** is formed with an external shoulder **37** which receives the rear end of the cylindrical casing **28** and is attached thereto by any appropriate securing medium. A propellant grain **38** having an axial bore **39** and disposed within the shell **32** fills the forward and intermediate portions of said shell but ends short of the nozzle **36** in a flat burning surface **41** which confronts said nozzle.

In order to provide gas transfer communication between the combustion chambers **18** and **33**, a tube **42**, made of a heat resistant material, communicates with the forward combustion chamber **18** and extends, axially of the missile, into the rear combustion chamber 33. More specifically, the tube 42 passes through the axial bore **26** of the grain **25** and the opening **22** into the casing **28**. It extends axially through the casing **28** and enters the rear combustion chamber **33** through the opening **35**, whereupon it extends along the bore **39** of the grain **38** to communicate with the chamber 33.

The rocket engine of this invention, therefore, comprises the forward combustion chamber **18** containing the propellant grain **25**, the rear combustion chamber **33** containing the propellant grain **38** and the gas transfer tube **42** providing communication between said combustion chambers. In operation, the grain **25** in the forward chamber **18** burns rearwardly, from the burning surface **27**, and the combustion gases thereby produced are conducted through the tube **42** to the rear chamber **33**. The grain **38** in the rear chamber **33** burns forwardly from the burning surface **41**. The combustion gases from both combustion chambers, **18** and **33**, combine and efflux through the exhaust nozzle **36**, thus providing a propulsive thrust.

By dividing the propellent grain of the rocket engine in the manner described above, it is possible to safely mount the wing sockets **29** or other wing attaching means at the center-of-gravity of said engine. In addition, such an arrangement also permits the convenient location of guidance and/or wing actuating equipment proximate to the wing sockets and wings.

It can be seen that the center-of-gravity **11** of the rocket engine of this invention will remain substantially in the same position during the burning of the propellant grains **25** and **38**. This feature is of utmost importance when the engine is used as the propulsive system in an aerial vehicle.

The propulsive thrust produced by this improved rocket engine is greatly increased over a conventional rocket engine of equal cross-sectional dimensions. This will be understood from the fact that two surfaces on the propellant grains, **27** and **41**, of the improved engine burn to produce a greater volume of combustion gases and a higher pressure. Thus, the combustion gases efflux through the exit nozzle with a greater momentum to produce an increased thrust, whereas the conventional rocket engine employs a propellant grain which burns from a single surface only, thereby producing a smaller volume of combustion gases at lower pressures, with the result that the thrust of the conventional rocket engine is smaller.

Obviously many modifications and variations of the present invention are possible in the light of the above teachings. It is therefore to be understood that within the scope of the appended claims the invention may be practiced otherwise than as specifically described.

What is claimed is:

1. In combination with an aerial vehicle having two pairs of wings arranged in a cruciform configuration and pivotally mounted on the body of said vehicle at the center of gravity thereof and two pairs of tail fins mounted on the rear end of said body of said vehicle, a rocket engine located in said vehicle body, said engine including spaced forward and rearward combustion chambers containing combustible material, said combustion chambers and combustible material being located substantially symmetrical with respect to the center of said vehicle, said combustible material of said forward combustion chamber being arranged to burn rearwardly of said vehicle and said combustible material of said rearward combustion chamber being arranged to burn forwardly of said vehicle, a rearwardly directed nozzle connected to said rearward combustion chamber, and means including a transfer tube for conducting combustion gases generated upon ignition of said combustible material from said forward combustion chamber to said rearward combustion chamber for efflux through said nozzle, whereby upon burning of said combustible material in said combustion chambers, the center of gravity of said vehicle remains substantially in the same position.

2. An arrangement as set forth in claim 1, wherein said combustion chambers are cylindrical in shape.

3. An arrangement as set forth in claim 2, wherein said combustion chambers have hemispherical end walls.

4. An arrangement as set forth in claim 1, wherein structure is provided between said combustion chambers to define an intermediate chamber for receiving guidance and control equipment, including means for attaching said two pairs of wings to said vehicle.

References Cited in the file of this patent

UNITED STATES PATENTS

D. 171,293	Boyd	Jan. 19, 1954
2,206,809	Denoix	July 2, 1940

FOREIGN PATENTS

580,598	France	Sept. 4, 1924
1,012,420	France	Apr. 16, 1952

OTHER REFERENCES

Popular Science, Dec. 1943, page 67.

(No Model.) 7 Sheets—Sheet 1.

W. S. BURROUGHS.

CALCULATING MACHINE.

No. 388,116. Patented Aug. 21, 1888.

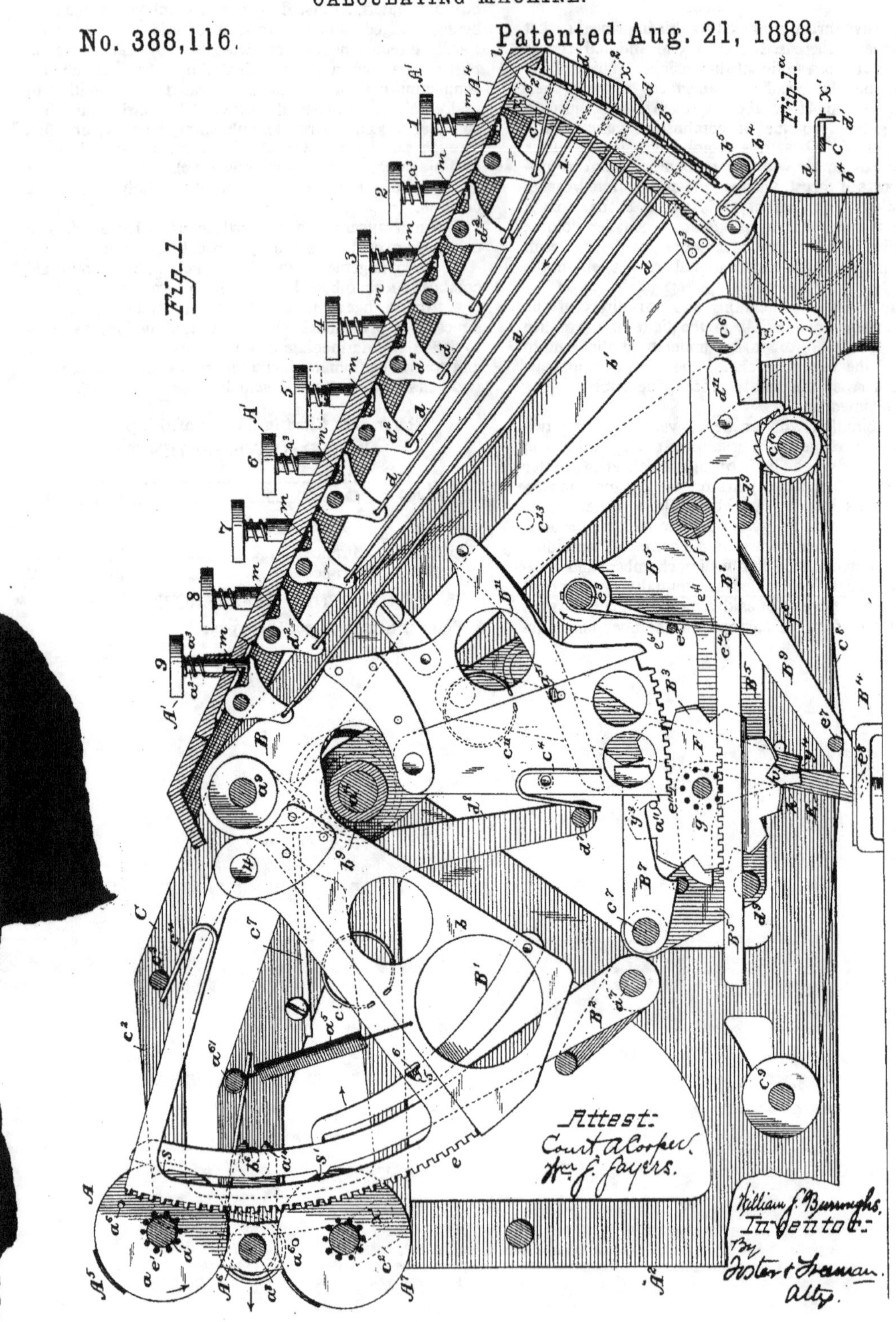

N. PETERS, Photo-Lithographer, Washington, D. C.

(No Model.)

7 Sheets—Sheet 2.

W. S. BURROUGHS.

CALCULATING MACHINE.

No. 388,116.

Patented Aug. 21, 1888.

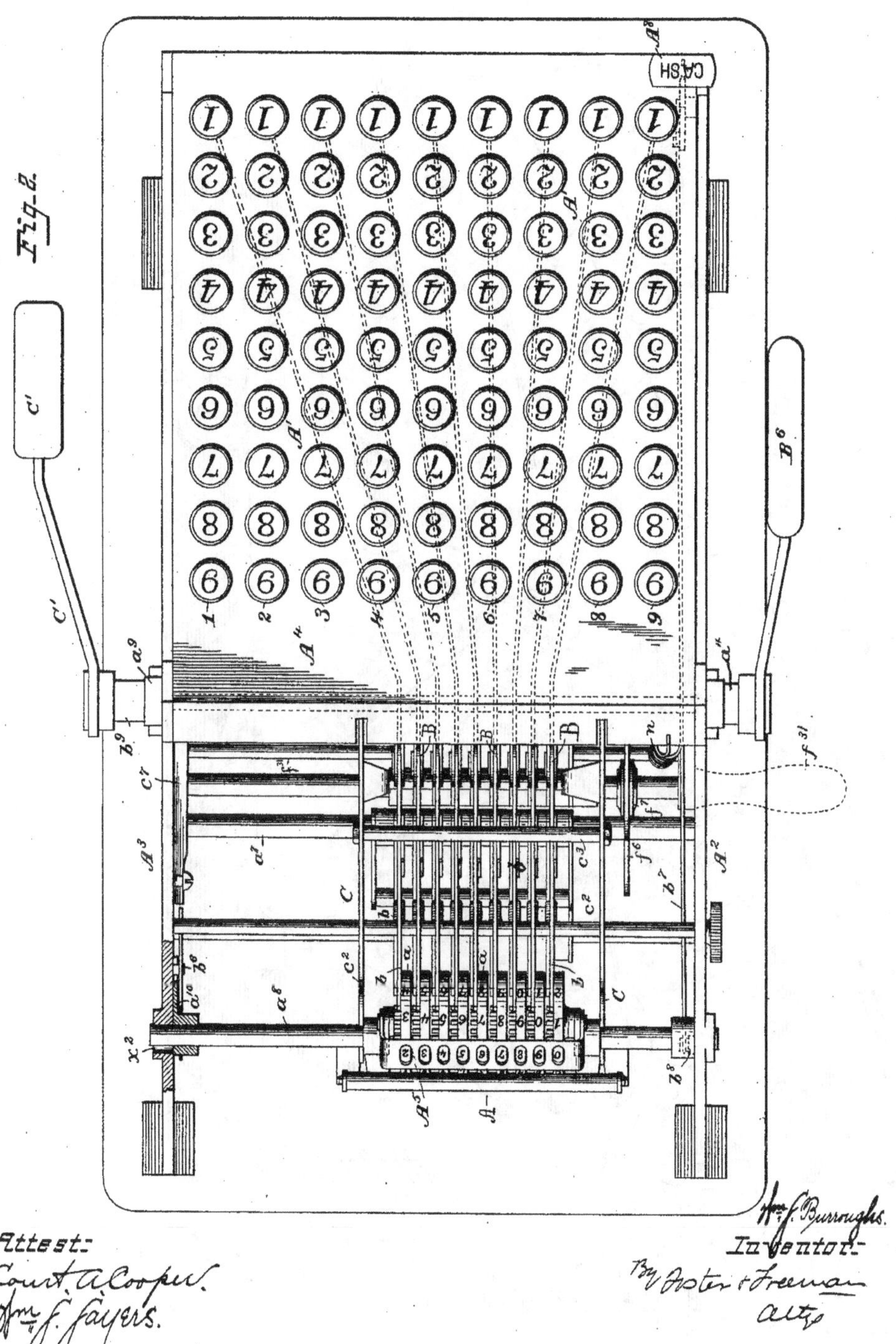

Attest:

Court A. Cooper.

Wm. J. Sayers.

Wm. S. Burroughs.

Inventor:

By Foster & Freeman

Attys

N. PETERS, Photo-Lithographer, Washington, D. C.

(No Model.)

7 Sheets—Sheet 3.

W. S. BURROUGHS.

CALCULATING MACHINE.

No. 388,116.

Patented Aug. 21, 1888.

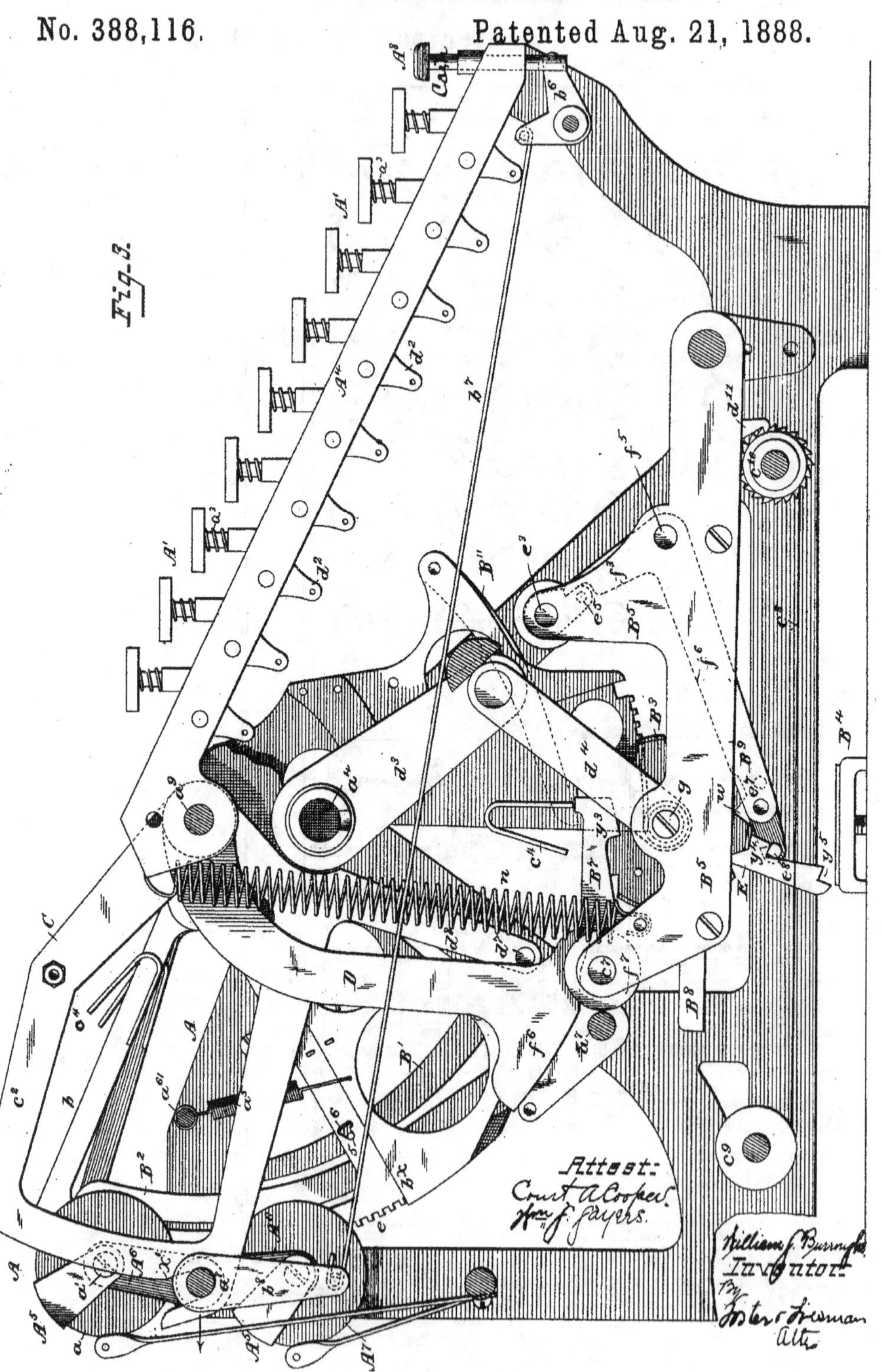

N. PETERS, Photo-Lithographer, Washington, D. C.

(No Model.)

7 Sheets—Sheet 4.

W. S. BURROUGHS.

CALCULATING MACHINE.

No. 388,116.

Patented Aug. 21, 1888.

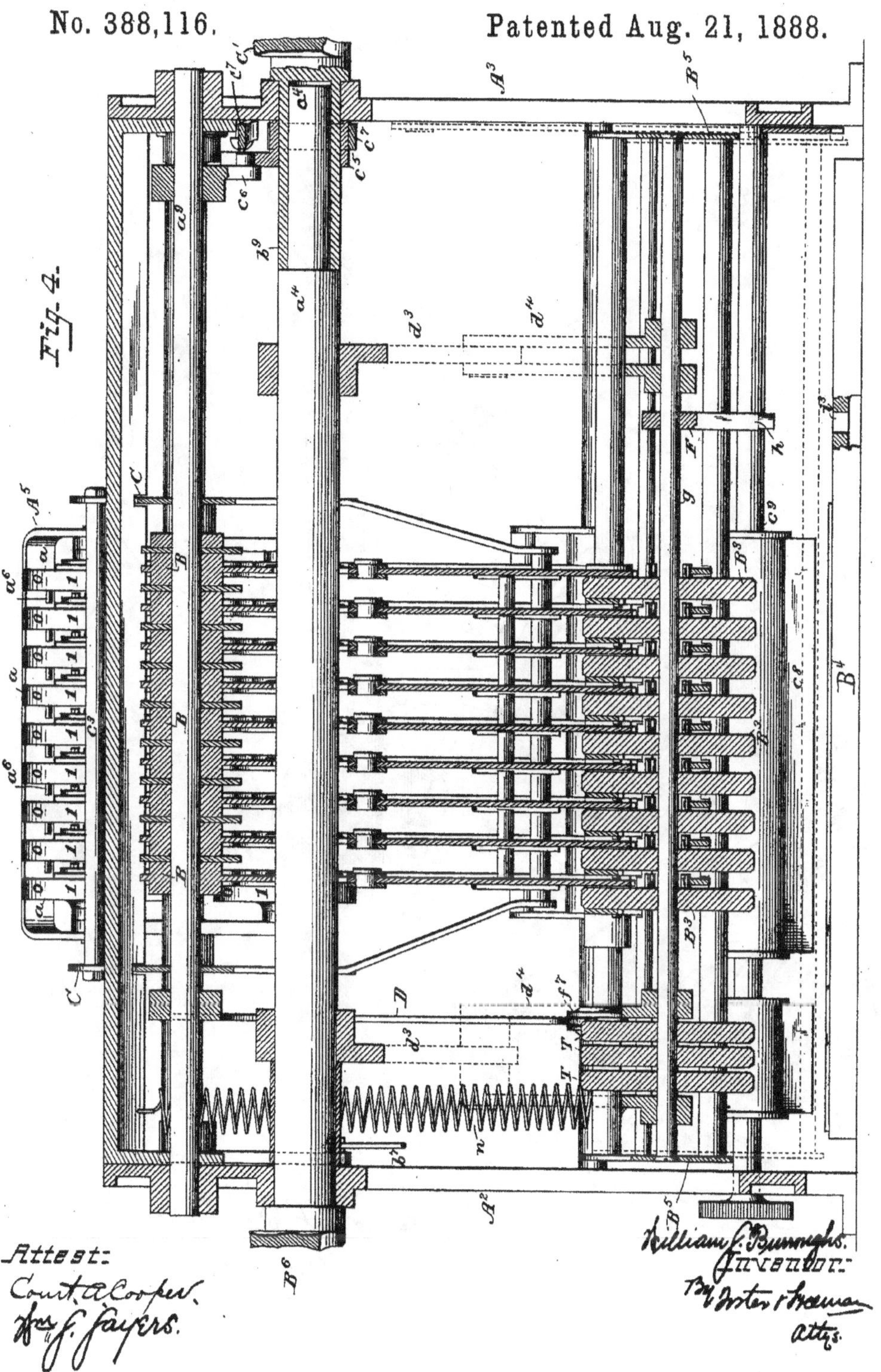

Attest:
Court A. Cooper.
Jas. J. Sayers.

William S. Burroughs.
Inventor.
By Foster & Freeman
Attys.

N. PETERS, Photo-Lithographer, Washington, D. C.

(No Model.) 7 Sheets—Sheet 5.

W. S. BURROUGHS.

CALCULATING MACHINE.

No. 388,116. Patented Aug. 21, 1888.

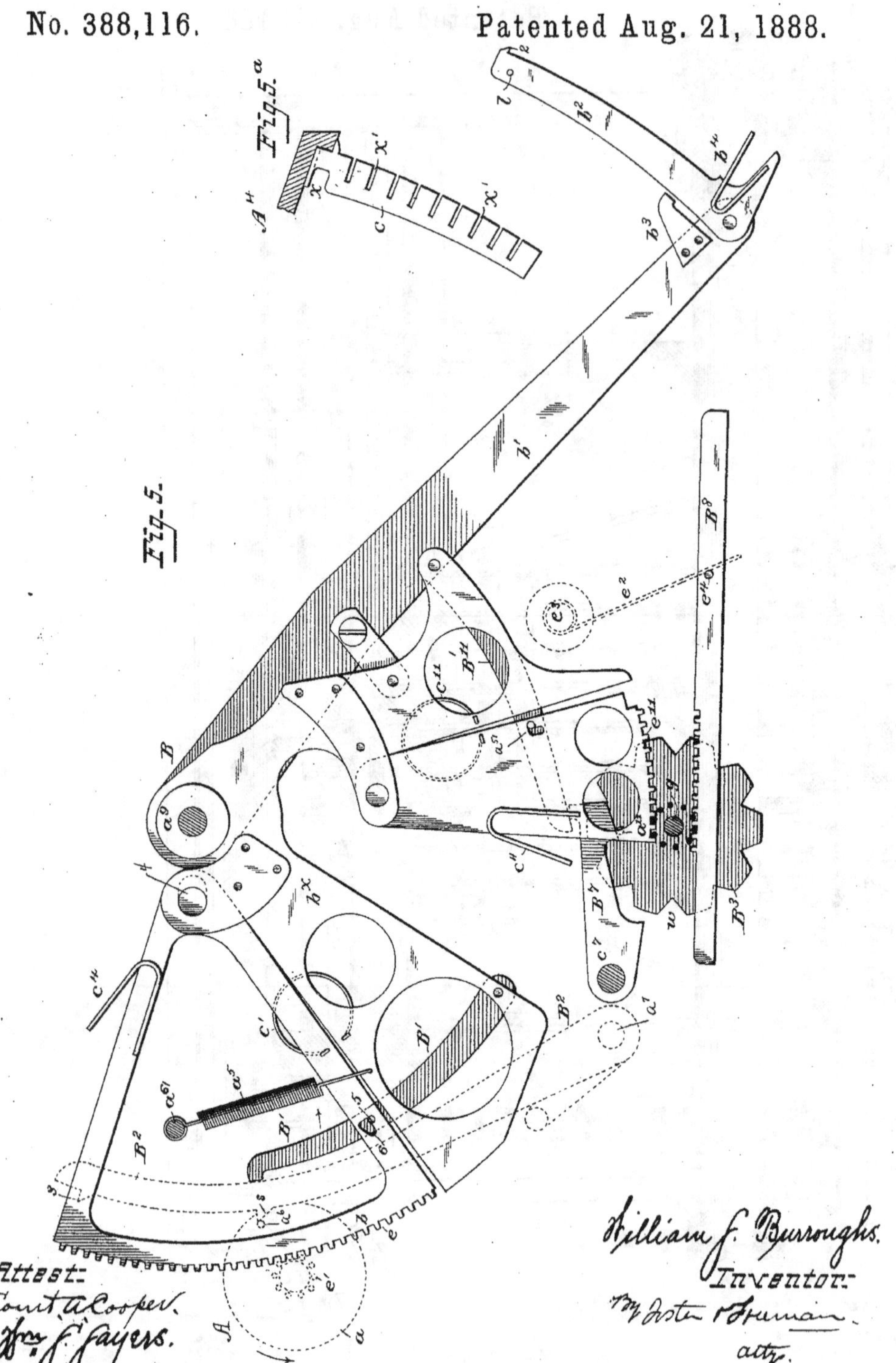

Attest:
Cont. A. Cooper.
Wm. J. Sayers.

William S. Burroughs.
Inventor.
By Foster & Freeman
atty.

N. PETERS, Photo-Lithographer, Washington, D.C.

(No Model.) 7 Sheets—Sheet 6.

W. S. BURROUGHS.

CALCULATING MACHINE.

No. 388,116. Patented Aug. 21, 1888.

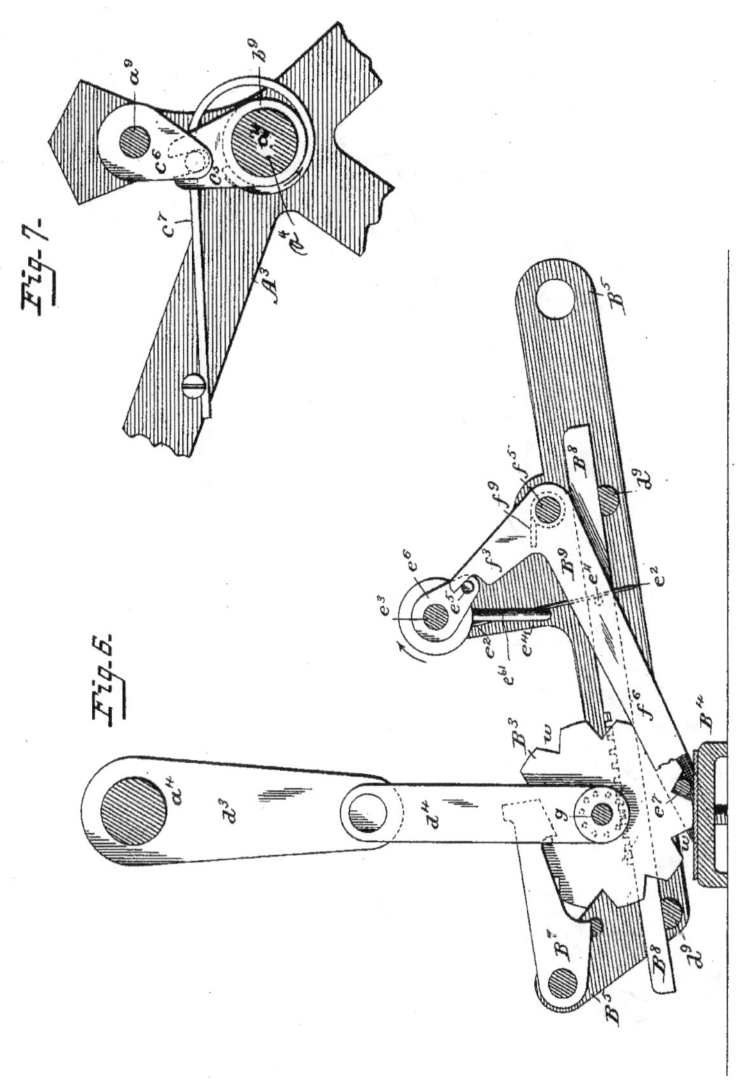

Attest:
Court A. Cooper.
Wm. J. Jaifers.

William S. Burroughs.
Inventor:
By Foster & Freeman.
Attys.

N. PETERS. Photo-Lithographer, Washington, D. C.

(No Model.) 7 Sheets—Sheet 7.

W. S. BURROUGHS.

CALCULATING MACHINE.

No. 388,116. Patented Aug. 21, 1888.

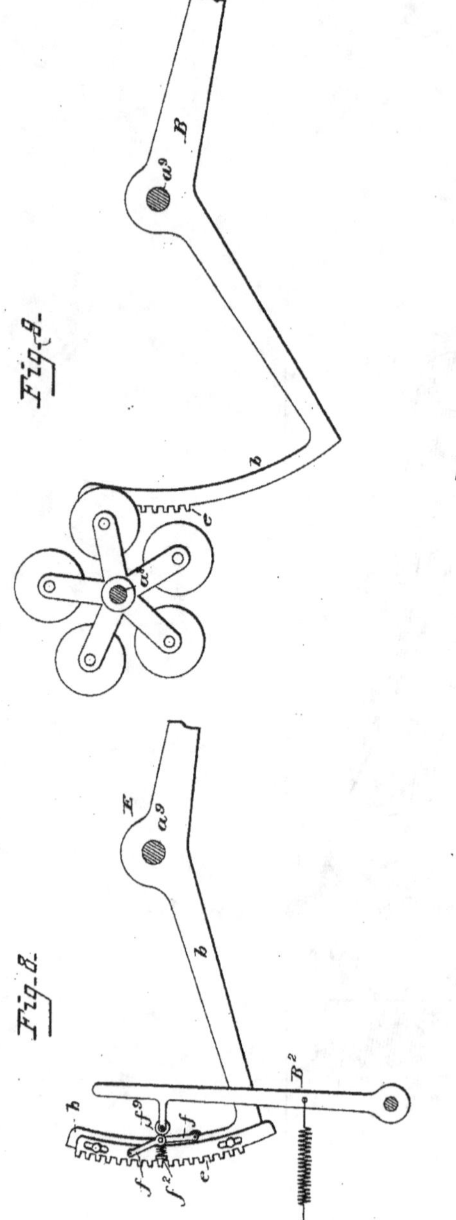

Attest:

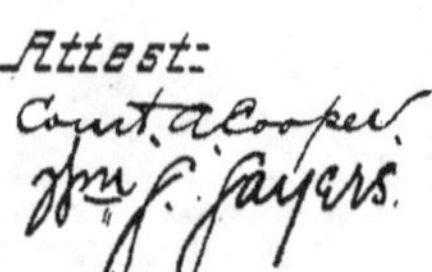

William S. Burroughs, Inventor:

By Foster & Freeman Attys.

N. PETERS, Photo-Lithographer, Washington, D. C.

UNITED STATES PATENT OFFICE.

WILLIAM S. BURROUGHS, OF ST. LOUIS, MISSOURI, ASSIGNOR, BY DIRECT AND MESNE ASSIGNMENTS, TO THE AMERICAN ARITHMOMETER COMPANY, OF SAME PLACE.

CALCULATING-MACHINE.

SPECIFICATION forming part of Letters Patent No. 388,116, dated August 21, 1888.

Application filed January 10, 1885. Serial No. 152,485. (No model.)

To all whom it may concern:

Be it known that I, WILLIAM S. BURROUGHS, a citizen of the United States, residing at St. Louis, in the State of Missouri, have invented certain new and useful Improvements in Mechanical Accountants, of which the following is a specification.

My invention relates to that class of apparatus used for mechanically assisting arithmetical calculations; and my invention consists in the combination, with one or more registers, of a series of independent keys and intervening connections constructed, arranged, and operating, as fully specified hereinafter, so as to indicate upon the register the sum of any series of numbers by the proper manipulation of the keys, and also so as to print or permanently record the final result.

In the drawings, Figure 1 is a longitudinal sectional elevation of an apparatus embodying my invention. Fig. 1ª is a cross-section on the line 1 2, Fig. 1. Fig. 2 is a plan view of the apparatus in part section. Fig. 3 is a longitudinal elevation in part section. Fig. 4 is a transverse sectional elevation. Fig. 5 is a detail view showing the connections between the keys and the registers, with the exception of the "regulating devices." Fig. 5ª is a detail view of part of the regulating devices. Fig. 6 is a detail view of the printing-register and its immediate connections. Fig. 7 is a detached view showing the connections between two of the shafts. Fig. 8 is a view showing a modified form of registering operating device. Fig. 9 is a view illustrating a mode of adjusting a series of registers. Fig. 10 illustrates a modification of the regulating devices.

The indicating-register A consists of a series of movable numbered pieces—as plates, disks, wheels, or segments—indicating by their position the sum added at each operation of the machine, and I therefore term them the "indicators." Each indicator is shown as consisting of a disk, a, revolving freely upon a shaft, a', and having upon its periphery a series of figures from 0 to 9; but it may be a plate or segment suitably supported and graduated in the same manner as the disk. Each indicator derives its movement from the operation of a series of nine keys, A', of any suitable construction, marked separately with a series of numbers from 1 to 9, and between the indicator and each key, and governed by the latter, intervene a series of devices, which may be of any suitable construction to transmit the motion of the key to the indicator, and regulate the adjustment of the latter, and which I include under the general terms of "connections" and "operating connections."

The registering device, the keys and connections, and other parts, hereinafter referred to, are supported in a suitable frame having side pieces, A^2 A^3, and connecting-bars, which latter also serve as supports for the intervening mechanism, and the frame supports a key-board, A^4, provided with recessed nipples, through which slide the shanks a^2 of the keys A', a spiral spring, a^3, tending to keep each key in its elevated position, and a lip at the bottom of the key striking a detachable strip, m, which bears against a flat side of the key and prevents the turning thereof, limits its movement, and facilitates its ready detachment when necessary.

The series of keys which operate in connection with each indicator is arranged upon a line parallel to the sides of the frame, and a lever, B, supported by a transverse shaft, a^9, when released by the action of any one of a series of keys operates automatically the indicator pertaining to that series.

Each lever B consists of a segmental "head," b, and an arm, b', the heads of all the levers being parallel to each other and in close proximity, and the arms diverging from the fulcrum a^9, so that the rear end of each arm will be below the key of one of the series farthest from the head.

Teeth at the edge of each head b constitute a rack, e, which engages with a pinion, e', at the side of one of the indicator-disks, and the movement of the lever is so limited that the indicator-disk will be turned nine-tenths of a revolution at each full movement of the lever, a perforated plate or gage, A^5, serving as an index to expose or designate the figure indicating the number added.

It is necessary that the indicator be moved

as each key is depressed to an extent proportioned to the position of the key and the number indicated upon the latter, and in order to effect this without varying the movements of the keys or the power applied to operate the latter I employ devices between each series of keys and the lever operated thereby, which constitute part of the "connections" between the keys and indicator, and which I term the "regulating devices."

The arms b' may be weighted so as to normally fall to their lowest position when released; but I prefer to secure a more positive action by means of springs a^5, connected to a cross-bar, a^{61}, of the frame and to the heads of the levers, so as to lift the latter.

The regulating devices are so constructed that when a key is depressed the lever will be released, and will then fall to an extent corresponding to the position of the operating-key, which, therefore, instead of acting directly upon the lever, operates the latter intermediately through said regulating devices, so that the key may recover its position at once after being struck.

In the construction shown the regulating devices connected with each series of keys consist of a lock which holds the arm b' of the lever normally in its highest position and a movable stop operated by the key, which limits, when it is set, the descent of the arm b' of the lever.

The lock, as shown, consists of a strip, b^2, pivoted to the end of each arm b' and limited in its rear movement by stop b^3 and carrying a lug, l. Adjacent to each strip b^2 a fixed plate, c, Figs. 1 and 5^a, is secured to the key-board A^4 and pendent therefrom and provided with a notch, x, near its upper end, into which the lug l enters as the arm b' is brought to its highest position and the strip b^2 falls forward. This forward movement of the strip b^2 is insured by the contact of a spring projection, b^4, extending from the strip b^2, with a stationary cross-bar, b^5, supported by the sides of the frame.

In the outer edge of the plate c is a series of notches, x', corresponding in number to the keys of the series, and through each notch extends the bent end d' of a rod, d, which is connected to the lower arm of a crank-lever, d^2, one of which is pivoted to a lug under the key-board beneath each key, and has its other end or arm projecting below the end of the key. The bent ends d' of the rods d constitute the stops. When one of the keys (say, the key 5, Fig. 1) is depressed, the rod d of the lever d^2, below said key, will be drawn in the direction of its arrow, and the end d', bearing upon the edge of the strip b^2, will carry the latter inward until the pin 1 is carried out of the notch x, when the action of the spring a^6 will cause the arm b' and strip b^2 to fall until a lip, 2, at the upper end of the strip b^2 strikes the bent end d' of the rod d, which acts as a stop, limiting the further movement of the lever downward. As the inward movement of each stop end d' carries the strip b^2 inward the lip 2 of the strip will escape contact with all of the stops above the one which has been so set by the movement of the key, so that the said stop only will be struck by the lip; and as the stops, beginning at the top, are connected with the keys in the order in which the latter are numbered, the movement of the lever will be greater in proportion as the stop operated is lower down, and the movement of the indicator will be proportionately increased.

As in other registering devices, it becomes necessary when the number added exceeds the highest number upon the indicator operated to also adjust the adjacent indicator accordingly. Thus, if one of the indicators is set to display the figure 8 and the number 5 or any number greater than 1 is to be added and indicated, the said indicator must be turned until it exhibits the figure 3, while the adjacent indicator must also be turned to disclose the figure 1, the sum of the two numbers being 13, which requires the adjustment of two indicators to show it.

To effect the turning of the second indicator without operating two keys, I construct the parts which I term the "connection" between each indicator and its key, so that a portion of said connection geared with the indicator is capable of a slight movement independently of the other parts of the connection. I lock these two parts together so that ordinarily they will operate as one, and provide means whereby the movable part is unlocked and moved to turn the indicator the extent of one figure whenever the adjacent indicator is brought to a position to necessitate such movement. Thus, the head of each lever B consists of the arm or portion $b^{\times}$, connected rigidly to the arm b', and the toothed segment b, pivoted at 4 to the arm $b^{\times}$, and a spring, c', is connected to the two parts, b $b^{\times}$, so as to tend to separate them, and a lever, B', is pivoted to the part $b^{\times}$, and is provided with a pin, 5, which enters an L-shaped slot, 6, in the part b and holds the latter in proximity to the part $b^{\times}$ until the lever B' is swung in the direction of its arrow, Figs. 1 and 5, when the pin will enter the vertical portion of the slot and permit the portion b to rise under the action of the spring c', and thereby turn the indicator to which said portion b is geared one-tenth of a revolution in the direction of its arrow. The lever B' is moved to unlock the part b by a pin, a^6, projecting from the side of the indicator next to that to which the part b is geared, which pin, as the figure 0 is brought toward the index-plate A^5, strikes a projection, s, on a lever, B^2, pivoted to a cross-bar, a^7, and swings it so as to bear against the end of the lever B' and carry the latter with it to unlock the swinging segment, which will then rise under the action of its spring c' and turn the indicator next adjacent to that indicator whose pin acted upon the lever B^2.

All the indicators are operated by the movements of the respective keys in the manner above described, and after a series of keys has been struck to register one of the numbers to be added together it becomes necessary to bring the parts to their normal positions, so as to permit any additional number to be added, inasmuch as after a lever, B, has been adjusted by the action of one key its head must be depressed before it can respond to the action of another key. Thus, if the number 72,842 is to be added, all the heads are depressed to bring the parts to an operative position, and the keys 7 2 8 4 2 in the series 5 4 3 2 1 are successively struck, thereby releasing the corresponding levers, which move to different degrees, according to the positions of the keys and the positions to which the stops d' are set, and the indicators show by the figures visible through the index-plate A^5 an increase corresponding to the sum added. If, now, the sum 234 is to be added, the heads of the levers are all again depressed to restore the parts to position, so that when the keys 2 3 4 in the series 3 2 1 are struck the corresponding levers can move each to an extent necessary to insure the required adjustment of the indicator and the exhibition of figures indicating an increase of 234.

The adjustment of the levers is effected by pressure upon the upper sides or edges of the segments b, which has the effect of moving such of the latter as have been separated from the portions $b^\times$ into contact therewith, when they will be locked in place by the arms B' falling back and carrying the pins 5 into the horizontal portions of the slots. The combined movement will then carry down all the segments until the arms b' are all elevated to their highest positions.

As the depression of the segments while in gear with the pinions e' would turn back the indicators, I provide means for throwing the segments and pinions out of gear prior to any downward movement of the segments. One means of effecting this is shown, and consists in journaling the shaft a' in a swinging frame, A^6, pivoted upon the cross-bar a^8, and in swinging the frame back to throw the pinions and racks out of gear prior to the descent of the segments. One means of effecting this adjustment is shown, and consists of a frame, C, carried by the shaft a^9, Fig. 4, and consisting of side pieces, c^2 c^2, and a cross-bar, c^3, each side piece having near its lower edge an inclined slot, x', Figs. 1 and 3, adapted to receive the cross-bar a^8, sliding in slots in the side pieces of the main frame, when the frame C is raised, the inclined edge x^3 of the slot serving to thrust the bar a^8 back in the direction of the arrow as the said frame C begins to swing downward, which backward motion of the cross-bar a^8 will be effected before the cross-bar c^3 is brought in contact with the upper edges of the segments.

The frame C is operated by a hand-lever, C', which is connected to operate the shaft a^9, as described hereinafter, and when the end of the lever which is toward the operator is depressed the frame C will be swung downward and will force out the cross-bar a^8, with the frame A^6 and the indicators, until the pinions e' are free from gear with the racks, and the edges of the frame will then remain in contact with the cross-bar a^8 and hold it in position, the cross-bar c^3 being then brought against the upper edges of the segments and depressing the latter. After the levers B are brought to their normal position the hand-lever C' is released, when it will rise and the frame C will swing upward and the cross-bar a^8 will enter the slots x', and the pinions will be brought into gear with the racks as the frame C reaches the limit of its upward movement.

To permit the return movement of the cross-bar a^8, the ends thereof enter slots x^2 in the side pieces of the frame, and an arm, a^{10}, projecting from the cross-bar a^8, is slotted to receive a pin, b^6, on one of the side pieces, which prevents the turning of the cross-bar as it slides back and forth.

To prevent objectionable shocks and jars in bringing the cross-bar c^3 against the segments, I place springs c^4 upon the segments so as to be struck by the cross-bar.

It will be seen that by the construction above described each indicator is operated upon the depression of any one of the keys of a single series, and that it has the effect when turned beyond a complete revolution of moving the adjacent indicator one step; that all the keys have the same extent of movement and are operated by the same amount of pressure, but that while the movements of the keys are the same, the extent of the movement of the connections between the keys of each series and each indicator will vary according to the position of the key which is operated, with a corresponding variation in the motion of the indicator. It will be apparent that these effects may be secured by the use of keys and intermediate connections differing to some extent from those described. For instance, each series of keys may operate upon a shaft suitably geared with the corresponding indicator instead of through the medium of a rocking-lever. The register may of course be of any usual or suitable character. The indicators may be differently connected, as is common in registering devices, so that each will be moved one step as the adjacent indicator completes its entire movement, in which case the rack must be thrown out of gear with the second indicator until the latter has moved one step.

Instead of withdrawing the register from engagement with the racks the latter may be hung to pivots sliding in elongated openings, so as to be withdrawn from the register, as shown in Fig. 5, dotted lines, or the register and the rack may be kept in constant connection, each pinion having a ratchet-connection with the indicator, so as to turn the latter when the rack is raised, but to revolve independently of the indicator when the rack descends.

It will be obvious that any suitable locking mechanism may be employed for connecting the rack portion of the segment with its support, so as to permit a limited independent movement of the rack portion, and that such locking mechanism may be actuated from the indicators in a different manner from that described. Thus the segmental rack e may be guided to slide upon the head b, as shown in Fig. 8, and be moved thereon to a limited extent by straightening or bending the toggle-levers $f f$, which may be effected by bringing a roller, f^9, to bear against the same in one direction and by a spring, f^2, forcing them in the opposite direction, the roller f^9 being carried by levers operated in like manner as the levers B^2, or in any other suitable manner.

Different forms of regulating devices may be employed for determining the extent of the movement of the connections according to the key operated. Thus the keys may be connected to operate arms B^2, Fig. 10, with shoulders x^3 arranged to hold the arms b' of the levers B in their elevated position until the arms B^2 are swung forward.

Instead of restoring the levers B to their position by means of the cross-bar c^3 bearing upon the levers, they may be moved by a cross-bar, c^{13}, extending beneath the arms b', as shown in dotted lines, Fig. 1, or in any other suitable manner.

In some kinds of calculations it is necessary to indicate the sums of different kinds of articles or money. For instance, in a bank it is sometimes necessary to ascertain the aggregate amount of a series of checks and also the amount of money represented by notes or coin and pertaining to the same transaction. In order to permit this to be done with facility, I provide two or more registering devices in connection with one set of keys and intermediate mechanism, and means whereby either register may be thrown into operative connection with the keys. Thus the second register, A^7, is hung to the same frame, A^6, that carries the first register, A, and this frame is vibrated so as to bring the pinions of either register into gear with the racks e. This vibration is effected by means of a key, A^8, actuating a crank-lever, b^6, connected by a rod, b^7, with an arm, b^8, upon the shaft a^8. The upper register is held in connection with the keys while the sum of the checks is being taken, and after this is done the key A^8 is depressed and the lower register will be swung into gear, and the amount of cash is registered thereon, and if additional cash or checks are then received the additional amounts may be added upon either register by swinging it into operative connection with the key and without any alteration of the other register.

Where, as in custom-houses and other places, it is necessary to indicate the value or number of a series of different articles, a series of registers may be employed to be operated from the same series of keys. One mode of arranging the registers in such case is shown in Fig. 9, which shows five registers carried by an armed frame, revolving upon or with the shaft a^8, and capable of being turned so that either register may be brought at will in operation with the connecting devices between the register and the keys.

When two registers are arranged one above the other, so that either may be brought into connection with the racks, the latter are necessarily longer than would be required if the arrangement shown in Fig. 9, or if but one registering device, was used.

When the arrangement shown in Figs. 1 to 7 is employed, the slot in the arm a^{10}, which receives the pin b^6, is widened, as shown in Fig. 1, so as to permit the swinging of the arm a^{10} required by the vibration of the shaft a^8.

As it is necessary that each indicator of the lower register shall upon the completion of its revolution move the succeeding indicator one step, as in the upper register, I effect this by extending the levers B^2 or projections s' thereof, so as to be struck by the pins a^6 of the lower indicators to unlock the segments in the same manner as they are unlocked by the upper indicators when the upper register is in use and with like effect.

The movement of the frame C is effected from the handle C′ by connecting the latter to a sleeve, b^9, receiving and turning upon the shaft a^4 and carrying a slotted arm, c^5, receiving a pin upon the end of an arm, c^6, extending from the shaft a^9, and a spring, c^7, secured to the side frame, is coiled around the hub of the shaft b^9 and secured thereto at the end and serves to turn the sleeve in the direction of the arrow, Fig. 7, the sleeve being turned in a reverse direction to depress the frame C whenever the handle C′ is depressed.

It is frequently desirable to secure a permanent indication of the sum shown upon the register, but this cannot always be well done without so covering the register as to prevent the figures upon the latter from being seen. In order to secure a visible representation as well as a permanent indication of the number registered, I employ, in addition to the registering devices described, recording devices, one arranged so as to be readily inspected and the other constructed and combined with means whereby to also print the numbers registered upon a strip of paper. Thus the second register, A^7, may be combined with gears throwing it into connection with the register A and with an inked ribbon and platen whereby the row of figures in line upon the lower register may be transferred to the paper, while those on the upper register are exposed. I prefer, however, instead of using the register A^7, to use an independent printing-recorder, B^3, preferably arranged beneath the shaft a^4 and above a platen, B^4, upon the base-plate of the machine. When this arrangement is employed the shaft g of the recorder B^3 is carried by a frame, B^5, hung to studs c^8 upon the side frames of the machine, and combined with devices whereby the said frame may be raised to bring the recorder into connection with the devices for operating it from the keys, and lowered to bring the lower row of type or figures against the paper upon the platen.

The adjustment of the indicators or wheels of the printing-register is effected by connections precisely similar to those employed for adjusting the upper register, each lever B carrying a second head or segment with a rack, e'', which gears with the corresponding pinion of the adjacent indicator of the lower register and operates the same in the same manner as has been described in connection with the register A. In the operating device for the printing-recorder the movable section of the head or segment is provided with an L-shaped slot to receive a pin, a^{51}, upon a locking-arm, B'', operating in the same manner as the locking-levers B' and operated from the pins a'' on the indicators through the medium of levers B^7, each of which is hung to a cross-bar, c^7, on the frame, and as a pin, a'', is brought beneath its inclined lower edge the lever is raised, strikes the end of the locking-lever B'', and thereby releases the adjacent segment to permit it to swing out under the action of a spring, c'', to move the adjacent indicator one step.

The frame B^5 may be depressed to bring the printing-recorder in contact with the platen by means of the arm f^{31}, (dotted lines, Fig. 2,) extending from the frame and adapted to be operated directly by hand, the frame being raised by a spring, n, or otherwise. I prefer, however, to employ devices operating more positively, and consisting, as shown, of toggle-levers d^3 d^4, the former secured to the shaft a^4 and the latter jointed to the levers d^3 and also to the frame B^5, as shown, so that by depressing a handle or arm, B^6, on the shaft a^4 to rock the latter in one direction the frame B^5 will be carried downward and the printing-indicators will be brought against an inked ribbon, c^8, while the spring n lifts the frame, when pressure upon the arm B^6 is removed. The inked ribbon is carried by rollers c^9 c^{10}, the latter provided with a ratchet with which a pawl, d^{11}, upon the frame B^5 engages, so as to move the ribbon slightly at each movement of the frame.

As in the devices operating with the register A, it is necessary to restore the movable parts of the segments to their position after each operation upon the keys; and this I effect by means of a cross-bar, d^7, carried by arms d^8 of the frame C and brought against springs at the edges of the heads of the segments of the recorder, as the cross-bar c^3 is brought against the corresponding parts of the segments operating the register A.

It is of course necessary to throw the pinions of the register B^3 out of gear with the racks e'' before each readjustment of the levers B. This is effected at the same time that a like operation is effected with the register A by means of an arm, D, Fig. 3, connected to the shaft a^3, having a cam end, f^6, which bears upon a grooved wheel, f^7, upon the cross-bar c^7 of the frame B^5, and when the frame C is depressed the cam end of the arm forces downward the end of the frame B^5 to carry the pinions from gear with the racks.

It is generally desirable that the printing-wheels shall be restored with all the indicators at zero after any number has been registered and printed, in order that it may be in position to be properly reset to indicate any number to be subsequently registered and printed, although the register A may indicate the sum of both numbers; and to secure this result I combine with each indicator a rack-bar, B^8, sliding in bearings d^9 d^9 upon the frame B^5, and each gearing with the pinion of one of the indicators, and a spring, e^2, upon a cross-bar, e^3, is arranged to be brought to bear upon a pin or bearing, e^4, upon each rack-bar B^8, so as to throw it inward to bring the pin a'' of the adjacent indicator against a shoulder, y^3, of one of the levers B^7, when the sign "0" will be the lowermost sign upon the indicator.

When the indicators of the printing-recorder are to be operated from the keys, they should be left perfectly free to turn without resistance, and the shaft e^3, which carries the springs e^2, is therefore hung in the frame B^5, so as to swing freely, and each spring e^2 will swing the rack-bar B^8 forward without resistance as the rack-bar B^8 is moved forward. When the indicators are to be restored to position, the shaft e^3 is turned in the direction of its arrow, Figs. 1 and 6, so as to cause the springs e^2 to bear against the pins e^4, when each bar B^8 will be moved until the indicator connected therewith is brought to the zero position, the extent of the movement of course depending upon the extent to which the indicator has been previously turned from such position. The movement of the shaft e^3 requisite to bring the springs to bear upon the pins e^4 results from the swinging by hand of a crank-lever, B^9, pivoted upon a shaft, f^5, carried by the frame B^5, and one arm, f^3, of which lever bears upon a pin, e^5, projecting from an arm, e^6, upon the shaft e^3, the long arm f^6 of the lever extending downward and forward and in conjunction with a similar arm at the opposite side carrying a cross-bar, e^7. A spring, f^9, Fig. 6, dotted lines, coiled upon the shaft f^5, tends to raise slightly the lower end of the arm f^6, and a lug, e^8, at the end of said arm is arranged to engage with shoulders y^4 y^5 upon an arm, E, pivoted to one of the side frames of the machine and swinging freely upon its pivot. Arms e^{61}, projecting from the shaft e^3, carry a cross-bar, e^{11}, Fig. 1, upon which the springs e^2 bear, and which when carried from the springs by the rocking of the shaft e^3 in the direction of the arrow permits the springs to move independently in acting upon the rack-bars B^8. When the frame B^5 is depressed to effect the printing, the lug e^8 is carried beneath the shoulder y^5 of the lever E, and when the frame B^5 again rises the lever B^9 will be retained in its position, Fig. 1, and its arm f, bearing upon the lug e^5, will swing the shaft e^3 in the direction of its arrow, Fig. 6, and carry the cross-bar e^{11} away from the springs e^2, which will then move inward the bars B^8 and restore the indicators to their zero positions. The parts remain in the position described until the frame B^5 is about horizontal; but as it rises higher the lever B^9 will be slightly retracted and the lug e^8 will be withdrawn from the shoulder y^5, and the lever B^9 will be lifted by the action of the spring f^9 until the lug e^8 strikes the shoulder y^4, Fig. 3, the lever B^9 being then free from contact with the pin e^5, so that the shaft e^3 can swing freely and the springs e^2 will exert no action upon the bars B^8.

The cross-bar e^7 acts as an equalizing-bar to bring the wheels into line and hold them in place. As the printing-recorder descends upon the said cross-bar the latter passes into the notches w of the wheels B^3, and, bearing against the inclined sides of the latter, brings all the indicators into line and holds them in place, and as the register rises (the bar e^7 being held in place by the action of the lever E) the wheels pass from the bar and are then free to turn under the action of the devices set in motion by the keys.

It is desirable in many instances to prevent duplicate printing—that is, after the recorder has once been pressed upon the paper to prevent it from again being forced down to make a print until a new number has been registered. To effect this an L-shaped dog, F, Figs. 1 and 4, is hung loosely to the shaft g of the printing-recorder, so that one arm will extend over the bar d^9, while the other arm, h, is pendent and is provided with an inclined edge, v, so arranged as to be struck by the cross-bar e^7 when the latter enters the notches w.

In the platen B^4 is an opening or notch, i^3, so arranged that when the cross-bar e^7 is in the lowest notch, w, to its greatest depth, the dog F will be held in such position that its end will enter the opening i^3, and the frame B^5 can descend to such an extent as to effect the printing. When, however, the printing has been effected and the cross-bar e^7 is held by the action of the lever E in the position shown in Fig. 1, the dog F will swing to such a position that its lower end will strike the face of the platen and prevent the contact of the type with the paper if the frame is depressed. This arrangement also prevents the battering of the type, which might result if any one or more of the wheels was turned so that the face of the type would not be presented absolutely parallel to that of the platen. If one of the wheels was thus out of adjustment, the cross-bar e^7 could not travel as far as the bottoms of all of the notches w, and the dog F could not therefore be moved by the cross-bar to its full extent, and consequently would not be in position to enter the opening i^3 and would prevent the descent of the frame and the battering of the type.

I do not limit myself to the mode described of restoring the indicators to their normal positions, as other means might be adopted. For instance, they might be weighted so as to normally hang with the figure 0 lowermost, to take this position whenever the indicators are free from contact with suitable friction devices. Other means than those described may be employed for bringing the springs to bear upon the rack-bars when the indicators are to be adjusted, leaving them free at other times, and other stop-motions may be used to prevent the full descent of the printing-indicators after one impression or when any of said wheels are out of adjustment.

In some instances it is desirable to print the date upon each slip upon which the number is printed. This I effect by arranging dating-wheels T upon the shaft g, or otherwise supporting them on the frame B^5, so as to operate in connection with the other printing-wheels when the frame B^5 is depressed.

I do not here claim any of the features shown herein and also shown and claimed in my applications Serial No. 174,593, filed August 17, 1885; Serial No. 195,583, filed March 17, 1886, and Serial No. 256,566, filed November 30, 1887; nor the printing devices herein described and forming the subject-matter of my application, Serial No. 279,609, filed July 11, 1888.

I claim—

1. The combination of a series of numbered independent indicators, a series of independent keys to each indicator, connections between each of the series of keys and each indicator, said connections being arranged to insure the movement of each indicator upon the movement of any key of its series and including a series of stops to each series of keys adjustable by but independent of the keys, arranged to vary the extent of movement of the indicator according to the position of the key struck, substantially as described.

2. The combination of a series of independent indicators, a series of keys to each indicator, connections whereby each indicator is operated on the movement of any key of its series, connections whereby each indicator on completing a revolution turns the adjacent indicator of higher order one step, and means for disconnecting the indicators from the connections after each number is registered to permit the connections to assume a position to operate the indicators to register another number, substantially as described.

3. The combination of the series of independent numbered indicators and a series of independent keys having uniform movements connected with each indicator and constructed to operate two or more of the indicators simultaneously when released by the action of two or more keys, and locking and releasing devices operated by but independent of the keys for releasing and regulating the movement of the indicator-operating devices, substantially as described.

4. The combination, with the series of indicators and with a series of keys connected with

each indicator, of a series of levers each connected to turn the indicator by its movement, and locking and releasing and regulating devices arranged between each lever and its keys, whereby the lever is released and its movement regulated according to the position of the key struck, substantially as set forth.

5. The combination, with the indicators and pinions and independent keys arranged in series, of actuating-levers carrying racks engaging with the pinions and regulating devices between each lever and each series of keys, the keys capable of movement independently of said devices, substantially as specified.

6. The combination, with the keys, indicators, and intermediate operating-connections between each key and each indicator, of means, substantially as described, for moving the indicators to throw them out of gear with the said connections upon their return motion, substantially as set forth.

7. The combination, with one or more keys, a series of levers, indicators, and pinions, of devices for throwing the indicators out of connection with the levers after the indicators have been operated by the movements of the keys, substantially as set forth.

8. The combination, with the indicators and pinions and with the operating levers and racks, of appliances for throwing the pinions and racks out of gear after the movement of the indicators, for the purpose specified.

9. The combination, with the indicators, a series of keys to each indicator, and a series of levers for operating the indicators, of appliances for throwing the indicators out of gear with the operating devices when the latter are moved in one direction, substantially as specified.

10. The combination, with the keys, a series of independent rack-levers and indicators, of a frame supporting the indicators and adjustable to and from the said levers, substantially as set forth.

11. The combination, with the series of operating rack-levers, the shaft a^8, and indicators supported by said shaft a^8, of a vibrating frame provided with edges bearing against the shaft and constructed to move the latter to and from the levers, substantially as specified.

12. The combination, with the indicators, keys, and a series of levers acting upon the indicators, of a cross-bar and means for moving the bar to restore the levers to their normal positions, substantially as specified.

13. The combination, with the indicators, a series of independent operating rack-levers, and series of keys, of a cross-bar arranged to move the levers to their normal position after they have been lifted by the action of the keys, substantially as set forth.

14. The combination, with the indicators and actuating-levers and independent keys, of a frame carrying a cross-bar arranged to strike the actuating-levers, and a handle connected to operate said frame, substantially as set forth.

15. The combination, with the indicators and a series of independent rack-levers, of a frame carrying a cross-bar for moving said levers, and devices whereby to throw the indicators in and out of gear with the levers, substantially as specified.

16. The combination, with the indicators, a series of independent keys to each indicator, and a series of independent intermediate connections, of a regulating device between the said connections and the keys, constructed to insure and determine the movement of the connections, substantially as set forth.

17. The combination, with the indicators and a series of independent keys to each indicator, of a separate connection for moving each indicator, and a lock connected to be operated by each key of the series, whereby each connection is held in its operative position, substantially as set forth.

18. The combination, with the series of keys, indicators, and intermediate connections, of a lock for securing each connection, and connections between each key and the lock, whereby said lock is operated by each key of the series, substantially as specified.

19. The combination, with the series of independent indicators, a series of keys to each indicator, and an operating-lever to each series of keys, of a locking-plate and connections between each key and said plate, substantially as set forth.

20. The combination, with the indicators, keys, operating-connections, and locks, of stops, each connected to be operated by one of the keys and arranged to limit the movement of the operating-connections according to the key depressed, substantially as specified.

21. The combination, with the indicators, keys, and operating-connections, of a series of stops for limiting the movements of said connections, each connected to and movable by one of the keys, substantially as set forth.

22. The combination, with the operating-lever and a series of keys, of a corresponding series of stops arranged to limit the movements of the lever, and connections between each key and one of the stops, substantially as specified.

23. The combination of the operating-lever carrying an arm provided with a lip, 2, a series of stops and connections between the stops and keys, whereby any one of the stops may be thrown into the path of the lip, substantially as specified.

24. The combination, with the operating-lever and a series of keys, of a lock for securing the lever in its elevated position, a series of stops for limiting the downward movements of the lever, and connections between each key and the lock and one of the stops, substantially as specified.

25. The combination of a frame having a

stationary shoulder, the operating-lever, keys, notched bar, stops connected to be operated by the keys, and a strip, b^2, pivoted to the lever, constructed to engage with said stationary shoulder on the frame, and provided with a lip, 2, arranged to engage with the stops, substantially as specified.

26. The combination, with each indicator and a series of keys to each indicator, of a series of independent intermediate connections, a spring for operating each connection to a limited extent independently of the key, a detent, and means for releasing the latter to permit the connection on one indicator to move independently of the key and operate its indicator when the adjacent indicator completes its revolution, substantially as specified.

27. The combination, with a series of indicators, a series of keys, and connections, of means for operating the latter upon the movement of any key or keys, the said connections being provided with parts capable of a limited movement independent of the other parts, with locking devices, and with means for releasing the latter as each indicator completes a revolution, substantially as set forth.

28. The combination, with a series of indicators, of a corresponding series of actuating devices, and connections whereby the actuating device of one indicator is moved one step, whether in motion or at rest, as the adjacent indicator completes its movement, substantially as set forth.

29. The combination, with the series of indicators and series of keys and series of independent actuating-connections between the keys and indicators, of means, substantially as described, for turning each indicator one step independently of the key action as the next lower indicator completes a revolution, substantially as described.

30. The combination, with the indicators and keys, of actuating-levers constructed to move the indicators under the action of the keys, and each lever consisting of two parts, one having a limited movement independent of the other under the action of a spring, a lock for holding the two parts in connection, and connections between the indicators and locks, whereby each movable portion is released to automatically actuate the adjacent indicator as the next indicator completes its revolution, substantially as set forth.

31. The combination, with the indicating-disks and keys, of levers, each provided with a part geared with one of the indicators and capable of a limited movement to turn the latter, with a locking-lever, and connections between the latter and the adjacent indicator, substantially as and for the purpose set forth.

32. The combination, with the indicators, of operating-levers in two parts, and locking-levers B′ and B^2, substantially as specified.

33. The combination, with a series of indicators, keys, and intermediate connections, of one or more additional series of indicators, and means for throwing either series into connection with the operating devices, substantially as set forth.

34. The combination of a series of indicators, a series of keys to each indicator, and connections whereby each indicator may be set by the action of any key of one series, and a device for restoring the connections to their normal positions at the will of the operator, substantially as set forth.

35. The combination, with the series of keys, of two or more registering devices, each consisting of a series of numbered indicators, and series of intermediate independent operating-connections, and means for turning the registering devices to bring either one of the same into connection with the operating devices, substantially as set forth.

36. The combination of a series of indicators, and operating-keys and connections for moving said indicators, and operating appliances independent of the keys and indicators, whereby each indicator is moved one step by said appliances independently of the keys as the adjacent indicator completes its revolution, and devices operated by the indicators for throwing said appliances into action as each indicator completes its revolution, substantially as described.

37. The combination, with two or more series of keys, of a series of printing indicators and independent connections, whereby each indicator is controlled by each key of one of the series, and means for throwing the indicators out of gear with the connections, substantially as set forth.

38. The combination, with the series of keys and registering device operated therefrom, of an independent printing-recorder, and connections whereby the latter is moved from the same keys and to the same extent as the said registering device, and means for throwing each register out of gear with the connections, substantially as described.

39. The combination, with the series of disks provided with lateral pins, of levers B, racks hung to said levers, and locking-levers B′, and springs c', substantially as described.

40. The combination of the disks provided with pins a^6, levers carrying racks pivoted thereto, springs c', locking-levers B′, and levers B^2, substantially as described.

41. The combination, with the register and the recorder, of levers each carrying two series of racks capable of independent movement, one gearing with the register and the other with the recorder, and rack-operating devices, substantially as described.

42. The combination, with the levers B, carrying racks pivoted thereto, and locking-levers B′, of two adjustable registers, and levers B^2, constructed to operate with the disks of each register, substantially as described.

43. The combination of the keys, indicating-register, intermediate connections, and printing-recorder, frame B^5, carrying the same,

and toggle-levers d^3 d^1, substantially as described.

44. The combination, with the independent keys arranged in series, and indicators and connections, of levers d^2, slotted plate c, and rods d, connected to the levers and having terminal stops, substantially as described.

In testimony whereof I have signed my name to this specification in the presence of two subscribing witnesses.

W. S. BURROUGHS.

Witnesses:
F. L. FREEMAN,
CHARLES E. FOSTER.

(No Model.)

W. L. JUDSON.

CLASP LOCKER OR UNLOCKER FOR SHOES.

No. 504,038. Patented Aug. 29, 1893.

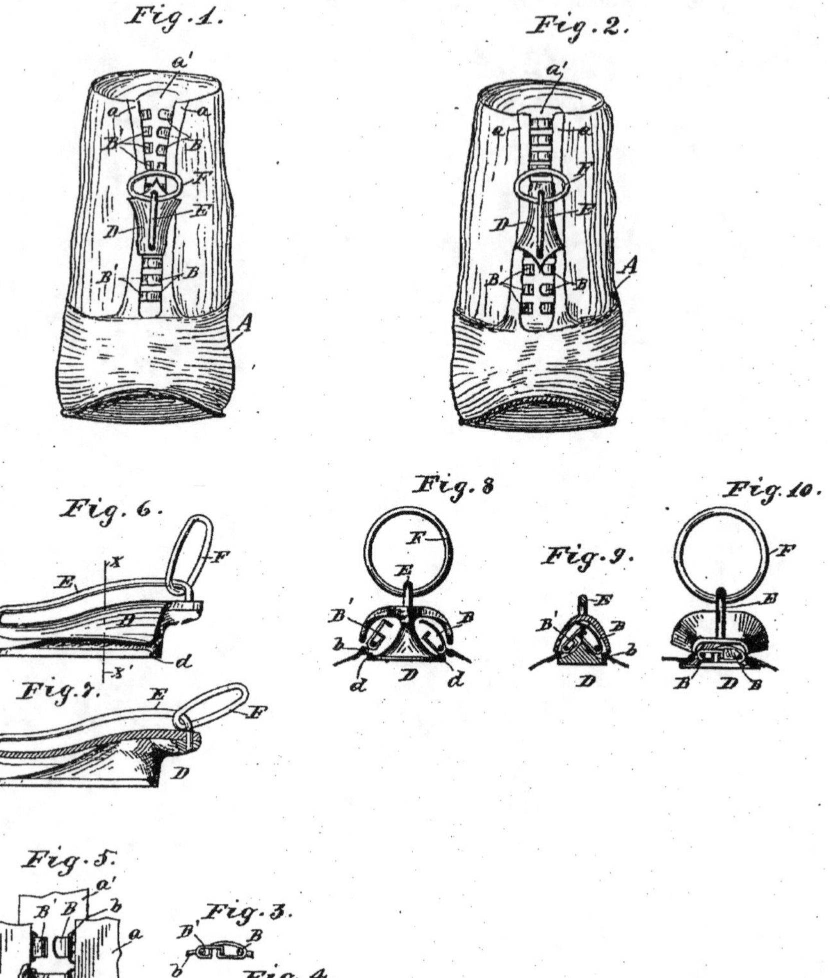

Witnesses.
A. U. Opsahl.
E. F. Elmore

Inventor.
Whitcomb L. Judson
By his Attorney.
Jas. F. Williamson

UNITED STATES PATENT OFFICE.

WHITCOMB L. JUDSON, OF CHICAGO, ILLINOIS.

CLASP LOCKER OR UNLOCKER FOR SHOES.

SPECIFICATION forming part of Letters Patent No. 504,038, dated August 29, 1893.

Application filed November 7, 1891. Renewed February 18, 1893. Serial No. 462,923. (No model.)

To all whom it may concern:

Be it known that I, WHITCOMB L. JUDSON, a citizen of the United States, residing at Chicago, in the county of Cook and State of Illinois, have invented certain new and useful Improvements in Clasp Lockers or Unlockers for Shoes, &c.; and I do hereby declare the following to be a full, clear, and exact description of the invention, such as will enable others skilled in the art to which it appertains to make and use the same.

My invention relates to clasp lockers or unlockers for automatically engaging or disengaging an entire series of clasps by a single continuous movement.

The invention was especially designed, for use as a shoe-fastener; but is capable of general application wherever clasps consisting of interlocking parts may be applied, as for example, to mail-bags, belts, and the closing of seams uniting flexible bodies. To these ends, the clasps are made with interlocking parts, which when in position, can only engage with each other when at an angle to the line of strain. The clasps have underreaching and overlapping projections or lips at their forward ends, which prevent the engagement or disengagement of the hook-portions of the clasps, except when thrown upward, so that the parts stand at an angle to each other of about ninety degrees. These clasps or fasteners, when in position on the flaps of a shoe or other adjacent parts which are to be united, may be engaged one at a time in succession, by bringing the two parts of the clasp into their proper angular relation to each other, by hand. But this is a tedious operation; and makes it difficult to draw the adjacent parts together, under the proper strain. I therefore provide a hand device, consisting of a movable guide, having cam-ways for permitting the passage of the clasps, by the movement of the guide from one end to the other of the series; and the cam-ways are so shaped and related that by the passage of the guide in one direction, the clasps will be drawn together and engaged, while by the passage of the guide in the other direction, the clasps will be disengaged and separated. In other words, one end of the guide has two channels or grooves, for receiving the parts of the fasteners when open or disengaged, and this may be called the forward end of the guide. The other or back end of the guide has a single channel or cam-way, into which the two channels from the forward end converge over an angular center ridge or instep. By moving the guide, so that the separate parts of the clasp enter the respective channels or cam-ways at the front of the guide the entire series of clasps will be delivered from the other or rear end of the guide properly engaged together. If the clasps be engaged and the united set be introduced at the rear end of the guide, and the guide moved over the same, the clasps will be delivered, disengaged from each other, at the forward end of the guide.

The invention, as applied to fasten shoes, is illustrated in the accompanying drawings, wherein, like letters referring to like parts throughout, Figure 1 is a front view of a shoe embodying my invention, showing the guide as applied to close the fastenings. Fig. 2 is a similar view, showing the guide, as applied to open the fastenings. Fig. 3 is an end view of one of the clasps detached, shown as in the locked position of the parts. Fig. 4 is a similar view, showing the angle which the parts of the clasps must assume, to lock or unlock. Fig. 5 is a plan, showing several clasps in position, on portions of the flaps of a shoe. Fig. 6 is a side elevation of the guide, the right hand end of the figure, being the front end of the guide. Fig. 7 is a longitudinal section of the guide. Fig. 8 is a front end view of the same, showing the entering position or delivering position of the unlocked clasps. Fig. 9 is a vertical cross-section on the line X X of Fig. 6, showing the clasps at their engaging or disengaging position, at the top of the ridge. Fig. 10 is a rear end view of the guide, showing the entering or delivering position of the clasps when engaged.

A is the body of the shoe, and *a a* are the flaps of the same. *a'* is the tongue underlying the flaps.

B B' are the two parts of the clasp, constructed with hook-portions and underreaching and overlapping parts, as before stated. The clasps are attached to the flaps of the shoe in any suitable way, shown as by wires *b* in Fig. 5, and as by lacing-strings C, in Fig. 11.

D is the guide. The base-piece of the guide is flat on its under surface, and has on its margin upturned lips *d*. The top-plate of the guide is concave or bell-shaped, and the connecting-body or center-piece uniting the two plates is angular in cross section, as before stated, and extends from the forward end of the guide to a point near the center of the same, and serving to divide the space between the base and top pieces into two channels or camways at the front end of the guide, which terminate as before stated, in the common channel or camway, at the back of the guide. A bail E is fixed to the top piece of the guide and carries a ring F, which serves as a finger-pull to operate the guide. In virtue of the bail, the ring or finger-pull may be shifted, so that the strain may be applied near the forward end of the guide, for moving the guide forward, and near the rear end of the guide, when moving the same backward.

The operation of this device has already been described.

The clasps may be easily and cheaply made of any suitable metal, and may be finished in any desired manner, so as to give an ornamental appearance. They may be very small in size and when properly applied to a shoe, will give the same a neat appearance and be comfortable to the wearer. It should be noted that the clasps are placed sufficiently close together on the flaps of the shoe, so that they cannot be disengaged by an endwise movement of the same.

The guide or hand-device may be made relatively small, as compared with the drawings, so that it may be readily inserted at the lower end of the series of fasteners, working on the tongue of the shoe, as a base or trackway. It should be noted that the guide acts not only to engage the clasps in its forward motion, but serves also to draw the flaps together, and the parts of the shoe tightly about the foot.

The practicability of the invention herein described has been demonstrated by actual usage of the same.

It will be noted, that in the construction shown in Fig. 11, the shoe is provided with top or overlapping flaps *c*, for concealing the fasteners from view.

What I claim, and desire to secure by Letters Patent of the United States, is as follows:

1. A device for engaging and disengaging a series of two-part clasps upon a shoe or other article, consisting of a guide-block, having two guide-ways, which are separated at one end thereof, and converge into a single guide-way, at the other end thereof, said guideways being adapted to engage and carry the interlocking parts of the clasps into or out of engagement with each other, as the block is moved forward or backward over the same, substantially as described.

2. A device for engaging and disengaging a series of two-part clasps, upon a shoe or other article, the interlocking members of which are engageable or disengageable by an angular movement of the same, the said device consisting of a guide-block having a pair of camways or guide-channels, in angular relation to each other at one end of the block, and converging and blending into a common cam-channel or guideway at the opposite end of the block, substantially as described.

3. A hand-device, for engaging or disengaging a series of two-part clasps, upon a shoe or other article, by a single continuous movement, the said device consisting of a movable guide-block having a pair of diverging camways or guide-channels, at one end of the block, which converge and blend into a single camway or guide-channel at the other end of the block, over an angular surface, located at the junction of the said ways, and extended outward on an incline toward the end of the single camway, substantially as and for the purpose set forth.

4. A device for engaging and disengaging a series of clasps upon a shoe or other article, the interlocking parts of which engage or disengage by an angular movement of the same, the said device consisting of a guide-block, having two bell-mouthed guideways in angular relation to each other, on opposite sides of the block, at one end thereof, and converging and blending over an angular surface into a single bell-mouthed guideway at the other end of the block, substantially as described.

5. A hand device for locking or unlocking a series of two-part clasps or similar interlocking parts, which engage or disengage by an angular movement, the said device consisting of a movable guide-block provided with a pair of divergent camways or guide-channels, at one end of the block with bell mouths at an angle to each other in the vertical plane, the said channel converging over an angular surface at the junction of the ways into a single camway or guide channel at the other end of the block having a bell mouth in the horizontal plane, whereby, under a single continuous movement of the hand device, the clasps may be drawn together and engaged or be disengaged and separated at will.

6. A device for engaging and disengaging a series of two-part clasps upon a shoe or other article, the same consisting of a block having two guideways which are separated at one end thereof and converge into a single guideway at the other end and a shifting finger-pull connected with the block and arranged to slide to either end for applying power to pull the block in either direction, substantially as and for the purpose set forth.

7. A device for engaging and disengaging a series of two-part clasps upon a shoe or other article, the same consisting of a block having two guideways which are separated

at one end thereof and converge into a single guideway at the other end, said block being provided with a staple extending from end to end thereof and a ring upon the staple constituting a shifting finger-pull to draw the block in either direction, substantially as described.

In testimony whereof I affix my signature in presence of two witnesses.

WHITCOMB L. JUDSON.

Witnesses:
JAS. F. WILLIAMSON,
CHARLES O. HENTHORN.

I. Hodgson,

Parlor Skate.

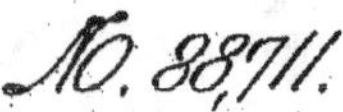
No. 88,711. Patented Apr. 6, 1869.

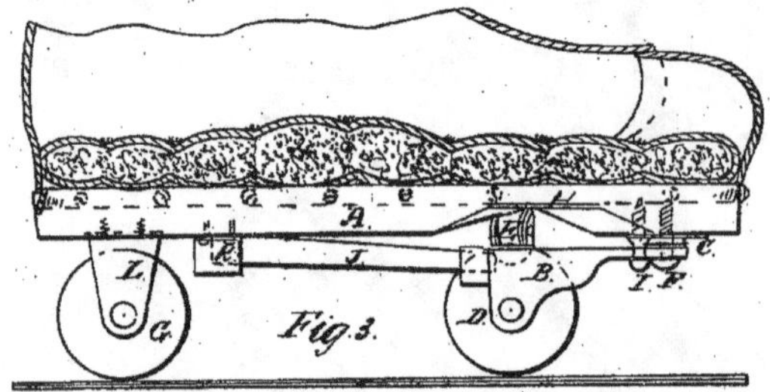

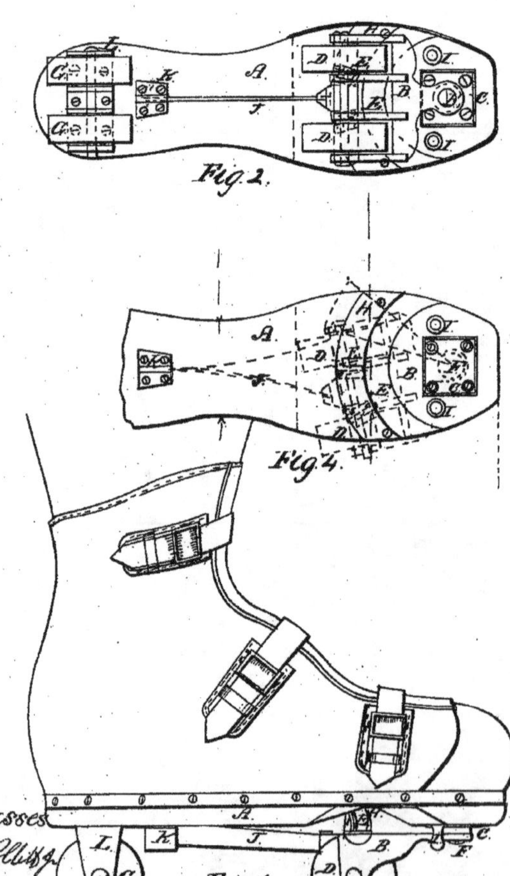

Witnesses
John Pollitt Jr.
Leon Beaver

Inventor
Isaac Hodgson

N. PETERS, PHOTO-LITHOGRAPHER, WASHINGTON, D. C.

UNITED STATES PATENT OFFICE.

ISAAC HODGSON, OF INDIANAPOLIS, INDIANA.

ROLLER-SKATE.

Specification forming part of Letters Patent No. **88,711**, dated April 6, 1869.

To all whom it may concern:

Be it known that I, ISAAC HODGSON, of Indianapolis, in the county of Marion and State of Indiana, have invented a new and useful Parlor-Velocipede; and I do hereby declare that the following is a full, clear, and exact description thereof, that will enable skilled artisans to make and use the same, reference being had to the accompanying drawings, and to the letters of reference marked thereon, making a part of this specification.

This invention relates principally to mode of operation of the forward wheel-frame; and it consists in the peculiar construction and manner of attaching the wheel-frame to the sole of the shoe, by which the operator is enabled at pleasure to run in a direct course, or, by the natural horizontal motion of the foot to the right or left, to change the course to a curved track without canting, rocking, or tipping the foot or sole to which the wheel-frame is attached, combined with the arrangement of a spring, which returns the wheel-frame to a direct line when the pressure upon the wheels is removed by raising the foot.

This invention further relates to construction and manner of attaching a padded shoe to the sole, by which the ordinary walking-shoe may be dispensed with, and the comfort of the wearer greatly enhanced.

Figure 1 is a profile of my invention. Fig. 2 is an inverted view, showing the mode of attaching the wheel-frame, spring, &c., to the sole. Fig. 3 is a longitudinal vertical section, showing the pad, &c. Fig. 4 is an inverted view of the front wheels, spring, &c., showing their alternating direction.

Similar letters of reference indicate like parts in the several figures.

A represents the sole of the shoe; and B the front wheel-frame, with arm extending forward and pivoted to the plate C, which is securely attached to the sole. Directly over the axis of the wheels D the friction-rollers E, which are connected with the wheel-frame B, are set in lines radiating from the pivot F, over the friction-rollers E.

A segmental way, H, is secured to the sole A, so that when the weight of the body is thrown on the rear wheels, G, and the toes turned horizontally to the right or left, the wheel-frame B will turn on the pivot F, as indicated by red lines in Fig. 4, and enable the operator to freely move the foot on the friction-rollers E, and describe any desired curve with the foot in a horizontal position.

Stops I are secured to the sole A, to limit the sweep of the wheel-frame B. The spring J, which is attached to the sole at K and extends to the wheel-frame B, where it freely enters a slit in the frame between the wheels D, is of sufficient strength and elasticity to force the wheel-frame B from the position shown by red lines in Fig. 4 to a direct line, as shown in Fig. 2.

The rear wheel-frame, L, is secured to the sole A, so that the wheels G will always be in a direct line; the pad M of the shoe to be of elastic material, of any desired thickness and elasticity to accommodate the foot.

The shoe may be of leather or canvas, lined with flannel, and securely attached to the edges of the sole, and provided with straps and buckles, for convenient and secure fastening.

I claim—

1. The wheel-frame B, provided with the forward-projecting arm, and furnished with the friction-rollers E, interposed between the frame and the sole, and attached by the arm to the forward part or toe of the sole, in the manner and for the purpose substantially as set forth.

2. The spring J, in combination with the wheel-frame B, constructed and arranged substantially as and for the purpose set forth.

ISAAC HODGSON.

Witnesses:
JOHN POLLITT, Jr.,
LEON BEAVER.

R. M. HOE.

3 Sheets—Sheet 1.

Printing Press.

No. 5,199.

Patented July 24, 1847.

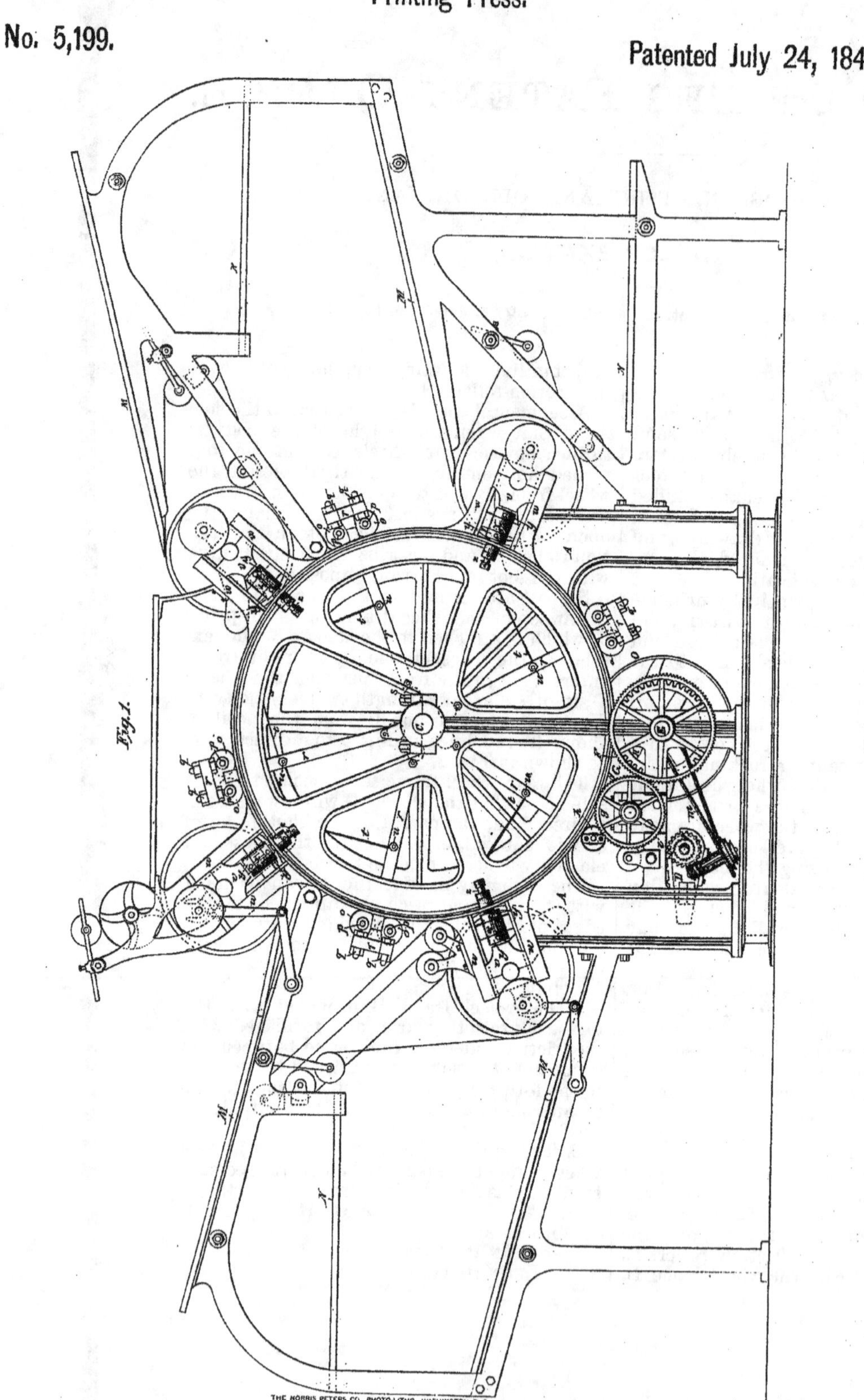

THE NORRIS PETERS CO., PHOTO-LITHO., WASHINGTON, D. C.

R. M. HOE.

3 Sheets—Sheet 2.

Printing Press.

No. 5,199.

Patented July 24, 1847.

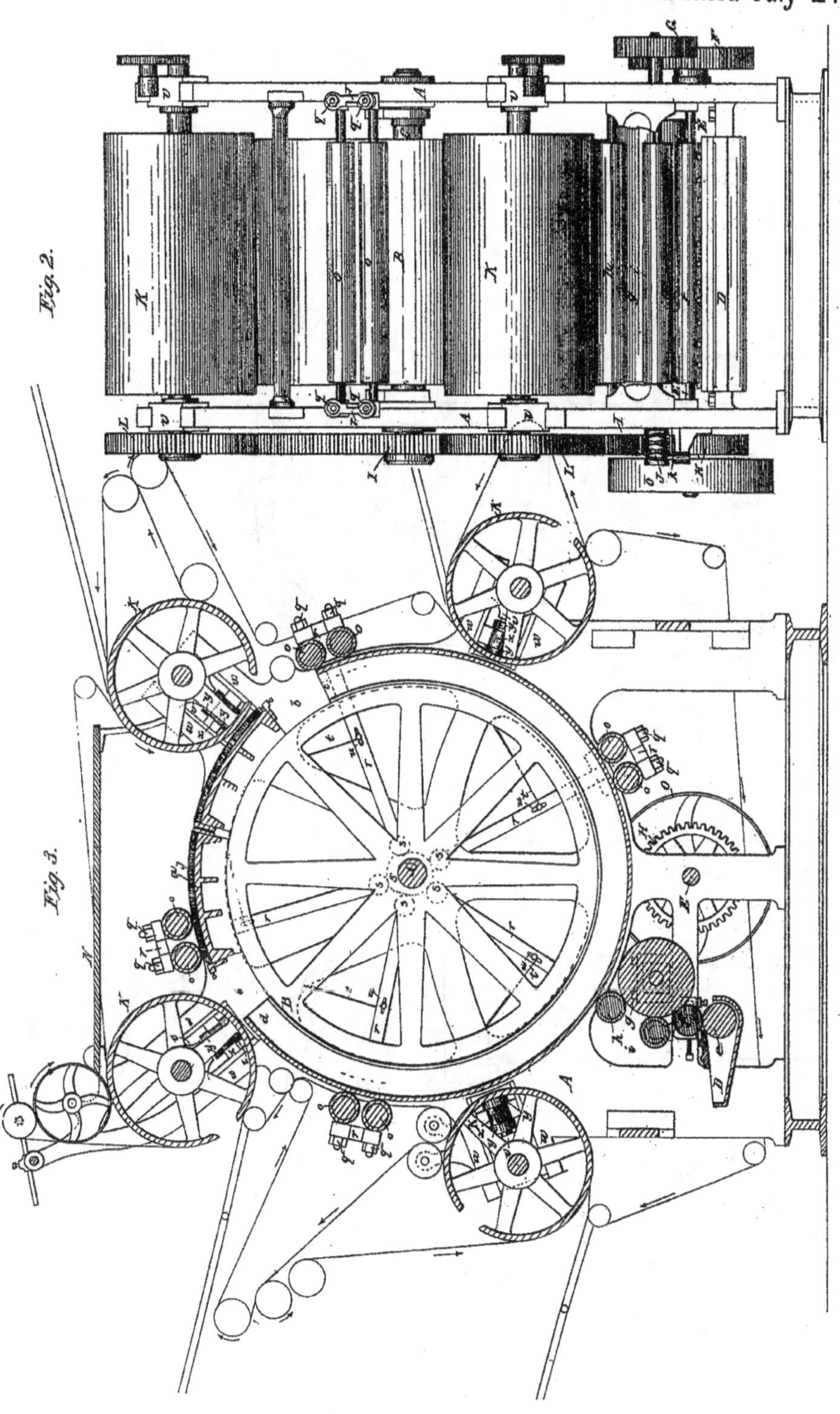

THE NORRIS PETERS CO., PHOTO-LITHO., WASHINGTON, D. C.

3 Sheets—Sheet 3.

R. M. HOE.

Printing Press.

No. 5,199. Patented July 24, 1847.

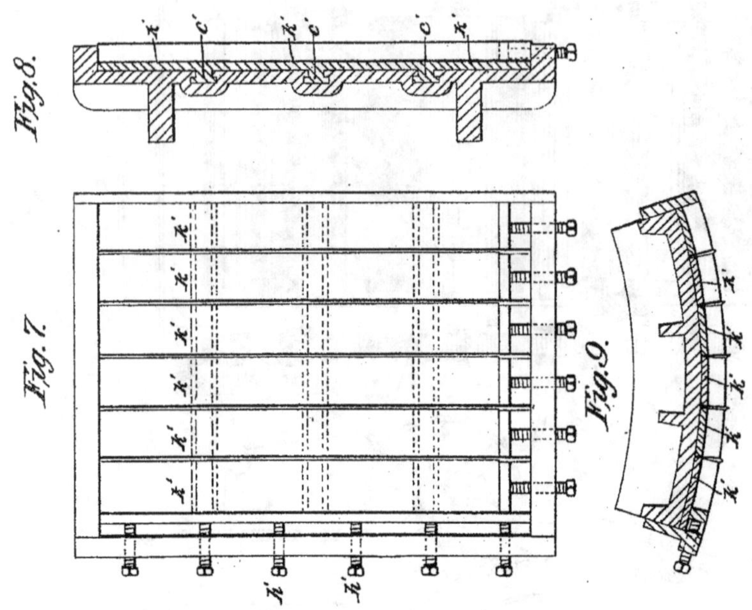

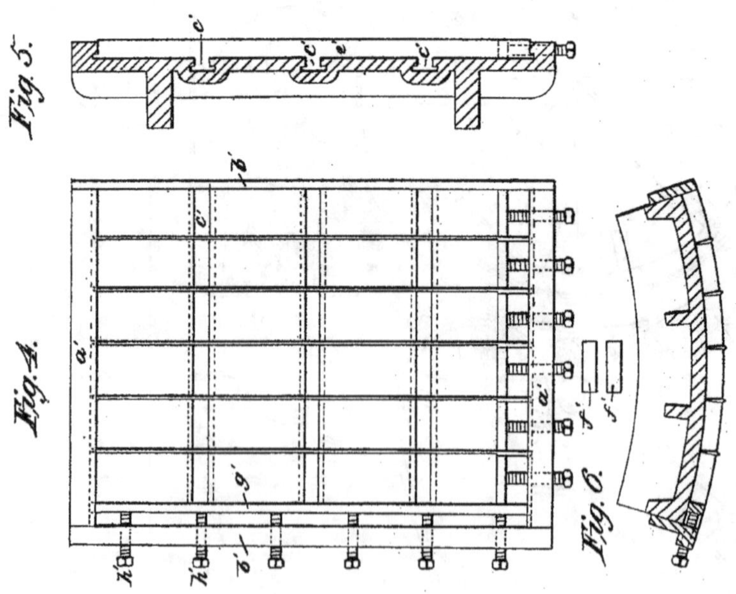

THE NORRIS PETERS CO., PHOTO-LITHO., WASHINGTON, D. C.

United States Patent Office.

RICHARD M. HOE, OF NEW YORK, N. Y.

IMPROVEMENT IN ROTARY PRINTING-PRESSES.

Specification forming part of Letters Patent No. **5,199**, dated July 24, 1847.

To all whom it may concern:

Be it known that I, RICHARD M. HOE, of the city, county, and State of New York, have invented new and useful Improvements in the Printing-Press which I denominate "Hoe's Cylindrical-Bed Press," and I do hereby declare that the following is a full, clear, and exact description of the principle or character which distinguishes it from all other things before known and of the manner of making, constructing, and using the same, reference being had to the accompanying drawings, making part of this specification, in which—

Figure 1 is a side elevation of the press; Fig. 2, a front elevation; Fig. 3, a longitudinal vertical section; Fig. 4, a plan of the cylindrical bed that receives the types to exhibit the method of securing them; Fig. 5, a longitudinal section, and Fig. 6 a cross-section thereof; and Figs. 7, 8, and 9 like views of a modification of the method of securing the types.

The same letters indicate like parts in all the figures.

My improvements are applied to that class of printing-presses in which the form of types is arranged on the surface of a cylinder with a series of impression-cylinders, inking-rollers, &c., arranged around it, so that by its rotation the types are successively inked and give their impression to the sheets of paper as they are fed in by the impression-cylinders, the number of sheets to be printed by one revolution of the cylindrical bed depending on the number of impression-cylinders arranged around it, the number of impression-cylinders being governed by the diameter of the cylinder that carries the types and the distance between the impression-cylinders.

The nature of the first part of my invention consists in arranging the form or forms of types on a segment of a cylinder, while the other portion of its surface is employed to distribute the ink, and therefore answering the purpose of a distributing-table.

The second part of my invention consists in giving to the inking-rollers, which are arranged in sets around the cylindrical form and distributing-table, and also to the transferring-roller of the inking apparatus, a motion in and out or toward and from the axis of the cylinder, around which they are arranged, so that they may make pressure on the cylindrical distributing-table as it passes under them to distribute the ink and be thrown out sufficiently far from the center to transfer the ink to the form of types, as the face of the types must be the segment of a larger cylinder than the segment that forms the distributing-table, that this (the distributing-table) may not ink the impression-cylinders when passing by them.

The third part of my invention relates to the inking apparatus; and it consists in giving to the ductor or fountain roller that takes the ink from the fountain a slow continuous rotary, instead of an intermitting motion, as heretofore, so that the ink shall be regularly transmitted to the taking-roller and thence to the distributing-roller, &c., and also in connecting the arbor of this ductor or fountain roller with the mechanism that gives to it the slow rotary motion by means of a ratchet, that it (the ductor-roller) may be turned forward when desired to alter the supply of ink.

The last part of my invention relates to the method of securing and retaining the types on the cylindrical bed by means of column-rules, which are thicker at the outer than at the inner edge, so that the faces of any two of them shall be parallel with each other, or nearly so, to hold the column of type as tight at the top as at the base, the said rules being made with projections from the lower edge to fit in rabbeted grooves in the bed, so that the columns of types, with the rules separating them, may be pressed together by screws at the side of the bed, in the usual manner of securing types, and thus secure and hold the form of types on a cylindrical surface as effectually as on a flat surface, this important object having long been essayed in various ways, but never before to my knowledge successfully attained.

In the accompanying drawings, A represents a frame properly adapted to the various parts of the press, and B a cylinder of large size mounted on a shaft C, running in appropriate bearings. About one-fourth of the circumference of this cylinder constitutes the bed *a* of the press, the periphery of which is of course the segment of a cylinder adapted to receive the form of types either in the manner to be presently described or in any other manner which may be desired. On each side of this bed there is a small open space *b b* to give free access to the ends of the bed for putting in and removing the types, and then the remnant of the periphery of the cylinder from *c* to *d* constitutes the cylindrical distributing-table, its surface being properly adapted to the distribution of ink, as distributing-tables for this purpose are generally made, except that it is cylindrical instead of flat. The diameter of this part of the cylinder should be less than that of the form of types, that it may pass by the impression-rollers without touching them.

The ink is taken from the fountain D, of the usual construction, by the ductor-roller *e*, transferred from this to the taking-roller *f*, thence transferred to the vibrating distributing-roller *g*, and taken from this by the transferring-roller *h* to the distributing-table *c d* of the cylinder B, one or more small distributing-rollers *i* being applied to the surface of the vibrating distributer and between the taking and transferring rollers for the purpose of more equally distributing the ink. This small distributing-roller may be composed of rings of cloth slipped onto an inclined cylinder or shaft. The vibrating distributing-roller *g* receives its rotary motion with considerable velocity (the surface moving with an equal velocity to the distributing-table *c d*) from the main shaft E by means of a cog-wheel F, which engages another cog-wheel G of less diameter on the arbor of the rollers, and these wheels are of sufficient thickness to allow of the vibration of the roller, with its arbor, in the direction of its axis without disengaging the cogs, and this vibrating motion is obtained by means of the double worm *j* on the end of the arbor, the two grooves crossing each other, so that by running on a swivel-feather *k* one of the grooves or worms will travel on the feather to the end, and then as it turns to run into reverse groove the feather is turned, which carries the arbor back, and so on back and forth.

The taking-roller *f*, the transferring-roller *h*, and the small distributing-roller *i* are carried by the rotating motion of the vibrating distributing-roller by contact of their surfaces, and the ductor or fountain roller receives a slow and continuous rotary motion to carry up the ink from the fountain by a worm *l*, that takes into the cogs of a worm-wheel *l′* on the arbor of the ductor, motion being communicated to the arbor of the worm by a belt *m* from a pulley (not seen in the drawings) on the main driving-shaft E. The worm-wheel *l′* on the arbor of the ductor turns freely thereon, and is connected by a ratchet-wheel and pawl *n*, so that the mechanism can carry the ductor in one direction, while the ratchet admits of turning it forward independently of the worm and its connections when it becomes necessary to alter the supply of ink.

The main cylinder B receives motion from the main shaft E by means of the pinion H, which engages with a cog-wheel I on the shaft C of the cylinder, and as the cylinder B rotates in the direction of the arrow the form of types J thereon is in succession carried to and under four impression-cylinders K K K K, arranged at proper distances around the cylinder to give the impression to four sheets of paper introduced between the form of types and the impression-cylinders, one sheet being introduced by each impression-cylinder in the same manner as in the well-known double-cylinder press. The impression-cylinders are constructed in the same manner as those employed in the class of presses just referred to, and they are either provided in the well-known manner with fingers for taking and liberating the sheets; or a system of tapes may be used for this purpose, and as these make no part of my invention it is deemed unnecessary to describe them.

The shaft of each of the impression-cylinders has a cog-wheel L on one end, which engages with the cog-wheel I on the shaft of the cylinder, by which the impression-cylinders receive their appropriate motion, and care must be taken to have the pitch-line of these cog-wheels so regulated that the surface of the form of types and that of the impression-cylinders shall move with the same velocity to prevent the slipping of one surface on the other, which would destroy the impressions.

Between every two of the impression-cylinders there is a set of inking-rollers, making one set to each impression-cylinder, each set consisting of two rollers *o o*, the journals of which run in boxes that are adjustable by screw-nuts *q q* in the ends of two sliding bars *r r*, one on each side of the press and moving in appropriate slides in the sides of the frame. These bars converge to the axis of the cylinder B, and are provided at the inner end each with a friction-roller *s*, (represented in Fig. 3 by dotted lines,) which run on the periphery of a cam *s′*, (also represented by dotted lines,) and this cam is so formed as to force out these bars with the inking-rollers just as the form of types approach them, that they may make a gentle pressure to ink the types, and as the form leaves them to permit the bars and rollers to be moved in by the tension of a spring *t*, which bears on an adjustable pin *u* on the bars, so that the inking-rollers may run on the distributing-table to receive the ink from it. There must of course be one spring for each bar.

The journals of the impression-cylinders run in boxes *v*, that slide in standards *w w*, and from the inner end of each there is a screw-stem with a nut *y* above and below a cross-bar *x*, through which the stem passes, by means of which the position of the impression-cylinders relatively to the form of types can be regulated, and below this and passing through the frame there is a set-screw *z*, (one for each sliding box,) which determines the depth to which the screw-stem of the sliding box shall move toward the axis of the form of types. This sets the impression-cylinders for the degree of pressure to be given in taking the impression, while by means of the screw-nuts on the stems of the sliding boxes the impression-cylinders can be raised at pleasure and thrown out of play.

Each impression-cylinder is provided with one feeding-table M and one delivery-table N to receive the printed sheets.

It will be obvious from the foregoing that the form of types can occupy more or less of the surface of the cylinder at the pleasure of the constructer, and that the number of impression-cylinders can be increased or decreased, as it may be desired, to make the press of greater or less capacity; but it must be observed that there must be one set of inking-rollers for each impression-cylinder, although one inking apparatus is sufficient for several impression-cylinders, although it is deemed advisable not to have more than four cylinders for one inking apparatus.

Power is applied to drive this press by a belt from some first mover running onto a belt-wheel O on the main shaft E, or in any other manner which may be preferred.

If desired, stereotype-plates may be secured to the cylinder instead of the form of types, a portion of the surface of the said cylinder being made and employed as a distributing-table for the distribution of the ink.

Having thus described my improvements in the press and the manner of constructing and using the same, I will proceed to describe my improved method of securing the form of types on a cylindrical surface. The bed *a* is a segment of a cylinder with flanges *a′ b′* at the ends and sides. In the direction of the periphery there are rabbeted parallel grooves *c′*, cut to receive correspondingly-formed tongues projecting from the lower edge of column-rules *e′*, the ends of which are made to fit in rabbets cut in the inside face of the flanges *a′ a′* of the bed. These column-rules are made thicker at the outer than at the inner edge; or, in other words, they are so formed as to present the form of a wedge in their cross-section, so as to bind the types near their upper end. As the types are set on a cylindrical surface and their sides are parallel instead of radiating from the center of the circle, if the rules were made of equal thickness—that is, with parallel sides—it will be obvious that the types, however tight they might be bound together at the base, would be loose at the top; but by making the rules thicker at the outer than the inner edge the types of each column are bound together just as tight at the top as at the base, and by this means are as firmly held on a cylindrical as they would be on a flat surface, for the rules are held down by the tongues fitting in the grooves and the ends in the rabbets of the side flanges of the bed, and so long as the rules are held in place so long will the types be. The grooves in the bed are fitted up with blocks *f′*.

In setting up the form the blocks *f′*, No. 1, are put in the grooves so as to be flush with the surface of the bed. The first column of types is set up. Rule No. 1 is then inserted; then blocks 2, the second column of types, rule 2, and so on to the end, and then finally a bar *g*, against which bear the ends of the screws *h′*, that pass through one of the end flanges *b′* of the bed to bind the form of types in one direction, the usual or any other method being used for binding them in the other direction.

It will be be obvious to every one skilled in this branch of the art that the principle which I have adopted for securing the types on a cylindrical bed may be variously modified without changing the character of my invention, and as an evidence of this it may be well to describe one of the modifications which I have contemplated, which is as follows, viz: Instead of making tongues to project from the lower edge of the rules, they (the rules) are attached to plates *k*, which are segments of a cylinder corresponding with the cylindrical bed and connected with it by tongues fitting in grooves *c′*, in the same manner as the tongues of the rules. By this modification the rules, which of necessity are made very thin, are sustained along their whole length by their attachment to the segment-plates, instead of being sustained at intervals by the tongues, as in the first modification.

What I claim as my invention, and desire to secure by Letters Patent, is—

1. Putting the form or forms of types on a movable or permanent segment of a cylinder which forms the bed and chase, substantially as described, and also when this is combined with the cylindrical distributing-table, which occupies another segment of the same cylinder, substantially as described.

2. Giving to the inking-rollers a movement toward and from the center of the cylinder that carries the form of types, substantially as described, when this is combined with the form of types and the distributing-table made on one and the same cylinder and of different radius, as described, whereby the inking-rollers are adapted to the different diameter of the form of types and the distributing-table, as described.

3. Giving to the ductor or fountain roller of the inking apparatus a slow continuous rotary motion, in combination with the ratchet-connection between the roller and the mechanism from which it receives its continuous rotary motion, substantially as described, whereby the ink is more regularly supplied, and by which, also, this supply may be altered when desired, as described.

4. The method of securing the form of types on a cylindrical surface with column-rules

made thicker toward their outer than their inner edge by connecting these with grooves in the bed, by which they are permitted to approach and recede from each other, and at the same time kept down to the same radius, substantially as described, whereby prismatic types can be secured and held on a cylindrical surface as effectually as on a flat surface, as described.

RICHD. M. HOE.

Witnesses:
CHAS. M. KELLER,
JAMES MACLYON, Jr.

May 20, 1958 R. C. BAUMANN 2,835,548

SATELLITE STRUCTURE

Filed Aug. 1, 1957 3 Sheets-Sheet 1

FIG. 1

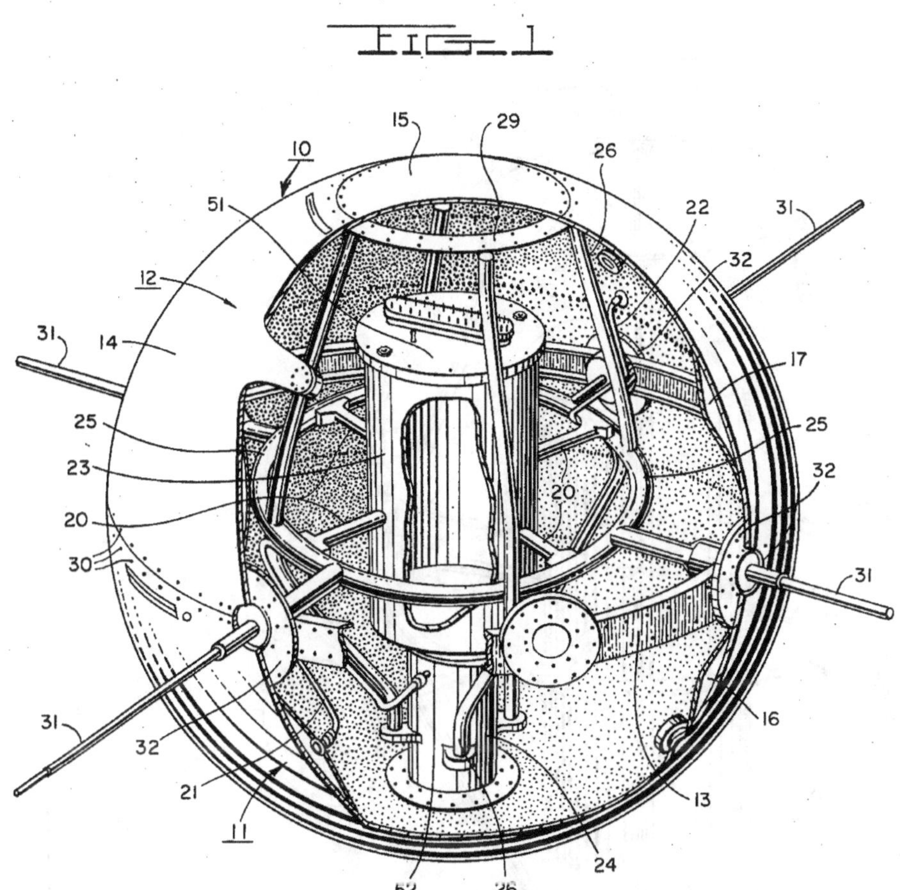

INVENTOR
ROBERT C. BAUMANN

BY W. R. Matthews
Richard E. Reed

ATTORNEYS

May 20, 1958 R. C. BAUMANN 2,835,548

SATELLITE STRUCTURE

Filed Aug. 1, 1957 3 Sheets-Sheet 2

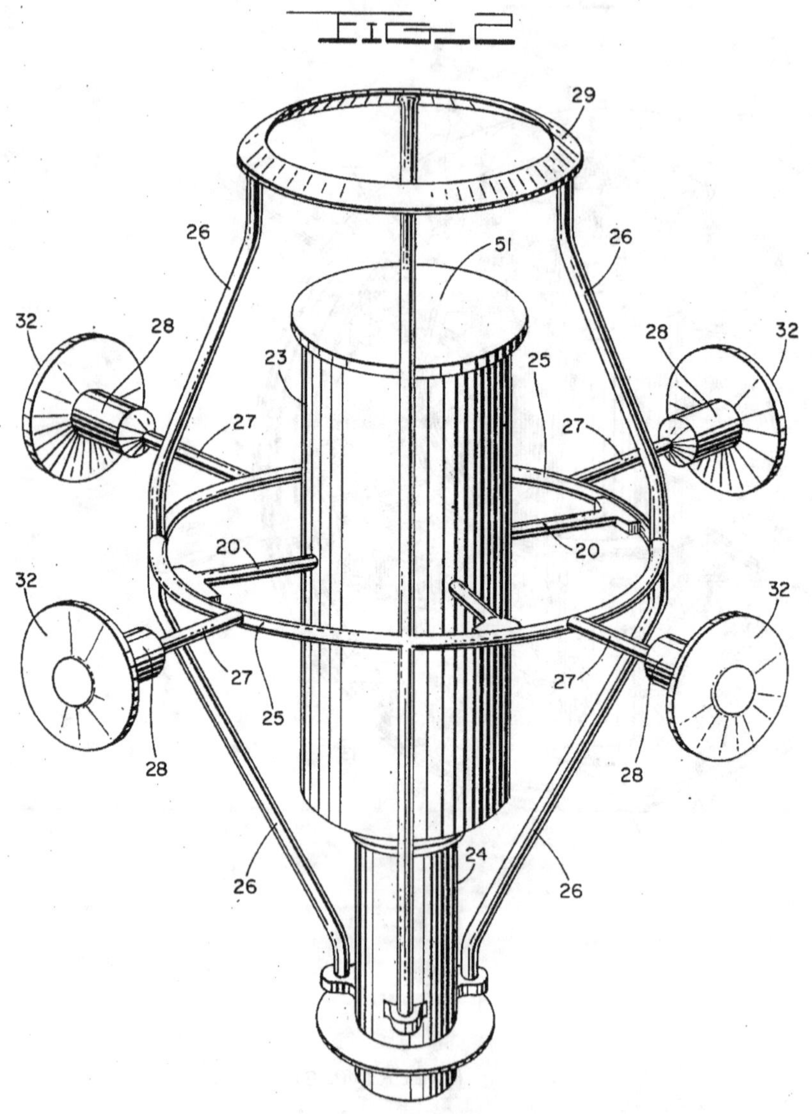

INVENTOR
ROBERT C. BAUMANN

BY W.R. Mattox
Richard C. Reed
ATTORNEYS

May 20, 1958 R. C. BAUMANN 2,835,548

SATELLITE STRUCTURE

Filed Aug. 1, 1957 3 Sheets-Sheet 3

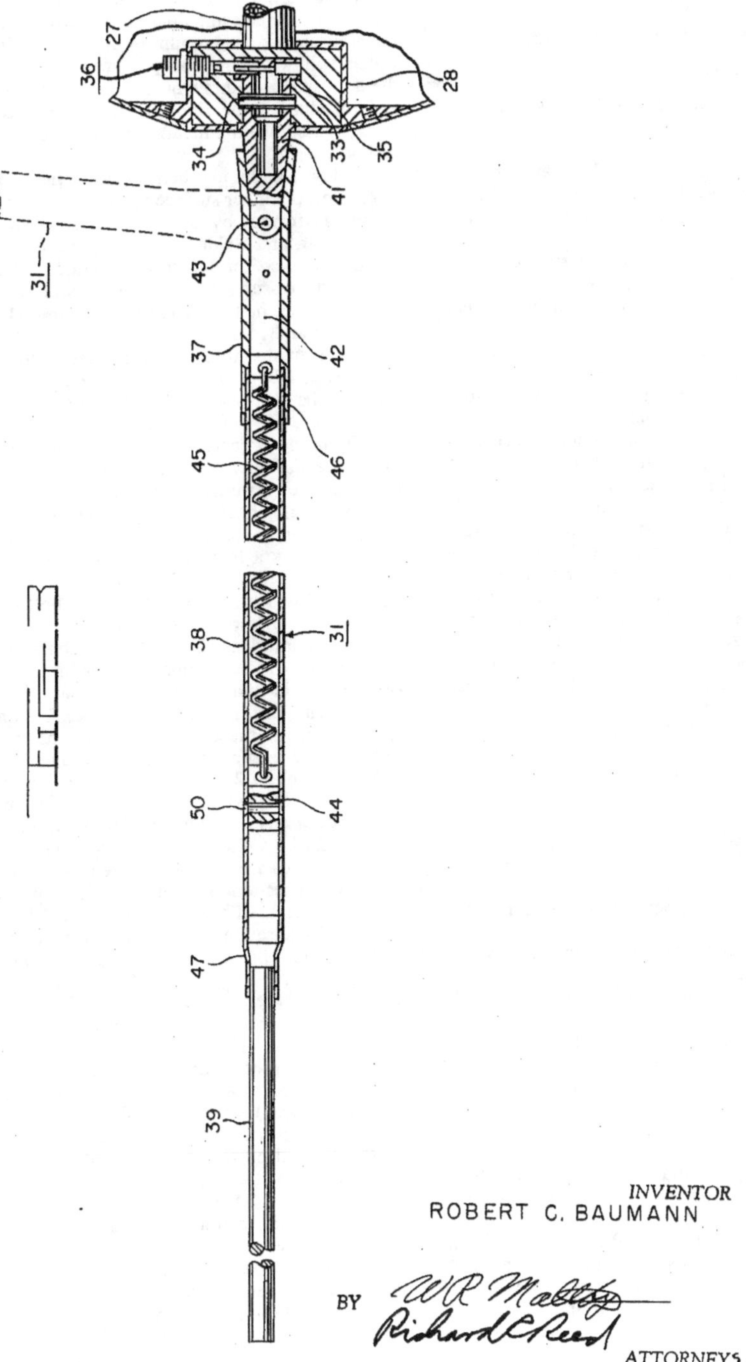

INVENTOR
ROBERT C. BAUMANN

BY W R Maltby
Richard C Reed
ATTORNEYS

United States Patent Office

2,835,548

Patented May 20, 1958

1

2,835,548

SATELLITE STRUCTURE

Robert C. Baumann, Alexandria, Va., assignor to the United States of America as represented by the Secretary of the Navy

Application August 1, 1957, Serial No. 675,787

9 Claims. (Cl. 312—352)

(Granted under Title 35, U. S. Code (1952), sec. 266)

The invention described herein may be manufactured and used by or for the Government of the United States of America for governmental purposes without the payment of any royalties thereon or therefor.

The present invention relates to the design of earth satellites and more particularly to the outer shell structure of the satellite and the supporting members therein which carry the instruments.

Heretofore scientific explorations of the upper atmosphere in order to obtain a better understanding of the physical phenomena in these regions has been carried out by the use of gas filled balloons and by rockets carrying specific instruments. These explorations have been limited to certain areas of the upper atmosphere and to short periods of time for taking the desired data.

Earth satellites made according to the present invention can be fired into the upper atmosphere to encircle the earth and to obtain data throughout the whole area about the earth. Observations can be made of electromagnetic radiation from the sun which does not penetrate the earth's atmosphere, and to study incoming radiations and relate them to the affected regions of the atmosphere such as ozonosphere and the ionosphere. The satellite will further provide new and unprecedented opportunities for scientific measurements of the upper atmosphere and will increase the observable time for taking measurements and provide a more widespread test area. Also more intelligent information about the size and shape of the earth can be obtained by such a satellite.

It is therefore an object of the present invention to provide a satellite structure which can be fired into the upper atmosphere and remain for relatively long periods of time.

Another object is to provide a structure which is adapted to carry instruments into the upper atmosphere for upper atmosphere observation.

Yet another object is to provide a structure which can be easily assembled and disassembled.

Other and more specific objects of this invention will become apparent upon a careful consideration of the following detailed description when taken with the accompanying drawings, in which:

Fig. 1 is a plan view of the satellite which is cut away to illustrate the inner structure;

Fig. 2 is a side elevation view of the inner structure of the satellite;

Fig. 3 is a sectional view of the antenna which illustrates the mechanism which operates the antenna to permit folding.

The present invention provides a spherical shell within which a supporting structure aids in maintaining the shape of the spherical shell and also provides easy access to the innermost part for securing and assembling the instruments. The inner structure has antennas connected thereto which are adapted to extend outwardly along the equator to provide the necessary function of sending and receiving signals.

Referring now to the drawings wherein like reference characters represent like parts throughout, the satellite

2

structure **10**, as illustrated, comprises a housing which has a lower hemispherical section **11** and an upper hemispherical section **12** which are fastened at the equator to a channeled structure **13**, by rivets and/or screws as appropriate. The upper hemispherical section **12** is formed in two parts **14**, **15** wherein the uppermost part **15** permits limited access to the inside thereof for adjusting the instruments and final assembly thereof. The housing is formed of magnesium or any other suitable material which will withstand the pressures and temperatures of the atmosphere within which the satellite structure travels, and in addition the rigorous vibration, acceleration and aerodynamic heating incurred during the ascending trajectory.

The lower and upper hemispherical housing sections are girdled on the inner surface respectively by pressure zones **16** and **17** formed by an annular band of metal similar to the housing and welded thereto. The band is somewhat rounded and so formed so as to afford equal strength to withstand pressures both internally and externally. The pressure zones have pressure lines **21** and **22** which extend therefrom and connect with a pressure gauge (not shown) in the inner structure. The pressure zones are adapted to withstand both positive and negative (vacuum) pressures. The zones are filled with unequal pressures for the purpose of determining if puncture occurs during the ascending trajectory and further to determine which of the hemispherical sections has been punctured in the event the shell is punctured during flight.

The housing has an inner supporting structure which comprises an inner cylindrical chamber **23** secured at the bottom by a low thermal conductivity support **52** of "Kel–F" or other suitable material, to a main support column **24** which is connected at the bottom to the inner surface of the lower hemispherical surface. Concentric with the cylindrical chamber is a tubular ring **25** which is connected to the chamber by four tubular rods **20** made of "Kel–F" or other suitable material with low thermal conductivity, said rods extending therefrom along equally spaced radii at the equator of the spherical shell. The tubular ring is supported vertically by four bow shaped tubular members **26** spaced 90 degrees apart with respect to a plane through the equator of the spherical shell. The bottom ends of members **26** are secured to the main support **24** by welding or any other suitable manner and the upper ends are likewise secured to an annular member **29** to which sections **14** and **15** of the upper hemispherical section are secured. The main support **24** is also designed to receive the satellite separation mechanism. Extending radially from the concentric ring **25** along radii in the equatorial plane are four tubular rods **27** spaced 90° apart and 45° with respect to members **26**. Each of the rods terminate in an enlarged cylindrical tubular portion **28** which supports an antenna **31** and a flanged portion **32** on the end thereof that aids in supporting the shell structure at the equator. As can be seen by illustration in Fig. 1, the shell structure is also supported at the north and south poles by the supporting frame structure.

The flanged portion **32** is curved to fit along the inner surface of the shell structure which is secured thereto by suitable screws **30** or any other suitable means and the cylindrical end portion of the antenna supporting structure in adapted to receive the end of the antenna and an insulating member **33** by which the antennas are secured to the tubular end portions. The insulating members may be made of Teflon or any other suitable material which is, cylindrical in shape and has a diameter such that it fits tightly into the cylindrical end piece. The insulating member is formed in two pieces and adapted to fit about the end of the antennas which is held thereto by a pin **34** and a rib **35** on the end of the antenna. A connector **36** makes contact with the antenna to provide connecting means through which signals may be sent or received and also to provide means for holding the insulating member and antennas in the cylindrical end piece 28.

The antennas are designed such that they may be folded at an acute angle with respect to the antenna support rods 27. The antennas are made in three sections 37, 38 and 39 of aluminum tubing having a wall thickness of 0.024 inch. The tubes provide a means by which suitable mechanism may be installed to permit folding and subsequent automatic return to a locked unfolded position as shown in Fig. 3. The mechanism includes a short stub end 41 which is secured to the cylindrical end pieces 28 by insulating member 33 and connector 36, and tapered on the other end to be received by a tapered end of antenna section 37. The stub end 41 is secured to an elongated cylindrical member 42 which is adapted to be inserted for free movement into the antenna end section 37 and pivotably connected to stub end 41 at 43. A fixed member 44 is secured in the outer end of section 38 by rivet 50 and a spring 45 is connected thereto and to the cylindrical member 42. The inner end of antenna section 38 is connected to the outer end of section 37 at 46 adjacent to cylindrical member 41 and the inner end of section 39 is secured at 47 to the outer end of antenna section 38 adjacent to the fixed member 44.

In order to position the antenna in its folded position, the antenna is pulled away from the spherical section until the inner tapered end of section 37 clears the pivot 43 which permits folding. During launching of the satellite the antennas will be in a folded position shown by dotted lines in Fig. 3 and resting upon suitable stops on the nose cone section, not shown, and upon release of the nose section, the spring 45 will pull the antennas into normal flight position as shown in Figs. 1 and 3. Such an arrangement affords protection for the antennas during the critical stages of launch as well as enabling the use of a relatively long antenna without modification to the launching vehicle.

In addition to the antennas equally spaced about the equator there are suitably spaced four microphones, a Lyman Alpha solar cell and a Lyman Alpha ion chamber. Further there are various gages and connections thereto from the shell structure such as erosion gages, temperature gages, pressure gages and any other attachment for suitable equipment.

The cylindrical chamber 23 is adapted for use as the power supply storage and for securing various instruments therein. These instruments do not constitute a part of the present invention, therefore, further discussion is not seen to be necessary. However, the top cover 51 for the cylindrical chamber provides the connections for most of the instruments in the chamber and is therefore designed to secure the connectors therein.

The internal structure, the internal cylindrical chamber and shell assembly are electro-plated with zinc, copper, silver and a coating of 0.00005 inch of gold to facilitate handling, reduce corrosion, and for thermal considerations. The outer surface of the magnesium sphere is further coated evaporatively with a silicon monoxide coating which has several underlying coatings of other metal substances as follows: a layer of chromium, a layer of silicon monoxide, and a layer of aluminum. The final silicon monoxide coating gives the desired thermal emissivity. These coatings are for the purpose of regulating, to some degree, the mean orbital temperature of the housing by setting the ratio between absorptivity and emissivity.

The above structure has been described for a satellite structure to be used in actual flights in the upper atmosphere. However, it is to be understood that applicant is not to be limited to the materials from which the structure is made since it is obvious that similar structures can be made of other materials. The materials from which the satellite structure is made will depend on the particular use to which it will be applied, that is, similar structures can be used for giving lectures, group discussions or even as a toy and will not require the particular materials for the structure as required for upper atmosphere flights.

Obviously many modifications and variations of the present invention are possible in the light of the above teachings. It is therefore to be understood that within the scope of the appended claims the invention may be practiced otherwise than as specifically described.

What is claimed is:

1. A satellite which comprises a thin shell spherical structure and an inner support structure, said inner support structure comprising a support column, a plurality of bow-shaped members and a first ring, all positioned concentrically about an axis through said spherical structure, said bow-shaped members being secured at one end to said support column and at the other end to said first ring, said support column and said first ring being secured to the inner surface of said spherical structure, a second ring secured to said bow-shaped members at the equator of said spherical structure and a plurality of radially extending members secured to said second ring and to the inner surface of said spherical structure at the equator.

2. A satellite as claimed in claim 1 wherein the ends of said plurality of radially extending members secured to said ring and to the inner surface of said spherical structure are adapted for mounting antennas that extend outwardly from said spherical structure, said antennas being adapted for pivotable movement for angularly positioning said antennas with respect to said mounting structure.

3. A satellite structure comprising an outer spherical structure and a supporting structure within said spherical structure, said supporting structure being formed of a plurality of bow-shaped members assembled about an axis of said spherical structure and secured respectively at opposite ends to a cylindrical chamber and a concentric ring each of which are secured to the inner surface of said spherical structure, and a ring secured to said bow-shaped members at points on a plane perpendicular to the axis about which said spherical structure is secured.

4. A satellite structure which comprises a thin shell spherical structure and a supporting structure within said spherical structure, said supporting structure comprising a plurality of bow-shaped members, a support column and a first ring, all assembled concentrically about an axis through said spherical structure perpendicular to a plane through the spherical section at the equator, said bow-shaped members being equally spaced and secured at one end to said support column and secured at the opposite end to said first ring, a second ring structure secured to said bow-shaped members along the plane at the equator, radially extending support members secured to said second ring structure about said bow-shaped members and to said support column, and other radially extending support members secured to said second ring structure about said bow-shaped members and to the inner surface of said spherical structure at the equator.

5. A satellite structure as claimed in claim 4 in which at least four bow-shaped members form a part of said supporting structure.

6. A satellite structure as claimed in claim 4 in which said spherical structure is formed by a plurality of sections.

7. A satellite structure as claimed in claim 4 in which said spherical structure is formed by one section from the equator and below and by two sections from the equator and above said equator, two of said sections being adapted to be secured at the equator to circularly extending channel sections and said two sections above said equator being adapted to be secured to said concentric ring of said supporting structure.

8. A satellite structure which comprises a thin shell three sectioned spherical structure and a supporting structure within said spherical structure, said supporting

structure comprising at least four bow-shaped members, a support column and a first ring all assembled concentrically about an axis through said spherical structure perpendicular to a plane through the spherical structure at the equator, said bow-shaped members being equally spaced and secured at one end to said support column and secured at the opposite end to said first ring, said support column being secured to the inner surface of said spherical structure, a second ring structure secured to said bow-shaped members on a plane through the equator of said spherical structure, radially extending support members secured to said ring structure and to said support column on said plane, other radially extending support members secured to said ring structure at one end and having a flanged end at the other end adapted to be secured to the inner surface of said spherical surface at the equator thereof, said flanged end of said other radially extending support members being adapted to support antennas on the outer surface of said spherical structure.

9. A satellite structure as claimed in claim 8 wherein one of said sections forms a part of the shell structure from the equator and below and the other two sections form the surface above said equator, the section below the equator and one of the sections above the equator being adapted to be connected to circularly extending channel sections at the equator and the two sections above the equator being adapted to be connected to said first ring of said support structure.

References Cited in the file of this patent

UNITED STATES PATENTS

731,394 Terwilleger ------------- June 16, 1903

766,643 Miniszewski ------------- Aug. 2, 1904

Sheet 1. 2 Sheets.

Sholes, Glidden & Soule.

Type Writing Mach.

No. 79,265. Patented Jun. 23, 1868

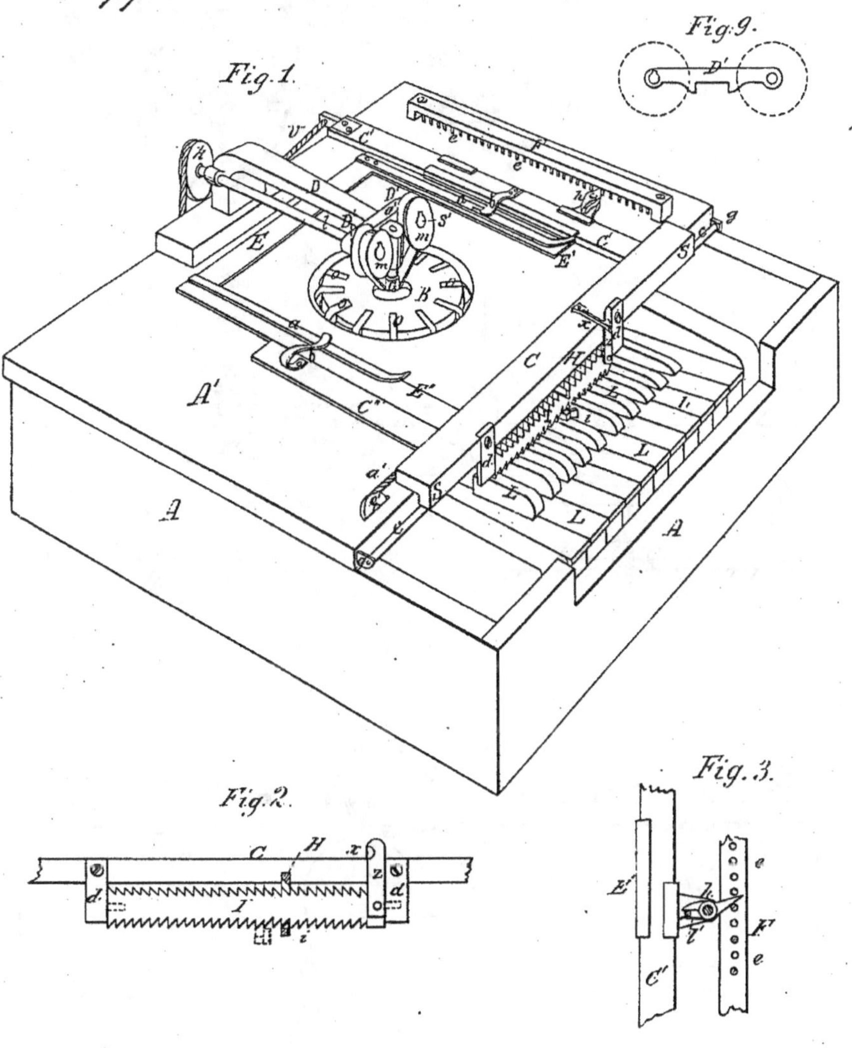

WITNESSES.

James Densmore.

L. Wailer.

INVENTORS.

C. Latham Sholes

Carlos Glidden

S. W. Soule

by Dodge & Munn

attys.

Sheet 2. 2 Sheets.

Sholes, Glidden & Soule.

Type Writing Mach.

No. 79,265. Patented Jun. 23, 1868.

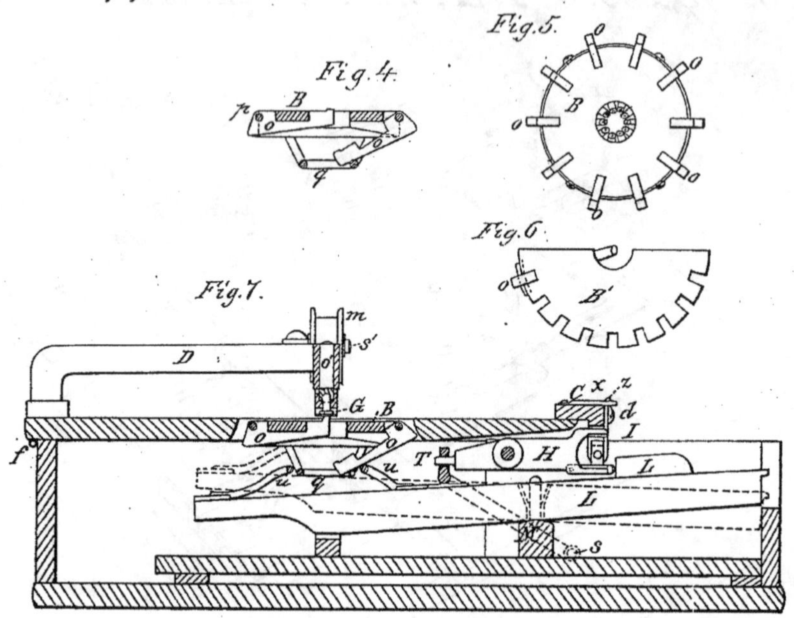

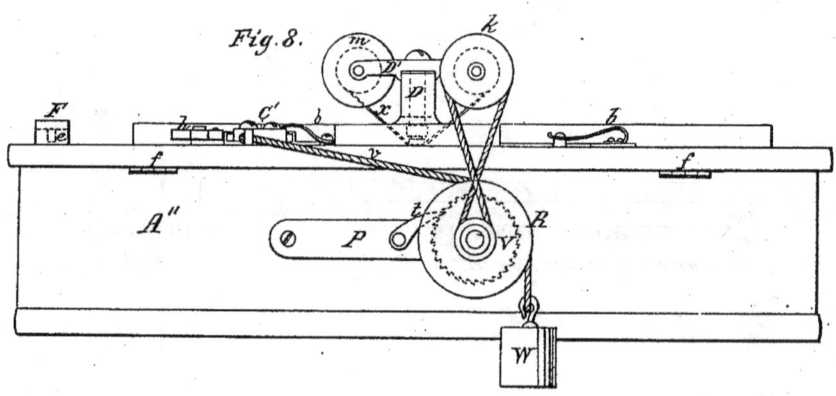

INVENTORS.

WITNESSES.

James Densmore.

L. Hailer.

C. Latham Sholes.

Carlos Glidden

J. W. Soule

by Dodge & Munn attys.

UNITED STATES PATENT OFFICE.

C. LATHAM SHOLES, CARLOS GLIDDEN, AND SAMUEL W. SOULE, OF MILWAUKEE, WISCONSIN.

IMPROVEMENT IN TYPE-WRITING MACHINES.

Specification forming part of Letters Patent No. **79,265**, dated June 23, 1868.

To all whom it may concern:

Be it known that we, C. LATHAM SHOLES, CARLOS GLIDDEN, and SAMUEL W. SOULE, of the city of Milwaukee, and county of Milwaukee, and State of Wisconsin, have invented new and useful Improvements in Type-Writing Machines; and we do hereby declare that the following is a full, clear, and exact description of the invention, which will enable those skilled in the art to make and use the same, reference being had to the accompanying drawings, forming part of this specification, in which—

Figure 1 is a perspective view of the machine; Figs. 2, 3, 4, 5, 6, and 9, views of detached parts thereof; Fig. 7, a view of a longitudinal vertical section thereof, and Fig. 8 a view of the rear elevation of the same.

This invention is of improvements to an invention of a type-writing machine, an application for a patent for which we filed October 11, 1867. Its features are a better way of working the type-bars, of holding the paper on the carriage, of moving and regulating the movement of the carriage, of holding, applying, and moving the inking-ribbon, a self-adjusting platen, and a rest or cushion for the type-bars to follow.

Make a case A, about two feet square, four to six inches deep, or of any requisite dimensions, of material and finish to one's taste, with the lid or cover A′ hinged to the back board A² by hinges *f*, as shown in Figs. 7 and 8. In the cover cut a circle, as shown in Fig. 7. Make a circular annular disk B, of any hard tough material (we use and prefer brass) four to five inches in diameter, or any required size, with a circle or hole in the center, one to one and a half or more inches in diameter, with the outer edge or periphery one-half to three-fourths of an inch or more thick, and the inner edge or circumference of the central circle two-eighths to three-eighths of an inch or more thick, with the top side planed level and smooth and the bottom side beveled, if preferred, from the outer to the inner edge with as many radial slots or grooves as types to be used cut in the bottom side from the central circle to the periphery, and deep to within an eighth of an inch of the top, less or more, with slots in the outer edge or periphery one-half to three-fourths of an inch or more deep toward the central circle to meet and fit exactly the radial grooves, and with a groove for pivot-wire cut in and circumscribing the periphery, as shown in Figs. 1 and 5.

Of any suitable material (we use and prefer steel) make as many type bars or hammers *o* as types to be used or slots in the disk. Pivot the outer ends of the type-bars in the slots in the outer edge by a wire laid in the groove in the periphery circumscribing the disk. On the upper sides of the inner ends of the type-bars cut in relief the types to be used. Make all the type-bars of the exact length of the radius of the circle of the disk, so each type on the inner ends, when thrown up into the radial grooves, will strike against the central point. (See Figs. 1, 5, and 7.) Fasten the disk thus combined with the type-bars in the circle in the cover of the case, as shown in Figs. 1 and 7, by any convenient means not interfering with the working of the type-bars. (We set it on wire posts fastened to the bottom of the case.) In the case, on a suitable frame, put a key-board similar to the key-board of a piano, having as many keys L, plus one, as types to be used, as shown in Fig. 1, each key reaching from the front in under or opposite the type-bars and pivoted to or vibrating on the fulcrum or beam M, as shown in Fig. 7. On the inner end of each key, excepting the space-key, fasten a finger *u*, made in any convenient way, (we use a stiff wire,) or bend the inner ends of the keys so the fingers will be part of the keys to reach the corresponding type-bar, so that when the front end of the key is pressed down it will strike and throw the type-bar up into its radial groove and its type-end against the central point, as shown in Fig. 7. The ends of the fingers will thus be in a circle corresponding to the circle of the disk and type-bars. Within and below the circle of the fingers and type-bars set a cushion or rest *q*, of any material for the type-bars to fall back and rest on after having been thrown up against the central point, as shown in Fig. 7. Over the central point of the inner circle of the disk suspend a solid anvil or post O′ in any firm manner, as by the arm D, fastened to the edge of the case and reaching out to the anvil, as shown in Figs. 1 and 7. In the bottom of the anvil make a spherical cavity or bowl. Make a platen G of any hard smooth substance (we use metal) with the bottom or face finished smooth and level, and with the top spherical to fit the bowl in the bottom of the anvil. Fit and attach the spherical end of the platen in and to the bowl of the anvil, thus making of the connection a universal joint, and making the platen self-adjustable. (See Figs. 1 and 7.) Hang the platen as near the plane of the surface of the cover of the case as will just admit the paper to be written on and the carbonized paper or inking-ribbon to pass easily under the platen and over the disk and case. This adjustable platen insures the types meeting the paper evenly and squarely, and giving a full and fair impression thereof when thrown against the paper.

Make an open frame C, C′, and C^2 with the bars C′ and C^2 as arms to the main bar C, as shown in Fig. 1, the arms projecting at a right angle to the main bar. Extend the arm C′ so that when the main bar C is laid flush and even with the front edge of the main part of the cover of the case it will reach entirely across to the back of the case and project so that the cord v may be attached to the open end, as shown in Fig. 1. To the front edge of the bar C attach a cleat S, to jut down against the edge of the cover of the case, as shown in Fig. 1. On the front edge of the top of the cover lay a rail, and on the under side of the bar C at each end, in the corner next to the cleat S, pivot a small flange-wheel to roll on the rail and enable the frame to move easily from right to left and back, or attach the ears g to the edge of the cover or table, as shown in Fig. 1, (two, next the keys, not being seen in the drawings, because of the cleat S,) and under the cleat S fasten two rings to serve as guides. To the ears g attach rods c, extending from the ears seen in Fig. 1 to the ears unseen next the keys and through the guides. This will enable the frame to slide easily from right to left and back, and be a guide to keep it always in place. To and within the frame C, C′, and C^2 attach another open frame E, E′, and E^2, as shown in Fig. 1, with the bar E opposite and parallel to the bar C, and the bars E′ and E^2 parallel with the bars C′ and C^2. To the bars C′ and C^2 attach springs b on a line through the center of the platen, parallel to the bar C, to press down on the bars E′ and E^2. Arrange the frame E, E′, and E^2 to slide to and from the bar C, the bars E′ and E^2 along the bars C′ and C^2, either by slots or grooves in the inner edges of the bars C′ and C^2, and tongues on the outer edges of the bars E′ and E^2 to fit and work therein, or by clasps on the bars E′ and E^2, reaching over and around the bars C′ and C^2, and fitted so as to slide readily or by any other obvious device. At the ends of the bars E′ and E^2, where they join the bar E, fasten two limber, thin, flat wire springs a, as long as the bars E′ and E^2, so that in sliding the frame E, E′, and E^2 to and from the bar C the springs a will be pressed close to the bars E′ and E^2 at every point in their length as they pass down and under the springs b, attached to the bars C′ and C^2. Rabbet the bars E, E′, and E^2 at their inner edges, so they may be as thin as practicable, and form a chase or bed for the paper to lie in. This combination of devices forms a simple and practicable paper-carriage, the larger and primary frame C, C′, and C^2, movable to and from in one direction—say east and west—carrying the smaller and secondary frame E, E′, and E^2 with it, and the latter frame movable in the transverse direction to and from north and south, while the former is stationary, thus furnishing a movement in one direction for a line of words and in the opposite direction for a series of lines.

On the edge of the cover of the case at the right of the paper-carriage attach the bar F, laid on stops or shoulders, so that the under side of the bar will be one-half inch or more above the table or cover of the case. In this bar set a series of pins e, running down into the table, so as to be fast and firm at regular and equal distances apart, the distance desired for the space (including the line) from one line of writing to another, as shown in Fig. 1. From the right-hand edge of the bar E′ of the paper-carriage project a lip out under the bar C′ or from the clasp attached to E′ and around C′, and on this lip pivot a pawl h, with a sharp incline on the side toward the front of the case running to a point, so arranged with a stop that it cannot be turned on the pivot in the direction of the back of the case, but readily turned in the opposite direction and held in position by a yielding-spring l, all as shown in Figs. 1 and 3. By moving the carriage to the right side of the case the point of the pawl h will just pass a pin e on the side from the front of the case. The incline of the pawl on the side next the pin being equal to the distance from one pin to another, and the pawl not being turnable on its pivot in the direction from the front to the back of the case, the frame E, E′, and E^2, with the paper, when on it, necessarily will be moved the proper distance from one line of writing to another.

Attach to the right-hand corner of the carriage-frame a cord a' and run it lengthwise of the bar under the bar C in a groove in the bar or table for that purpose, or it may be close to and inside of the bar, over a pulley e', fitted in and below the top surface of the table, as shown in Fig. 1, and fasten to the other end of it a weight under the case, but unseen in the drawings. To the other end of the bar C′ fasten a cord v, and run it down over a large pulley R on the back side of the case A^2, and to the other end of the cord hang the weight W, as shown in Fig. 8. These cords v and a', attached one to each corner of the carriage on one side, running over the pulleys R and e' and fastened to the weight W and the weight W′ (unseen in the drawings) are the force and means of moving the carriage and paper while writing.

Under the table or cover of the case, behind the beam or fulcrum M, between the fulcrum and the disk a suitable distance, on and across all the keys, lay a bar T, with the ends bent at a right angle and extended and pivoted to the frame below and in front of the fulcrum, as shown at *s*, Fig. 7, so that when the front ends of the keys are pressed down the rear ends will strike against and raise the bar an extent in proportion to the distance from the fulcrum. Connect a lever H to the middle of the bar T, midway of the key-board, extending directly over and parallel with and between the middle keys, and pivoted in the middle on a suitable support, as shown in Fig. 7. Bifurcate the front end of this lever and make the right-side faces of the forks perpendicular and the left-side faces inclined, the upper one to the left upward and the under one to the left downward, with the under edge of the upper fork and the upper edge of the under fork sharp like saw-teeth, as shown in Figs. 1 and 2, particularly in Fig. 2. Fasten to the bar C of the carriage-frame two holders or arms *d*, extending down through the cleat S, or fasten them directly to the cleat, and pivot in their lower ends the ends of the ratchet-bar I, as shown in Figs. 1 and 2. Serrate the bar I on both sides with notches like saw-teeth, as shown in Figs. 1 and 2. Make these notches, teeth, or cogs regular and equidistant apart, the exact distance required for a letter in writing or printing on the paper. Make the left side of the faces of the teeth or cogs perpendicular, both above and below, and the right-side faces inclined exactly alike, but the reverse of the teeth or cogs of the inner edges of the forks of the lever H, so that of the lever H, with its forks embracing the ratchet-bar I, in moving up and down first one and then the other forks will strike and fit into the notches of the bar I, as shown in Figs. 1 and 2. Make the forks of the lever H so far asunder as just to allow the ratchet I in its widest way to pass between. At the right side, considered from the front of the under fork of the lever H, attach a thin yielding spring *i*, as shown in Figs. 1 and 2. Make the upper and sharp edge of the under fork stand a hair-breadth or slight distance to the right of the under and sharp edge of the upper fork, and then, as the weights W and W′, attached to the cords *v* and *a′*, over the pulleys R and *e′*, as shown in Figs. 1 and 8, (excepting that the weight W′ is unseen in the drawings,) are constantly pulling at the carriage to draw it from the right to the left of the table or case. When the upper fork is thrown up out of an upper notch in the ratchet, the carriage will move to the left till the left-side perpendicular face of the tooth or cog next to the right and below meets and strikes against the right-side perpendicular face of the under fork, and the carriage is thereby stopped. Fix the thin yielding spring *i* so that when the upper fork is pressed down into an upper notch of the ratchet I the spring will fly back against and up into the next tooth and notch to the right below. The office of this yielding spring is to assist the under fork to catch every under tooth and not let one slip by. As the ratchet is moved along by the carriage till the face of the tooth to the right below strikes and stops against the spring and under fork, the left perpendicular face of the tooth directly above is moved to and directly in line up and down into a hair-breadth with the perpendicular face of the fork above, so that when the front end or forks of the lever are moved or pressed down and the under fork lets go its hold of an under tooth the upper fork falls into the notch and against the tooth directly above and prevents the ratchet from moving; but when the forks are thrown up and the upper fork lets go its hold of the tooth above the ratchet moves to the left the space of one notch till the next tooth to the right below, with the yielding spring in the notch at its perpendicular face, strikes against the perpendicular face of the under fork. In this way the ratchet and carriage are held firmly still, while the front or bifurcated end of the lever H is thrown and held down, but moves to the left one notch, a regular, exact, and equal distance every time the bifurcated end of the lever is thrown up, and as striking down the front end of each key, as at L in Fig. 1, raises the bar T laid across the key at the rear of the fulcrum M, and raises the rear end of the lever H, attached to the bar T, it therefore necessarily throws down the bifurcated or front end of the lever, and as the key rises to its place of rest again all these movements are reversed and necessarily throw up again the front end of the lever. Thus the working of the keys L, in combination with the weights W and W′, (the latter unseen,) the cords *v* and *a′* the pulleys R and *e′*, the bar T, the lever H, the ratchet I, and the carriage inevitably moves the paper a regular, uniform, and exact distance—any distance desired for a type or letter every time a key is struck—and the paper is moved while the type-bar is falling to the cushion, and stopped and held firmly stationary while the type is struck against it and the platen.

On the end of the ratchet I, to the right, attach the lever *z*, to turn it down flatwise when desired, as shown in Figs. 1 and 2. To the bar C of the carriage attach a yielding spring *x* to hold the lever and ratchet in perpendicular position, while the carriage is moving from right to left, as shown in Figs. 1 and 2. Turn the lever *z* forward and down, and therewith the ratchet, to a horizontal position, and the ratchet and carriage can be moved from left to right, the ratchet through and between the embracing-forks of the lever H readily and without obstruction. This can be done by the hand or by any obvious device by a foot-treadle, thus completing the means for the right and left movement of the carriage and paper.

On the front end of the arm D, just behind the anvil, put a cross-beam D′, as shown in Figs. 1, 8, and 9. In the end of the cross-beam, at the right, put a gudgeon *s*′, and through the end at the left run a shaft *l*, and through a box at the left side of the back end of the arm D, as shown in Figs. 1 and 8, make two ribbon-spools *m*, of any adequate size, with holes in their centers, to slip on and revolve on the gudgeon *s*′, as shown in Figs. 1, 8, and 9. At the circumference of the holes in the spools *m* in the inner edge of each spool, through from side to side, cut a slot to fit on a key or cog or spur on the front end of the shaft *l* forward of the cross-beam D′, so that whichever spool is put on the shaft will be fast thereto and cannot revolve thereon. On the hind end of the shaft *l* fasten a pulley *k*, as shown in Figs. 1 and 8. Make the pulleys R and *k*, as shown in Fig. 8, cone pulleys—that is, make each R and *k* a series of pulleys, decreasing in size in regular conical order. Pivot the pulley R on a bar P, and pivot the bar P to the back side of the case A^2, so that the pulley may rise and fall freely, as shown in Fig. 8. Attach a ratchet-wheel V, with a pawl *t*, pivoted to the bar P, as shown in Fig. 8, to follow and fall into the notches of the ratchet-wheel to prevent the wheel turning toward the bar F, as seen in Fig. 8, or from left to right, considered from the front. Connect the pulleys R and *k* with a cord or band *v*′, as shown in Fig. 8. The pulley R, pivoted to the loose-pivoted bar P, with the weight W pulling down on the pulley, will always keep the band *v*′ tight, so that it will not slip in working. Upon the spool *m* on the gudgeon *s*′ wind the inking-ribbon, and run one end under the platen G and attach it to the spool *m* on the shaft *l*, as shown in Fig. 1. Then, as striking each key L permits the weight W, by means of the cord *v* over the pulley R, to pull or move the carriage the space of one notch of the ratchet-bar I, it will necessarily roll the pulley R a corresponding distance, and, as the pulleys R and *k* are connected by the band *v*′, and the pulley *k* and the left spool *m*, considered from the front, both being fast to the shaft *l*, rolling the pulley R necessarily will roll the spools *m* and draw the ribbon from the loose spool on the gudgeon *s*′ under the platen G and onto the spool attached to the shaft *l*, and thus give a fresh place of the inking-ribbon every time for every type to strike against, and by means of the series of conical pulleys at R and *k* the feed of the inking-ribbon can be regulated as may be desired.

Thus made, the type-writer is the simplest, most perfectly adapted to its work—the writing of ordinary communications with types instead of a pen—and in every way the best of all machines yet designed for the purpose, particularly as to the cost of making the machine and the neatness and labor-saving quality of its work.

Fig. 6 of the drawings represents a crescent or the segment comprising half a disk. By making the circumference large enough to receive the requisite number of radial grooves the crescent may be substituted for the disk, or, in other words, the segment comprising one-half for the whole disk.

What we claim as new and useful in our invention, and desire to secure by patent, is—

1. The key-levers L, vibrating on the fulcrum M, with the inner ends or fingers *u* reaching under the type-bars, so that the keys will act directly on the types, substantially as and for the purpose described.

2. The spacer or ratchet I, combined with the bifurcated lever H, connected with the bar T, pivoted at *s* and resting on and across the arms of the keys L behind the fulcrum M, so that striking the faces of the keys will work the teeth of the forks of the lever up and down and into the notches of the spaces and give a certain uniform and regular space movement to the paper-carriage in line of the types, when made substantially as described.

3. The pins *e*, fastened to the table A′, combined with the pawl *h* and the spring *l*′ to give the paper-carriage a certain and regular cross-line movement at a right angle to the space movement from line to line, when made substantially as described.

4. The clasps or springs *b*, attached to the bars C and C′ on a line through the middle of the platen G, combined with the springs *a*, attached to the bar E to hold the paper to the carriage and press it down smooth and tight in passing under the platen, when made substantially as described.

5. The spools *m*, combined with the gudgeon *s*′, the shaft *l*, the pulleys *k* and R, the band *v*′, the cord *v*, the weight W, the ratchet-wheel V, the pawl *t*, and the bar P, pivoted to the back of the case A^2 to feed a fresh part of the inking-ribbon under the platen to each type successively, when made substantially as described.

This specification signed this 1st day of May, A. D. 1868.

C. LATHAM SHOLES.
CARLOS GLIDDEN.
SAMUEL W. SOULE.

Witnesses:
G. E. WEISS,
F. J. CROSBY.

A. S. HALLIDIE.

ENDLESS WIRE ROPEWAY.

No. 110,971. Patented Jan. 17, 1871.

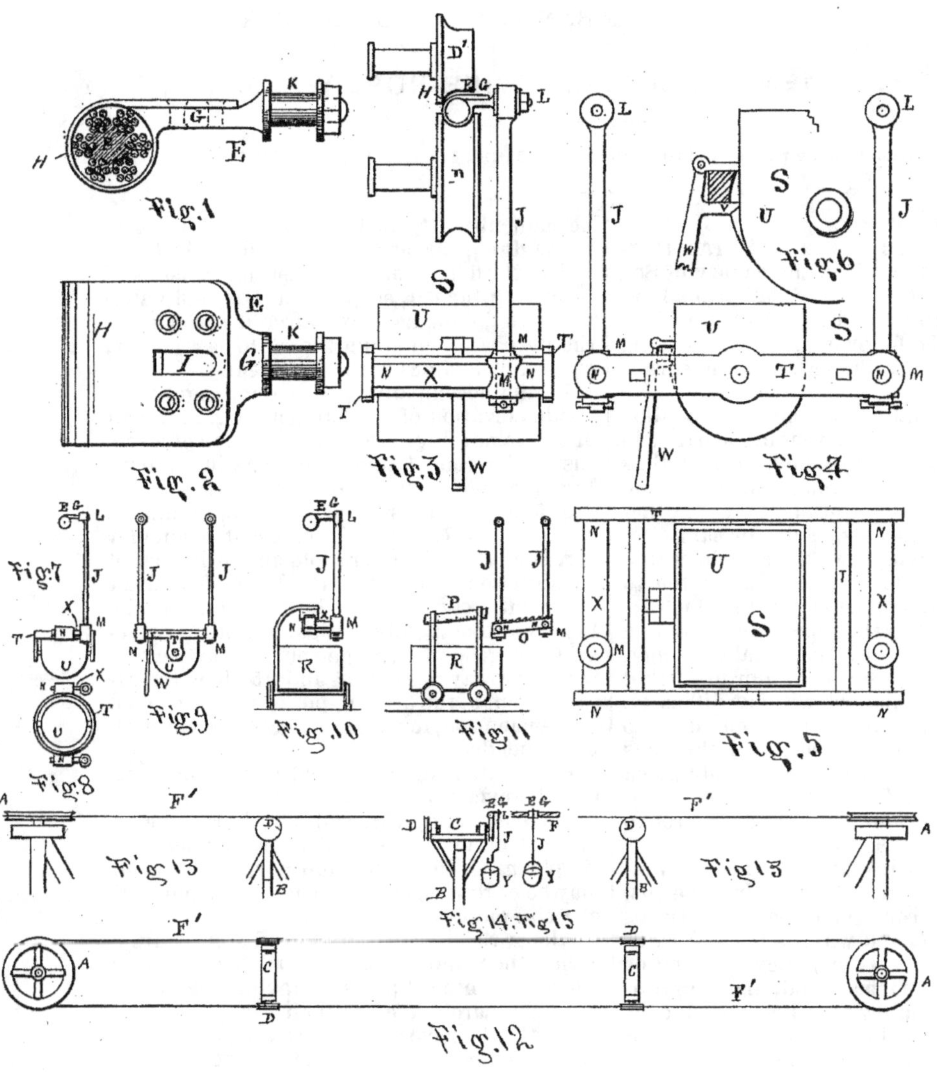

Witnesses

T. J. Thibault

David R. Smith

Inventor

Andrew Smith Hallidie

THE NATIONAL LITHOGRAPHING COMPANY,
WASHINGTON, D. C.

UNITED STATES PATENT OFFICE.

ANDREW SMITH HALLIDIE, OF SAN FRANCISCO, CALIFORNIA.

IMPROVEMENT IN ENDLESS-WIRE-ROPE WAYS.

Specification forming part of Letters Patent No. **110,971**, dated January 17, 1871.

To all whom it may concern:

Be it known that I, ANDREW SMITH HALLIDIE, of San Francisco, in the county of San Francisco, and in the State of California, have invented an Improved Endless-Wire-Rope Way, of which the following is a specification, reference being had to the accompanying drawing.

My invention relates to an improved method of obtaining power from weights carried in buckets, sacks, or cars attached to an endless wire-rope moving over or around sheaves or pulleys, and to the transportation of ores or other materials in said buckets, sacks, or cars.

In endless wire-rope ways or "wire tramways," as they are sometimes called, hitherto constructed, it has been necessary to detach the cars or buckets from the rope before passing the end pulleys, because no suitable hanger or attachment has been known or used that could pass the end pulleys without difficulty.

Another difficulty that has been hitherto considered as perhaps insurmountable has arisen from the tendency of the well-known wave motion that is frequently communicated to the rope to throw said rope out of the groove in the bearing-pulleys.

The object of my invention is, first, to provide a hanger or device for carrying the suspension-rods that can remain permanently attached to the rope, and will readily and without difficulty pass all the pulleys around or over which the rope may lead, and admit of the placing of a pulley or other guard over the bearing-sheaves, to prevent the escape of the rope from the groove of said bearing-sheaves.

The second object of my invention is to provide an improved method or system of constructing and operating the cars, buckets, and other apparatus to be used in the connection with my improved hanger; and

My invention consists, first, of permanently attaching the suspension-rods to the rope by means of a hanger consisting of a horizontal arm permanently secured to the upper and outer quarter of a rope, in a manner hereinafter more fully described.

The second part of my invention consists of the employment, in connection with a hanger permanently attached to the rope, of an automatic dump-car, as hereinafter described.

The third part of my invention consists of providing the car, suspension-rods, and hangers with joints, as hereinafter described, for the purpose of enabling the car to pass angles, curves, or inclines without difficulty.

In the accompanying drawing, Figure 1 is a side elevation of the hanger; Fig. 2, a plan of the same; Fig. 3, an end elevation of a car, suspension-rods, hanger, and bearing-pulley; Fig. 4, a side elevation of same car and rods; Fig. 5, a plan of said car; Fig. 6, details of same. Figs. 7, 8, and 9 represent a circular car, similar in its operation and principle of construction to the above. Figs. 10 and 11 represent a car that may be attached and detached automatically. Figs. 12 and 13 represent the general arrangement of the endless wire-rope way. Figs. 14 and 15 show the arrangement of the bearing-pulleys and a hooked suspension-rod, to which buckets or sacks may be suspended.

Each part is distinguished by the same letter wherever it appears in the drawing.

The end pulleys, A, will usually revolve on vertical or nearly vertical axes, and may be simply grooved in the ordinary manner, or may be of that class known as "gripe-pulleys" or "clip-pulleys."

Along the line, at suitable distances from each other and from the end pulleys, are placed well-braced posts B, having transverse arms C securely fixed to their tops.

To the transverse arms C the sheaves or bearing-pulleys D are attached, overhanging the arms C, as shown in Figs. 12, 13, and 14, and free to revolve on their axes.

It will be seen by reference to Fig. 3 that the depth of the groove in the sheaves D should be about half the diameter of the rope, and that the radius of said groove should be equal to half the diameter of the rope, increased by the thickness of that part of the hanger E surrounding the lower part of the rope.

The axis of each sheave or pulley should always be perpendicular to the plane in which the rope approaches and recedes from the sheave or pulley. When the line is straight all the bearing-sheaves on either side of the way will be in the same vertical plane; but when an angle occurs in the line the sheaves must be inclined to correspond to the rule above given. A guard of some kind should be placed over each bearing-pulley to prevent the escape of the rope from the groove of said pulley.

In Fig. 3 a pulley, D′, of a suitable form for this purpose, is shown.

F is the rope. The lines F′ represent the rope in Figs. 12 and 13.

It is evident that any brakes or any apparatus or gearing required for communicating or transmitting power may be attached to either of the end pulleys, A, or to their axles.

The hangers E are permanently attached to the rope at suitable distances.

The hangers E consist, essentially, of an arm, G, and its fastenings, the arm G being always retained in a horizontal position by the weight of the car or suspension-rods, or both, and secured to the rope in such a manner that its upper surface shall be on a tangent, or nearly on a tangent, to the upper part of the circumference of the rope, and its lower surface horizontal, or nearly so, while its vertical thickness is only sufficient to secure the required strength and stiffness, and said thickness should never exceed one-half the diameter of the rope. The fastenings of the hanger by which the arm is secured to the rope should increase the diameter of the rope as little as possible. A hanger constructed and secured to the rope in this manner will pass readily over the bearing-pulleys and around the end pulleys.

The arm G, Figs. 1, 2, and 3, is formed in one piece with the band H. To place this hanger the band may be heated, or may be made of steel and so thin as not to require heating, and bent around the rope and secured by rivets, as shown.

The rivet-holes may be made in the shop with the aid of a gage or mandrel, so that when made to coincide by means of a key driven through the keyway I, the rope will be so tightly compressed as to prevent the hanger from slipping along the rope.

The arm G may be secured to the rope by thin steel clasps passing through between the strands of the rope and turned back and clinched. The suspension-rods J swing freely on the journal K of the hanger.

When no cars are used, and buckets or sacks Y, Figs. 14 and 15, are hooked on and off the lower ends of the suspension-rods, said lower ends should have sufficient weight to retain the arm G in a horizontal position even when the sacks are removed.

When cars are employed they should be suspended by rods and hangers in pairs, and jointed in a manner substantially as shown at L, M, and N, Figs. 3, 4, 5, 7, 8, 9, 10, and 11, in order that the cars may pass readily around the end pulleys or around the sheaves or bearing-pulleys at any angles in the line.

Figs. 10 and 11 represent an arrangement by which removable cars may be employed.

O is an inclined bar or beam, attached to the suspension-rods, and having suitable teeth or projections on its upper surface. P is a beam, inclined to the same angle and rigidly attached to the car, having corresponding teeth or recesses on its under side. The rope carrying the hangers, rods, and the bar O along in the direction of the car, the bar O will pass under and engage the bar P, carrying the car R with it.

It is evident that floors or tracks can be so arranged as to lift the car off at any desired point, and place it in a position to be engaged again at any other desired point.

The car S has a frame, T, in which is pivoted the dump-bucket U. The axle of the bucket U is so situated as to be below and at one side of the center of gravity when the bucket is full, and above the center of gravity when the bucket is empty, so that the bucket will turn on its axle and empty itself when released from the catch V and right itself when empty.

The catch V is attached to the lever W in such a manner that when the car is in motion, if the lever W comes in contact with any suitable stop or projection placed for that purpose, the bucket will be released and the weight of the lower end of the lever will cause the catch to re-engage the bucket when it rights itself, and hold it in its proper position until its return to the place at which it is desirable to deposit the load.

The suspension-rods of the cars, besides swinging freely on the journal K, forming the joint L, are free to turn in the sockets M, forming the joints M.

The horizontal arm X, to which the socket M is attached, is free to oscillate in the frame T, forming the joint N.

As the center of gravity of whatever depends from the arm G will always seek a position in a vertical plane passing through the center of the rope in the hanger, it is necessary to so proportion the parts of the car that the axle of the buckets will always be found in a horizontal position when said center of gravity has attained its said position in said plane.

The construction of the car represented in Figs. 7, 8, 9, and 10 is exactly the same in principle as the car represented in Figs. 3, 4, 5, and 6, only the forms of the buckets and frames vary, those of the former being circular and of the latter rectangular.

Having thus described my invention, what I claim as new, and desire to secure by Letters Patent, is—

1. The method herein described of attaching the suspension-rods to the rope, by means of a hanger proceeding or projecting horizon-

tally from the upper and outer quarter of the rope, in a manner substantially as hereinbefore described, and for the purpose hereinbefore set forth.

2. The hanger E, substantially as and for the purposes set forth.

3. The hanger E, in combination with the rods J and frame T, having the joints L M N, or their equivalents, substantially as described, and for the purposes set forth.

4. The dump-car S, constructed and operated substantially as described, in combination with the rods J and hanger E, substantially as and for the purposes set forth.

In testimony whereof I have hereunto set my hand this 22d day of November, A. D. 1870.

ANDREW SMITH HALLIDIE.

Witnesses:
F. J. THIBAULT,
DAVID R. SMITH.

L. YALE, Jr.

POST-OFFICE DRAWER LOCK.

No. 31,278. Patented Jan. 29, 1861.

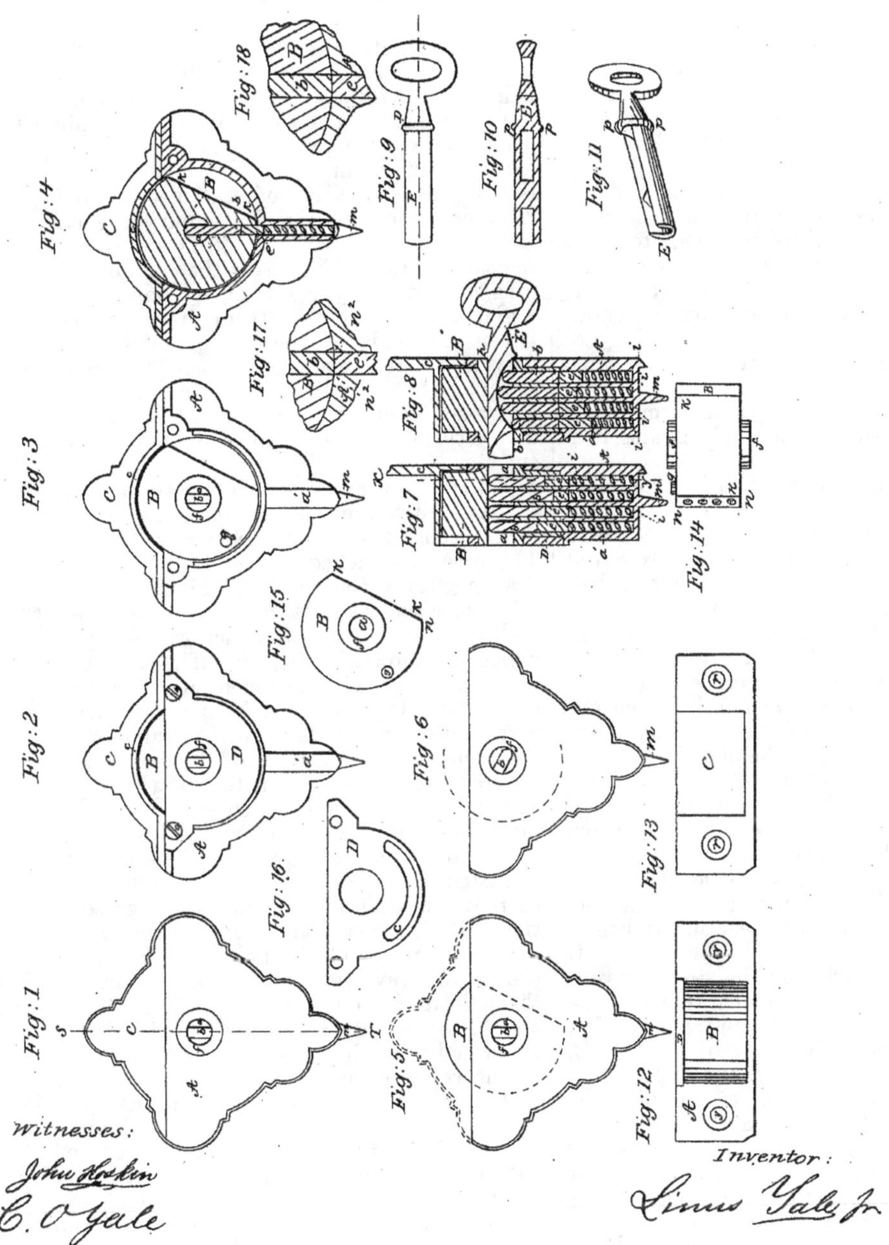

Witnesses:

John Hoskin

C. O. Yale

Inventor:

Linus Yale Jr

UNITED STATES PATENT OFFICE.

LINUS YALE, JR., OF PHILADELPHIA, PENNSYLVANIA.

LOCK.

Specification of Letters Patent No. 31,278, dated January 29, 1861.

To all whom it may concern:

Be it known that I, LINUS YALE, Jr., of the city and county of Philadelphia, State of Pennsylvania, have invented a new and useful Improvement in Locks for Drawers, Closets, Cupboards, &c.; and I do hereby declare the following to be a true and exact description of its construction.

To enable others skilled in the art to make and use my invention I will proceed to describe its construction and operation, reference being had to the drawings hereunto annexed, in which and in the following description the same letters refer to the same parts of the lock.

A is the case, B the revolving tumbler or bolt, C is the keeper, D is the back and E is the key.

Figure 1 is a front view of the complete lock, locked, with the keeper in place. Fig. 2 is a rear view of the same. Fig. 3 is a rear view of the lock with the back removed showing the bolt B in place in the locked position. Fig. 4 is a section of the lock through the red line X Y Fig. 7, showing as also in Figs. 7 and 8, the key hole *a*, the pistons *b b b b*, their drivers *e e e e*, and their actuating springs *h h h h*, which are placed in cylindrical holes *i i i i*, drilled for that purpose in the pin-chamber in the case *a'*, and also in the bolt B. Fig. 5 is a front view of the lock with the keeper removed, but indicated by the dotted lines, to show the bolt in the locked position. Fig. 6 shows the lock as it appears with the keeper removed and the bolt in the unlocked position. Figs. 7 and 8 are vertical sections of Fig. 1 through the line S, T. In Fig. 8 the key E is shown in its place, and the stops, or pins *b b b b* and *e e e e*, set ready to allow the bolt to rotate. Figs. 9, 10 and 11 are plan section and perspective view of the key E. Fig. 12 is an edge view of the lock, showing the case, bolt and back, and also the screw holes *y*, *y*, by which it is attached. Fig. 13 is an edge view of the keeper C showing its screw holes whereby it is attached. Figs. 14 and 15 are two views of the bolt showing the key hole *a*, the shoulders *f f* on which it rotates in the case and back, the stop pin *g*, the piston holes *i i i i* and the slightly flattened portion of its periphery at *n*, and the slabbed off or cut away portion of its periphery at *k k*, which when the bolt is in the unlocked position, disengages it from the keeper, which till then has fastened it. Fig. 16 is the back of the lock D, exhibiting that side which when it is in place, is toward the bolt, thereby showing the recess *c* made to receive the stop pin *g* on the bolt whereby it is stopped in the locked or in the unlocked position. Fig. 17 is an enlarged view of a "piston" and "driver" showing the flattening of their abutting ends at n^2. Fig. 18 is a view on the same scale as Fig. 17 of the ends of the pins as ordinarily constructed showing at a glance the greater width of the joints between the ends of the pistons and their drivers as ordinarily constructed, thereby enhancing the facility of picking the lock, and also lessening the power of making a diversity of locks, each opened by a different key.

The case A and keeper C, are so constructed that their front faces as shown Fig. 1, constitute the escutcheon of the lock. The case is attached to the lock rail of a drawer &c. by the spur *m* which is thrust into the wood when it is fitted to its place, and also by screws at the screw holes *y y* which are shown in Fig. 12. The keeper C Fig. 13 is likewise fastened at the screw holes *r r*. The case and keeper are chambered out to receive the revolving tumbler B, which acts also as the bolt of the lock. The general form of this bolt is that of a frustum of a cylinder, or from which cylinder a segment has been removed or slabbed off as shown at *k k* Figs. 14 and 15. The back D is fastened to the case by the screws *o*, *o*, thereby retaining the bolt in its proper position. Both the case and back are drilled to fit the shoulders of the bolt *f*, *f*, on which the latter revolves. The bolt is drilled through to receive the key E Figs. 9, 10 and 11, by which it is actuated. It is also furnished with the stop pin *g*, which works in the recess *c* made for it in the back D, and which recess is so fitted as to stop the bolt in either its locked or unlocked position. In the locked position a segment of the bolt projects above the edge of the case as shown Figs. 2, 3, 4, and 5, entering the keeper which opposes the opening of the drawer or door to which it is attached, until it is unlocked: the latter is accomplished by revolving the bolt until its slabbed off surface *k k* is brought into the same plane with the upper portion of the case as at Fig. 6, and thus no longer projecting above the case into the keeper, the latter no longer opposes any obstruction to the opening of the drawer door &c., in other words it is unlocked.

To prevent the bolt from being rotated, or unlocked by any key except its own, both the bolt and pin chamber *a'* of the case are drilled with one or more holes in the same right line, as shown at *i i i i* in the plan of the bolt Fig. 14 and also in Figs. 4, 7 and 8, in which their relative positions are more clearly shown. These holes are each furnished with a pair of "stops" or "bolts", those which come in contact with the key being designated as "pistons" *b b b b*—and those which lie behind them lower in the pin chamber as "drivers" *e e e e;* behind each of these drivers is a helical spring *h h h h*, which forces the pistons and drivers into and across the keyhole. The pistons and drivers are of such a length and so proportioned, that before the key is inserted, the drivers have their lower ends in the pin chamber of the case, while their upper ends rest in the bolt, and thus prevent it from revolving as shown in section Fig. 7. The length of the pistons *b b b b* are such that when the key is thrust into the key hole as far as the ring *p* around it will admit, see Fig. 8, the upper ends of the pistons being forced against the planes cut in the key, by the force of the springs *h h h h*, then the lower ends should at this time terminate at the intersection of the bolt and the case; in other words the pistons are contained in the bolt, while the drivers are carried back into the pin-chamber *a'* in the case, as shown Fig. 8. They now oppose no resistance to the revolution of the bolt, which may be revolved by turning the key to its unlocked position. When the bolt is once more revolved to its locked position, on withdrawing the key, the springs force up the pistons and drivers, the latter entering the bolt and thus fastening the case and it together.

The key E as shown Figs. 9, 10 and 11, is a cylindrical barrel, which has a groove or channel before spoken of cut longitudinally in it, this groove is somewhat wider than the diameter of the pistons, and may be cut on different planes, varied in depth to make variations in different keys. This groove or channel is shown in the perspective drawing Fig. 11 and also in section Fig. 8.

An inspection of Figs 2 and 3 will show that the bolt chamber of the case may be made very thin and still be sufficiently strong, allowing great economy in material.

What I claim as my invention and desire to secure by Letters Patent is—

1. Using a "revolving tumbler" for a "bolt" when the same is used in the described manner, or in an equivalet manner, with "jointed pins" which are the "stops" or "guards".

2. Reversing the main plate of a "pin lock" to answer the purpose of an escutcheon, to protect the drawer &c. from injury by the key.

3. The flat plane *n n* on the revolving tumbler for the purpose described.

4. The part *m* for the purpose described.

5. The use of a metal keeper C, when the same has a front plate to complete the design of the front of the lock.

6. The thin curb around the tumbler instead of the ordinary case of this class of lock thereby saving metal and cheapening its construction.

7. Placing the projection *a'* or spring chamber opposite the bolt hole, so that the drilling thereof may be done from the inside without making an outward opening.

8. The use of, in this class of lock, flat ended and close jointed pins with the least possible waste or rounding of corners.

LINUS YALE, Jr.

Witnesses:

John Hoskin,

C. O. Yale.

March 8, 1932. I. SIKORSKY 1,848,389

AIRCRAFT, ESPECIALLY AIRCRAFT OF THE DIRECT LIFT AMPHIBIAN TYPE AND MEANS OF CONSTRUCTING AND OPERATING THE SAME

Original Filed Feb. 14, 1929 8 Sheets-Sheet 1

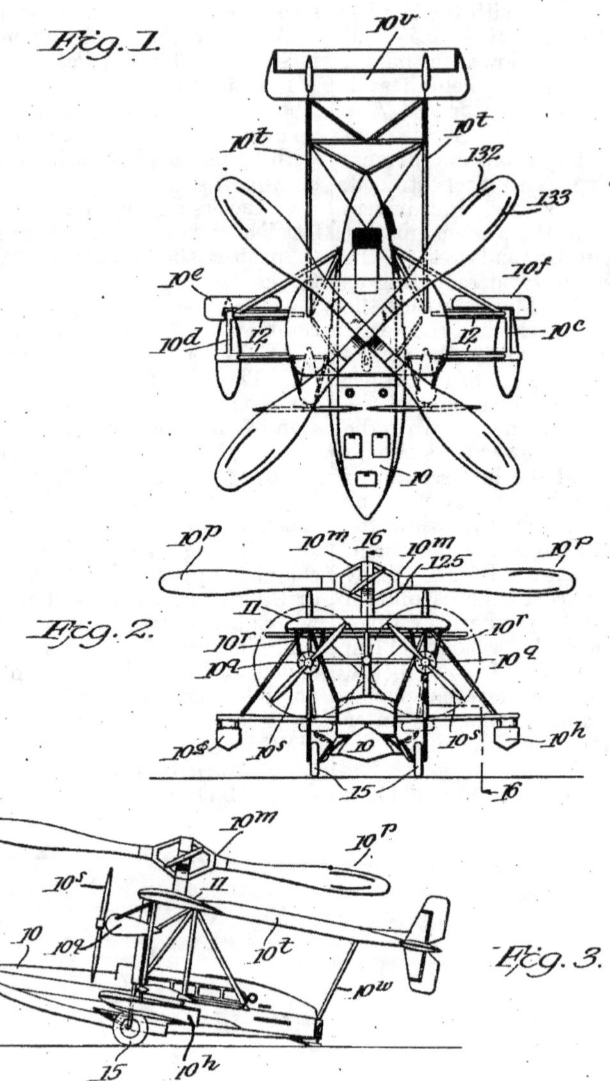

INVENTOR

BY

ATTORNEY

March 8, 1932. I. SIKORSKY 1,848,389

AIRCRAFT, ESPECIALLY AIRCRAFT OF THE DIRECT LIFT AMPHIBIAN TYPE AND MEANS OF CONSTRUCTING AND OPERATING THE SAME

Original Filed Feb. 14, 1929 8 Sheets-Sheet 2

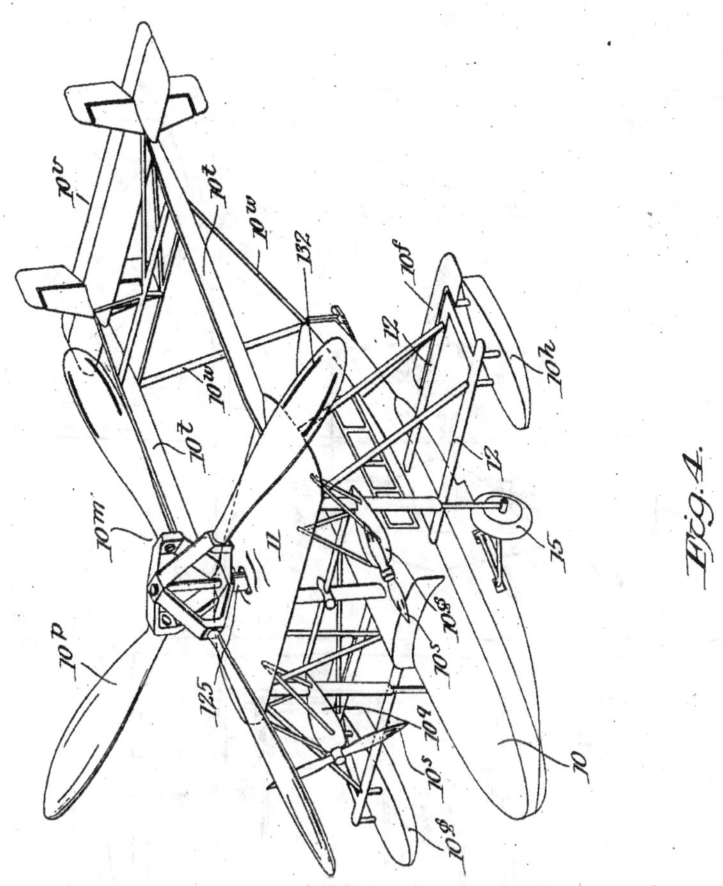

Fig. 4.

INVENTOR.

BY

ATTORNEYS.

March 8, 1932. I. SIKORSKY 1,848,389

AIRCRAFT, ESPECIALLY AIRCRAFT OF THE DIRECT LIFT AMPHIBIAN TYPE AND MEANS OF CONSTRUCTING AND OPERATING THE SAME

Original Filed Feb. 14, 1929 8 Sheets-Sheet 3

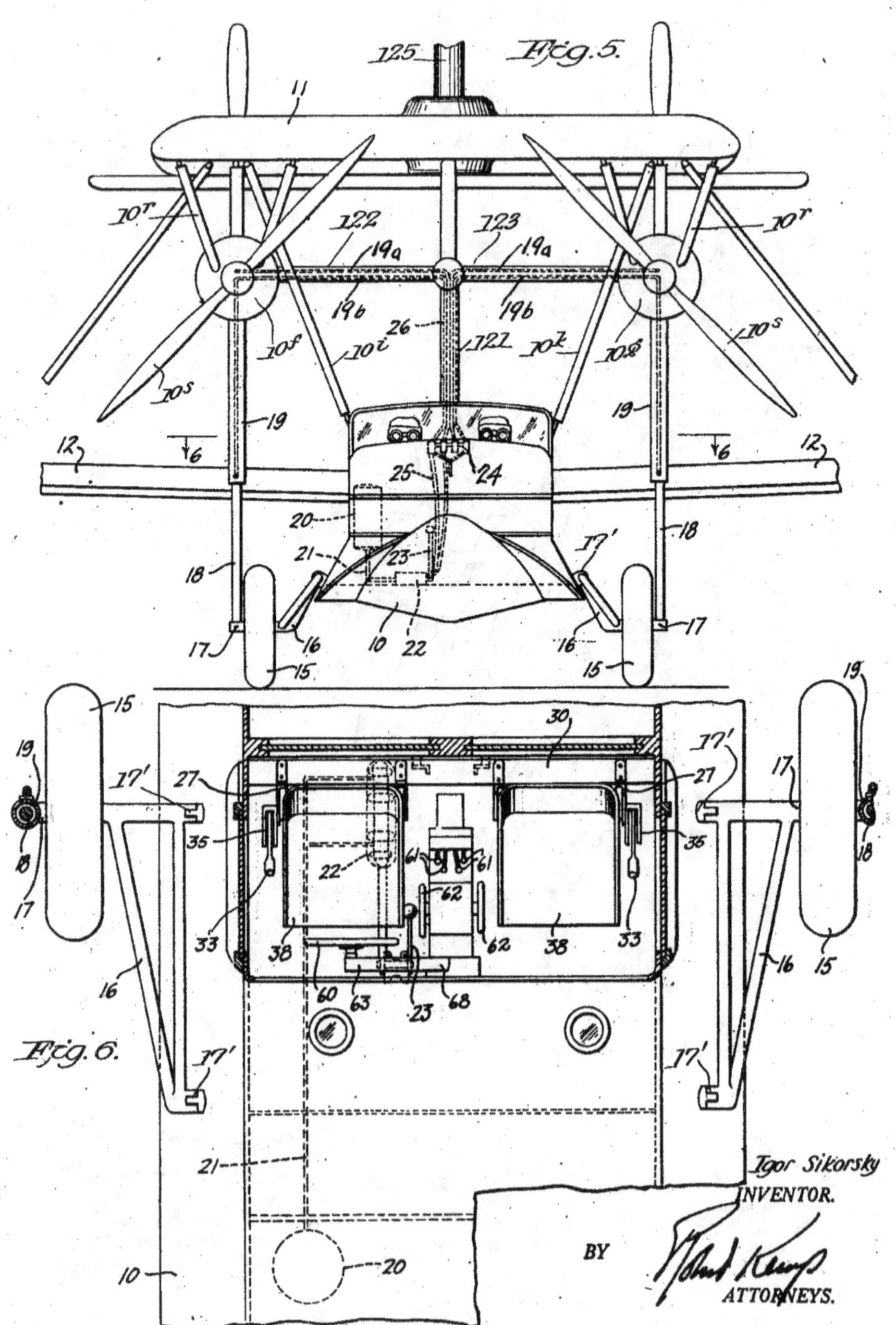

Igor Sikorsky
INVENTOR.

BY

ATTORNEYS.

March 8, 1932. I. SIKORSKY 1,848,389

AIRCRAFT, ESPECIALLY AIRCRAFT OF THE DIRECT LIFT AMPHIBIAN TYPE AND MEANS OF CONSTRUCTING AND OPERATING THE SAME

Original Filed Feb. 14, 1929 8 Sheets-Sheet 4

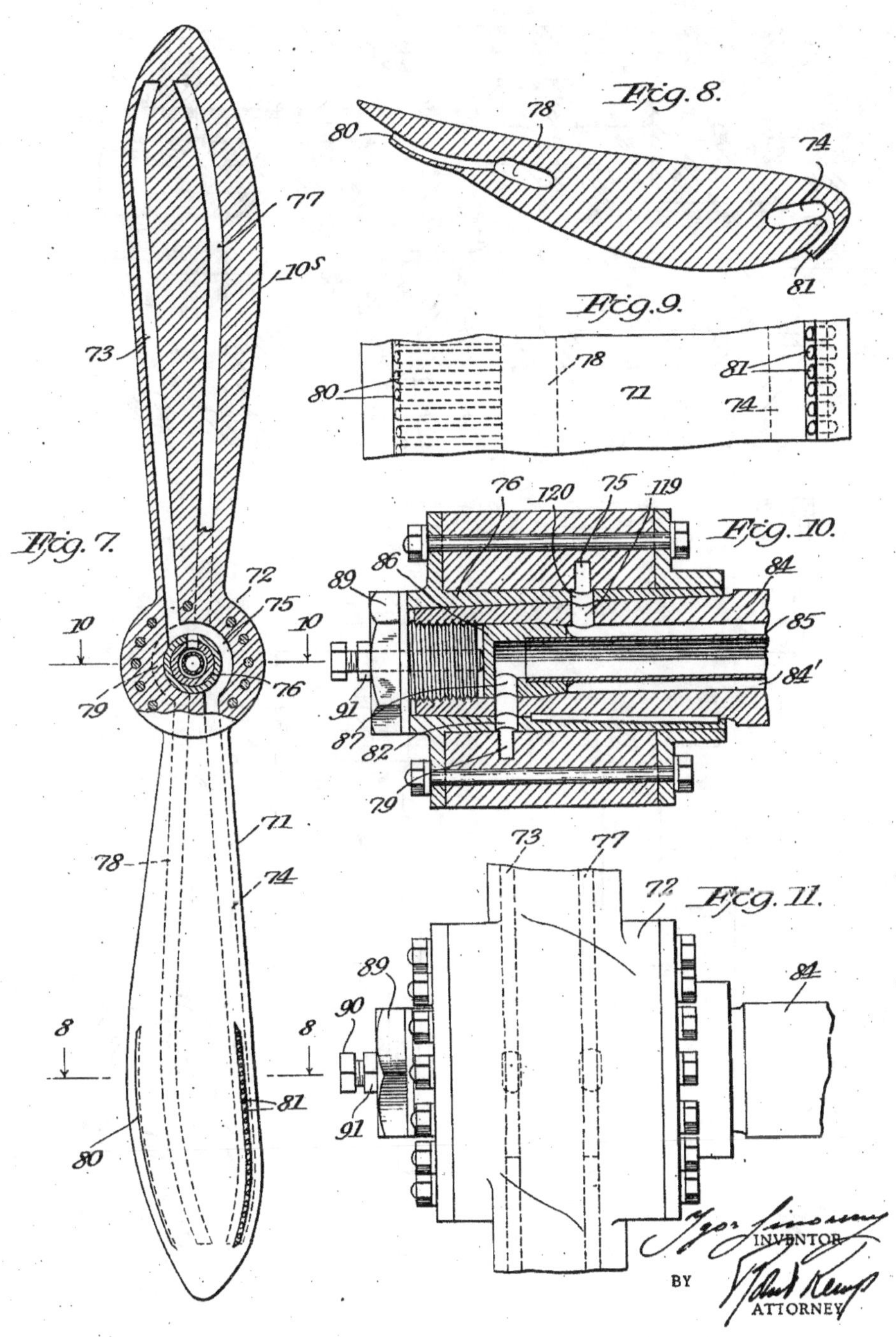

March 8, 1932. I. SIKORSKY 1,848,389

AIRCRAFT, ESPECIALLY AIRCRAFT OF THE DIRECT LIFT AMPHIBIAN TYPE AND MEANS OF CONSTRUCTING AND OPERATING THE SAME

Original Filed Feb. 14, 1929 8 Sheets-Sheet 5

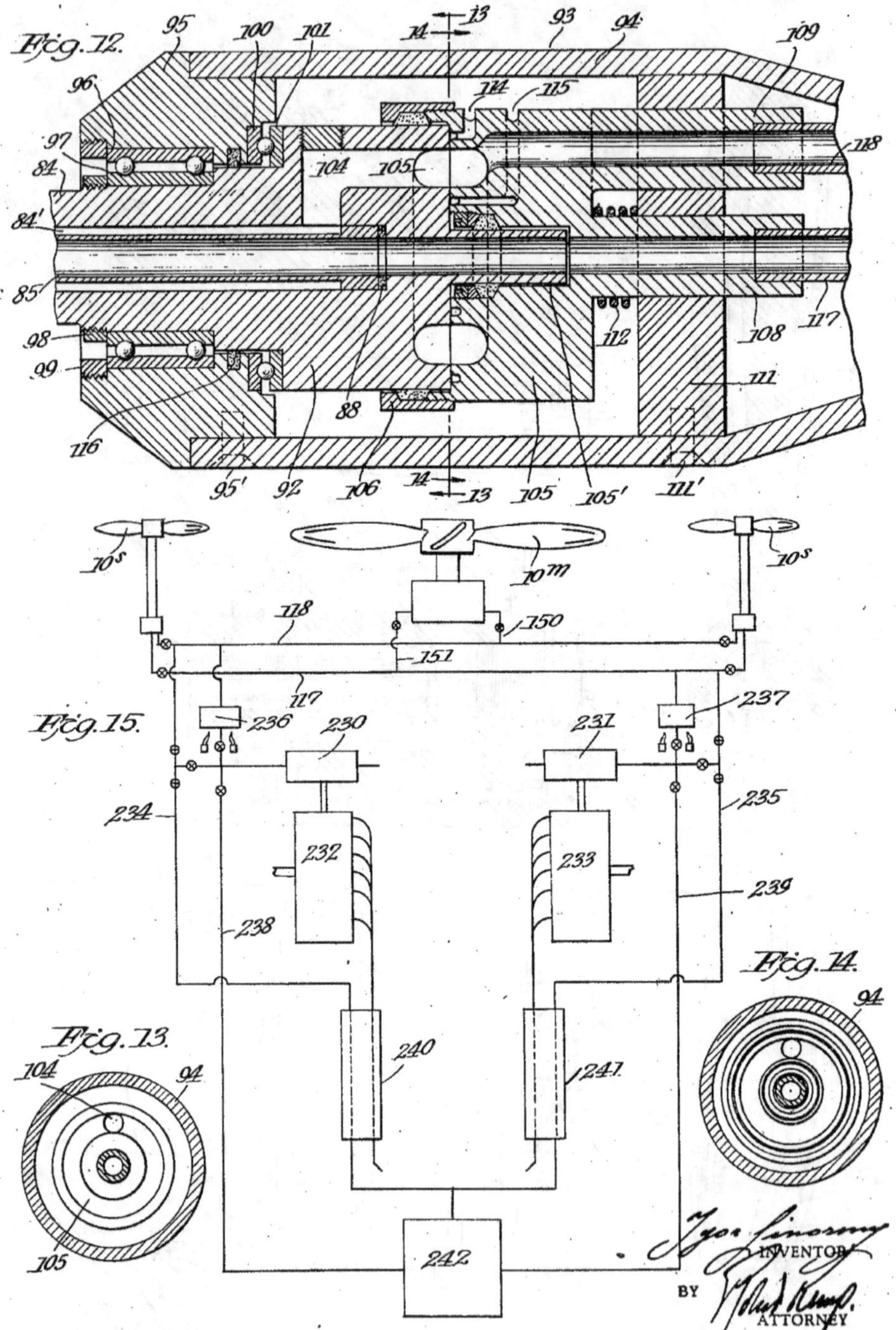

March 8, 1932. I. SIKORSKY 1,848,389

AIRCRAFT, ESPECIALLY AIRCRAFT OF THE DIRECT LIFT AMPHIBIAN TYPE AND MEANS OF CONSTRUCTING AND OPERATING THE SAME

Original Filed Feb. 14, 1929 8 Sheets-Sheet 6

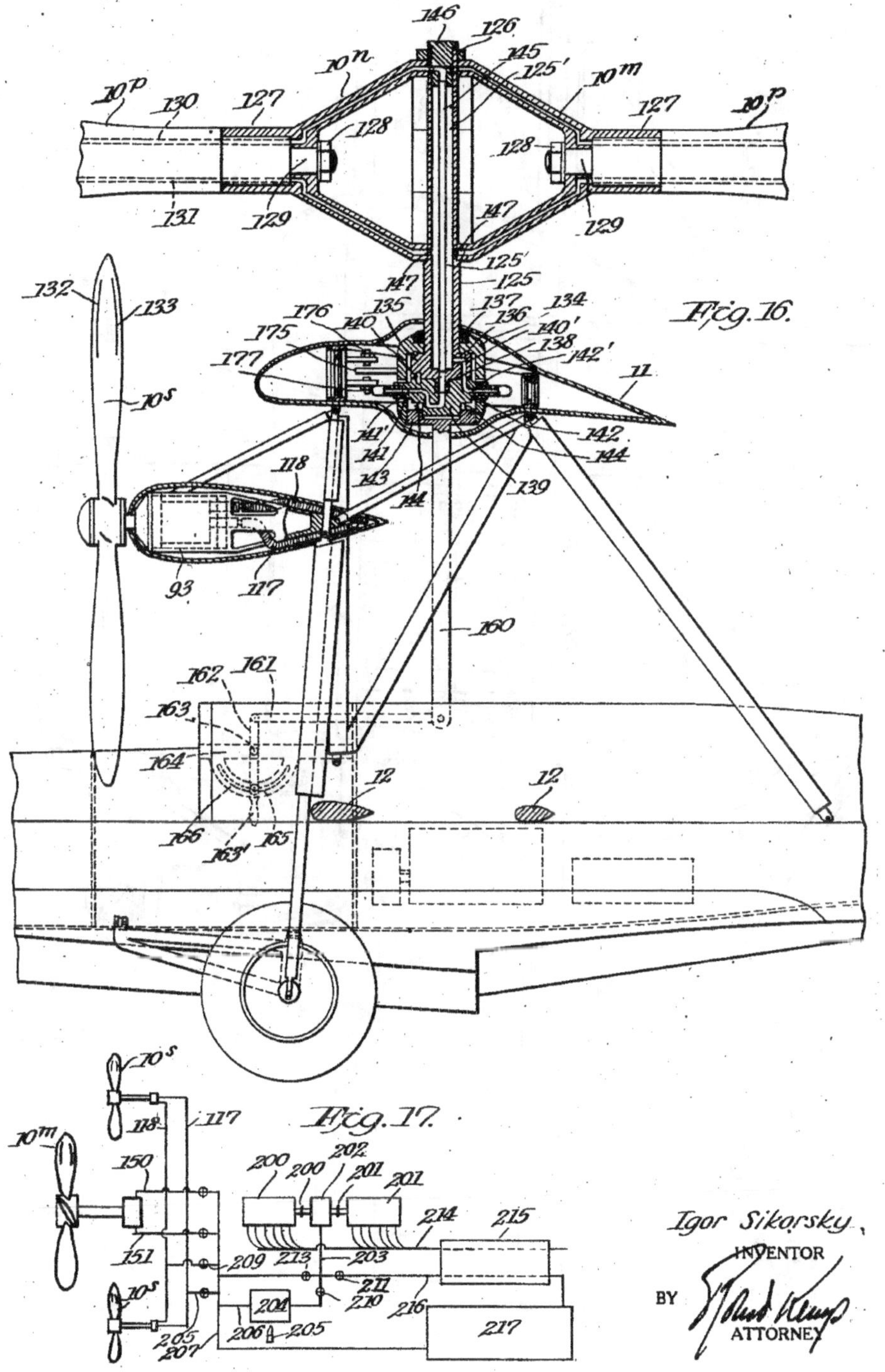

Igor Sikorsky, INVENTOR

BY [signature] ATTORNEY

March 8, 1932. I. SIKORSKY 1,848,389

AIRCRAFT, ESPECIALLY AIRCRAFT OF THE DIRECT LIFT AMPHIBIAN TYPE AND MEANS OF CONSTRUCTING AND OPERATING THE SAME

Original Filed Feb. 14, 1929 8 Sheets-Sheet 7

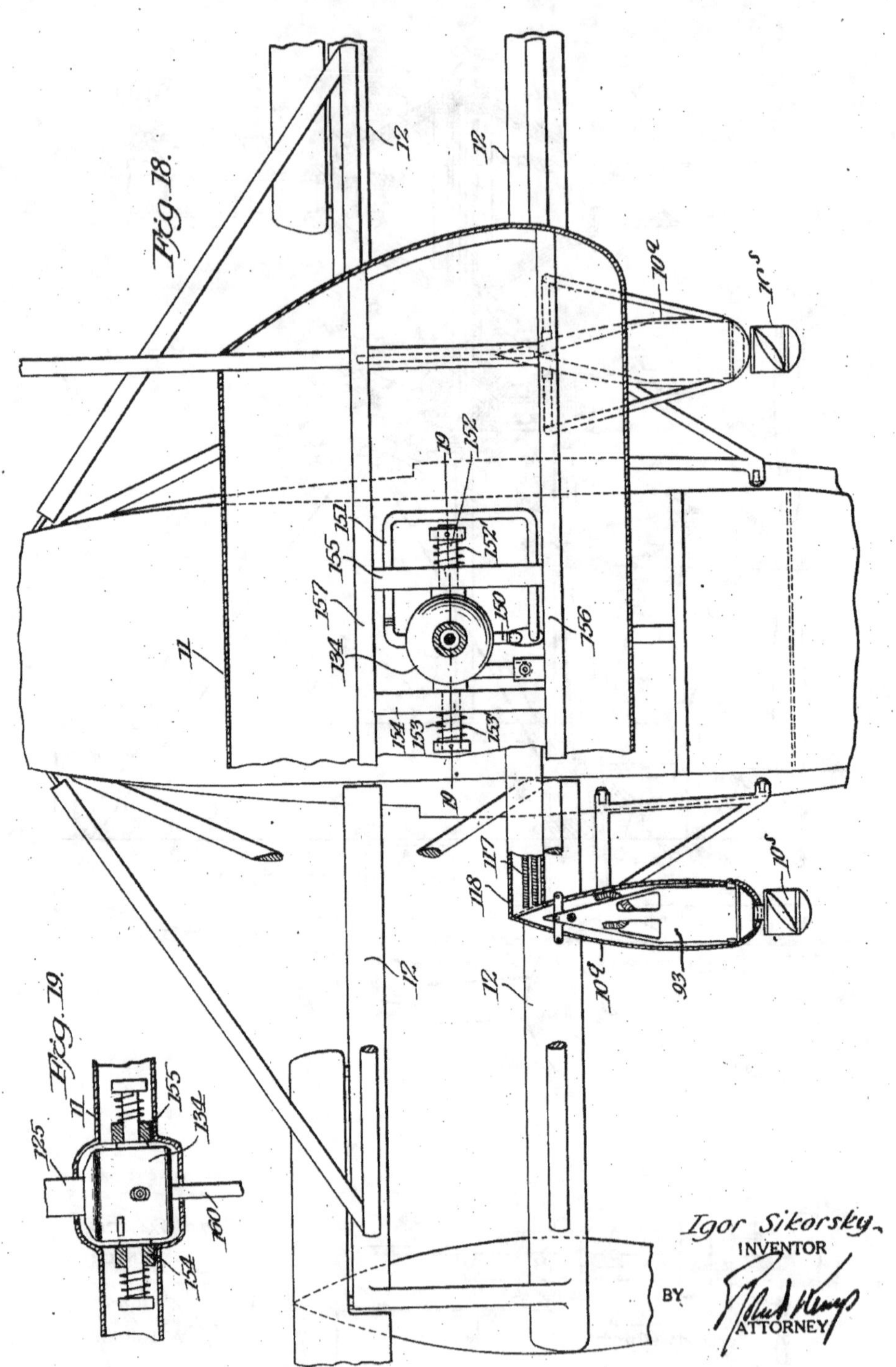

March 8, 1932. I. SIKORSKY 1,848,389

AIRCRAFT, ESPECIALLY AIRCRAFT OF THE DIRECT LIFT AMPHIBIAN TYPE AND MEANS OF CONSTRUCTING AND OPERATING THE SAME

Original Filed Feb. 14, 1929 8 Sheets-Sheet 8

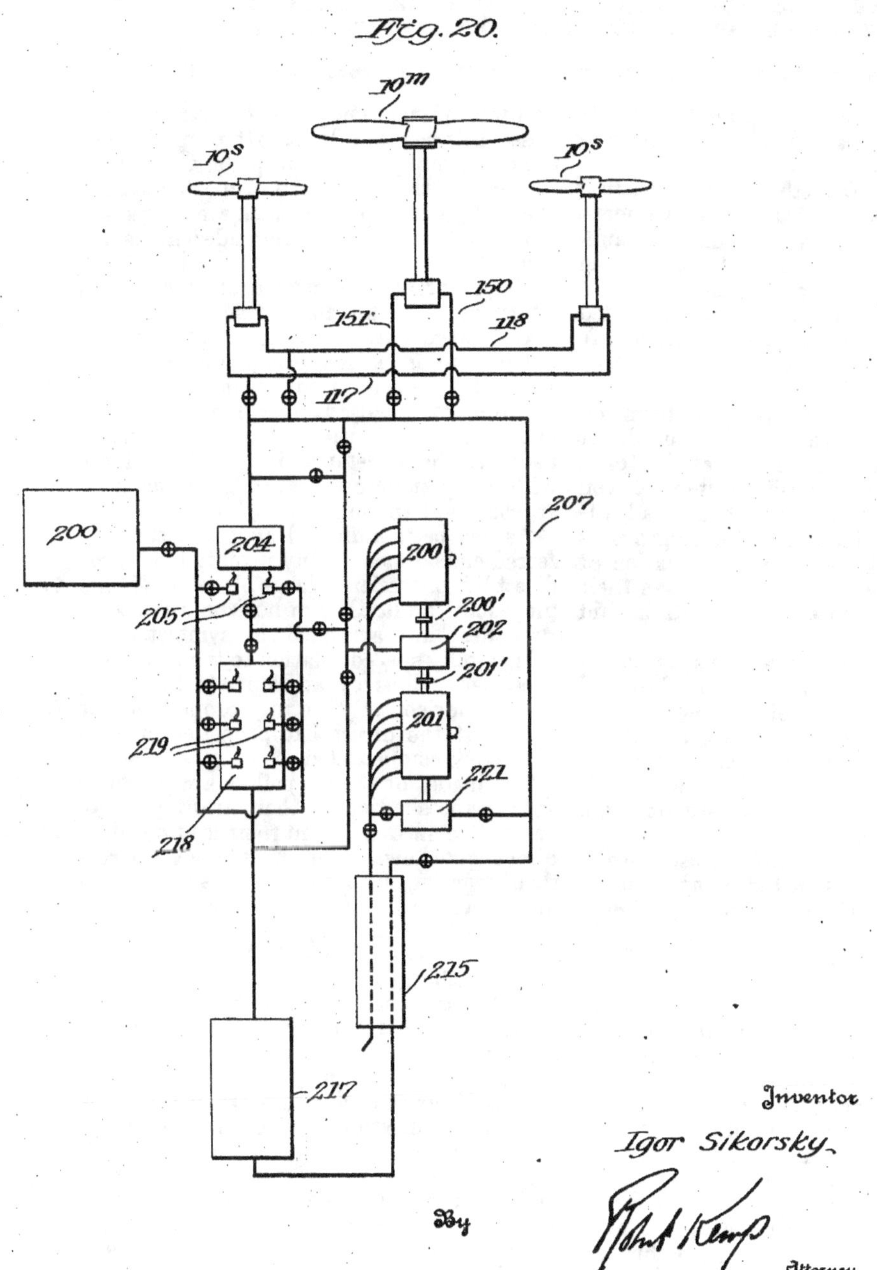

Inventor

Igor Sikorsky

By Robert Kemp

Attorney

Patented Mar. 8, 1932 **1,848,389**

UNITED STATES PATENT OFFICE

IGOR SIKORSKY, OF NICHOLS, CONNECTICUT, ASSIGNOR TO SIKORSKY AVIATION CORPORATION, OF WILMINGTON, DELAWARE, A CORPORATION OF DELAWARE

AIRCRAFT, ESPECIALLY AIRCRAFT OF THE DIRECT LIFT AMPHIBIAN TYPE AND MEANS OF CONSTRUCTING AND OPERATING THE SAME

Application filed February 14, 1929, Serial No. 339,784. Renewed May 29, 1931.

The present invention relates broadly to aircraft and more particularly to aircraft of the direct lift type.

It concerns a new method of constructing and operating direct lift aircraft in a manner to insure increased safety, maneuverability, comfort and reliability during taking off, flying and alighting operations.

It further contemplates aircraft provided with direct lift driving means which are driven by air or other gases issuing from reaction jets in said driving means.

It also has to do with the position, form and arrangement of the various parts of direct lift aircraft structures, as well as the form, arrangement, relative position, interrelation and details of fixed, movable and adjustable fixtures contained in or forming a part of direct lift aerial vehicles, the relation of said fixtures to each other as well as their relation to other structural elements in said aerial vehicles.

The invention further pertains especially to various combinations of any or all of said above improvements, their application to, or their use on, in, or in connection with individual heavier-than-air aircraft units of the multimotor, land-water-air type, capable of navigating with equal facility on land, water or in the air.

Where the term "aircraft" is used in the present application it includes any form of aerial vehicle capable of navigating through the air. The invention has application to aircraft of the heavier-than-air, lighter-than-air or combination type.

The term, "heavier-than-air aircraft unit of the multimotor, land-water-air type" designates a form of aircraft sometimes known as a "multimotor amphibian."

In one of its aspects the invention has to do specifically with an amplibian provided with a direct lift propeller and a plurality of horizontally acting propellers, all of the propellers being reaction-driven, that is, being rotatable under the reactive force of a fluid medium under pressure discharged through orifices in their blades toward the following edges.

As another feature of the invention an aerofoil of relatively short span is interposed between the direct lift propeller and the horizontally acting propellers. Further, the body of the amphibian supports laterally extending outriggers to which are hinged ailerons and at whose lower sides floats are secured.

The inherent structural characteristics of multimotor amphibians are such as to give rise to large parasitic drag. Consequently, the present invention has particular reference to this type of machine, since by the substitution of small transfer boxes for fluid under pressure in place of the usual engine nacelles, the resistance is greatly reduced with proportionate benefit to the flying characteristics of the machine.

As has been indicated above, in the preferred embodiment of my invention, a single direct lift propeller of large blade area is employed and a plurality of horizontally acting propellers are arranged in symmetrically horizontally spaced relation to the direct lift propeller. Thus by appropriate throttling of the horizontally acting propellers, the torque of the direct lift propeller may be effectively compensated.

The blades of the propellers are provided with orifices adjacent their trailing edges and with orifices adjacent their leading edges discharging toward the trailing edges across their negative or depression sides. The fluid under pressure discharged through the first mentioned orifices imparts rotation to the propellers in which effect the pressure medium discharged through the other set of orifices participates, although this latter is of greater importance as performing a lifting function.

The medium used may be air or air mixed with other gases such as the exhaust gases of internal combustion engines used to drive the air pump. However, I do not limit myself with respect to the nature of the fluid medium which may be any such as may be convenient and practicable.

According to one phase of the invention, the air which is to be supplied to the reaction propellers is first of all mechanically placed under pressure by means of a positive action air pump or blower, and during its subsequent passage to the propellers, is expanded by application of heat thereto with consequent increase in work capacity. The heating of the air may be accomplished in a number of different ways which will be hereinafter described.

In addition to the mechanical advantages obtained by thus expanding the air, the discharged air serves the purpose of preventing, to a large extent, the formation of ice on the aircraft members.

An embodiment of the invention will be described in detail with reference to the accompanying drawings in which:

Fig. 1 is a plan view of an amphibian constructed in accordance with the present invention.

Fig. 2 is a front elevation of the amphibian.

Fig. 3 is a side elevation of the amphibian.

Fig. 4 is an isometric perspective of the amphibian.

Fig. 5 is an enlarged view in front elevation of the central portion of the amphibian.

Fig. 6 is a section on the line 6—6 of Fig. 3.

Fig. 7 is an elevation of a propeller back, partly in section.

Fig. 8 is a section on line 8—8 of Fig. 7.

Fig. 9 is an enlarged elevation of a portion of the propeller shown in Fig. 7.

Fig. 10 is a section on line 10—10 of Fig. 7.

Fig. 11 is a side elevation of the propeller boss.

Fig. 12 is a longitudinal section through a transfer box for the fluid pressure medium.

Fig 13 is a partial section along line 13—13 of Fig. 12.

Fig. 14 is a partial section along line 14—14 of Fig. 12.

Fig. 15 is a diagram showing one form of arrangement of the units comprising the propulsion system.

Fig. 16 is a section on line 16—16 of Fig. 2.

Fig. 17 is a diagram showing another arrangement of the units comprising the propulsion system.

Fig. 18 is a plan view of a portion of the amphibian with parts in section.

Fig. 19 is a section on line 19—19 of Fig. 18, and

Fig. 20 is a diagram showing a modified arrangement of units comprising the propulsion system.

Referring first to Figs. 1 to 4, 10 designates the body-boat of an amphibian, while 12 designates streamline outriggers, springing laterally from the body-boat and joined together at their outer ends by means of members 10*c* and 10*d*. The rear outriggers have hinged thereto ailerons 10*e* and 10*f*, while floats 10*g* and 10*h* are secured beneath the outer ends of the outriggers. An aerofoil 11 of relatively short span is disposed above the body-boat and interconnected therewith as by struts 10*i*, 10*k* etc.

A shaft 125, normally lying in a vertical line passing through the center of gravity of the machine, supports a direct lift propeller 10*m* here shown as comprising an open framework boss 10*n* and blades 10*p*. Streamline housings 10*q* supported from suspension members 10*r* secured to the lower surface of plane 11 adjacent its ends, encase transfer boxes for the fluid pressure medium, which boxes in turn support shafts on which propellers 10*s* are mounted. Outriggers 10*t* spring rearwardly from plane 11 and support empennage assembly 10*v* at their outer ends. The outriggers 10*t* have a direct connection with the rear end of body-boat 10 through struts 10*w*.

Landing wheels 15, Figs. 2 to 6, may be moved from the operative position shown to an inoperative position in which they lie in horizontal planes immediately below outriggers 12.

Referring particularly to Figs. 5 and 6, it will be seen that the wheel axles 17 are supported on brackets 16 pivotally connected to the body 10 at points 17′ for movement about normally substantially horizontal axes. The outer ends of wheels 17 are pivotally connected to rods 18 guided in tubular members 19 which are in turn pivoted at their upper ends to the frame structure. Rods 18 are provided interiorly of tubular members 19 with piston heads which are adapted to be suitably influenced to move the wheels from operative to inoperative position.

According to the illustrated arrangement, hydraulic pressure is employed for this motive function, the particular fluid medium preferably being oil, glycerine or the like. A supply of the pressure medium is contained in a tank 20 and a conduit 21 leads from the tank to a pressure generating device 22 which is controlled by means of a lever 23. The lines to and from pressure generator 22 are indicated at 25 and lead to a distributor 24 disposed within convenient reach of the pilot in the pilots' compartment 30. Distributor 24 is connected to tubular members 19 by means of pipes 19*a* and 19*b*, the former debouching into the tubular members above the upper limit of travel of the piston heads associated with rods 18, and the latter debouching into said members below the lower limit of travel of the piston heads. Distributor 24 is suitably provided with valves so that fluid may be supplied at will to one or both of conduits 19*a* or to one or both of conduits 19*b* so that the wheels 15 may be correspondingly raised or lowered.

The fluid medium, in addition to acting as an operating agent, has the function of serving as cushioning means when the machine is landed on its wheels.

It will be noted that the pilots' compartment 30 is provided with two chairs 38 placed side by side. Since the arrangement of this compartment and the control devices therein have been particularly described and claimed in my above named applications, it will suffice here to state that the reference numerals 61 indicate the engine control levers, 33 indicates levers cooperating with segments 35 to adjust the seats to various heights and retain them in adjusted position, 62 indicates the stabilizer controls and 60 indicates the aileron control member which has a pivoted extension 63 so supported that it may be swung to bring wheel 60 in front of either chair 38.

In Fig. 7 a propeller 10*s* is shown comprising blades 70 and 71 and boss 72. The leading edge of each blade is provided with a longitudinal passage 73 and 74 respectively, these being joined by an arcuate recess 75 surrounding the propeller hub 76. Each blade has also a longitudinally extending passage 77 and 78 respectively adjacent the following edge, these passages being connected by an arcuate recess 79 similar to 75. For a distance adjacent the blade tips, discharge orifices 80 and 81 are formed, which communicate with passages 78 and 74 respectively. As particularly shown in Fig. 8, the propeller back is stepped downwardly toward its following edge, there being two steps determining respectively the position of orifices 81 and 80. Referring to Fig. 10, hub 76 is provided with an aperture 82 communicating with recess 79. Aperture 82 likewise registers with an aperture 83 formed in the wall of a hollow shaft 84 (see also Fig. 12) to the end of which the propeller hub is keyed. A tube 85, Figs. 10 and 12, is supported concentrically of the bore 84′ of shaft 84 and is closed at its front end, Fig. 10, by means of a cap 86 having an aperture 87 in register with aperture 83. The rear end of tube 85, Fig. 12 is provided with a circumferential flange for the purpose of maintaining the tube in concentric relation to bore 84′, the rear end of tube 85 seating against washer 88. The propeller hub is secured to the end of shaft 84 by means of a cap screw 89 cooperating with threads formed in the end of bore 84′. Cap screw 89 is provided with a threaded axial bore in which engages cap screw 90 which abuts with its inner end cap 86 to hold the latter and tube 85 in proper position. A nut 91 serves to lock cap screw 90 in adjusted position.

The end of shaft 84, remote from the propeller, is provided with an expanded head 92 extending within transfer box 93, Fig. 12. Transfer box 93 comprises a hollow cylindrical member 94 supporting at its end adjacent the propeller a ring 95 which latter and shaft 84 are appropriately shouldered to receive the members 96 and 97 of a ball bearing assembly constituting a journal bearing for the propeller shaft, members 96 and 97 being respectively held in position by means of rings 98 and 99.

The adjacent vertical faces of ring 95 and head 92 support members 100 and 101 of a ball bearing assembly which constitutes the thrust bearing. Ring 95 is rigidly secured to casing 94 by means of screws such as shown at 95′.

The rear face of head 92 is provided with an annular recess 105 concentric with bore 84′. This recess communicates by means of an angular passage 104 with bore 84′ forward of the flanged head of tube 85. A fitting 105 has a face adapted to contact with the rear face of head 92, this member being provided with a circumferential flange circumscribing the rear margin of head 92 and cooperating with ring 106 to form a packing gland. Member 105 is provided with a bore 105′ into which a central tubular extension 102 of head 92 projects and a packing gland 106ᵃ is provided to effect a tight fit between the extension and bore 105′. Member 105 is provided with a rearward tubular extension 108 concentric with bore 105′ and also with an eccentric rearward extension 109 provided with a bore which communicates with an annular recess 110 registering with recess 103. Extensions 108 and 109 pass through closely fitting apertures in a wall 111 secured within casing 94 by means of screws such as shown at 111′, the casing tapering off rearward of wall 111. Rotation of fitting 105 about extension 108 is prevented by the eccentric extension 109, although axial movement of the fitting relative to wall 111 is permitted. A spring 112 yieldingly urges fitting 105 against head 92.

In practice, the interior of the casing will be filled with oil, the contacting faces of members 92 and 105 receiving lubricants through ducts 114 and 115. The escape of the oil from the forward end of the casing is prevented by a gasket 116. The fluid medium under pressure is supplied to the hollow stem 108 through a preferably flexible tube 117 which leads from air air pump preferably disposed in the body-boat. The air thus supplied passes through head 92, tube 85, apertures 87, 83, 82, recess 79 and passages 77 and 78 to discharge orifices 80. Air conducted to tubular extension 109 through pipe 118 flows into the registering annular recesses 110 and 103, passage 104, bore 84′, an aperture 119 in shaft 84, an aperture 120 in hub 76, recess 75 and passages 73 and 74 to discharge orifices 80.

The transfer boxes 93, as has been mentioned above, are disposed in streamline housings 10*q* and rigidly secured in position through struts 10*r*. Pipes 117 and 118 (see also Figs. 16 and 18) are preferably led to the transfer boxes through the streamline conduits 121, 122 and 123, Fig. 5, which likewise house tubes 19*a* and 19*b*. All conduits, connections and other passages arranged to lead air to horizontal left side propeller 10s are also duplicated preferably in a symmetrical fashion for the same purpose of supplying air to right side propeller 10s.

The direct lift propeller 10m will now be described with particular reference to Figs. 1 to 4, 16, 18 and 19.

Referring particularly to Fig. 16, it will be seen that the propeller boss 10n consists of crossed loop-shaped members mounted at their crossed portions on a hollow vertical shaft 125. The lower members of the loops bear against a shoulder 125′, while a nut 126 secures them in position on the shaft. Blades 10p are secured in ferrules 127 integral with the loops and extending radially relatively to shaft 125. The propeller blades are secured in position by means of nuts 128 screwed to their stems 129. It will be seen that each blade is supported at an apex of a substantially triangular frame, the opposite triangular frames forming an integral frame of symmetrical substantially rhomboidal shape, these closed frames having major and minor axes, the propeller blades being in alignment with the major axis, while the frame is supported for rotation about its minor axis. The described arrangement gives a particularly rigid construction in that the boss has a large axial extent and supports the blades at a considerable distance from its rotational axis.

As shown in Fig. 16, the frame members of the boss are hollow and communicate with longitudinally extending passages 130 and 131 formed in the blades. Passage 130 communicates with orifices 132 in the leading edge of the blades, while passages 131 communicate with orifices 133 in the trailing edges in the same manner as has been described in connection with propeller 10s.

The lower end of shaft 125 is disposed in a transfer box 134 which is supported in aerofoil 11, the latter having a central vertically expanded portion for the accommodation of the box. Shaft 125 has an expanded head 135 interiorly of the transfer box and forms a thrust member cooperating with the latter through balls 136. A ball bearing journal support is indicated at 137. The lower face of head 135 is provided with an extension 138 which mates with a recess formed in the face of a fitting 139, the extension and recess being concentric with bore 125′ of shaft 125. The contacting faces of head 135 and member 139 are provided with registering annular recesses concentric with extension 138 and forming together a tubular duct 140. At diametrically opposite points member 139 is provided with arms 141 and 142 which extend through slots formed in the side walls of box 134, these slots extending in the axial direction of shaft 125. The engagement of extensions 141 and 142 with the walls of the slots prevents rotation of member 139 relative to box 134, although axial movement is permitted. The lower end of the box is closed by means of a cap 143 and between the latter and member 35 are interposed compression springs 144 which yieldingly urge member 139 against head 135.

Extension 141 has formed therein a passage 141′ which through extension 138, a tube 145 and passages in a block 146 in register with the passages formed in the upper frame members, communicates with blade passages 130. Extension 142 has a passage 142′ communicating with chamber 140 and thence through a passage 140′ with bore 125′. Bore 125′ communicates by means of apertures 147 with the passages of the lower frame members of boss 10n and thence with the blade passages 131. Extensions 141 and 142 are connected by means of flexible tubing 150 and 151 with the blower, tubes 150 and 151 being passed downwardly to the blower through the streamline conduit 121.

Box 134, Fig. 18, is provided on diametrically opposite sides with trunnions 152 and 153 supported in members 154 and 155 which are mounted between the front and rear spars 156 and 157 of aerofoil 11. Oppositely acting torsion springs 152′ and 153′ tend to maintain shaft 125 in a constant position relative to members 154 and 155, this position being vertical when the machine is in operation.

Extending downwardly from cap 143, Fig. 16, is a rod 160 which at its lower end is connected by means of a link 161 to a lever 162 pivotally mounted at 163 to a frame 164 disposed in the pilots' compartment. Lever 162 is provided with a handle 163 within easy reach of the pilot and may be locked in adjusted position by means of a threaded stud and nut 165 which cooperate with a slotted segment 166. Movement of handle 163 causes a corresponding movement of box 134 about its trunnions and consequently a tilting movement of shaft 125 in a fore and aft direction. It will be noted that the covering of aerofoil 11 is provided with top and bottom slots to permit free movement of shaft 125 and rod 160.

According to the described arrangement, shaft 125 may be adjusted from a vertical position in which propeller 10m exerts a purely vertical force, to a forwardly tilted position in which a forwardly acting resultant is obtained. The tilting movement of box 134 is limited by means of an arm 175 fixed thereto and cooperating with adjustable abutments 176 and 177.

In Fig. 17, I have shown a layout of one form of an entire propulsion system. In this figure numerals 200 and 201 denote internal combustion engines operating a positive action pump or blower 202 through the intermediary of clutches 200′ and 201′. A line 203 connects the outlet of the blower with a chamber 204 adapted to be heated by means of a burner 205. Chamber 204 communicates by means of a line 206 with a line 207 from which branches 208 and 209 lead respectively to lines 117 and 118, which feed the horizontally acting propellers 10*s*. Branches 150 and 151 feed the direct lift propeller 10*m*.

By closing valves 210 and 211, the air may be passed directly through line 212, valve 213 being open, to line 207. The exhaust of both engines is led into a line 214 which passes through a heat exchanger 215 through which likewise passes a line 216, which through an air expansion chamber or reservoir 217 is in communication with line 207. By opening valve 211 and closing valves 210 and 213, air from the blower may be passed through the heat exchanger, the reservoir and line 207 to the propellers.

Thus, it will be seen that blower 202 may be operated by one or the other of motors 200 or 201 or by both of them and that the air therefrom may be led directly to line 207 and thence to the propellers, or may be warmed by passage through chamber 204, or may be warmed by passage through chamber 215 and then led to the propellers.

The layout according to Fig. 20 is generally similar to that of Fig. 17, although according to this figure, the air may be passed additionally through a chamber 218 in direct contact with burners 219, both the latter and burners 205 being supplied with fuel from a tank 220. Further, according to this showing, the exhaust may be by-passed through a blower 221 into line 207. It is obvious that various combinations are possible through the proper manipulation of the illustrated valves and it is not believed that further description is necessary.

According to Fig. 15, two blowers 230 and 231 are provided, each being driven by an independent internal combustion engine 232 and 233. Blowers 230 and 231 may discharge directly into lines 234 and 235 and thence to lines 117, 118, 150 and 151 to supply the propellers. By suitable manipulation of the valves, however, the blowers may be connected directly through heated chambers 236 and 237, to lines 117 and 118. By a further manipulation of the valves, the air from the blowers may be diverted into lines 238 and 239 to reservoir 242 and thence through heat exchangers 240 and 241 to lines 234 and 235. According to this arrangement, one blower alone may supply line 117 and the other line 118, or one blower alone may supply both lines.

While I have described my invention with some particularity, it is to be understood that I do not intend to restrict myself except as determined in the following claims.

I claim:

1. In an aircraft, a direct lift reaction driven propeller, a rotatable vertical shaft on which the propeller is mounted, an aerofoil beneath said propeller and a transfer box for a fluid medium under pressure and including bearing means for said shaft disposed in said aerofoil.

2. In an aircraft, a direct lift reaction driven propeller, a rotatable vertical shaft on which the propeller is mounted, an aerofoil beneath said propeller, a transfer box for fluid medium under pressure and including bearing means for said shaft disposed in said aerofoil, and mounting means for said box to enable it to tilt to move the shaft relative to said aerofoil out of its normal vertical position.

3. In an aircraft, a direct lift reaction driven propeller, a rotatable vertical shaft on which the propeller is mounted, an aerofoil beneath said propeller, a transfer box for fluid medium under pressure and including bearing means for said shaft disposed in said aerofoil, and mounting means for said box to enable it to tilt to move the shaft relative to said aerofoil out of its normal vertical position in a fore and aft direction.

4. In an aircraft, a direct lift reaction driven propeller, a rotatable vertical shaft on which the propeller is mounted, an aerofoil beneath said propeller, a transfer box for fluid medium under pressure and including bearing means for said shaft disposed in said aerofoil, trunnions supporting said box for tilting relative to said aerofoil in a fore and aft direction, and means operable to tilt said box.

5. In an aircraft, a reaction driven direct lift propeller comprising an open framework boss and hollow blades secured thereto, said framework comprising hollow members adapted to conduct a fluid medium under pressure to said hollow blades.

6. In an aircraft, a direct lift propeller comprising a boss in the form of a plurality of crossed loop-like frames, a shaft supporting said frames at their crossed portions, and blades extending radially relative to said shaft and secured to said frame.

7. In an aircraft, a body, an outrigger extending laterally from each side of the body, an aileron hinged to each outrigger, an aerofoil of relatively short span above the body and secured thereto and to the outriggers, a rearwardly extending outrigger springing from said aerofoil and supporting an empennage assembly, a direct lift propeller above the aerofoil, and a horizontally-acting propeller intermediate the aerofoil and body.

8. In an aircraft, a body, an outrigger extending laterally from each side of the body, an aileron hinged to each outrigger, an aerofoil of relatively short span above the body and secured thereto and to the outriggers, a rearwardly extending outrigger springing from said aerofoil and supporting an empennage assembly, a direct lift propeller above the aerofoil, and a plurality of longitudinally

spaced horizontally-acting propellers intermediate the aerofoil and body and symmetrically disposed relative to the direct lift propeller.

9. In an amphibian, a body, an outrigger extending laterally from each side of the body, an aileron hinged to each outrigger, a float secured to each outrigger, an aerofoil of relatively short span above the body and secured thereto and to the outriggers, a rearwardly extending outrigger springing from said aerofoil and supporting an empennage assembly, a direct lift propeller above the aerofoil, and a horizontally-acting propeller intermediate the aerofoil and body.

10. In an amphibian, a body, an outrigger extending laterally from each side of the body, an aileron hinged to each outrigger, a float secured to each outrigger, an aerofoil of relatively short span above the body and secured thereto and to the outriggers, a rearwardly extending outrigger springing from said aerofoil and supporting an empennage assembly, a direct lift propeller above the aerofoil, and a plurality of horizontally spaced horizontally-acting propellers intermediate the aerofoil and body and symmetrically disposed relative to the direct lift propeller.

11. In an aircraft, a direct lift propeller comprising a blade provided with reaction jets adjacent its following edge and jets adjacent its leading edge discharging over the top of the blade toward its following edge, said blade having separate passages therein leading to the respective jets, a substantially triangular frame element supporting said blade at an apex, the legs of said frame adjacent said apex being hollow and communicating respectively with said passages.

12. In an aircraft, a reaction-driven propeller, a blower, a plurality of internal combustion engines adapted to drive said blower either singly or in conjunction, a common conduit for the exhaust of said engines, and an airline connecting said blower and propeller and passing in heat-exchanging relation with said conduit.

13. In an aircraft, a reaction-driven propeller, a motor-driven air blower, an externally heated chamber and an internally heated chamber, and means to connect said blower and propeller through either of said chambers or through both of said chambers in either order.

14. In an aircraft, a reaction driven propeller, a blower, an internal combustion engine driving said blower, an externally heated chamber and an internally heated chamber, and means to connect said blower and propeller through either of said chambers or through both of said chambers in either order and in heat-exchanging relation with the engine exhaust.

15. In an aircraft, a reaction-driven propeller, a blower, a plurality of internal combustion engines adapted to drive said blower either singly or in conjunction, a common conduit for the exhaust of said engines, an airline connecting said blower and propeller, a second blower driven by one of said engines and having an inlet connectible with said conduit, the outlet of said second blower being connectible with said line whereby the exhaust gases may be injected into the latter.

Signed at College Point, Long Island, in the county of Queens and State of New York this 21st day of December, A. D. 1928.

IGOR SIKORSKY.

Nov. 14, 1972 M. A. FAGET 3,702,688

SPACE SHUTTLE VEHICLE AND SYSTEM

Filed Jan. 4, 1971 6 Sheets-Sheet 1

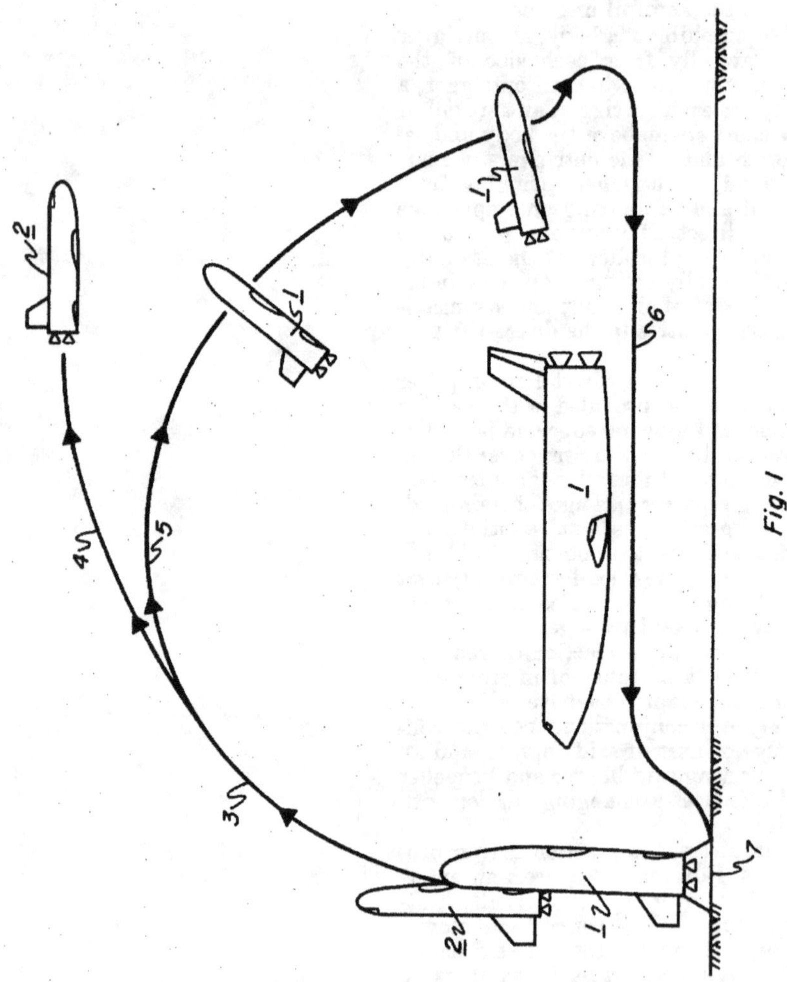

Fig. 1

MAXIME A. FAGET
Inventor

by W.A. Marcontell
Attorney

Nov. 14, 1972 M. A. FAGET 3,702,688

SPACE SHUTTLE VEHICLE AND SYSTEM

Filed Jan. 4, 1971 6 Sheets-Sheet 2

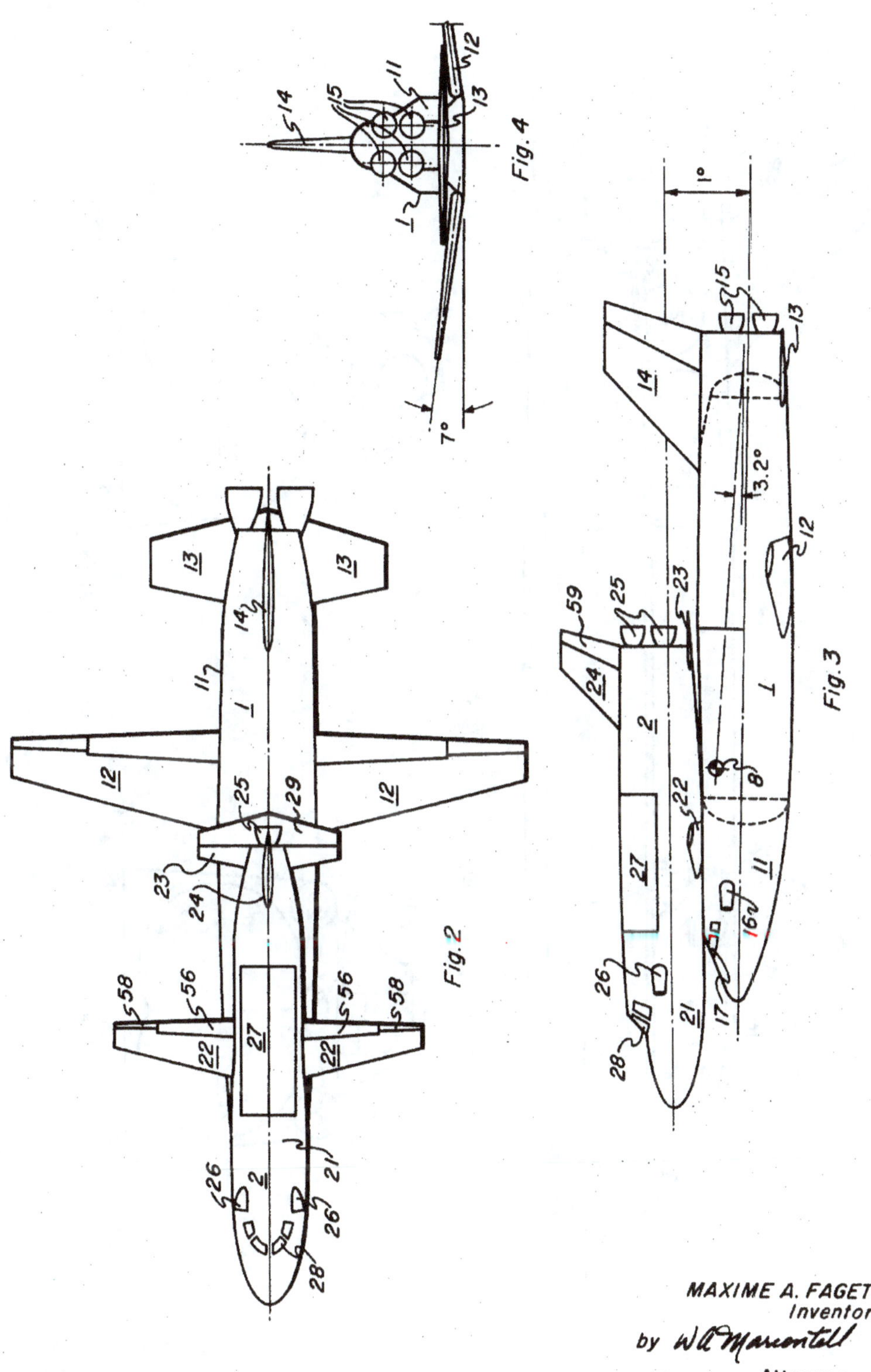

MAXIME A. FAGET
Inventor
by W. A. Marcontell
Attorney

SPACE SHUTTLE VEHICLE AND SYSTEM

Filed Jan. 4, 1971 6 Sheets-Sheet 3

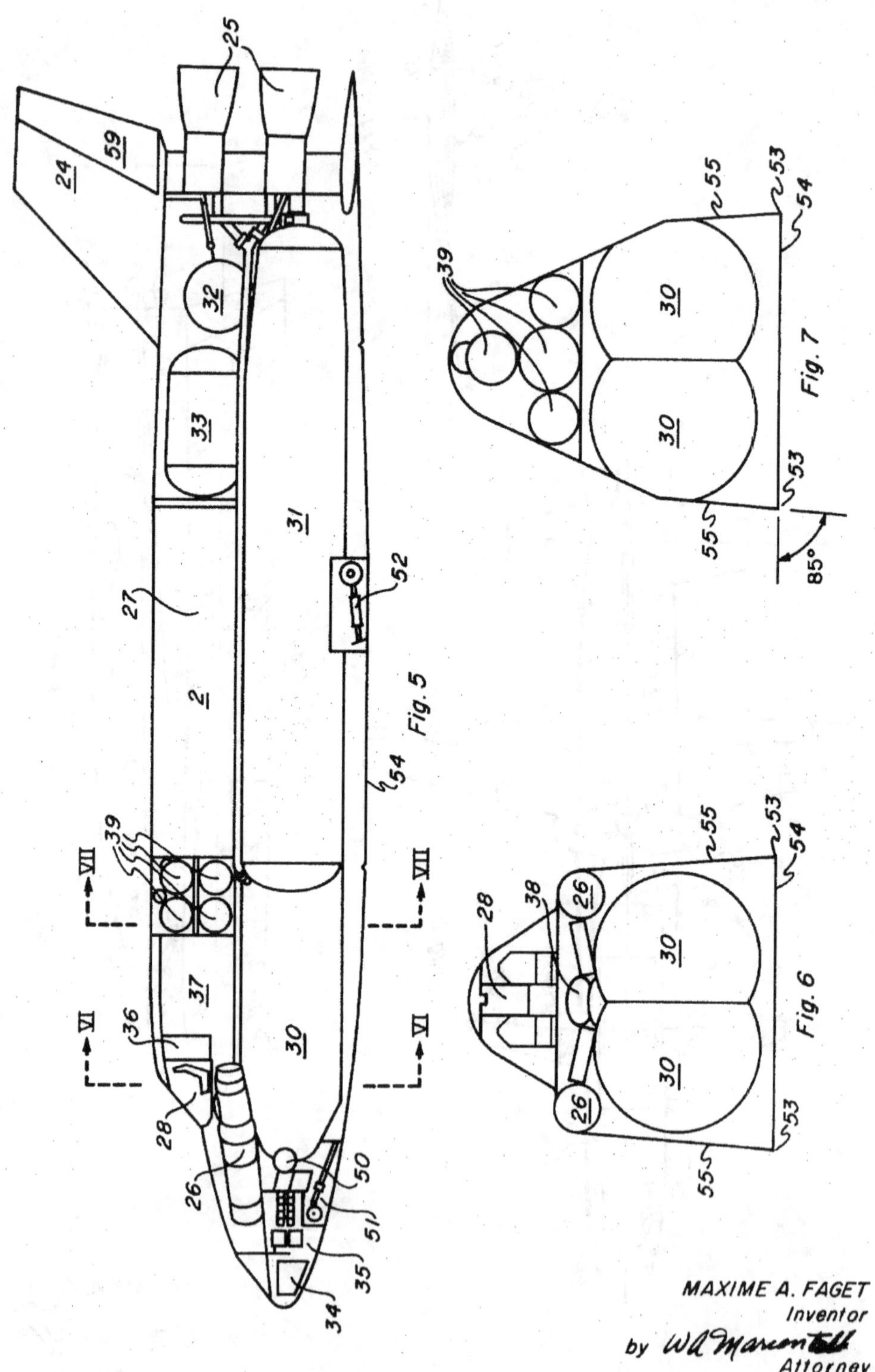

MAXIME A. FAGET
Inventor

by W. A. Marcontell
Attorney

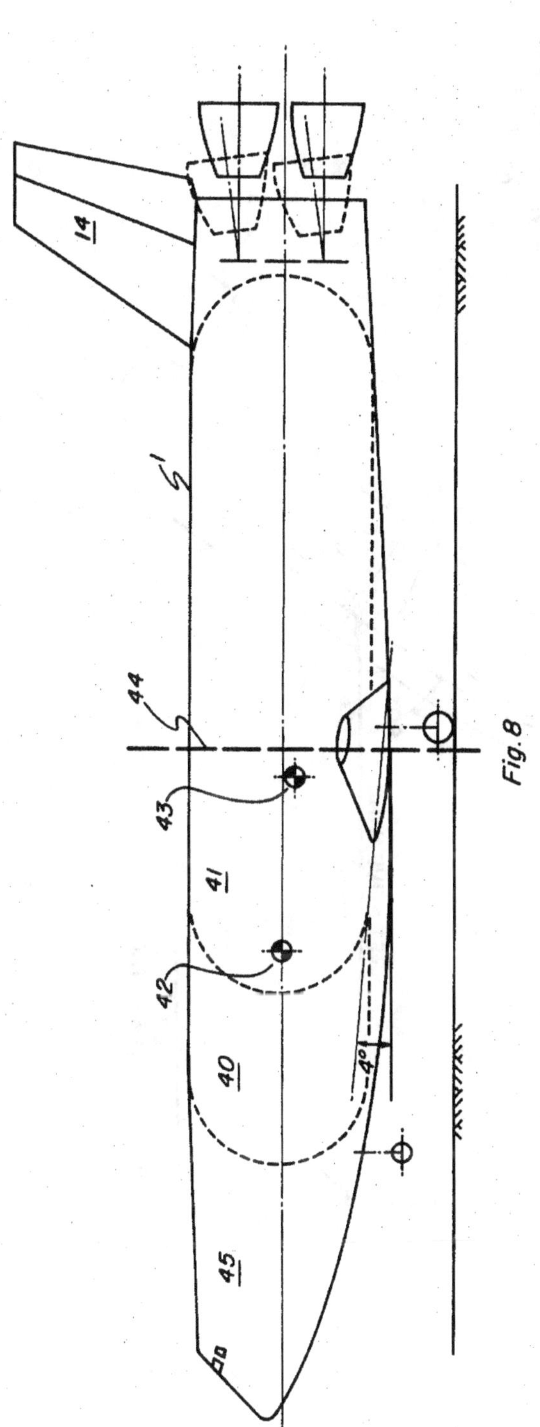

Fig. 8

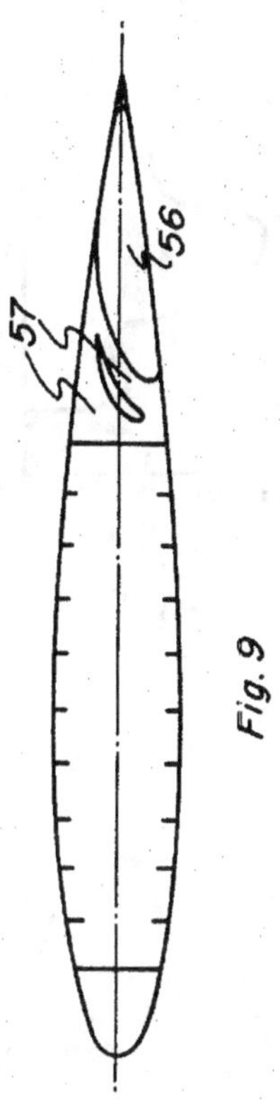

Fig. 9

MAXIME A. FAGET
Inventor

by W.A. Marcontell
Attorney

Nov. 14, 1972 M. A. FAGET 3,702,688

SPACE SHUTTLE VEHICLE AND SYSTEM

Filed Jan. 4, 1971 6 Sheets-Sheet 5

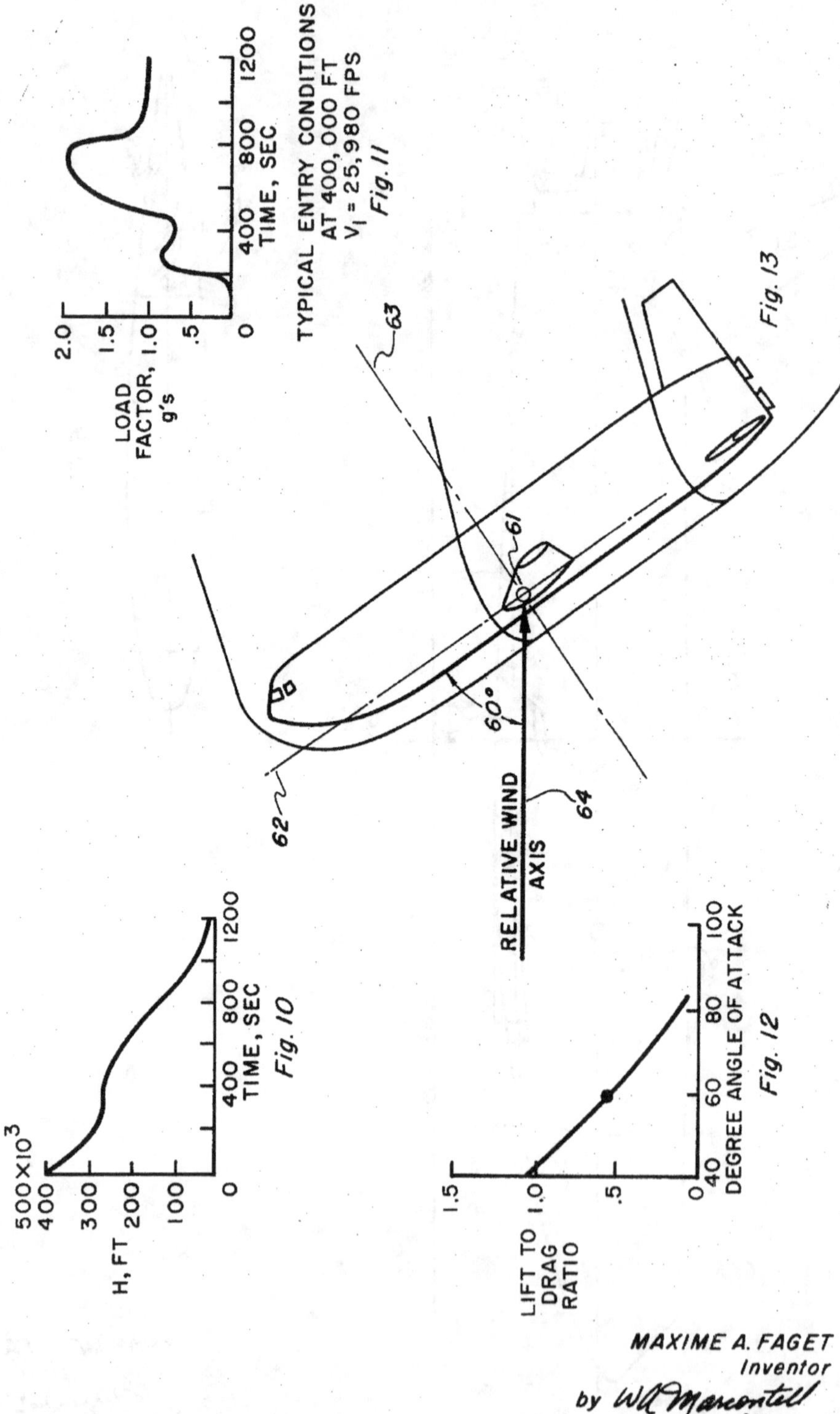

MAXIME A. FAGET
Inventor
by W.A. Marcontell
Attorney

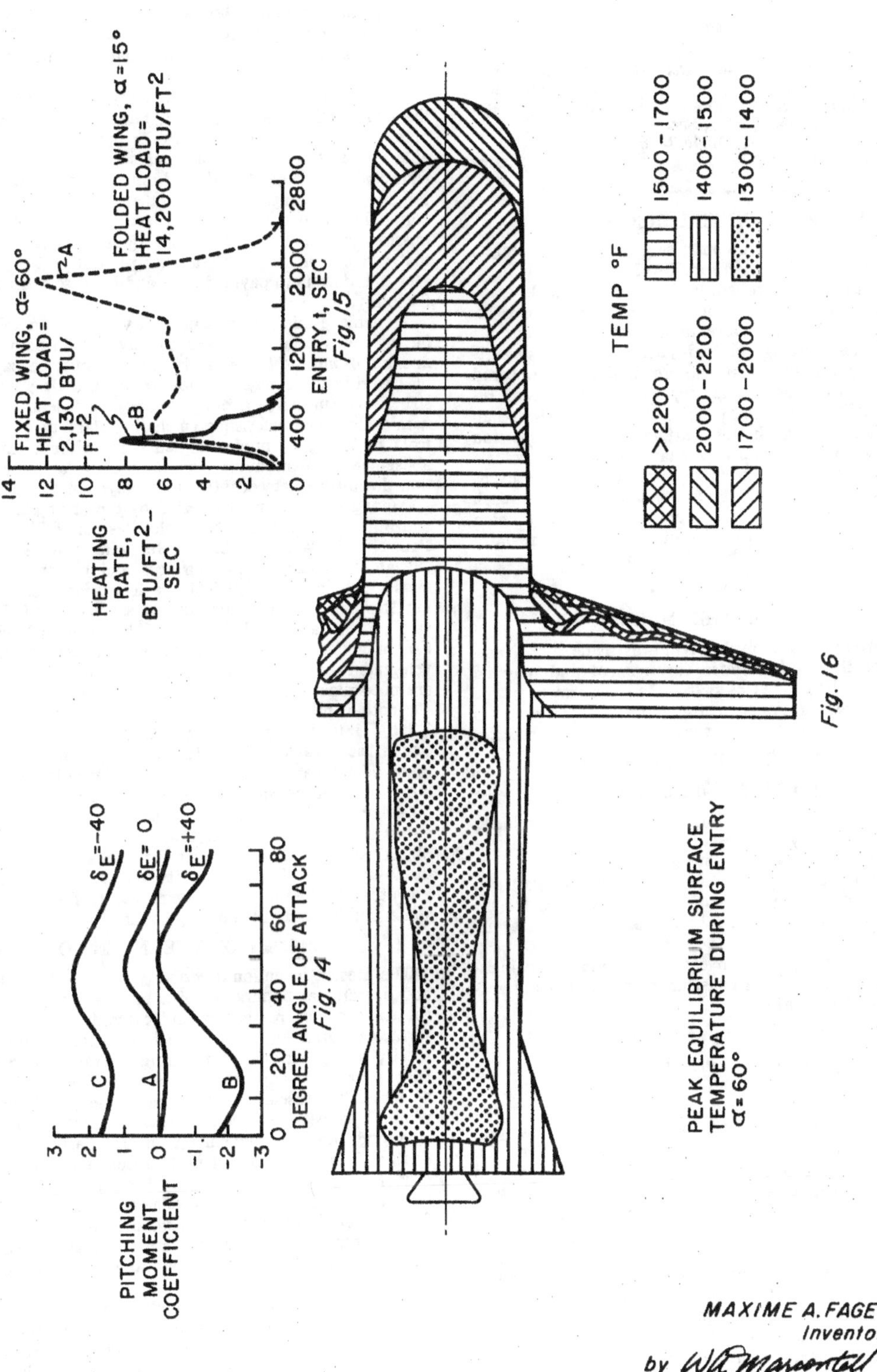

Fig. 15

Fig. 16

Fig. 14

MAXIME A. FAGET
Inventor

by W.A. Marcontell
Attorney

United States Patent Office

3,702,688
Patented Nov. 14, 1972

3,702,688

SPACE SHUTTLE VEHICLE AND SYSTEM

Maxime A. Faget, Dickinson, Tex., assignor to the United States of America as represented by the Administrator of the National Aeronautics and Space Administration

Filed Jan. 4, 1971, Ser. No. 103,551
Int. Cl. B64c *37/02*
U.S. Cl. 244—155 17 Claims

ABSTRACT OF THE DISCLOSURE

A space shuttle system comprising two reusable stages, joined "piggyback" fashion for lift-off, each stage being a manual attitude controlled vehicle having fixed, aerodynamic support and control surfaces for horizontal atmospheric flight and conventional, near stall, aircraft landings. The fuselage bottom surface of each stage is transversely relatively flat and longitudinally cambered to provide dynamic lift at hypersonic atmosphere re-entry velocities and high angles of attack. Newtonian fluid flow states over bottom surfaces and the dispersion of flow stagnation regions over large areas hold vehicle surface temperatures during re-entry to tolerable levels. Other aerodynamic criteria are balanced so that each stage will have stable flight characteristics at both high and low attack angle attitudes during both the re-entry and the subsonic atmospheric flight phases.

ORIGIN OF THE INVENTION

The invention described herein was made by an employee of the United States Government and may be manufactured and used by or for the Government of the United States of America for governmental purposes without the payment of any royalties thereon or therefor.

BACKGROUND OF THE INVENTION

(1) Field of the invention

The present invention relates to aerospace vehicles suitable for carrying substantial payloads beyond the earth's atmosphere and for return therefrom.

Also disclosed is a technique of combining at least two reusable aero-space vehicles of the type described for the launch of large payloads to earth orbital altitudes and beyond.

Moreover, the present invention contemplates a logistical support method for constructing and sustaining artificial satellites in orbit.

(2) Description of the prior art

Historically, rocket propelled vehicles have been, for the most part, unmanned ballistic devices. Flight path control has been largely limited to the ascent trajectory; the sensory function of control being performed by preprogrammed, automatic, inertial guidance systems. So long as the exclusive purpose of a rocket shot was to occasionally deliver an inanimate and expendable article to the outer reaches of the atmosphere or the depths of space, such one-way transit was satisfactory and practicable.

With the advent of human activity in space, however, it became necessary to devise reliable return devices and techniques. Such techniques include that of U.S. Pat. No. 3,093,346 to M. A. Faget et al., which discloses a space capsule for human occupancy of the type used in the manned ventures beyond the earth's atmosphere to date. Although vehicles of the Faget et al. type have proven most satisfactory for early space missions of limited scope, such craft have relatively little lateral maneuverability after re-entry into the atmosphere and are in-

capable of controlled, horizontal landing. Accordingly, precise landing points are impossible to predetermine and parachutes are necessary for the final few hundred feet of descent.

The next generation of space activity will be focused on the construction and support of large manned satellites or interplanetary vehicles. Since the present and foreseeable future states of the space vehicle propulsion art are restricted by practical payload limits and costs, multiple freight and passenger sorties are therefore necessary. Accordingly, the past practice of a large inventory of single mission, expendable vehicles in various sizes, is a luxury that can no longer be tolerated. Specifically, it is necessary to reduce both payload costs and increase operational flexibility.

To achieve this end, it is necessary to develop vehicles having operational efficiencies equally high in both space and atmospheric phases. Such vehicles must be capable of carrying large payloads to orbital altitudes; non-destructively surviving re-entry stresses; sufficient aerodynamic flight characteristics (including stability and control) to land at a predetermined point without external assistance; and minimum service, fueling, and preparation down-time antecedent to re-deployment.

Among the concepts suggested by the prior patent art and other literature are the disclosures of: Phillips, U.S. Pat. No. 3,104,079, a fixed delta wing vehicle having stowable auxiliary lifting surfaces and elevons; Kehlet et al., U.S. Pat. No. 3,090,580, a lenticular vehicle having stowable aerodynamic control surfaces; and Eggers et al., U.S. Pat. No. 3,276,722, a "lifting body" vehicle. Common limitations of these vehicles, however, are development and operational costs and the fact that they are not suitable for first stage booster service.

Another prior art concept is that of U.S. Pat. No. 3,369,771 to Walley et al., which is directed to a particular recoverable booster design and system of multiple stage deployment. The Walley et al. invention includes a delta winged booster vehicle having a lifting body fuselage that may be used in one of several launch modes as either a booster or orbiter craft. Although the mission objective of Walley et al. is similar to that of the present disclosure, the distinctions between respective vehicles are of a substantial and primary nature for reasons to subsequently be made more apparent.

SUMMARY OF THE INVENTION

The present invention describes primary vehicle design and operational parameters of a reusable shuttle system for transport of passengers and cargo from the earth's surface to orbit and return. Such spacecraft parameters specify a conventionally appearing aircraft having fixed wing panels and empennage secured in fixed position to a voluminous fuselage. Total area of the vehicle bottom profile is substantially equally divided between the fuselage and the wing-empennage group. Particularly distinctive from the prior art is the fuselage bottom geometry. For reasons of stability and surface heat distribution, the fuselage bottom is relatively transversely flat, forming relatively small radius chines at the juncture with the side panels. Moreover, the interior angle formed between the mean planes of said bottom and side panels is less than 90°.

A unique aerodynamic characteristic of the fixed geometry vehicle disclosed herein is that the stabilizing forces about the three stability axes remain relatively constant throughout the re-entry regime from high hypersonic, high angle of attack to low subsonic, low angle of attack without anomalous moments and interactions in the middle Mach numbers and transonic flight regions. Such moment anomalies and instability interactions are common to all winged and lifting body vehicles previously proposed as re-entry configurations. This characteristic is exploited by the subject vehicle to provide simple, straightforward control—one control means for re-entry and one for subsonic flight.

Contributing criteria to the aforedescribed characteristic are that the subject vehicle have both high and low angle of attack trim stability in both hypersonic and subsonic flight phases and sufficient pitch control to drive the vehicle from one stable attitude to the other. More particularly, as determined about the vehicle axes, the present invention provides for positive static roll, pitch, and yaw stability throughout the operational velocity regime. As resolved about the re-entry trajectory axis, however, i.e., an axis parallel with the relative wind, the subject vehicle has neutral static yaw stability.

Longitudinally, the fuselage bottom is ski-shaped. At the high attack angle trim point, where the wings and empennage are fully stalled, lift per unit area of fuselage is substantially equal to that of the wing-empennage group.

When used as a space transport system, two vehicles having the foregoing characteristics are joined together as first and second stages. More appropriately, the larger, first stage vehicle will hereafter be identified as the booster and the smaller, second stage vehicle as the orbiter, either of which may be characterized as a "shuttle vehicle." Relative assembly of the two vehicles for launching is to secure the orbiter bottom to the booster top, well forward of the booster pitch axis. Such disposition provides straightforward staging with good provisions for launch-load transmission between stages.

Initial ascent of the compound vehicle from an earth base is from the vertical launch position. Ascent power is provided by rocket engines positioned in the booster fuselage extreme aft end. For a launch weight of approximately 2.5 million lbs. and 25,000 lbs. orbit payload weight, staging should occur at approximately 10,000 f.p.s. While the orbiter vehicle continues under independent rocket power, the booster re-enters the atmosphere, decelerates to subsonic velocity, and makes a transitional maneuver into subsonic flight. It then either glides or cruises under turbojet power, back to land at the launch site.

The orbiter ascent trajectory delivers the loaded vehicle to the altitude and velocity vector definitive of the desired orbit where the primary engines are extinguished. While in orbit, the orbiter payload is removed. If a previously orbiting payload is to be returned to earth, said returning payload is positioned within the orbiter bay or appropriately attached to the orbiter top side. Retro engines are momentarily started to decelerate the vehicle to an acceptable re-entry velocity. Re-entry, transition and return flight of the orbiter is similar to that of the booster.

In this manner, many payload types such as satellites or sub-components of large structures such as permanent space stations or interplanetary vehicles may be placed in orbit by repeated sorties of the same shuttle vehicles. Of paramount significance is the fact that no structural portion of the delivery system is expended or abandoned. Moreover, substantial savings over prior systems may be gained in the developmental phase of the present invention by virtue of the fact that subsonic flight tests may be initiated from conventional airport takeoffs. Expendable prototypes for high altitude, vertical launch tests are eliminated. It is entirely possible that an original prototype may evolve into an operational "line" item.

BRIEF DESCRIPTION OF THE DRAWINGS

Relative to the drawings wherein like reference characters designate like or corresponding parts throughout the several views;

FIG. 1 is a schematic representation of the invention vehicles and method of deployment;

FIG. 2 is a plan view of the booster and orbiter vehicles combined in ascent configuration;

FIG. 3 is a side elevation view of the booster and orbiter vehicles combined in ascent configuration;

FIG. 4 is an aft end elevation of the booster vehicle;

FIG. 5 is a schematic cross-sectional side elevation of the orbiter vehicle;

FIG. 6 is a schematic cross-sectional end elevation of the orbiter vehicle taken from plane VI—VI of FIG. 5;

FIG. 7 is a schematic cross-sectional end elevation of the orbiter vehicle taken from plane VII—VII of FIG. 5;

FIG. 8 is a schematic cross-sectional side elevation of the booster vehicle;

FIG. 9 is a cross-sectional end elevation of a typical shuttle vehicle wing;

FIG. 10 is a graph representing the locus of altitude-time coordinates during a typical atmospheric re-entry of a shuttle vehicle;

FIG. 11 is a graph representing the locus of deceleration force-time coordinates during a typical atmospheric re-entry of a shuttle vehicle;

FIG. 12 is a graph representing the locus of lift/drag-attack angle coordinates of a typical shuttle vehicle;

FIG. 13 is a schematic representation of a shuttle vehicle in the initial atmospheric entry phase depicting pitch attitude and fluid flow pattern;

FIG. 14 is a graph representing the locus of pitching moment coefficient-attack angle coordinates respective to three positions of the vertical stabilizer;

FIG. 15 is a graph representing the locus of shuttle vehicle surface heating rate-time coordinates during a typical atmospheric re-entry of two shuttle vehicle designs having different *L/D* characteristics; and

FIG. 16 is a schematic bottom plan of a shuttle vehicle depicting the re-entry heat distribution.

DESCRIPTION OF THE PREFERRED EMBODIMENT

Due to mission objectives of the next generation of manned space activity, the present invention comprehends the logistical transport and support system shown in FIG. 1 wherein a second stage or orbiter vehicle **2** is secured piggyback fashion to a first stage or booster vehicle **1** for launch from a vertical position.

Numerical criteria of the system described herein are predicated on the following booster and orbiter specifications. It should be understood, however, that the following specifications are neither limiting or exclusive but merely representative of a possible combination and are stated only as an example.

	Orbiter	Booster
	Weight × 1000 lb.	
Launch weight	360.0	1,132.0
Weight at insertion/burnout	97.2	180.0
Payload	12.5	
Combined liftoff weight	1,492.0	
Main engine thrust	468.0	1,940.0
Crew	2	2
Passengers	10	
Payload volume, ft.^{3}	2,000	

The ascent trajectory **3** of the composite vehicle carries the orbiter **2** to a selected staging velocity. Since the booster engines provide all thrust from lift-off to staging, both vehicles being functionally independent, all fuel consumption up to the staging point is from the booster **1** tankage.

When staging occurs, the orbiter **2** is thrust free and clear of the booster **1** whereupon the orbiter main engines are started. Thereafter, the orbiter **2** is propelled along the orbit injection trajectory **4**.

After separation from the orbiter **2**, the main engines of the booster **1** are extinguished thereby allowing the earth gravitational field to pull the booster **1** back into the atmosphere along the re-entry path **5**. Re-entry attitude of approximately 60° angle to the relative wind is sustained throughout entry at the end of which the angle of attack is reduced to about 20°, the normal subsonic flight attitude.

After the booster is returned to normal subsonic flight attitude, air breathing auxiliary engines are started to sustain altitude with aerodynamic lift for the base return flight **6**.

Final recovery at base **7** is a conventional, near stall, wheel landing.

Since the booster **1** is designed, as hereafter explained, to suffer no structural degradation due to the preceding sub-orbital flight and re-entry, only minimum maintenance is necessary for deployment in addition to vertical erection, loading, fueling, and replenishment of crew life-support consumables.

The orbiter **2**, being the payload carrying vehicle of the system, delivers same to the desired station for removal from a payload bay. Thereafter, retro engines, not shown, are briefly started to reduce the orbiter velocity below a critical orbit velocity so that re-entry and return to base **7** in the same manner as booster **1** may be effected.

Relative to the booster and orbiter vehicle characteristics and construction details, reference is first made to FIGS. 1–3. Booster **1** comprises a fuselage **11**, fixed positioned wings **12**, horizontal stabilizer **13**, vertical stabilizer **14**, primary propulsion rocket engines **15**, and auxiliary propulsion turbojet engines **16**.

The orbiter **2**, in many respects, is a smaller version of the booster vehicle having fuselage **21**, fixed wings **22**, horizontal stabilizer **23**, vertical stabilizer **24**, primary propulsion engines **25**, and auxiliary engines **26**. In addition, orbiter **2** is provided with a top opening, large volume, payload receiving bay **27**.

Both booster and orbiter vehicles may be manually controlled and provided with crew compartments **17** and **28** respectively.

Some internal details of booster **1** are illustrated by dashed lines in FIG. 8 where the main engine propellant tanks **40** and **41** are shown to occupy most of the internal fuselage volume. Appreciable weight savings may be gained by integrating the tank and primary vehicle structure where possible. For a liquid oxygen-hydrogen propellant system, the greater density component, oxygen, would be stored in the forward tank **40** with the lighter propellant, hydrogen, in the aft tank **41**. This arrangement allows aerodynamic stability of the vehicle during the boost flight phase. When all tanks are full, the resultant vehicle center of gravity will be positioned ahead of the center of pressure located in the proximity of plane **44**.

It should be recalled that "center of pressure" is a theoretical aerodynamic concept wherein all aerodynamic forces acting upon a vehicle may be resolved about a point as a resultant lift and drag vector and without resultant aerodynamic moment.

In the booster fuselage section **45** forward of the oxygen tanks **40** is disposed the crew compartment **17**, navigational equipment, auxiliary propulsion engines **16**, jet fuel, and fuel for attitude control engines. The attitude control engines (not shown) are small, low thrust reaction engines located in the tips of opposite wing panels and the fore and aft ends of the fuselage **11** for controlling the vehicle flight attitude during re-entry when mechanical surface controls are ineffective.

The orbiter vehicle **2**, having a more complex interior, is shown with greater detail by FIGS. 5–7. As in the booster, oxygen tanks **30** are located ahead of the center of pressure with the hydrogen tanks **31** occupying the lower aft volume of the fuselage. Propellants from tanks **30** and **31** fuel the main engines **25** during the orbit insertion trajectory **4** (FIG. 1). For on-orbit and interorbit propulsion, additional fuel is stored in tanks **32** and **33** which are high efficiency cryogenic storage vessels suitable for maintaining oxygen and hydrogen in the low temperature liquid state for long time periods.

In the nose of the orbiter fuselage **21**, ample space is provided for guidance and control equipment **34** and electric power supply equipment **35**. The space **36** immediately

aft of the crew compartment **28** is convenient for tracking, telemetry, and communication equipment. Passenger space may be provided at **37** thereby limiting the environmentally controlled volume to the composite of spaces **28**, **36**, and **37**. Environmental control equipment **38** may be conveniently positioned beneath the crew compartment **28**.

Between the passenger space **37** and payload bay **27**, auxiliary tankage **39** for attitude control engine propellants and the environmental control system may be secured with additional auxiliary tankage positioned in the region **50**.

Current designs of the present system contemplate positioning air-breathing, auxiliary propulsion, turbine engines **26** in the fuselage nose section as illustrated, but an attractive alternative is to place the engines **26** on the upper or "shaded" face of wings **22** in protective nacelles. Other embodiments of the present invention omit the air-breathing engines entirely and rely upon the gliding cross-range of the vehicle throughout the subsonic return flight **6**. This glider version of the invention accepts certain risks of loss, however, due to the fact that landings must be executed successfully in the first approach, no go-around capacity being available. Glider and sailplane experience have proven such risks to be entirely reasonable and acceptable, though.

Since the capacity for horizontal landings on conventional airport runways is a primary objective of the present shuttle system, retractable landing gear and wells therefor are provided in the fuselage structure at **51** and **52**.

The following structural and aerodynamic design parameters are relevant to both vehicles of the invention. To reduce descriptive redundancy therefore, further references will be limited to the orbiter **2**.

Aerodynamically, the factors of positive and neutral static stability is the initial tendency of a body to return to an equilibrium attitude following a disturbance. Representative of positive static stability is a spherical mass disposed within a bowl of spherical radius greater than that of the mass. Equilibrium position for the mass in a gravity environment is at rest in the bottom of the bowl. An external force disturbance to displace the mass from said equilibrium position is resisted by gravitational forces.

Neutral static stability may be defined as the tendency of a body to remain in equilibrium in a new position following a disturbance from an initial equilibrium position. This quality is represented by a spherical mass placed on a level surface in a gravity environment. The sphere will remain in any position on the surface to which it is displaced.

About the vehicle pitch axis, in the low or 20° angle of attack attitude, positive static stability is conventionally achieved by a discretely arranged relation between the vehicle center of gravity and the center of pressure as dictated by longitudinal positionment of the wing. Decalage between the wing and horizontal stabilizer provides an inherently correct stabilizing moment tending to restore the vehicle to the low angle of attack equilibrium attitude.

At the high (60°) re-entry attack angle attitude, the present vehicle represents a substantial departure from the prior art to achieve positive static pitch stability. In the first place, most of the bottom profile surface area—fuselage, wing and empennage—is arranged to "feel" substantially uniform pressure from the relative wind of re-entry. Accordingly, the vehicle fuselage **2** (FIG. 5) is constructed almost transversely flat along the bottom **54**. Since, as stated previously, the total bottom wetted area of the wing-empennage group is substantially equal to that of the fuselage, it therefore follows that lift per unit bottom wetted area is substantially equally distributed thereover.

As a note of departure, concerning operational angles of attack, it should be understood that 60° is merely a convenient optimum for the high angle attitude. A more accurate, but elusively variable, definition of high attack angle for present purposes would be that pitch attitude at which the wings **22** are aerodynamically stalled. For the particular wing **22** described herein, stalling may occur at pitch attitudes exceeding 40° angle of attack.

Conversely, a low angle of attack is defined as a pitch attitude of less than 40° angle of attack or an operational condition whereat fluid flow over the wings **22** generates aerodynamic lift.

In view of such distinct fluid flow patterns respective to the two positions of pitch equilibrium, it is also important to point out the nature of lift relevant to said pitch attitudes. For this purpose, it is convenient to relate the functional distinctions between impulse and reaction turbine motors. The first absorbs the impact of the energizing fluid whereas the latter is driven by the reaction from an expanding, hence, accelerating fluid mass. Relating the above to this description, lift forces on the vehicle due to fluid impact at the high angle of pitch attitude will be characterized as dynamic lift whereas lift due to fluid acceleration will be characterized as aerodynamic lift.

Returning to the explanation of positive static pitch stability, it may now be seen that more static stability in the high angle attitude is a function of uniform pressure movement distribution about the pitch axis (shown as a circle **61** in FIG. 13 and extending perpendicular to roll axis **62** and yaw axis **63**). This kind of stability is inherent with a flat plate of symmetric profile. Without additional control devices, however, such stability is statically negative. Imposition of any disturbing force will elicit an accelerating departure from the equilibrium position. For this reason, the present vehicle design provides a longitudinal curve to the bottom of the fuselage, at least at either extremity thereof. Other portions of the bottom of the fuselage may be substantially longitudinally flat, or may also have a gentle curvature. In the preferred embodiment, the curvature is greatest at the bow section, then substantially flat until the extreme aft section which exhibits a curve lesser than that of the bow. This design feature can best be seen in FIGS. 3, 5, and 13. The effect of this curve is to alter the local angle of attack to the relative wind along said curved portion at a rate differential to that of the remaining vehicle bottom area. Accordingly, the coefficient of presure, a function of local angle of attack, is altered to provide restoring forces effective to return the vehicle to the equilibrium position. In other words, if the vehicle pitch attitude departs in the nose up direction, the coefficient of pressure on the flat bottom surfaces aft of the pitch axis increases at a greater rate than the cumulative result of those along the more greatly curved bow. Hence, a relative force increase is imposed on the aft side of the pitch axis **61** to cause a counteracting moment thereby driving the vehicle pitch attitude back to the equilibrium position.

Pitch departures in the nose down direction produce the opposite result to increase relative moment forces on the bow side of the pitch axis **61**.

In summary, therefore, the curved bow of the present invention provides pitch angle restoring forces for positive pitch stability in the high angle operational mode with fixed structure and with no expenditure of control power.

The rate of curvature effects the rate of restoration, a function of dynamic stability, and should be analyzed in the context of other pitch motion damping forces.

Positive static roll stability relative to the vehicle axis **62** is more conventionally achieved by a dihedral relationship between the wing panels **22** and the fuselage **21**. The 7° dihedral shown in FIG. 4 with respect to booster **1** is deemed adequate for both, re-entry and subsonic flight conditions. A dihedral angle to the horizontal stabilizer **23** will further contribute to positive static stability.

Conventional design practice is also exploited for positive static yaw stability about axis **63** in the low angle flight mode. Accordingly, the present design provides for an extreme aft mounted vertical stabilizer **24** and sweep to the wing leading edge.

The significance of positive static yaw stability, as resolved about the vehicle axis **63** in the high angle flight mode is nominal due to the near alignment of said vehicle yaw axis with the relative wind of re-entry. An angular displacement about the vehicle yaw axis **63** necessarily induces a roll disturbance in reference to the relative wind **64**. There is, therefore, no need for a restoring force to return the vehicle attitude to an equilibrium position relative to a fixed yaw axis **63** reference. Since the vehicle has positive static stability about the vehicle roll axis, such vehicle axis related yaw merely induces a corrective roll to reposition said vehicle yaw axis.

As determined about an axis of reference **64** parallel with the relative wind, however, the present vehicle may be considered neutral in static yaw stability. This is to say that the vehicle will stabilize in any angular position of a plane including both the relative wind axis **64** and the vehicle roll axis **62** as said plane revolves about said relative wind axis **64**. This may be better understood by envisioning the vehicle roll axis **62** as a straight line surface element of a regular cone revolved about the relative wind axis **64**. By such analogy, the vehicle will remain in equilibrium at any position on the cone where all three axes; roll **62**, yaw **63**, and relative wind **64**; lie in the same plane.

An interesting consequence of the foregoing is that relative to an earth gravitational reference, the vehicle may stabilize in an upside down re-entry attitude as well as right side up.

Width of the fuselage nose bottom **54** also contributes to neutral static yaw stability by attenuating or preventing destabilizing moments caused by fluctuating fluid flow fields and resultant side forces on the nose.

Selective attitude control of the present vehicle in the high angle flight mode is accomplished conventionally by small reaction engines (not shown) disposed in the wing tips and fuselage ends. Due to the inherent stability of the present vehicle, however, power and fuel reserve for such attitude control engines is substantially reduced from that required for comparable prior art designs.

Adjustable area structural panels such as flaps and elevons are not only ineffective in the early re-entry phases but use thereof in the high-heating re-entry phase would raise substantial thermodynamic difficulties.

As a special note of high angle attitude control, due to the strong interrelationship of roll stability and the relative wind axis **64** vehicle flight path may be selectively altered by sustaining an induced roll moment long enough to re-establish stable fluid flow against the new directional face of the bottom surface area.

In the low angle flight mode, ailerons **58** (FIG. 2), elevator **29**, rudder **59** (FIG. 3) are the devices of selective attitude control. As to the elevator **29** in particular, sufficient area and deflection must be provided to drive the vehicle pitch attitude from the re-entry positive static equilibrium position of approximately 60° to a positive static equilibrium pitch attitude of approximately 20° for subsonic flight. Representative design parameters are illustrated by FIG. 14 where the two positions of static pitch stability are shown by curve A which corresponds a zero pitching moment coefficient with a zero elevator deflection angle at attack angles of 30° and 70° respectively. Reasonably symmetric responses are derived from the elevator **29** at deflection angles of 40° down (curve B) and 40° up (curve C) respectively, for smooth handling characteristics and firm response rates throughout the subsonic flight region.

Other design relationships between fuselage, wing and empennage of the present vehicle are consistent with conventional aircraft design practice and are selected for proven superior flying qualities at subsonic velocities and in approach and landing maneuvers.

For example, the airfoil section of wing **22** may be selected almost exclusively on the basis of optimization for subsonic cruise and landing. An NACA 0012–64 airfoil provided with low speed lift augmenting devices such as flaps **56** and slots **57** is shown in FIG. 9. The 40° angle of incidence for the wing shown in FIG. 8 is also conventional practice.

The straight wing shown in FIG. 2, is clearly the lightest planform for these requisite purposes but a delta planform offers other advantages.

Two functions of the broad, flat fuselage bottom **54** have been recited, i.e., positive pitch and neutral yaw static stabilities. At least two more flat bottom functions are to be added, i.e., re-entry rate control and heat management.

A body having the general physical characteristics of a space vehicle, entering earth bound atmosphere at a velocity of approximately 26,000 f.p.s., encounters sufficient atmospheric density at approximately 400,000 ft. altitude to start hypersonic heating. Vehicles having a Lift/Drag ratio (L/D) of 1.5 entering the atmosphere at said velocity and at a 15° angle of attack, generate a heating rate profile as is illustrated by dotted line curve A of FIG. 15. At this rate, temperatures on the surface of a folded wing, lifting body vehicle may reach as high as 3900° F. Such temperatures are beyond the present art state for load-carrying structural members and can be tolerated only by the use of ablative heat shields which not only impose weight penalties against the total lift-off payload but require replacement after each use. Moreover, vehicles with high L/D ratios, e.g., 1.5, must sustain high heating rates for much longer time periods thereby compounding weight penalties due to the insulation required to maintain the primary interior structure temperatures within the useful range of material strength as the re-entering vehicle literally soaks in the blast furnace environment.

The large, uniformly loaded bottom surface area of the present invention enables the designer to capitalize on the advantages of a much lower L/D ratio, e.g., 0.5, and thereby avoid the above problems of a higher L/D ratio. By selecting vehicle design parameters so as to yield as L/D-attack angle relationship as represented by FIG. 12, a 60° re-entry angle of attack will produce the desired results. Not only is the re-entry heating profile substantially reduced in intensity and duration as shown by curve B of FIG. 15, but the re-entry deceleration profile is held well within the tolerable limits of most healthy persons as shown by FIG. 11. The slope of the FIG. 10 curve represents the altitude rate of descent.

It should be pointed out, however, that the thermal advantages of a low L/D ratio are not purchased without consideration. In this case, the price paid is that of operating range. The vehicle to which FIG. 15, curve A, relates, having an $L/D=1.5$ for a re-entry angle of attack of 15° will have an operational cross-range of 2000 N.M. The present invention, having an $L/D=0.5$ and a re-entry angle of attack of 60°, follows the much shorter re-entry duration as depicted by curve B which allows an operational cross-range of only 200 N.M. However, such a sacrifice is not considered controlling in most circumstances since even 200 N.M. is adequate to reach a suitable airport for landing.

In determining the L/D ratio for the re-entry flight phase, it should be recalled that at the 60° pitch attitude, the wing **22** is substantially if not completely stalled. Therefore, total vehicular lift in this attitude is attributable exclusively to the dynamic pressure against the bottom wetted surfaces of the fuselage, wing and empennage. Not until the vehicle descends to approximately 40,000 ft. and 300 f.p.s. velocity is the elevator **29** deflected down 30° (FIG. 14, curve C) to rotate the vehicle into a low angle of attack flight for generation of aerodynamic lift by fluid flow over the wing.

The 60° re-entry angle of attack has another desirable product in addition to consistency with an advantageous L/D ratio. That pitch attitude in correlation with the flat bottom profile is also responsible for widely distributed flow stagnation regions over the vehicle bottom wetted area. Large areas of flow stagnation in relation to the total wetted area is a manifestation of Newtonian (laminar) free stream flow states across said bottom wetted surface.

The mechanics of restricting fluid flow across the bottom area to the laminar flow states include the generation of one or more shock stable waves of maximum magnitude preceding the vehicle throughout the hypersonic flight phase. Since the propagation of shock waves consumes large quantities of energy, it is possible by the afore-described configuration to dissipate a large percentage of the altitude and velocity energy potential of the present vehicle by such device. Those skilled in the art will appreciate the practical advantage of converting this energy to heat along the wave front at a discrete distance ahead of the vehicle structure per se.

Consistent with the laws of gas dynamics and conservation of energy, diffusion of atmospheric gas past the shock wave front converts a determinable percentage of the gas total energy from the dynamic state to the static state. In this respect, the shock wave phenomena functions as a compressor to convert the compressible medium from high relative velocity, low temperature and low static pressure to a condition of low relative velocity, high temperature and high static pressure. Unlike a compressor, however, total energy of the medium before and after passing the shock wave front is substantially unchanged. It is the post wave front high temperature consequence of these physical laws that cause re-entry vehicle designers greatest concern.

The present re-entry vehicle attack profile is designed to propagate one or more shock waves, the last of which sustains a relatively stable position at a discrete distance removed from the vehicle surface. Moreover, subsequent to diffusion past said last wave front, the flow velocity of the atmospheric gas should be reduced to near sonic or less. In this velocity region, gas flow states may be limited to the Newtonian (laminar) condition thereby minimizing heat transfer from the hot, post shock wave atmospheric gas as it flows over the vehicle bottom surface. There being little mixing of flow streams across the vehicle surface, the vehicle structure need only absorb that heat of the boundary layer wetting the vehicle surface which in turn, constitutes an effective insulating layer to inhibit heat transfer from flow streams more remote from the wetted surface. The substantially flat transverse profile of the fuselage bottom minimizes the distance traversed by such insulating flow streams from the inner or centered stagnation region laterally to the sharp chines **53** which cause the flow to thereafter separate from the vehicle surface. Accordingly, less opportunity is given the flow stream to accelerate to turbulent or mixing flow velocities thereby supporting the critical necessity of maintaining a laminar condition until the gas is completely separated from all vehicle surfaces. Of course, all heat energy released by the shock wave and not transferred back to the vehicle structure is left behind the vehicle to diffuse into the general atmosphere.

Results of the foregoing physics, as applied to the present vehicle, are illustrated by FIG. 16. Leading edges of aerodynamic surfaces having small radii of curvature are outside the stagnation regions where most of the heat transfer to the vehicle structure occurs. Therefore, no unreasonable heat concentration problems are presented. Under these circumstances, the maximum expected temperature is about 2400° F. along the wing root leading edge. Structural materials are presently available to accommodate this temperature without appreciable degradation of strength.

The 5° angle of relief (FIG. 7) at the sharp planar intersection of fuselage sides **55** with bottom **54** "shades" the vehicle side and top skin structure by causing fluid flow from the stagnation region to "peel" off sharply and cleanly, as explained previously. At the same time, the relatively small radius of the "hard chines" **53** minimizes the effect of side forces on the vehicle nose due to minor deviations in vehicle alignment with the flight path to cause destabilizing moments about the yaw axis **63**.

Considering both, booster **1** and orbiter **2**, as a single vehicular unit as illustrated by FIGS. 2 and 3, it is important that the unit be aerodynamically stable throughout the boost trajectory **3** (FIG. 1). Accordingly, the unit center of gravity **8** must be positioned ahead of the unit center of pressure, a theoretical function of lift, drag, and angle of attack. By attaching the orbiter **2** to the booster **1** topside well forward of the booster center of mass with a 1° centerline convergence angle, the resultant unit c.g. will fall along a 3.2° intersection line with the booster center line at the engine **15** thrust center. An engine gimbal envelope of only a few degrees is necessary to maintain thrust vector alignment with the unit c.g. **8**.

This "piggyback" arrangement provides a number of other features not available in a tandem or side-by-side shuttle configuration. The interconnecting structure necessary for securing the two vehicles together may be compact and contained within the aerodynamic shape of respective vehicles. Separation requirements to preclude booster **1** damage prior to firing the orbiter **2** engines **25**, are less stringent. Large dimension payload pods may be attached to the orbiter with minimum geometrical interference with the booster, a capacity also having re-entry advantages by protecting a recovery load from severe re-entry heating.

It should be understood that the drawings and specification set forth hereinabove present a detailed disclosure only of a preferred embodiment of the invention and that therefore the invention is not to be limited by the specific form disclosed.

What is claimed is:

1. An aircraft having fuselage and wing means:

said fuselage having a longitudinally cambered, continuous bottom along the full length thereof, said bottom being substantially flat in the transverse direction whereby substantial impact lifting forces are generated against said bottom at flight angles of attack at and exceeding full stall;

said wing means providing aerodynamic lifting forces at flight angles of attack of less than full stall and impact lifting forces at flight angles of attack at and exceeding full stall;

the impact lifting force per unit area of said bottom at said flight angles of attack at and exceeding full stall substantially equaling the impact lifting force per unit area of said wing.

2. An aircraft as described in claim **1** wherein sides of said fuselage join said bottom at small angles of relief.

3. An orbital payload delivery system comprising:

first and second vehicles, each having aerodynamic wings, attitude control surfaces, and fuselage; the bottom of said fuselage being longitudinally cambered and substantially transversely flat, the area of said fuselage bottom being substantially equal to the bottom area of said wings and control surfaces, collectively;

said second vehicle having a payload receiving bay and removable cover therefor on the top side of said fuselage;

said first and second vehicles adapted for releasable interconnection, the bottom of said second vehicle being secured to the top of said first vehicle for vertical ascent from a static base; and

the gravity center of an interconnected unit comprising said first and second vehicles disposed ahead of the center of pressure of said unit.

4. The orbital payload delivery system of claim **3** wherein said unit gravity center lies along the thrust center of first vehicle primary propulsion engines.

5. The orbital payload delivery system of claim **3** wherein said second vehicle fuselage is adapted for releasable interconnection with a payload to the external topside thereof, said payload being protected by said fuselage from environmental extremes during atmospheric re-entry.

6. An aircraft having a fuselage and aerodynamic wings and control surfaces;

the bottom of said fuselage being substantially transversely flat with full stall pitch stabilizing means comprising fixed structure of cambered profile;

large area fluid flow stagnation regions formed on said fuselage, wing and control surface bottoms at flight angles of attack at and in excess of full stall;

relatively small linear distances between the peripheral boundaries of said stagnation regions and the peripheral boundaries of said bottom surface areas; and

substantially laminar fluid flow states prevailing adjacent said bottom surfaces between said stagnation regions and said bottom area pheripheral boundaries.

7. An aircraft as described in claim **6** wherein the sides of said fuselage are relatively convergent from said bottom and fluid flow separates from said fuselage, wing and control surfaces at said bottom area peripheries under said full stall attack angle flight conditions.

8. A reaction propulsion aerospace vehicle having:

fixed position aerodynamic lifting and control surfaces;

two positively statically stable trim attitudes at subsonic velocities, one stable trim attitude at an angle of attack exceeding 40 degrees and the other at an angle of less than 40 degrees; and

sufficient pitching moment control means to selectively drive said vehicle pitch attitude between said two stable trim attitudes.

9. An aerospace vehicle as described in claim **8** wherein said vehicle is positively statically stable about pitch and roll fuselage reference axes and neutrally statically stable about a yaw axis coinciding with a relative wind axis in flight at said attack angle in excess of 40 degrees.

10. An aerospace vehicle as described in claim **9** wherein said vehicle is positively statically stable about a vehicle reference yaw axis in flight at said attack angle of less than 40 degrees.

11. A method of recovering a vehicle having fixed aerodynamic support and control surfaces from high altitudes and hypersonic velocities comprising:

regulating the flight attitude of said vehicle throughout the hypersonic velocity region to attack the relative wind with a substantially maximum area profile;

providing positive static pitch stability throughout said hypersonic velocity region by relating the lifting force from said relative wind to the angle of incidence of said vehicle over one portion thereof at a rate different from the remaining portion;

generating a shock wave between the undisturbed atmospheric fluid and said attack area profile at a discrete distance ahead of said vehicle;

conducting fluid flow behind said shock wave over said attack area surface at laminar flow states throughout said hypersonic velocity region;

regulating the flight attitude of said vehicle throughout the subsonic velocity region to attack the relative wind with a substantially minimum profile;

generating aerodynamic lift from said fixed aerodynamic support and control surfaces throughout said subsonic velocity region; and

horizontally landing said vehicle in a substantially stall attitude.

12. A high attitude vehicle recovery method as described in claim **11** additionally comprising:

substantially separating said fluid from said attack area surface at the peripheral boundary thereof throughout said hypersonic velocity region.

13. An aircraft having fuselage means fixed position wings and attitude control surfaces:

said fuselage having roll and pitch and first yaw axes relative thereto;

a second yaw axis parallel with the relative wind against said aircraft in flight and substantially intersecting the vehicle center of gravity; and

said aircraft being positively statically stable about said roll and pitch axes and neutrally statically stable about said second yaw axis in flight at full stall attack angle attitudes.

14. An aircraft as described in claim **13** also being positively statically stable about said first yaw axes in flight at angles of attack less than full stall.

15. An aircraft having fuselage and wing means:

the bottom of said fuselage being substantially flat over a first portion thereof, both longitudinally and transversely, whereby substantially uniform impact lifting pressures are imposed against said first portion of the bottom in full stall flight attitudes; and

said fuselage bottom having rigid structural means over a second portion thereof for effecting positive static pitch stability restoring moments on said aircraft in said full stall flight attitudes.

16. An aircraft as described in claim **15** also having means for positive static roll stability about a longitudinal fuselage axis and neutral static yaw stability about an axis parallel with the relative wind and substantially intersecting the aircraft center of gravity in said full stall flight attitudes.

17. An aircraft as described by claim **15** wherein said rigid structural means comprises longitudinally cambered surface means disposed along the leading end of said fuselage bottom.

References Cited

UNITED STATES PATENTS

3,104,079	9/1963	Phillips	244—155
3,132,825	5/1964	Postle et al.	244—155
3,369,771	2/1968	Walley et al.	244—155
3,058,691	10/1962	Eggers et al.	244—2
3,276,722	10/1966	Eggers et al.	244—155
3,090,580	5/1963	Kehlet et al.	244—155

EVON C. BLUNK, Primary Examiner

B. H. STONER, JR., Assistant Examiner

U.S. Cl. X.R.

244—2, 36, 90

J. F. GLIDDEN.

Wire-Fences.

No. 157,124. Patented Nov. 24, 1874.

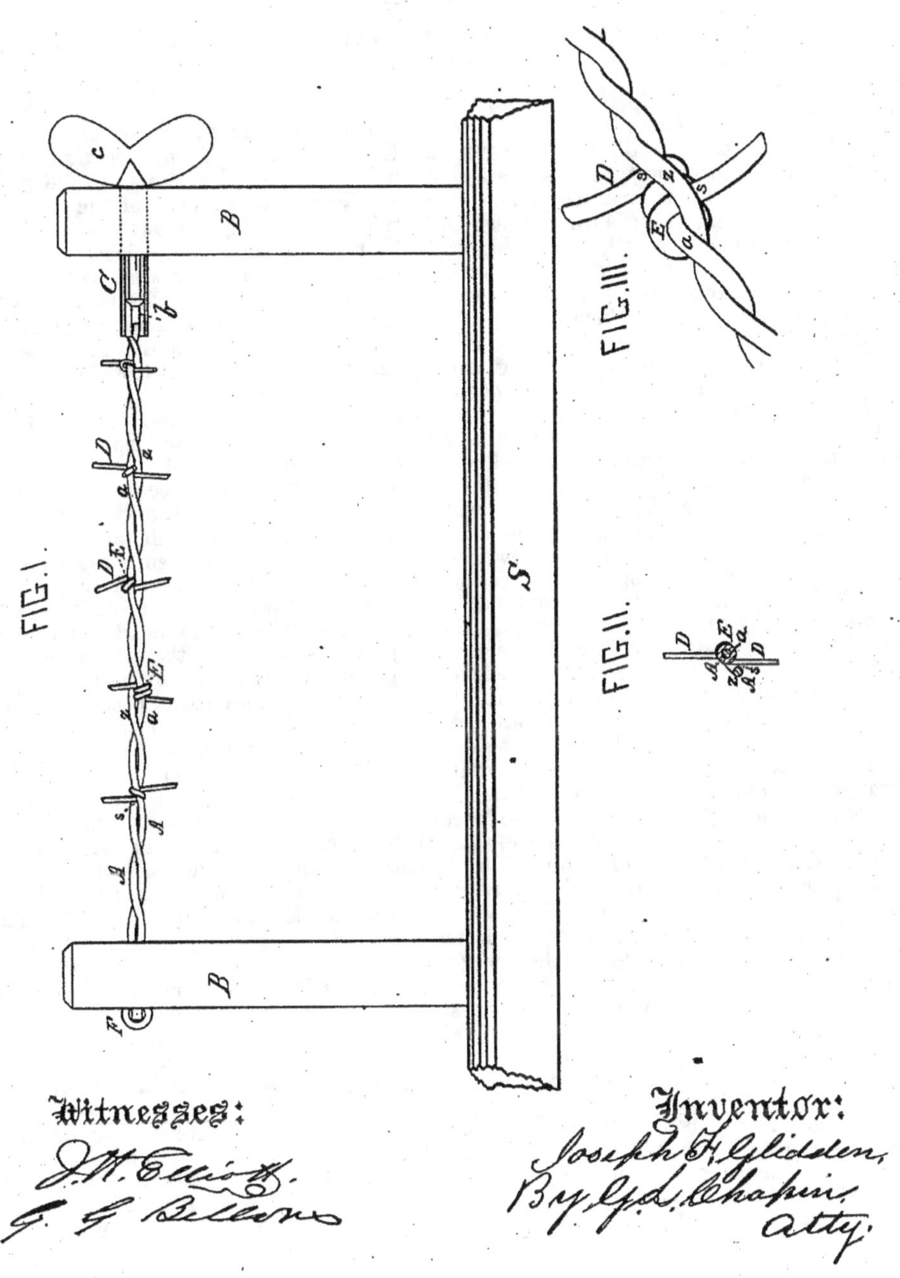

THE GRAPHIC CO. PHOTO-LITH, 39 & 41 PARK PLACE, N.Y.

UNITED STATES PATENT OFFICE.

JOSEPH F. GLIDDEN, OF DE KALB, ILLINOIS.

IMPROVEMENT IN WIRE-FENCES.

Specification forming part of Letters Patent No. **157,124**, dated November 24, 1874; application filed October 27, 1873.

To all whom it may concern:

Be it known that I, JOSEPH F. GLIDDEN, of De Kalb, in the county of De Kalb and State of Illinois, have invented a new and valuable Improvement in Wire-Fences; and that the following is a full, clear, and exact description of the construction and operation of the same, reference being had to the accompanying drawings, in which—

Figure 1 represents a side view of a section of fence exhibiting my invention. Fig. 2 is a sectional view, and Fig. 3 is a perspective view, of the same.

This invention has relation to means for preventing cattle from breaking through wire-fences; and it consists in combining, with the twisted fence-wires, a short transverse wire, coiled or bent at its central portion about one of the wire strands of the twist, with its free ends projecting in opposite directions, the other wire strand serving to bind the spur-wire firmly to its place, and in position, with its spur ends perpendicular to the direction of the fence-wire, lateral movement, as well as vibration, being prevented. It also consists in the construction and novel arrangement, in connection with such a twisted fence-wire, and its spur-wires, connected and arranged as above described, of a twisting-key or head-piece passing through the fence-post, carrying the ends of the fence-wires, and serving, when the spurs become loose, to tighten the twist of the wires, and thus render them rigid and firm in position.

In the accompanying drawings, the letter B designates the fence-posts, the twisted fence-wire connecting the same being indicated by the letter A. C represents the twisting-key, the shank of which passes through the fence-post, and is provided at its end with an eye, *b*, to which the fence-wire is attached. The outer end of said key is provided with a transverse thumb-piece, *c*, which serves for its manipulation, and at the same time, abutting against the post, forms a shoulder or stop, which prevents the contraction of the wire from drawing the key through its perforation in said post.

The fence-wire is composed at least of two strands, *a* and *z*, which are designed to be twisted together after the spur-wires have been arranged in place.

The letter D indicates the spur-wires. Each of these is formed of a short piece of wire, which is bent at its middle portion, as at E, around one only of the wire strands, this strand being designated by the letter *a*. In forming this middle bend or coil several turns are taken in the wire, so that it will extend along the strand-wire for a distance several times the breadth of its diameter, and thereby form a solid and substantial bearing-head for the spurs, which will effectually prevent them from vibrating laterally or being pushed down by cattle against the fence-wire. Although these spur-wires may be turned at once around the wire strand, it is preferred to form the central bend first, and to then slip them on the wire strand, arranging them at suitable distances apart. The spurs having thus been arranged on one of the wire strands are fixed in position and place by approaching the other wire strands *z* on the side of the bend from which the spurs extend, and then twisting the two strands *a z* together by means of the wire key above mentioned or otherwise. This operation locks each spur wire at its allotted place, and prevents it from moving therefrom in either direction. It clamps the bend of the spur-wire upon the wire *a*, thereby holding it against rotary vibration. Finally, the spur ends extending out between the strands on each side, and where the wires are more closely approximated in the twist, form shoulders or stops, *s*, which effectually prevent such rotation in either direction.

Should the spurs, from the untwisting of the strands, become loose and easily movable on their bearings, a few turns of the twisting-key will make them firm, besides straightening up the fence-wire.

What I claim as my invention, and desire to secure by Letters Patent, is—

A twisted fence-wire having the transverse spur-wire D bent at its middle portion about one of the wire strands *a* of said fence-wire, and clamped in position and place by the other wire strand *z*, twisted upon its fellow, substantially as specified.

JOSEPH F. GLIDDEN.

Witnesses:
G. L. CHAPIN,
J. H. ELLIOTT.

Ives W. McGaffey.

Sweeping-Machine.

No. 91,145.

Patented June 8, 1869.

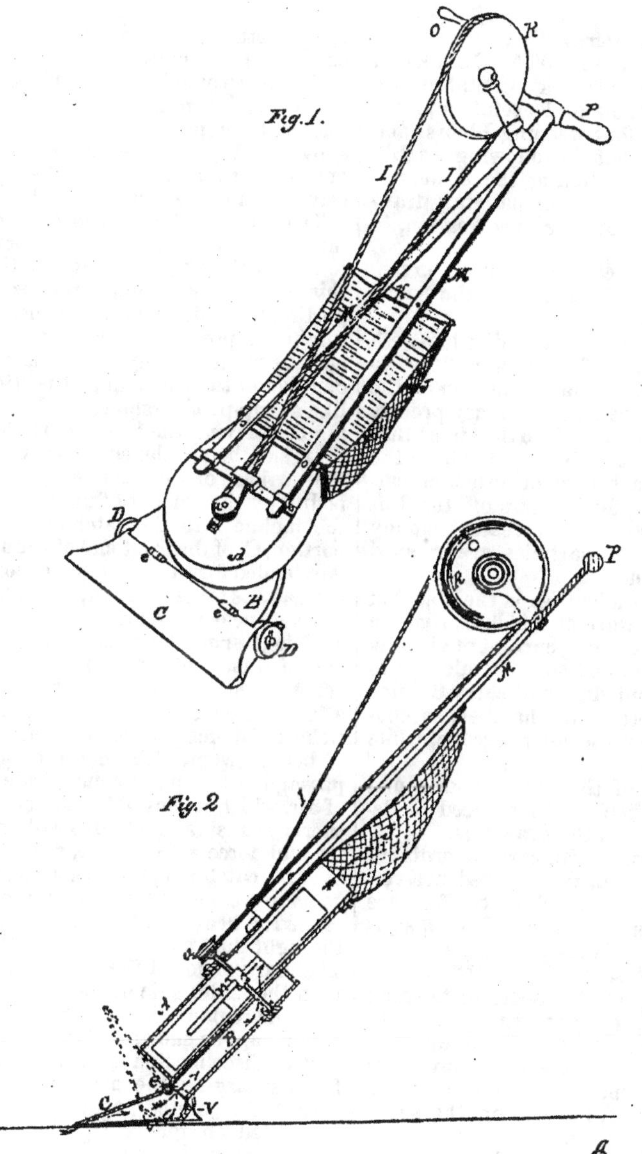

Witnesses,
L. Hailer
P. T. Dodge

Inventor, Ives W. McGaffey
by Dodge & Munn
his attys

UNITED STATES PATENT OFFICE.

IVES W. McGAFFEY, OF CHICAGO, ILLINOIS.

IMPROVED SWEEPING-MACHINE.

Specification forming part of Letters Patent No. **91,145**, dated June 8, 1869.

To all whom it may concern:

Be it known that I, IVES W. MCGAFFEY, of Chicago, in the county of Cook and State of Illinois, have invented new and useful Improvements in a Machine for Sweeping Floors; and I do hereby declare that the following is a full, clear, and exact description of the same, reference being had to the accompanying drawings, making a part of this specification, in which—

Figure 1 is a perspective view of my invention; Fig. 2, a vertical section of the same, taken through the center.

The accumulation of dust and dirt in dwelling-houses is a source of great annoyance to all good housekeepers, a large portion of the dust being so light that the ordinary process of sweeping sends it flying into the air, so that it is difficult to control or expel it from the room. The use of a broom or brush on carpets is objectionable, for it wears off the lint and fiber in fine particles, and creates a cloudy dust, while the heavier particles are brushed down into the carpet.

To obviate these difficulties is the object of my invention, the nature of which consists in the employment of a strong current of air, produced by mechanism, and so controlled as to take up the dust and dirt, and carry the fine particles into a porous air-chamber, so constructed as to allow the air to escape, while the dust is retained.

The construction of the several parts and the mode of operation I will proceed to describe with reference to the drawings.

A represents a rotary fan-case of ordinary construction, with a fan, W, mounted on a vertical shaft, H, which is fixed in the fan-case by suitable bearings, as shown in Fig. 2, so as to allow it to revolve freely upon its axis. A handle, M M, is attached to the fan-case, of suitable length to allow the operator to stand or move in an erect position while using the machine. A grooved wheel or pulley, R, is fixed on the handle, near the end, convenient to the hand of the operator, provided with a handle, O, on one side, for turning the same. From this hand-wheel motion is communicated to the fan by the use of a belt or cord, I, connecting it with a small pulley, *a*, on the end of the fan-shaft.

The air to supply the fan-blast is admitted at the axis through an air passage or conductor, B, of peculiar construction, which is shown in Fig. 2 extended forward, widening out at its front and tapered down, so as to receive the air in a thin, broad sheet.

The front portion of the air-conductor C is intended to rest upon the floor or carpet when in use, in the position shown in the drawings, this front portion being attached to the rear portion by a hinge-joint, *e*, which allows it to retain its proper position, while the handle may be carried at any desired angle. The under side of the front hinged portion is formed like a drip-pan, as shown in Fig. 2, for retaining or carrying the heavy particles of dirt, its front side being sloped down to an edge, so that articles of dirt too heavy to be taken up bodily by the current of air may be drawn up the incline into the catch-pan *o*. The upper portion, C, of the front end of the air-conductor is extended forward over and beyond the catch-pan, and is tapered down, so that when in use it comes near the floor or carpet, leaving only sufficient space between to allow small particles of dust and dirt to pass under it, and so constructed that the current of air is received at the extreme front, and compelled to move a short distance parallel with the surface which is being swept. The contraction of the air-passage at this point concentrates the volume of air, which is moved by the action of the fan, thereby causing it to move with greater rapidity and force. In the rear of the front, and over the catch-pan, the air-passage is enlarged, so as to weaken the current of air, and thereby cause the heavy particles to be separated from the light dust and deposited in the catch-pan. The rear portion of the catch-pan is connected with the under side of the air-conductor B by a semicircular joint, V, so that when it is turned up, as shown by dotted lines X, Fig. 2, it is disconnected and an opening formed for discharging the accumulated dirt. The rear of the jointed portion C, when in use, is supported on rollers D, while its front portion slides upon the surface.

The blast from the fan is discharged into an air-chamber, J, which is constructed of material sufficiently porous to allow the air to gradually pass through, while the dust and dirt are retained.

This air-chamber may be constructed in a variety of forms, and of different kinds of material, according to fancy or circumstances. For ordinary purposes I use a bag or sack, made of fine and firm cotton cloth, with an elastic band or cord fixed in a hem around its mouth, which is drawn closely over the discharge-opening of the fan, so as to receive the blast. Its opposite end is suspended to a cross-bar, K, on the handle.

To use this machine, the operator grasps the handle P with one hand and turns the wheel R briskly with the other, and moves forward over the floor or carpet to be swept, with the machine in the position shown in the drawings.

Having thus fully described the nature and object of my invention, what I claim is—

1. A machine, substantially such as is herein described, having a rotating fan arranged to draw or suck a current of air, and with it the dust or dirt from the floor, up into the machine, as set forth.

2. The combination of the case A with its fan, the conductor B, and mouth-piece C, when arranged to operate as and for the purpose set forth.

3. The combination of a fan for drawing or sucking up the dirt by a current of air and a porous receptacle, which serves to retain the particles of dirt while permitting the air to escape.

IVES W. McGAFFEY.

Witnesses:

W. M. HOWLAND,

C. C. CLARKE.

2 Sheets—Sheet 1.

A. G. BELL.

TELEGRAPHY.

No. 174,465. Patented March 7, 1876.

Fig. 1

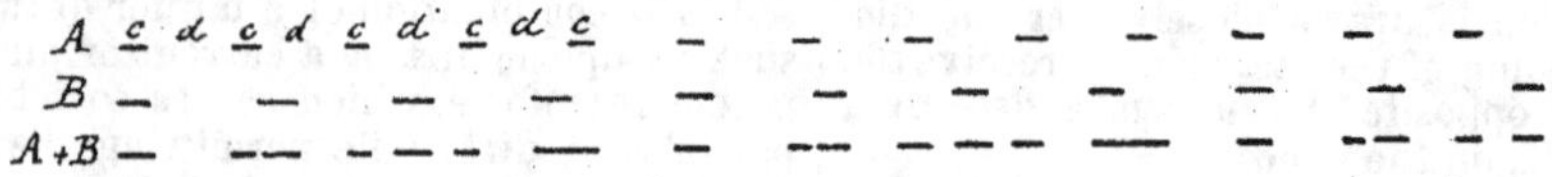

A c d c d c d c d c d c

B

A+B

Fig. 2.

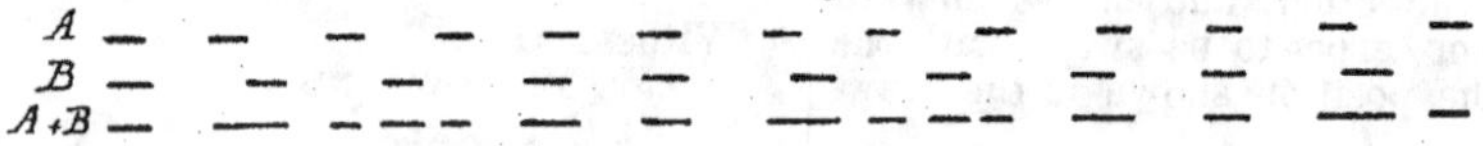

A

B

A+B

Fig. 3.

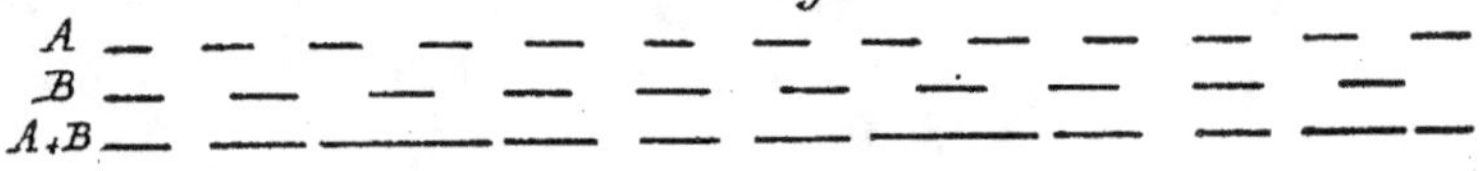

A

B

A+B

Fig. 4.

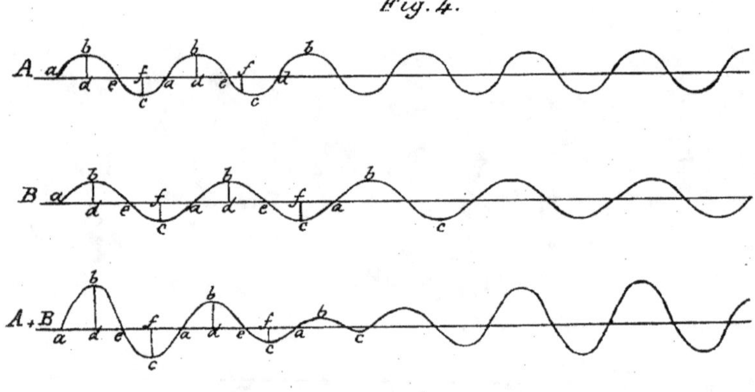

Fig. 5.

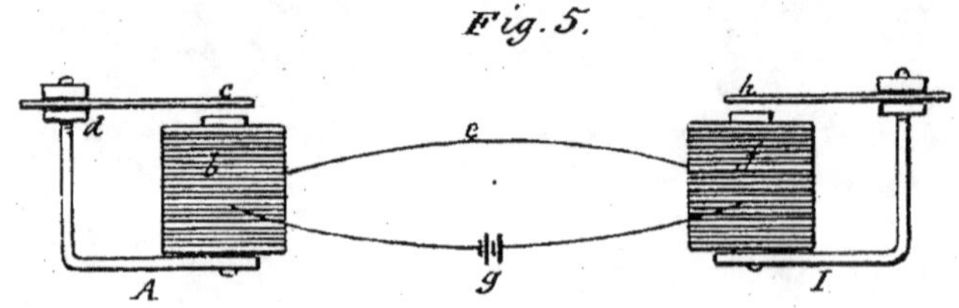

Witnesses

Ewell A. Dick.

W. J. Hutchinson

Inventor:

A. Graham Bell

by atty Pollok & Bailey

2 Sheets—Sheet 2.

A. G. BELL.

TELEGRAPHY.

No. 174,465. Patented March 7, 1876.

Fig 6.

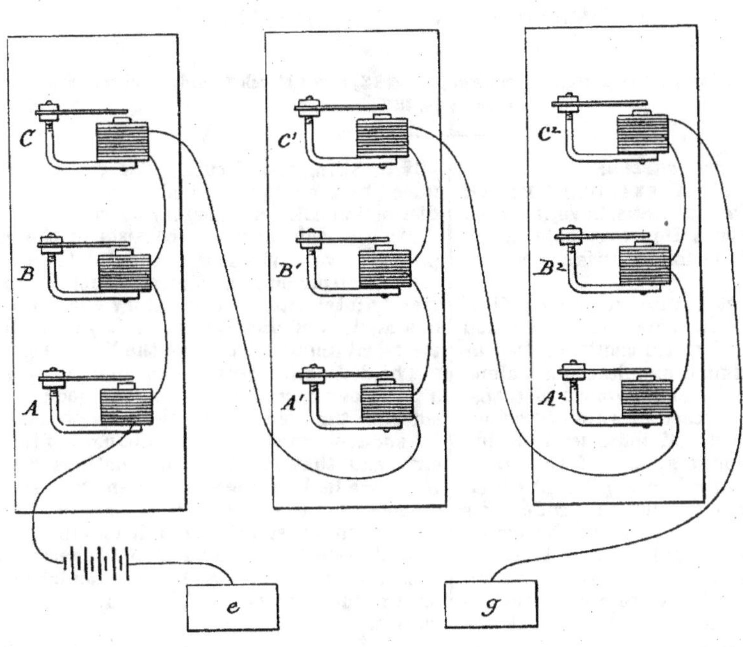

Fig. 7

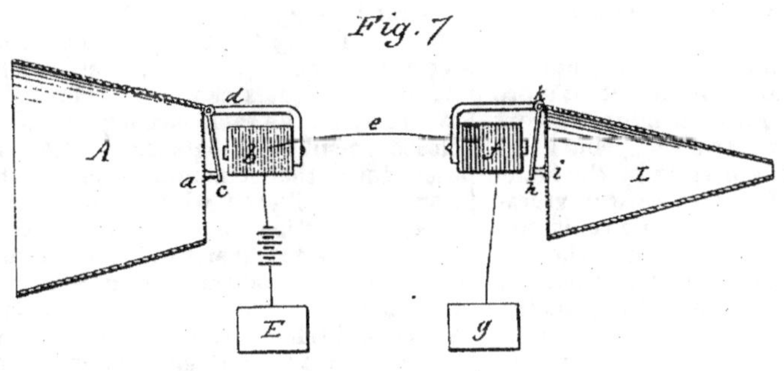

Witnesses

Ewell A. Dick

W. J. Hutchinson

Inventor:

A. Graham Bell

by attys Pollok & Bailey

UNITED STATES PATENT OFFICE.

ALEXANDER GRAHAM BELL, OF SALEM, MASSACHUSETTS.

IMPROVEMENT IN TELEGRAPHY.

Specification forming part of Letters Patent No. **174,465**, dated March 7, 1876; application filed February 14, 1876.

To all whom it may concern:

Be it known that I, ALEXANDER GRAHAM BELL, of Salem, Massachusetts, have invented certain new and useful Improvements in Telegraphy, of which the following is a specification:

In Letters Patent granted to me April 6, 1875, No. 161,739, I have described a method of, and apparatus for, transmitting two or more telegraphic signals simultaneously along a single wire by the employment of transmitting-instruments, each of which occasions a succession of electrical impulses differing in rate from the others; and of receiving-instruments, each tuned to a pitch at which it will be put in vibration to produce its fundamental note by one only of the transmitting-instruments; and of vibratory circuit-breakers operating to convert the vibratory movement of the receiving-instrument into a permanent make or break (as the case may be) of a local circuit, in which is placed a Morse sounder, register, or other telegraphic apparatus. I have also therein described a form of autograph-telegraph based upon the action of the above-mentioned instruments.

In illustration of my method of multiple telegraphy I have shown in the patent aforesaid, as one form of transmitting-instrument, an electro-magnet having a steel-spring armature, which is kept in vibration by the action of a local battery. This armature in vibrating makes and breaks the main circuit, producing an intermittent current upon the line-wire. I have found, however, that upon this plan the limit to the number of signals that can be sent simultaneously over the same wire is very speedily reached; for, when a number of transmitting-instruments, having different rates of vibration, are simultaneously making and breaking the same circuit, the effect upon the main line is practically equivalent to one continuous current.

In a pending application for Letters Patent, filed in the United States Patent Office February 25, 1875, I have described two ways of producing the intermittent current—the one by actual make and break of contact, the other by alternately increasing and diminishing the intensity of the current without actually breaking the circuit. The current produced by the latter method I shall term, for distinction sake, a pulsatory current.

My present invention consists in the employment of a vibratory or undulatory current of electricity in contradistinction to a merely intermittent or pulsatory current, and of a method of, and apparatus for, producing electrical undulations upon the line-wire.

The distinction between an undulatory and a pulsatory current will be understood by considering that electrical pulsations are caused by sudden or instantaneous changes of intensity, and that electrical undulations result from gradual changes of intensity exactly analogous to the changes in the density of air occasioned by simple pendulous vibrations. The electrical movement, like the aerial motion, can be represented by a sinusoidal curve or by the resultant of several sinusoidal curves.

Intermittent or pulsatory and undulatory currents may be of two kinds, accordingly as the successive impulses have all the same polarity or are alternately positive and negative.

The advantages I claim to derive from the use of an undulatory current in place of a merely intermittent one are, first, that a very much larger number of signals can be transmitted simultaneously on the same circuit; second, that a closed circuit and single main battery may be used; third, that communication in both directions is established without the necessity of special induction-coils; fourth, that cable dispatches may be transmitted more rapidly than by means of an intermittent current or by the methods at present in use; for, as it is unnecessary to discharge the cable before a new signal can be made, the lagging of cable-signals is prevented; fifth, and that as the circuit is never broken a spark-arrester becomes unnecessary.

It has long been known that when a permanent magnet is caused to approach the pole of an electro-magnet a current of electricity is induced in the coils of the latter, and that when it is made to recede a current of opposite polarity to the first appears upon the wire. When, therefore, a permanent magnet is caused to vibrate in front of the pole of an electromagnet an undulatory current of electricity is induced in the coils of the electro-magnet, the undulations of which correspond, in rapidity of succession, to the vibrations of the magnet, in polarity to the direction of its motion, and in intensity to the amplitude of its vibration.

That the difference between an undulatory and an intermittent current may be more clearly understood I shall describe the condition of the electrical current when the attempt is made to transmit two musical notes simultaneously—first upon the one plan and then upon the other. Let the interval between the two sounds be a major third; then their rates of vibration are in the ratio of 4 to 5. Now, when the intermittent current is used the circuit is made and broken four times by one transmitting-instrument in the same time that five makes and breaks are caused by the other. A and B, Figs. 1, 2, and 3, represent the intermittent currents produced, four impulses of B being made in the same time as five impulses of A. *c c c*, &c., show where and for how long time the circuit is made, and *d d d*, &c., indicate the duration of the breaks of the circuit. The line A and B shows the total effect upon the current when the transmitting-instruments for A and B are caused simultaneously to make and break the same circuit. The resultant effect depends very much upon the duration of the make relatively to the break. In Fig. 1 the ratio is as 1 to 4; in Fig. 2, as 1 to 2; and in Fig. 3 the makes and breaks are of equal duration. The combined effect, A and B, Fig. 3, is very nearly equivalent to a continuous current.

When many transmitting-instruments of different rates of vibration are simultaneously making and breaking the same circuit the current upon the main line becomes for all practical purposes continuous.

Next, consider the effect when an undulatory current is employed. Electrical undulations, induced by the vibration of a body capable of inductive action, can be represented graphically, without error, by the same sinusoidal curve which expresses the vibration of the inducing body itself, and the effect of its vibration upon the air; for, as above stated, the rate of oscillation in the electrical current corresponds to the rate of vibration of the inducing body—that is, to the pitch of the sound produced. The intensity of the current varies with the amplitude of the vibration—that is, with the loudness of the sound; and the polarity of the current corresponds to the direction of the vibrating body—that is, to the condensations and rarefactions of air produced by the vibration. Hence, the sinusoidal curve A or B, Fig. 4, represents, graphically, the electrical undulations induced in a circuit by the vibration of a body capable of inductive action.

The horizontal line *a d e f*, &c., represents the zero of current. The elevations *b b b*, &c., indicate impulses of positive electricity. The depressions *c c c*, &c., show impulses of negative electricity. The vertical distance *b d* or *c f* of any portion of the curve from the zero-line expresses the intensity of the positive or negative impulse at the part observed, and the horizontal distance *a a* indicates the duration of the electrical oscillation. The vibrations represented by the sinusoidal curves B and A, Fig. 4, are in the ratio aforesaid, of 4 to 5—that is, four oscillations of B are made in the same time as five oscillations of A.

The combined effect of A and B, when induced simultaneously on the same circuit, is expressed by the curve A+B, Fig. 4, which is the algebraical sum of the sinusoidal curves A and B. This curve A+B also indicates the actual motion of the air when the two musical notes considered are sounded simultaneously. Thus, when electrical undulations of different rates are simultaneously induced in the same circuit, an effect is produced exactly analogous to that occasioned in the air by the vibration of the inducing bodies. Hence, the coexistence upon a telegraphic circuit of electrical vibrations of different pitch is manifested, not by the obliteration of the vibratory character of the current, but by peculiarities in the shapes of the electrical undulations, or, in other words, by peculiarities in the shapes of the curves which represent those undulations.

There are many ways of producing undulatory currents of electricity, dependent for effect upon the vibrations or motions of bodies capable of inductive action. A few of the methods that may be employed I shall here specify. When a wire, through which a continuous current of electricity is passing, is caused to vibrate in the neighborhood of another wire, an undulatory current of electricity is induced in the latter. When a cylinder, upon which are arranged bar-magnets, is made to rotate in front of the pole of an electro-magnet, an undulatory current of electricity is induced in the coils of the electro-magnet.

Undulations are caused in a continuous voltaic current by the vibration or motion of bodies capable of inductive action; or by the vibration of the conducting-wire itself in the neighborhood of such bodies. Electrical undulations may also be caused by alternately increasing and diminishing the resistance of the circuit, or by alternately increasing and diminishing the power of the battery. The internal resistance of a battery is diminished by bringing the voltaic elements nearer together; and increased by placing them farther apart. The reciprocal vibration of the elements of a battery, therefore, occasions an undulatory action in the voltaic current. The external resistance may also be varied. For instance, let mercury or some other liquid form part of a voltaic circuit, then the more deeply the conducting-wire is immersed in the mercury or other liquid, the less resistance does the liquid offer to the passage of the current. Hence, the vibration of the conducting-wire in mercury or other liquid included in the circuit occasions undulations in the current. The vertical vibrations of the elements of a battery in the liquid in which they are immersed produces an undulatory action in the current by alternately increasing and diminishing the power of the battery.

In illustration of the method of creating electrical undulations, I shall show and describe one form of apparatus for producing the effect. I prefer to employ for this purpose an electro-magnet, A, Fig. 5, having a coil upon only one of its legs *b*. A steel-spring armature, *c*, is firmly clamped by one extremity to the uncovered leg *d* of the magnet, and its free end is allowed to project above the pole of the covered leg. The armature *c* can be set in vibration in a variety of ways, one of which is by wind, and, in vibrating, it produces a musical note of a certain definite pitch.

When the instrument A is placed in a voltaic circuit, *g b e f g*, the armature *c* becomes magnetic, and the polarity of its free end is opposed to that of the magnet underneath. So long as the armature *c* remains at rest, no effect is produced upon the voltaic current, but the moment it is set in vibration to produce its musical note a powerful inductive action takes place, and electrical undulations traverse the circuit *g b e f g*. The vibratory current passing through the coil of the electro-magnet *f* causes vibration in its armature *h* when the armatures *c h* of the two instruments A I are normally in unison with one another; but the armature *h* is unaffected by the passage of the undulatory current when the pitches of the two instruments are different.

A number of instruments may be placed upon a telegraphic circuit, as in Fig. 6. When the armature of any one of the instruments is set in vibration all the other instruments upon the circuit which are in unison with it respond, but those which have normally a different rate of vibration remain silent. Thus, if A, Fig. 6, is set in vibration, the armatures of A^1 and A^2 will vibrate also, but all the others on the circuit will remain still. So if B^1 is caused to emit its musical note the instruments B B^2 respond. They continue sounding so long as the mechanical vibration of B^1 is continued, but become silent with the cessation of its motion. The duration of the sound may be used to indicate the dot or dash of the Morse alphabet, and thus a telegraphic dispatch may be indicated by alternately interrupting and renewing the sound.

When two or more instruments of different pitch are simultaneously caused to vibrate, all the instruments of corresponding pitches upon the circuit are set in vibration, each responding to that one only of the transmitting instruments with which it is in unison. Thus the signals of A, Fig. 6, are repeated by A^1 and A^2, but by no other instrument upon the circuit; the signals of B^2 by B and B^1; and the signals of C^1 by C and C^2—whether A, B^2, and C^2 are successively or simultaneously caused to vibrate. Hence by these instruments two or more telegraphic signals or messages may be sent simultaneously over the same circuit without interfering with one another.

I desire here to remark that there are many other uses to which these instruments may be put, such as the simultaneous transmission of musical notes, differing in loudness as well as in pitch, and the telegraphic transmission of noises or sounds of any kind.

When the armature *c*, Fig. 5, is set in vibration the armature *h* responds not only in pitch, but in loudness. Thus, when *c* vibrates with little amplitude, a very soft musical note proceeds from *h*; and when *c* vibrates forcibly the amplitude of the vibration of *h* is considerably increased, and the resulting sound becomes louder. So, if A and B, Fig. 6, are sounded simultaneously, (A loudly and B softly,) the instruments A^1 and A^2 repeat loudly the signals of A, and B^1 B^2 repeat softly those of B.

One of the ways in which the armature *c*, Fig. 5, may be set in vibration has been stated above to be by wind. Another mode is shown in Fig. 7, whereby motion can be imparted to the armature by the human voice or by means of a musical instrument.

The armature *c*, Fig. 7, is fastened loosely by one extremity to the uncovered leg *d* of the electro-magnet *b*, and its other extremity is attached to the center of a stretched membrane, *a*. A cone, A, is used to converge sound-vibrations upon the membrane. When a sound is uttered in the cone the membrane *a* is set in vibration, the armature *c* is forced to partake of the motion, and thus electrical undulations are created upon the circuit E *b e f g*. These undulations are similar in form to the air vibrations caused by the sound—that is, they are represented graphically by similar curves.

The undulatory current passing through the electro-magnet *f* influences its armature *h* to copy the motion of the armature *c*. A similar sound to that uttered into A is then heard to proceed from L.

In this specification the three words "oscillation," "vibration," and "undulation," are used synonymously, and in contradistinction to the terms "intermittent" and "pulsatory." By the terms "body capable of inductive action," I mean a body which, when in motion, produces dynamical electricity. I include in the category of bodies capable of inductive action—brass, copper, and other metals, as well as iron and steel.

Having described my invention, what I claim, and desire to secure by Letters Patent is as follows:

1. A system of telegraphy in which the receiver is set in vibration by the employment of undulatory currents of electricity, substantially as set forth.

2. The combination, substantially as set forth, of a permanent magnet or other body capable of inductive action, with a closed circuit, so that the vibration of the one shall occasion electrical undulations in the other, or in itself, and this I claim, whether the permanent magnet be set in vibration in the neighborhood of the conducting-wire form-

ing the circuit, or whether the conducting-wire be set in vibration in the neighborhood of the permanent magnet, or whether the conducting-wire and the permanent magnet both simultaneously be set in vibration in each other's neighborhood.

3. The method of producing undulations in a continuous voltaic current by the vibration or motion of bodies capable of inductive action, or by the vibration or motion of the conducting-wire itself, in the neighborhood of such bodies, as set forth.

4. The method of producing undulations in a continuous voltaic circuit by gradually increasing and diminishing the resistance of the circuit, or by gradually increasing and diminishing the power of the battery, as set forth.

5. The method of, and apparatus for, transmitting vocal or other sounds telegraphically, as herein described, by causing electrical undulations, similar in form to the vibrations of the air accompanying the said vocal or other sound, substantially as set forth.

In testimony whereof I have hereunto signed my name this 20th day of January, A. D. 1876.

ALEX. GRAHAM BELL.

Witnesses:
THOMAS E. BARRY,
P. D. RICHARDS.

No. 708,553. Patented Sept. 9, 1902.

J. P. HOLLAND.

SUBMARINE BOAT.

(Application filed Aug. 7, 1901.)

(No Model.) 2 Sheets—Sheet 1.

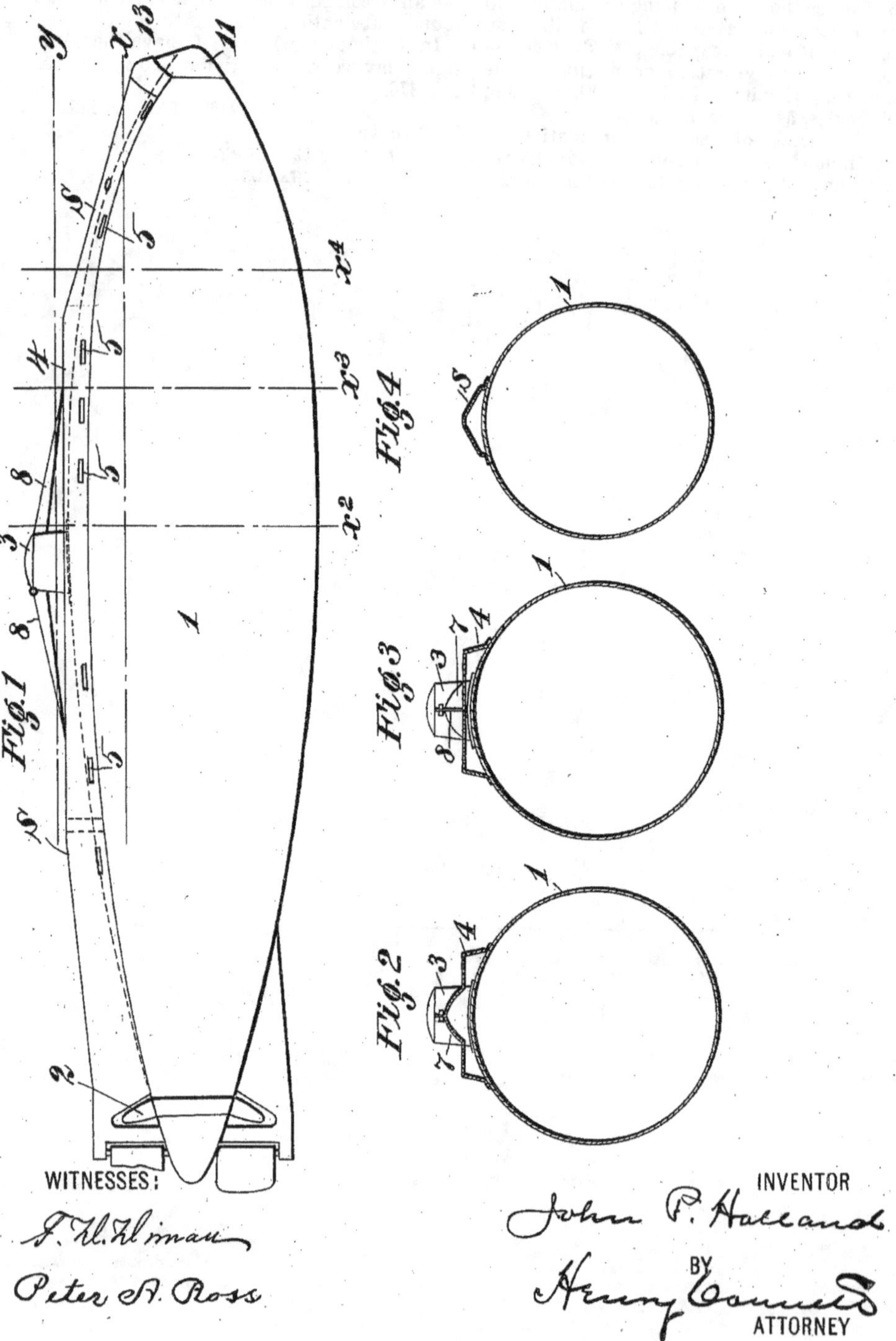

WITNESSES:

F. W. Wiman

Peter A. Ross

INVENTOR

John P. Holland

BY

Henry Connett

ATTORNEY

THE NORRIS PETERS CO., PHOTO-LITHO., WASHINGTON, D. C.

No. 708,553. Patented Sept. 9, 1902.

J. P. HOLLAND.

SUBMARINE BOAT.

(Application filed Aug. 7, 1901.)

(No Model.) 2 Sheets—Sheet 2.

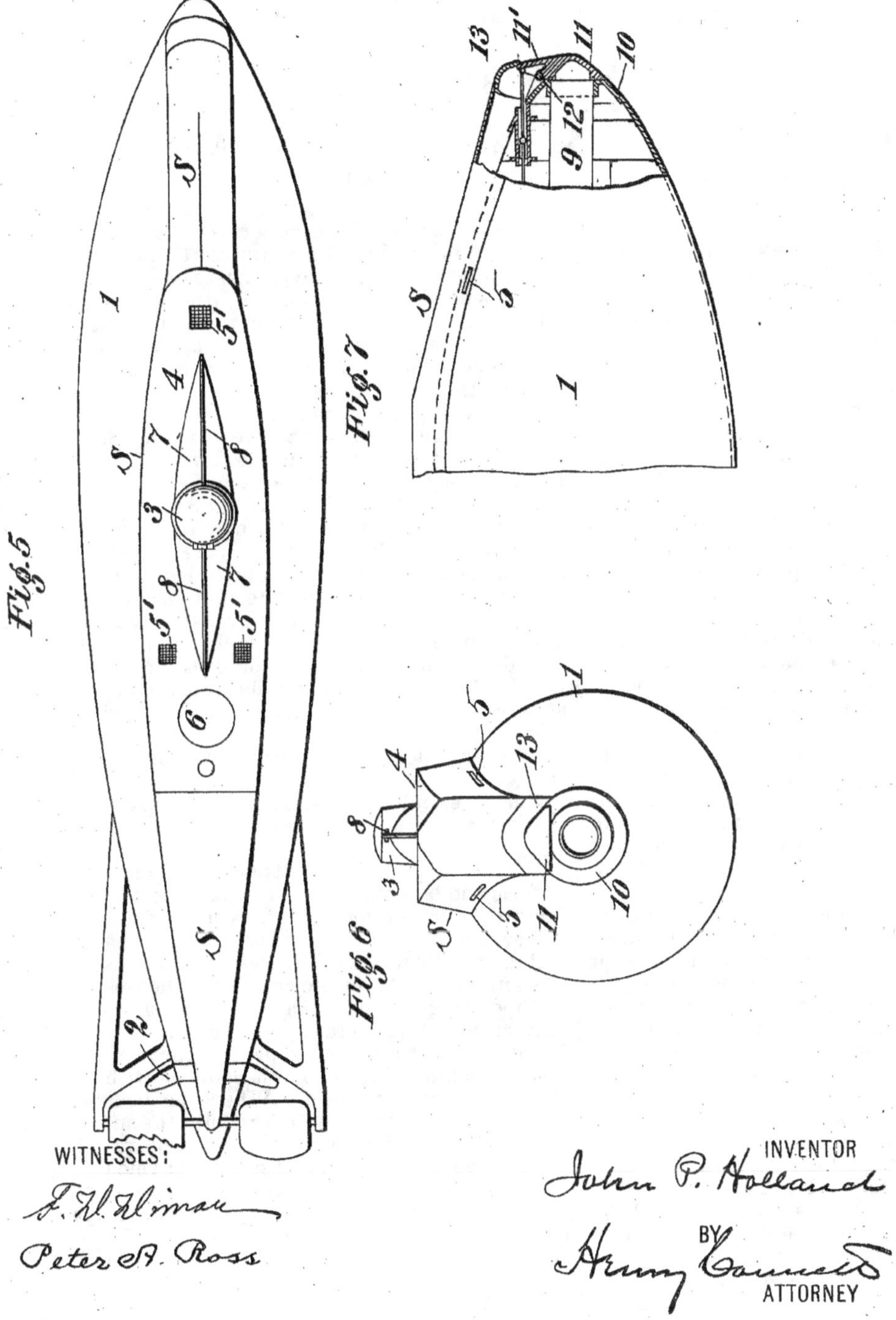

F. W. Ulman

Peter A. Ross

Henry Connett

THE NORRIS PETERS CO., PHOTO-LITHO., WASHINGTON, D. C.

UNITED STATES PATENT OFFICE.

JOHN P. HOLLAND, OF NEWARK, NEW JERSEY.

SUBMARINE BOAT.

SPECIFICATION forming part of Letters Patent No. 708,553, dated September 9, 1902.

Application filed August 7, 1901. Serial No. 71,130. (No model.)

To all whom it may concern:

Be it known that I, JOHN P. HOLLAND, a citizen of the United States, residing at Newark, in the county of Essex and State of New Jersey, have invented certain new and useful Improvements in Submarine Boats, of which the following is a specification.

This invention relates to the class of boats or vessels which are adapted to be operated or maneuvered both on the surface of the water and submerged; and the object is, in the main, to provide the boat, which will have the known spindle form or contour, with a deck or promenade for the crew when the boat is operating on the surface and such a superstructure for this purpose as will permit the boat to dive and operate submerged without impediment or hindrance.

In the accompanying drawings, which illustrate the invention embodied, Figure 1 is a side elevation of a submarine boat. Figs. 2, 3, and 4 are transverse sections of the same at the points in Fig. 1 indicated, respectively, by the lines x^2, x^3, and x^4. Fig. 5 is a plan of the boat. Fig. 6 is a bow end view of the boat, showing the cap of the expulsion-tube open. Fig. 7 is a sectional side elevation of the bow portion of the boat, showing the construction of the hinged cap of the expulsion-tube and the superstructure at the bow.

1 designates the spindle-shaped hull or body of the boat, 2 the propeller, and 3 the turret or conning-tower. These are or may be of the usual or known construction. In Fig. 1 the line x designates the water-level when the boat is adapted for surface running, and the line y designates the water-level when the boat is ballasted to put it awash or in diving condition.

On the rounded back or top of the boat is built and rigidly secured a hollow superstructure S, having a flat deck or promenade 4 extending both forward and aft of the conning-tower. This superstructure toward the bow has the form in cross-section seen in Fig. 4—that is, its top slopes off at each side from a central ridge in a vertical plane coincident with the axis of the boat; but at the bow it is arched, as seen in Fig. 6. Along the sides of the superstructure are scuppers 5 for the free escape of the water from the superstructure when the boat comes to the surface, and in the deck 4 are gratings 5′, one or more, to permit the air to escape as the boat sinks beneath the surface. In the superstructure is shown a well 6 for a coil of rope, and about the conning-tower, fore and aft, is a rounded structure 7 above the deck 4 and sloping from the tower down to the deck. In a vertical plane passing through the longitudinal axis of the boat are thin fins 8, which abut against the conning-tower. The upper edges of these fins slope from the top of the tower down to the deck 4 and serve as skids to carry a hawser or line over the tower and prevent fouling. The purpose of this superstructure is in part to provide a deck and promenade for the crew, to afford a cover and protection for ventilators, relief-valves, exhaust-pipes, and mufflers, which are on the outside of the upper surface or back of the hull of the boat, and to provide a convenient stowage-space for the anchor, cable, and mooring-lines.

The construction at the bow of the boat is seen in Figs. 6 and 7.

9 is the expulsion-tube, and 10 the muzzle-casting, where the said tube and the boat-hull are joined.

11 is the cap of the expulsion-tube, coned to form the bow-tip or nose of the boat and hinged to the casting 9 at 12. On the front or bow end of the superstructure is a strong metal hood 13, which takes over and protects the shield-plate 11′ of the cap 11 and houses the operating-gear of the cap. This hood not only protects the operating-gear of the cap, but it serves as a fender to prevent the accidental opening of the cap from collision with a dock or floating object. When the boat sinks, the water freely enters and fills the superstructure, and when the boat rises the water flows out freely. This construction avoids the necessity of providing an excess of water-ballast space in the interior of the boat to overcome or neutralize the buoyancy if the superstructure were made water-tight. The superstructure S extends the entire length of the boat from stem to stern, and the purpose in giving to it the inverted-V form seen in Fig. 4 is to reduce resistance in moving through the water. This also is the object of the

rounded structures 7 in front and rear of the conning-tower. They serve to part the water as the boat is running.

Having thus described my invention, I claim—

1. A submarine boat provided on its top or rounded back with a hollow superstructure having a flat, level promenade both forward and aft of the conning-tower, and having sloping portions extending from said level promenade down to the stem and stern of the boat, said hollow superstructure having at its sides always-open scuppers for the flow of water in and out, and having in its top always-open gratings for the flow of air in and out, substantially as set forth.

2. A submarine boat having on its top or rounded back a hollow superstructure with apertures for the free flow of water into and out of same, a conning-tower which extends up through said superstructure, the inclined structures, and the sloping fins 8 on the superstructure and abutting against the said tower to prevent the fouling of lines, substantially as set forth.

3. The combination with a submarine boat of spindle form and having an expulsion-tube, a muzzle-casting, and a cap 11 hinged to said casting and provided with a shield-plate 11′, of the hollow superstructure on the boat, provided with a hood 13 which houses the operative mechanism of the cap and the shield thereof, substantially as set forth.

4. A submarine boat having on its rounded top or back a hollow superstructure extending the entire length of the boat and open at all times for the outflow and inflow of air or water, said superstructure having at its middle part a flat, level, promenade-deck, and at its forward end an inverted-V form so as to reduce resistance in moving through the water, as set forth.

In witness whereof I have hereunto signed my name, this 30th day of July, 1901, in the presence of two subscribing witnesses.

JOHN P. HOLLAND.

Witnesses:
PETER A. ROSS,
K. M. CAPLINGER.

Erteilt auf Grund der Verordnung vom 12. Mai 1943
(RGBl. II S. 150)

DEUTSCHES REICH

AUSGEGEBEN AM
13. DEZEMBER 1944

REICHSPATENTAMT

PATENTSCHRIFT

№ 745930

KLASSE **63c** GRUPPE 35

P 77944 II/63c

Karl Fröhlich in Stuttgart und Erwin Komenda in Stuttgart-Korntal
sind als Erfinder genannt worden

Dr.-Ing. h. c. F. Porsche K.-G. in Stuttgart-Zuffenhausen

Kraftfahrzeug

Patentiert im Deutschen Reich vom 23. September 1938 an

Patenterteilung bekanntgemacht am 16. November 1944

Die Erfindung bezieht sich auf ein Kraftfahrzeug mit einem im Heck angeordneten, gegenüber dem Fahrgestell nachgiebig gelagerten Antriebsmotor, dessen im Motorraum liegender oberer Teil mit Hilfe einer Abdeckung gegen außen abgeschlossen ist, die in der Richtung vom Antriebsmotor gegen das Fahrgestell unterteilt ist und aus einem starren und einem nachgiebigen Teil besteht.

Es ist bei Schienenfahrzeugen bekannt, den Motor nachgiebig zu lagern und den Raum, in dem der Motor angeordnet ist, staubdicht gegen außen abzuschließen. Bei diesen bekannten Bauarten ist die Abdeckung des Motorraumes gegen unten in Richtung vom Antriebsmotor gegen die den Motor einhüllende Wandung unterteilt und besteht aus einem starren und einem nachgiebigen Teil, jedoch weisen diese Bauarten den Nachteil auf, daß sie keine starke Relativbewegung des Motors gegenüber der ihn umschließenden Wandung gestatten, ohne daß die Gefahr des Entstehens einer Lücke vorhanden wäre.

Diesen Nachteil vermeidet die Erfindung durch die besondere Art der Abdichtung, und zwar dadurch, daß der nachgiebige Teil der Abdeckung die Form einer im wesentlichen U-förmigen Leiste besitzt, zwischen deren Schenkeln der Rand des starren Teiles der Abdeckung zur Anlage kommt. Erst diese Maßnahme ermöglicht eine zuverlässige Abdichtung in jenen Fällen, bei welchen starke Relativbewegungen des Motors gegenüber der ihn umschließenden Wandung gegeben sind und ein Richtungswechsel der Bewegung stattfindet.

Die bisher bekannten Abdichtungen wiesen außerdem den Mangel auf, daß das Abdichtungsmittel an den Teilen, welche gegeneinander abgedichtet werden sollten, befestigt war, so daß ein rascher Ausbau des Motors aus dem Fahrgestell verhindert wurde. Auch dieser Nachteil erscheint durch die Maßnahmen der Erfindung vermieden, nach welchen der nachgiebige Teil der Abdeckung für sich am Fahrgestell befestigt ist und in seiner Lage verbleibt, wenn der starre Teil der Abdeckung beim Ausbau des Antriebsmotors entfernt wird.

Durch die im wesentlichen U-förmige Gestalt des nachgiebigen Teiles der Abdeckung, nämlich der Gummileiste, wird mit Sicherheit erreicht, daß bei jeder im Betrieb auftretenden Stellung des Antriebsmotors immer mindestens einer der Schenkel der Leiste abdichtend wirkt, und zwar unabhängig von den an sich unvermeidlichen Ungenauigkeiten, die sich beim Zusammenbau an den Dichtungsstellen ergeben. Dem Eindringen von Staub in den Motorraum ist auf diese Weise wirksam vorgebeugt, so daß der in ihm liegende Teil des Antriebsmotors mit allen Hilfseinrichtungen stets sauber gehalten wird; ein weiterer Vorteil besteht darin, daß die Ansaugluft des Antriebsmotors rein bleibt, wodurch dessen Instandhaltungskosten, ob er nun mit einem Luftfilter versehen ist oder nicht, gesenkt werden. Mit Vorteil ist hierbei die Ausbildung derart getroffen, daß der starre Teil der Abdeckung auf der Seite des Antriebsmotors liegt, da hierdurch hohe Temperaturen von dem immerhin empfindlichen nachgiebigen Teil der Abdeckung, der beispielsweise aus Gummi, gummiähnlichen Stoffen oder besonders imprägnierten Geweben besteht, ferngehalten werden können und der Aufbau der ganzen Abdeckung besonders einfach ist.

Bei Kraftfahrzeugen, deren Antriebsmotor im Fahrgestell schwenkbar gelagert ist, wird der nachgiebige Teil der Abdeckung im wesentlichen in einer durch die Schwenkachse des Antriebsmotors gehenden Ebene angeordnet. Hierdurch erfährt nicht nur der starre Teil der Abdeckung eine besonders einfache Gestaltung und kann der nachgiebige Teil der Abdeckung in einfacher Weise verlegt werden, sondern es werden vom letzteren auch ungünstige Beanspruchungen ferngehalten.

Von besonderem Vorteil ist es endlich, wenn der starre Teil der Abdeckung in bezug auf das Auspuffrohr des Antriebsmotors derart verlegt ist, daß dieses außerhalb des Motorraumes zu liegen kommt. Das Auspuffrohr braucht dann den starren Teil der Abdeckung nicht mehr zu durchsetzen, und es ist daher an dieser Stelle eine besondere, umständliche Abdichtung des Motorraumes nach außen nicht mehr erforderlich. Außerdem wird noch erreicht, daß der Motorraum durch das Auspuffrohr auch nicht mehr mittelbar durch die am Auspuffrohr vorbeistreichende und in ihn eintretende Luft erwärmt werden kann.

Die Erfindung ist in der Zeichnung an Hand eines Ausführungsbeispieles näher erläutert, und zwar zeigt

Fig. 1 eine Rückansicht eines Kraftfahrzeuges mit Heckantrieb, mit teilweise aufgeschnittenem Wagenkasten und aufgeschnittener Abdeckung, und

Fig. 2 den einen Rand des starren Teiles der Abdeckung mit dem nachgiebigen Teil derselben in etwa natürlicher Größe.

In Fig. 1 ist mit 1 der als Gegenläufer ausgebildete Antriebsmotor bezeichnet, der mit seinem Getriebeblock mit den Gummilagern 2, 3 um die zur Fahrzeuglängsachse parallele und damit zur Zeichenebene senkrechte Schwenkachse *S* im Fahrgestell im Sinne der Pfeile *B-B* schwenkbar gelagert ist; an ihm sind die den starren Teil der Abdeckung bildenden Abdeckbleche 4, 5 festgeschraubt, wobei die Formgebung und Lage derselben derart ist, daß die Auspuffrohre 6, 7 unterhalb und damit außerhalb des Motorraumes 8 zu liegen kommen. Gegen die mit dem Fahrgestell fest verbundenen Wagenkastenwände 9, 10 erfolgt die Abdichtung des Motorraumes 8 nach außen über die den nachgiebigen Teil der Abdeckung bildenden Gummileisten 11, 12, die in der durch die Schwenkachse *S* gehenden waagerechten Ebene *E-E* angeordnet sind.

Nach Fig. 2 weist die Gummileiste 11 einen U-förmigen Querschnitt auf. Sie ist mit ihrem Steg 13 mit Hilfe der diesen einklemmenden Blechleiste 14, die beispielsweise an der Wagenkastenwand 9 angepunktet ist, an dieser befestigt und umfaßt mit ihren Schenkeln 15, 16 den hochgestellten Rand 17 des Abdeckbleches 4 derart, daß bei starken Schwenkbewegungen des Antriebsmotors und damit des Abdeckbleches 4 immer mindestens einer der beiden Schenkel 15, 16 gegen den Rand 17 zur Anlage kommt und damit die einwandfreie Abdichtung des Motorraumes nach außen bewirkt.

Die Erfindungsmaßnahmen sind von der besonderen Art der Lagerung des Antriebsmotors unabhängig, können also auch dann angewendet werden, wenn dieser im Fahrgestell nur an zwei oder an vier Stellen gelagert ist.

Patentansprüche:

1. Kraftfahrzeug mit einem im Heck angeordneten, gegenüber dem Fahrgestell nachgiebig gelagerten Antriebsmotor, dessen im Motorraum liegender oberer Teil mit Hilfe einer Abdeckung gegen außen abgeschlossen ist, die in der Richtung vom Antriebsmotor gegen das Fahrgestell unterteilt ist und aus einem starren und einem nachgiebigen Teil besteht, dadurch gekennzeichnet, daß der nachgiebige Teil (11, 12) der Abdeckung die Form einer

im wesentlichen U-förmigen Leiste besitzt, zwischen deren Schenkeln (15, 16) der Rand (17) des starren Teiles (4, 5) der Abdeckung zur Anlage kommt.

2. Kraftfahrzeug nach Anspruch 1 mit im Fahrgestell schwenkbar gelagertem Antriebsmotor, dadurch gekennzeichnet, daß der nachgiebige Teil (11, 12) der Abdeckung im wesentlichen in einer durch die Schwenkachse (S) des Antriebsmotors (1) gehenden Ebene (E-E) angeordnet ist.

3. Kraftfahrzeug nach Anspruch 1, bei dem das Auspuffrohr (6, 7) des Antriebsmotors unterhalb des starren Teiles (4, 5) der Abdeckung und damit außerhalb des Motorraumes (8) liegt.

Hierzu 1 Blatt Zeichnungen

BERLIN. GEDRUCKT IN DER REICHSDRUCKEREI

Zu der Patentschrift 745 930
Kl. 63c Gr. 35

Zu der Patentschrift 745 930
Kl. 63c Gr. 35

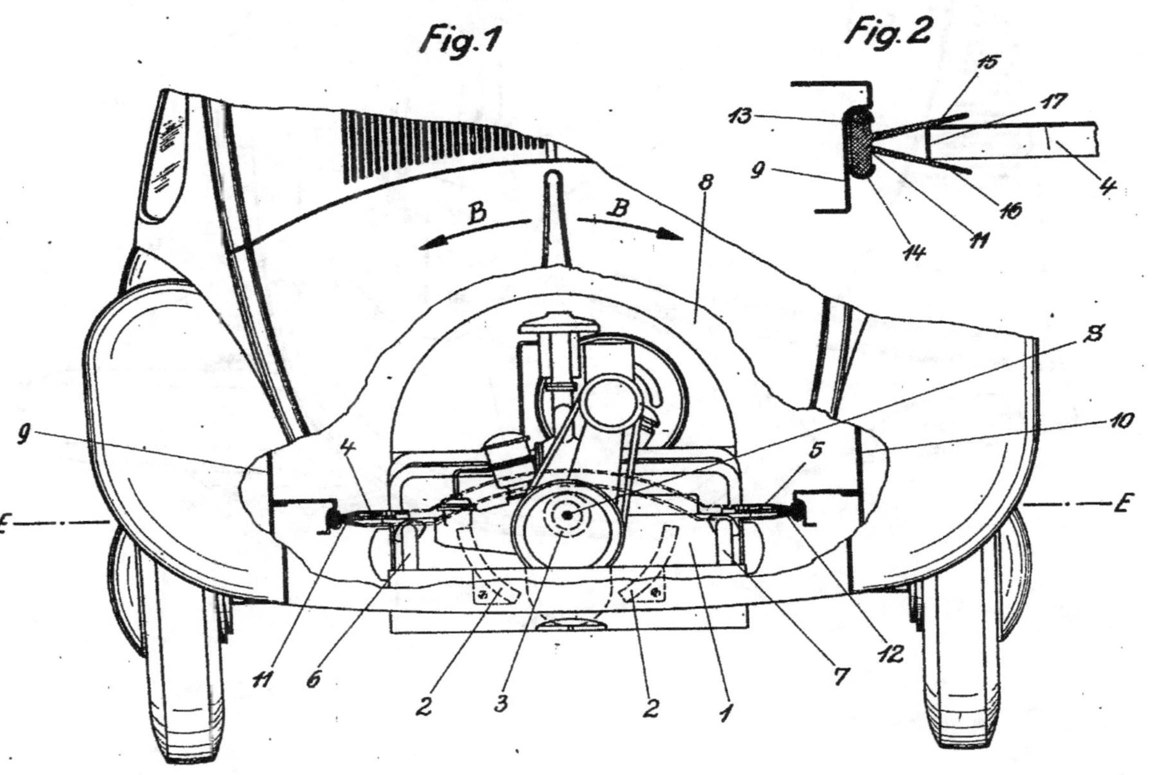

Zu der Patentschrift 745 930
Kl. 63 c Gr. 35

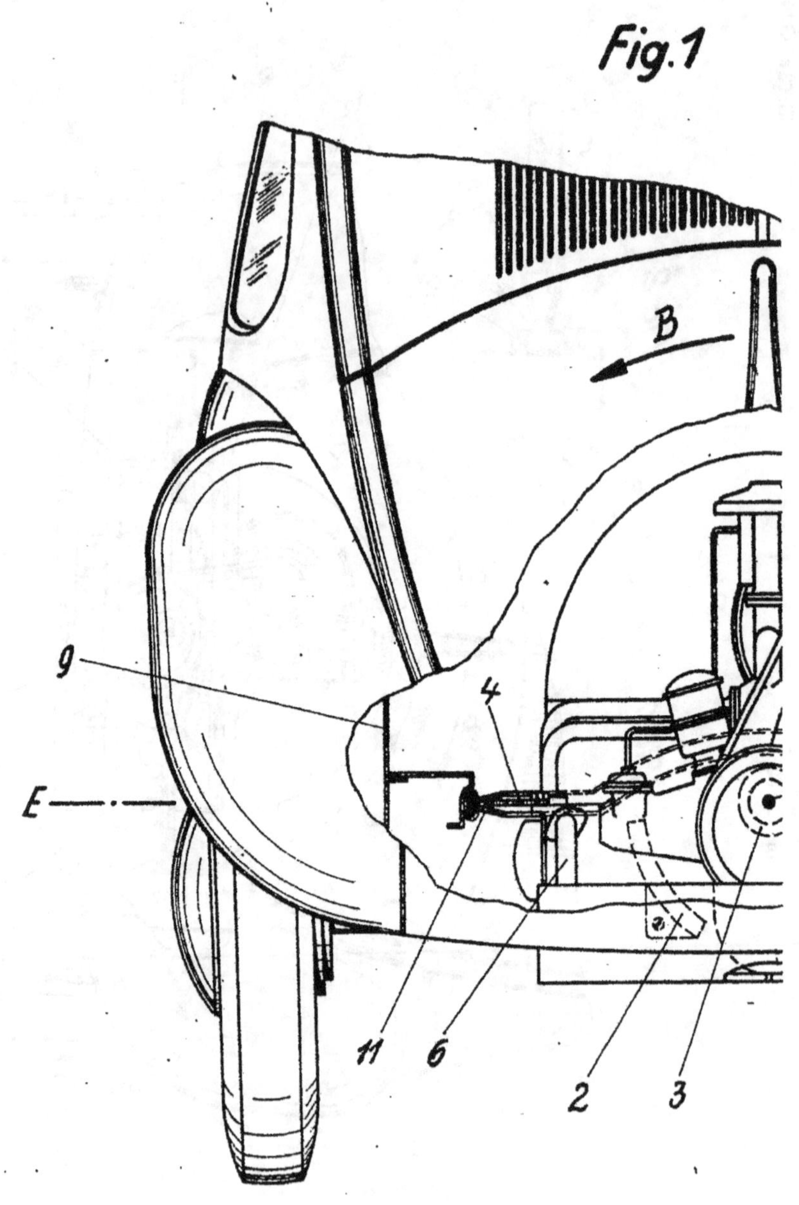

Zu der Patentschrift 745 930
Kl. 63 c Gr. 35

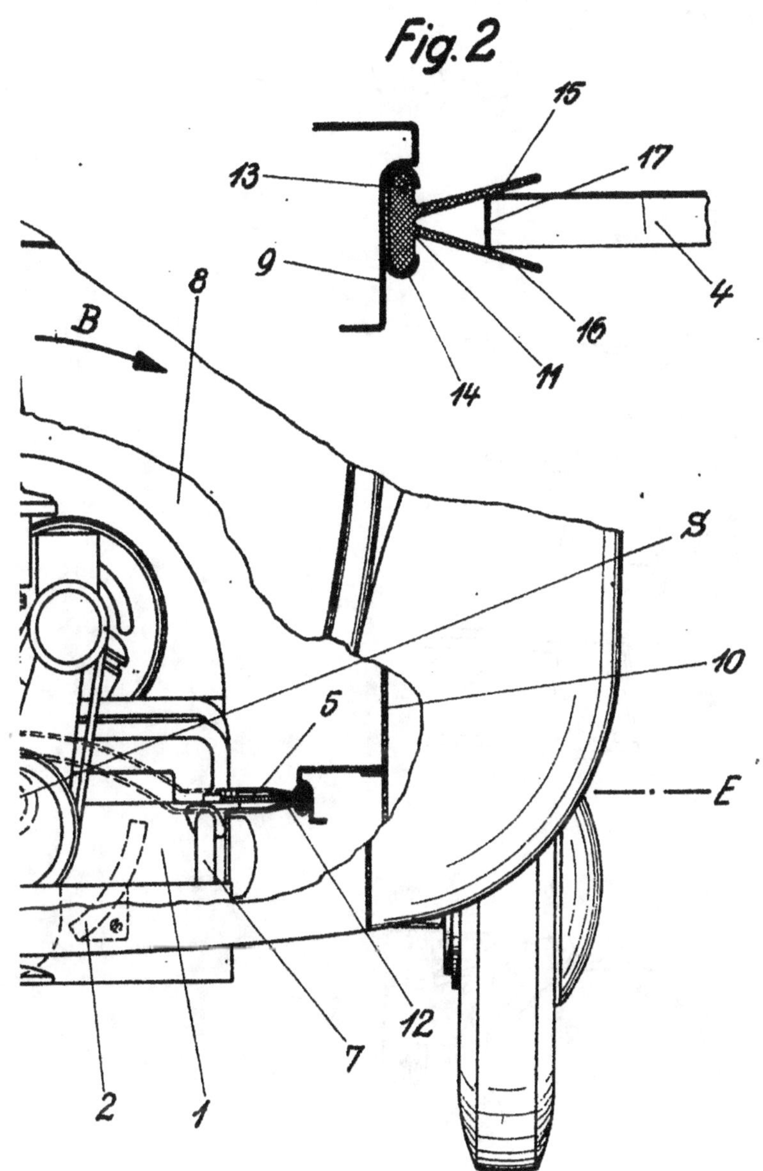

W. Hovey,

Washing Machine,

Patented Feb. 4, 1837.

No. 117,

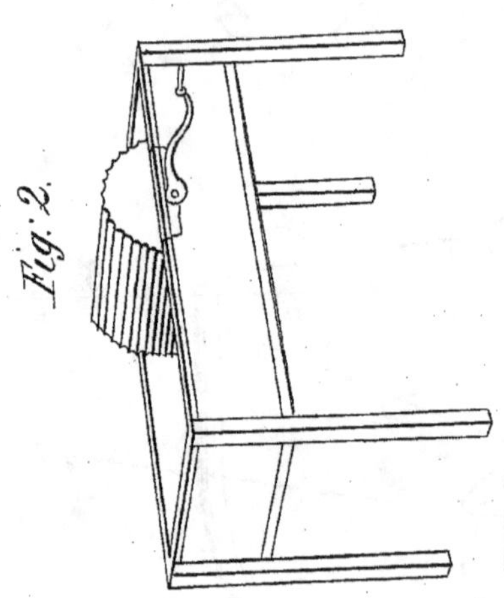

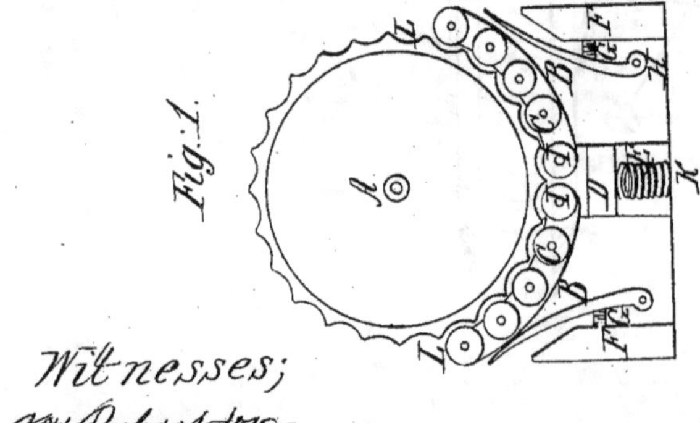

Witnesses;

Inventor;

N. PETERS, PHOTO-LITHOGRAPHER, WASHINGTON, D. C.

UNITED STATES PATENT OFFICE.

WILLIAM HOVEY, OF WORCESTER, MASSACHUSETTS.

CYLINDER WASHING-MACHINE.

Specification of Letters Patent No. 117, dated February 4, 1837.

To all whom it may concern:

Be it known that I, WILLIAM HOVEY, of Worcester, in the county of Worcester, in the State of Massachusetts, have invented a new and useful Improvement in the Construction of Machines for Washing Clothes, called the "cylinder washing-machine," of which the following is a full and exact description.

This machine consists of a box or sink about three feet long and from ten to twelve inches deep and from twelve to eighteen inches wide. In this box or sink, near the middle from each end, I cut grooves in the side pieces of the box running up and down in a perpendicular manner about three-eighths of an inch deep, more or less, and two or three inches wide, as the builder may choose. This box or sink is then ready to receive the application of the washing apparatus, which is constructed as follows, viz: Fitted into the groves cut in the sides of the box or sink are two brass stands or slides, one on each side of the box opposite each other, calculated and fitted to slide up and down in the grooves cut in said box; they are let in flush and even with the inside of the box and are connected together near the bottom by a piece of wood the same width as the slides and about one inch thick running transversely across the box. This connects the two slides in such a manner that they will both rise or fall in the grooves as this bottom piece that connects them is raised or lowered. This horizontal piece of wood that connects the upright slides of brass or other metal not subject to corrode stands on two or more spiral or other springs which are supported by the bottom of the box or sink in such a manner that this frame, which is composed of one bottom piece of wood and the two metallic slides as above described, will sink down by pressure and rise as it is relieved. This frame I shall call the spring frame, which supports the first division of the washing apparatus, viz, the small rollers, which are fitted to come in contact with one large one and in conjunction with it constitute the whole washing apparatus or principle by which the washing is done; all other parts I consider merely as necessary appendages. This first division I describe as follows: I have as many small rollers about one inch in diameter, more or less, as will when laid close and parallel to each other make about five-twelfths of a circle of from eight to twelve inches diameter. These rollers are placed horizontally and parallel to each other with gudgeons of metal in each end, which run in circular pieces of brass or other metal not subject to corrode. These circular pieces of metal are so fitted that the small rollers, when fitted into them, form horizontally a surface conforming to the segment of a circle equal to that which I intend the outer surface or circle of the large cylinder shall be; which cylinder constitutes the second division of the washing apparatus. These circular pieces of metal in which the small rollers revolve are four in number, making two independent segments of the small rollers when fitted in their proper place. These segments of small rollers are hung in the spring frame on metallic pins projecting from near the bottom of the metallic slides or upright part of the spring frame in such a manner that the two segments of small rollers when hung in said spring frame will form a segment of a circle about five-twelfths of its circumference, more or less, with the concave side on the upper surface and running across the box. These segments of small rollers are hung on the pins in the upright slides in the spring frame by inserting the pins on which they hang into female centers made directly in a line with the centers of the two bottom rollers where the two segments meet and form the center hinge of the whole segment when they are united in the spring frame in such a manner that the upper edge of each of these wings or segments of small rollers can be moved out or in from the true circle of the whole segment without altering the lower rollers as they are hung in the spring frame. I then attach springs at each end of the upper edge of these segments near the top rollers, which springs are connected with the box or sink and serve to press the segments or rollers toward the center of the large cylinder which is fitted into the box so as to revolve in these segments of small rollers, its convex surface corresponding with the concave surface of the segment of small rollers. This roller or cylinder is fluted, so as to correspond with the small rollers in such a manner that when the springs have pressed them up they will all lie in the flutes of the large cylinder, and as this large cylinder is turned on its axis the springs will yield and let them out of those flutes, and as they are pressed in and out by the springs and the fillets between the flutes alternately the washing is produced as the clothes are passed through between the surface of the large and small rollers. This large roller or cylinder runs by means of a crank and is turned around or backward and forward at will; it runs on a shaft or axis bearing on each side of the box; the best method of operating it is to turn it by hand backward and forward so as not to let the clothes run entirely out of the first division or small rollers nor should they wind around the large roller.

I disclaim all right to the principle or method of simply applying a fluted roller to a single segment of small rollers corresponding except such right as may be held in common with the public.

What I claim as my improvement consists in—

The method of dividing and applying the segment of small rollers as above described so as more fully to equalize the pressure between the surface of the small rollers and the fluted cylinder between which the washing is effected. I do not confine my claim to a single division of the small rollers any farther than simplicity and convenience may require to produce the desired effect. A division of each roller with springs pressing them toward the center of the fluted cylinder will more fully equalize the pressure between the small rollers and the fluted cylinder as they recede from or approach each other and will answer nearly as good a purpose as the one above described. I therefore claim the method of dividing and applying these segments of small rollers as above described or any other division that will serve to equalize the pressure between the surface of the small rollers and the fluted cylinder while in the operation of washing clothes of various thicknesses.

More fully to illustrate the construction of this machine reference is hereby made to the accompanying drawings, in which—

Figure 1 is a sectional view of the apparatus. Fig. 2 is a perspective view of the same.

In Fig. 1 A is an end view of the fluted cylinder. B B are the circular pieces of brass in which the small rollers run. I I are the center hinges or pins on which the brass pieces are hung to the spring frame. C C are the segments of small rollers divided between I I. D is a bottom section of the upright slides and end of the cross piece that connects them which constitute the spring frame. E is the spiral spring which supports the spring frame. F F are two cross partitions in the sink. G G are spiral springs which are let into the cross partitions F F and press against the dogs H H which in connection serve to press the upper part of the segments of small rollers C C toward the center of the fluted cylinder A as they hang on the hinges or pins I I. K is the bottom of the sink. L L are the places where the clothes are entered for washing between the fluted cylinder and the segment of small rollers.

In testimony whereof I, the said William Hovey, hereto subscribe my name in the presence of the witnesses whose names are hereto subscribed, on the tenth day of January, A. D. 1837.

WILLIAM HOVEY.

Signed in presence of—
J. H. Richardson,
Geo. W. Richardson.

Sheet 1-3 Sheets.

E. B. Bigelow.
Loom.

No. 169. Patented Apr. 20, 1837.

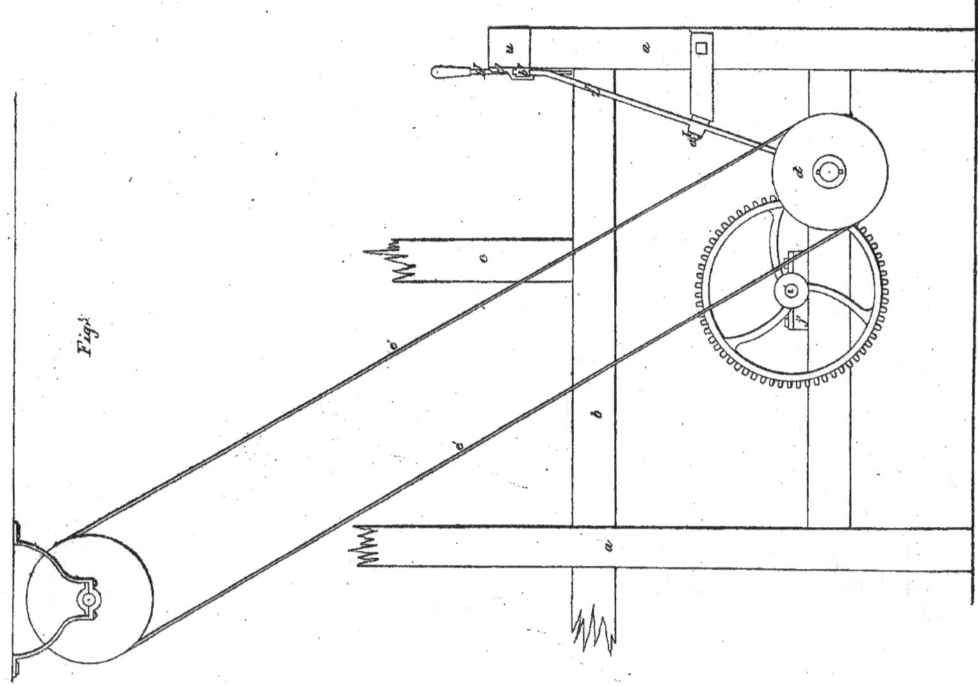

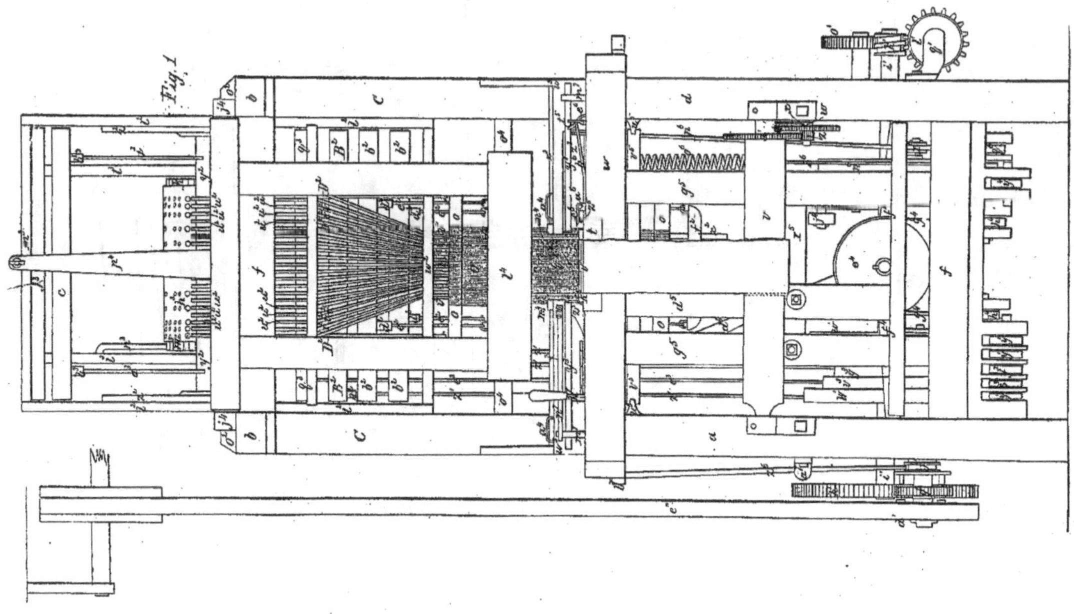

N. PETERS, PHOTO-LITHOGRAPHER, WASHINGTON, D. C.

E. B. Bigelow.
Loom.

Sheet 2-3 Sheets.

No. 169.

Patented Apr. 20, 1837.

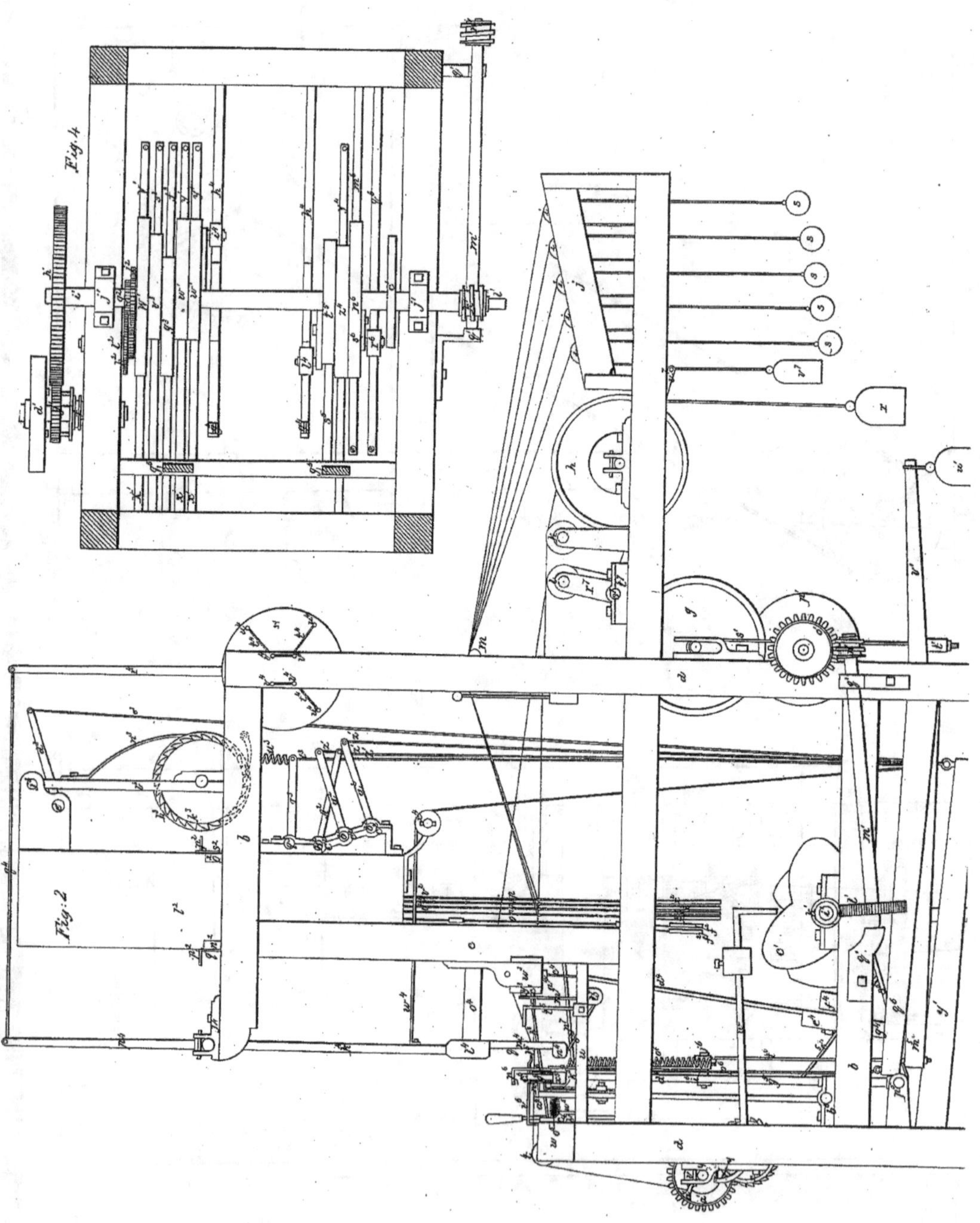

N. PETERS, PHOTO-LITHOGRAPHER, WASHINGTON, D. C.

Sheet 3-3 Sheets.

E. B. Bigelow.
Loom.

No 169. Patented Apr. 20, 1837.

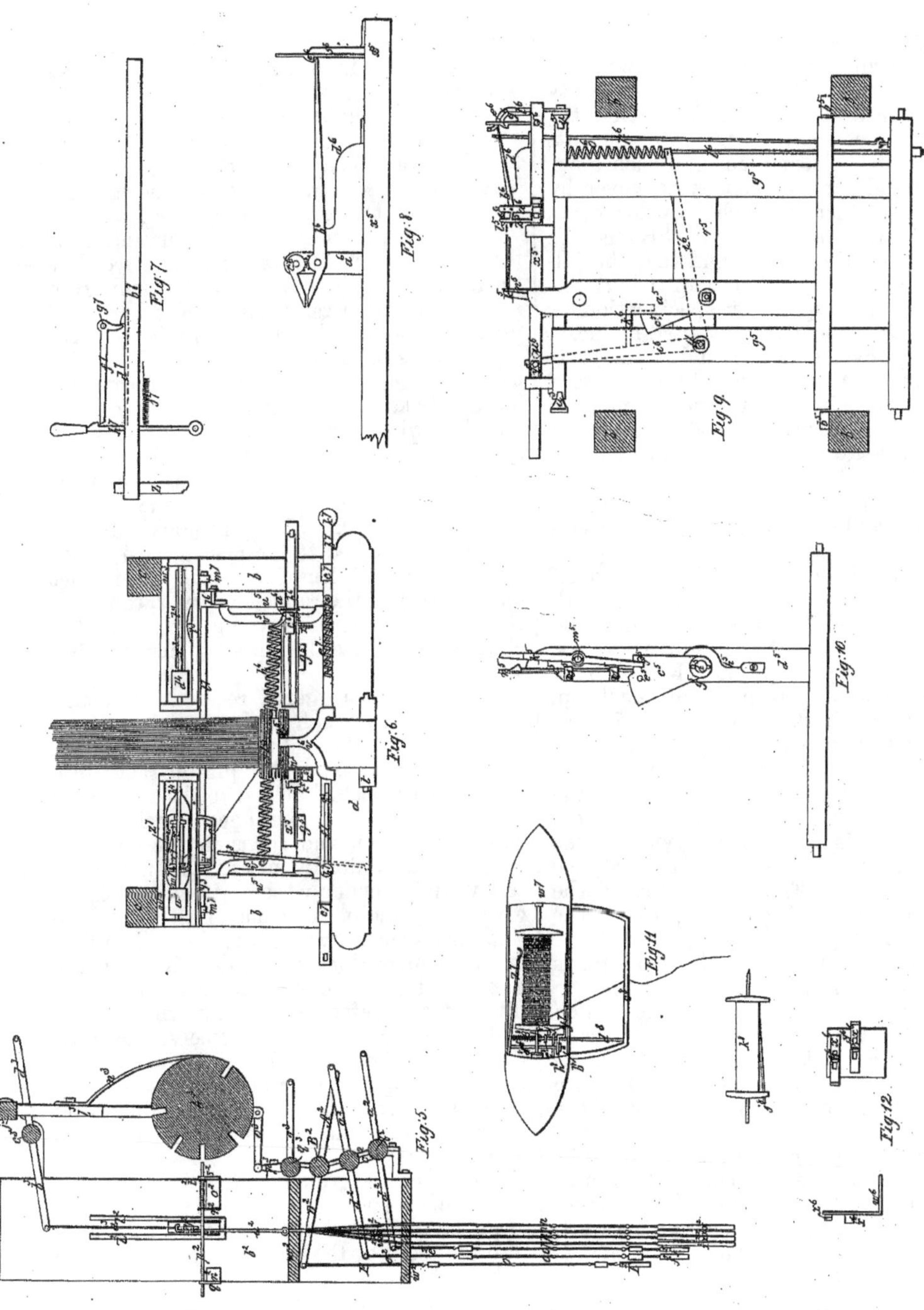

N. PETERS, PHOTO-LITHOGRAPHER, WASHINGTON, D. C.

UNITED STATES PATENT OFFICE.

ERASTUS B. BIGELOW, OF WEST BOYLSTON, MASSACHUSETTS.

POWER-LOOM FOR WEAVING COACH-LACE AND OTHER SIMILAR FABRICS.

Specification forming part of Letters Patent No. 169, dated April 20, 1837; Reissued September 26, 1846, No. 87.

To all whom it may concern:

Be it known that I, ERASTUS BRIGHAM BIGELOW, of West Boylston, in the county of Worcester and State of Massachusetts, have invented new and useful improvements in looms to weave coach-lace and such other similar wrought fabrics as may be woven by the said machinery, which improvements combined are denominated "Boylston's Power Coach-Lace Loom," and that the following description, with the drawings annexed thereto, compose my specification of the said improvements, as invented by me.

These improvements consist in constructing, combining, and applying to use, certain mechanical contrivances in such a manner as to perform the entire operation of weaving coach-lace, &c., by water, steam, or other rotary power.

Figure 1, is a front elevation of the loom. Fig. 2, is a profile elevation of the right end of the loom. Fig. 3, shows the driving parts of the loom on the left end. Fig. 4, is a horizontal view of the cams and treadles, with the upper part of the loom removed. Fig. 5, is a sectional view of the figuring works, detached from the machine. Fig. 6, is a horizontal view, of the parts of the loom, forward of the shuttle boxes. Fig. 7, represents the stop motion. Fig. 8, is a side view of the pliers or forceps, with one standard removed. Fig. 9, is a front view of the apparatus to shift the wires, detached from the machine. Fig. 10, is a back view of the apparatus to push the wires toward the pliers or forceps. Fig. 11, represents the shuttle and bobbin. Fig. 12, represents the temples or cloth guides.

The same letters refer to the same parts when they occur in any of the figures.

a, *a*, *a*, *a*, are four upright posts, which are connected together, by horizontal rails *b*, *b*, *b*, *b*. *c*, *c*, are two other upright posts resting on the cross rails *b*, *b*. *f*, *f*, *f*, *f*, are cross horizontal rails for connecting the ends of the framework.

The warp beam *g*, containing the linen or ground warp, and the warp beam *h* containing the worsted or that warp which is raised above the ground warp to conceal the weft, are mounted on axles turning in bearings attached to the framework.

The bobbins *i i i* contain the figuring warp and are supported by the creel *j*. From these beams and bobbins, the warp threads pass over their respective whip-rollers *k*, *l*, *m*, thence through the raddle *n*, headles *o*, *o*, and *p*, *p*, and reed *q*, in the ordinary way. *r* is a weight suspended by a friction cord passing around the warp-beam *h*, to keep the warp at a proper degree of tension. The weights *s*, *s*, *s*, *s*, in the same manner preserve the tension of the figuring warp.

The cloth produced by the intervention of the warp and weft threads in front of the reed, passes over the roller *t*, turning in bearings in the breast beam *u*, and is wound upon the cloth roller *v*. *w*, is a lever turning on the stud of the ratchet wheel, *x*, with the clicks *y*, *y*, jointed to its short arm. These clicks act on, and turn the ratchet wheel *x*, when the lever *w*, is moved by the cam c'. The pinion gear *z* on the side of the ratchet wheel *x* takes into the teeth of the cogged wheel a', affixed to the axle of the cloth roller *v*. b', b', are clicks playing on a stud attached to the post *a*, (being hid in the drawing behind the post *a*) which act on the teeth of the ratchet wheel *x*, to secure the lace that has been wound up, when the clicks *y*, *y*, release their hold to act on other teeth.

The pulley or rigger d' receives its motion from the mill work, by the belt e' and moves the loom when the clutch f' connects it with the pinion gear g'. This pinion takes into the cogged wheel h' affixed to the main axle i'. The axle i' is arranged horizontally across the loom, and turns in the bearings j', j'. This axle, by means of an endless screw, and cams or wipers attached thereto, gives the principal motions of all the operations performed by the machine. The first of these operations to be described, is the giving out of the linen warp as fast as it is filled by the weft; second, that which raises and depresses the warp, to receive the shuttle; third, regulating the variations in the pattern or figure; fourth, driving the shuttle to and fro; fifth, beating up the weft; sixth, shifting the wires over which the figure is wrought.

The linen or ground warp, is given out as follows: The worm or endless screw k', is affixed to the end of the shaft i', and takes into the teeth of the cogged wheel l', on the end of the axle m'. To the reverse end of the axle m', another worm or endless screw n', is affixed, which takes into the teeth of the cogged wheel o', and revolves the cylinder p', on its axle in the bearings r', r'. The axle m', turns in bearings q', q', attached to the frame-work.

The diameter of the cylinder p', and the number of teeth of the cogged wheels l', o', are so proportioned, as to move the disk or circumference of the cylinder p', at each throw of the shuttle, through a space equal to the length of the linen warp filled by the thread of the weft, thus introduced.

The upper extremities of the wires s', s', hook on to the ends of the axle of the warp-beam g, and are connected at their lower-extremities by the cross-bar t'. The weight w', is suspended from the lever v', resting on the cross-bar t', and presses the warp on the warp beam g, against the disk of the cylinder p', thus causing the beam g to turn with the cylinder p', and deliver a certain length of warp at each revolution of the cylinder, however much the diameter of the warp around the beam may vary.

The whip-roller l over which the linen warp passes, turns in bearings in the arms r^7, r^7, extending from the axle s^7 which turns in the bearings t^7, t^7. u^7 is an arm extending from the axle s^7, from which the weight v^7, is suspended; which weight serves to tighten the warp, and at the same time allow the whip-roller l, to move forward and prevent too great tension of the warp when the heddles are raised to form the sheds.

The raising and depressing of the warp to receive the shuttle are effected as follows: The cams or wipers w', w', W', are affixed to the axle i', and act on the levers or treadles y', y', Y', which play on the fulcra x', x', X'. The straps or cords z', z', Z', connect the treadles y, y', Y', to the arms a^2, a^2, A^2, Fig. 5, extending from the axles b^2, b^2, B^2, which turn in the bearings c^2, c^2, C^2. d^2, d^2, D^2, are arms extending from the reverse sides of the axles b^2, b^2, B^2, from which the heddles o, o, O are suspended by the cords e^2, e^2, E^2. f^2, f^2, F^2, are weights connected to the bottom of heddles o, o, O.

As the shaft i, revolves the eccentric parts of the cams come around at proper intervals, and force down the treadles y', y', Y', which by means of the cords z', z', Z', turns the axles b^2, b^2, B^2, raises the heddles and forms the sheds of the warp. When the cams relieve their action on their respective treadles, the headles are again depressed by the weights f^2, f^2, F^2.

The two leaves of heddles o, o, are sufficient to weave laces with plain grounds, or fabrics, but in weaving those laces in which certain of the warp threads are overlaid, the third leaf O is required, which leaf being raised relatively at such distant intervals of time, require a graduated motion of the cam which moves it. This is effected as follows;—the cogged wheel g^2, Fig. 4, affixed to the axle i', takes into the teeth of the largest part of the pinion wheel, h^2, which turns on the stud i^2 attached to the frame-work. The teeth of the small part of the cogged pinion h^2, takes into the teeth of the cogged wheel j^2, affixed to the cam W' which revolves loose on the shaft i'. The number of teeth of these cogged wheels may be so calculated as to give different degrees of relative motion to the cam W', according to the number of threads of the weft, overlaid by the warp;—thus when the warp overlays three threads of the weft, two revolutions of the axle i' will be required to one of the cam.

The third operation or that which regulates the variations in the pattern or figure, is next described. l^2, l^2, are upright post or side-pieces (see Fig. 5), which are connected together by the cross bars, or rails m^2, n^2, o^2. When this part is connected with the main frame-work, the ends of the cross-bars n^2, o^2, rest on the horizontal rails b, b, and are bolted or otherwise affixed thereto. p^2, p^2, p^2, are horizontal wires, sliding with a reciprocatory motion in holes through the plates of iron q^2, r^2, s^2, which are screwed to the cross rails n^2, o^2. Between the plates of iron r^2, s^2, a spiral spring is encircled around each horizontal wire p^2, p^2, p^2, one end of which is attached thereto by the pin t^2, inserted in the wire, the other abuts against the plate of iron r^2. This spring yields to any gentle pressure, made on that extremity projecting beyond the plate of iron s^2, and returns again when that pressure is removed till the pin t^2 strikes against the plate s^2. Eyes are formed in the central parts of these horizontal wires, through which the vertical wires u^2, u^2, pass. The wires u^2, u^2, are hooked at the upper extremity and arranged vertically over the lifting bar. In the lower extremities of the vertical wires, eyes are formed, to which the cords v^2 v^2 are attached. From these eyes the cords v^2, v^2, pass through the guide holes in the cross-pieces w^2, w^2, to the heddles p, p, suspended therefrom.

x^2, x^2, are weights attached to the heddles p, p, to depress them after they have been raised by the lifting-bar, and preserve their tension during the operation of the other parts of the loom. The lifting bar y^2 has a vertical reciprocating motion in the guides r^2, r^2, which are attached to the side-pieces of the frame t^2, t^2. The wires a^3, a^3, connect the lifting bar y^2, to the arms b^3, b^3, extending from the axle c^3.

d^3, is the reverse arm of the axle c^3, to which is attached the upper extremity of the cord e^3, Fig. 2 which connects it with the treadle f^3, Fig. 4. The lifting-bar y^2, Fig. 5 is raised by the cam g^3 acting on the treadle f^3, Fig. 4 and when the cam relieves its action thereon, it falls again by its own gravity.

It is evident from what has been described that when the lifting bar y^2 is raised it will carry up with it all the vertical wires u^2, u^2, and heddles connected with them, and no variation will be produced in the pattern or figure; but, if before we raise the lifting bar, we push back certain of the horizontal wires p^2, p^2, p^2, and thus withdraw the vertical wires connected with them, from its action, it will carry up with it, those vertical wires and heddles only, the horizontal wires of which have not been pushed back, and a corresponding variation will be produced in the pattern or figure.

The apparatus to push back certain of the horizontal wires is thus described. The cylinder h^3, is mounted on an axle turning in bearings in the frame l^3, which oscillates on the axis j^3. In the circumference of this cylinder as many longitudinal rows of holes are formed, directly opposite to the ends of the horizontal wires p^2, p^2, p^2, as there are variations to be made in the pattern or figure.

The number and position of the holes in each row, are varied according to the variation to be made in the figure; that is, holes are bored in any one of the rows, opposite to the ends of those horizontal wires only, which connect with the heddles required to be raised when the said row swings against the horizontal wires.

The ratchet wheel k^3 Fig. 2 having as many teeth as there are longitudinal rows of holes in the cylinder h^3, is attached to the axel of the said cylinder, and is acted on by the click l^3, which (being concealed in the drawing behind the horizontal rail b) is attached to the framework. m^3 Fig. 1, is another ratchet wheel, which has as many teeth and is affixed to the axle like the former, and is acted on by the spring n^3, attached to the frame i^3 Fig. 5.

The bars o^3, o^3, connect the frame t^3, with the arms p^3, p^3, extending from the axle q. From the reverse side of this axle another arm r^3, extends which is connected with the treadle s^3, Fig. 4 by the cord t^3, Fig. 2. u^3, is a spring one extremity of which is attached to the arm r^3, the other to the framework. v^3 is a cam or wiper attached to the axle i^1. This apparatus operates as follows:

The cam v^3 forces down the treadle s^3, and turns the axle q^3, which being connected with the frame i^3, carries the cylinder away from the horizontal wires.

As the cylinder is thus moved back, one tooth of the ratchet wheel k^3, strikes against the click l^3, which overcomes the elastic force of the spring n^3, and turns the cylinder on its axis the distance of one tooth of the ratchet wheel, or in other words, the distance between the centers of any two rows of holes in its circumference; at this instant the spring n^3, takes into a new space of the ratchet wheel m^3, and secures the cylinder in its proper position.

When the cam v^3, relieves its action on the treadle s^3, the spring u^3, raises the arm r^3, and forces the cylinder against the wires, and pushes back those which have no holes opposite to them in the cylinder, while, those wires which have corresponding holes in the cylinder, enter therein, and remain at rest, with the hooked wires connected with them over the lifting-bar.

The cylinder thus presenting a new row of holes, and swinging against the ends of the horizontal wires, and pushing certain of the hooked wires, from the action of the lifting-bar every time it is raised, produces variations in the pattern, or figure corresponding with the variations of the holes in its circumference.

The fourth operation is that which drives the shuttle to and fro, and may be understood as follows. The shuttle boxes w^3 w^3 Fig. 6 are affixed to the posts c, c, at a suitable distance to allow the rod to pass between them; x^3, x^3, are picker-rods; y^3, y^3, represent the shuttle binders; z^3, z^3, Fig. 1 are openings or mortices in the front side of the shuttle boxes, to receive the guide wire of the shuttle.

The pickers a^4, a^4, Fig. 6 slide on the rods x^3, x^3, with their lower extremities playing in mortices b^4, b^4, in the bottom of the boxes; c^4, c^4, Fig. 2 are picker-strings connecting the pickers a^4, a^4, to the picker-staff d^4, which is affixed to the pulley e^4, Fig. 1. This pulley turns on a stud attached to the cross-bar f^4. The strap g^4 passing over the pulley connects it to the treadle h^4, h^4. i^4, i^4, are cam-bolts attached to the cams on the main axle i^1, and as they revolve with the said axle force down the treadle h^4, h^4, alternately, and by means of the strap g^4, vibrates the picker-staff and throws the shuttle to and fro.

The weft is beat up as follows:—the lay is mounted on an axle turning in bearings j^4, j^4 Fig. 2, attached to the framework; k^4, k^4, are the swords of the lag; l^4 is the top shell, which receives and supports the upper edge of the rod.

The bars of iron m^4, m^4 are affixed to the top shell and extend downward to the lower shell of the lay n^4, and supports the lower edge of the rod. o^4 o^4 are straps with one extremity of each attached to the top shell of the lag t^4, the other end of each to the framework. The upright arm p^4, extending from the axle of the lag, is connected by the cord q^4, to the lever r^4, which is attached to the pulley s^4. The pulley s^4, is suspended between the posts a, a, by the cord t^4, passing through the holes u^4, u^4, in the pulley, and the holes v^4, v^4, in the posts a, a. The pulley is turned to twist the cord, which offers a degree of resistance in proportion to the tension of the twist, and serves to bring forward the lag to beat up the weft. The cord w^4, attached to the lag, passes from them over the pulley y^4, to the treadle y^4. The cam z^4 on the main axle i^1, forces down the treadle y^4, which by means of the cord w^4, draws back the lag, and increases the tension of the cord t^4, which cord, as the cam relieves its action on the treadle, throws the lag forward and beats up the weft. The degree of motion given to the lag by the cord t^4 is determined by the straps o^4, o^4.

We now come to the sixth operation, or that which shifts the wires over which the figure is wrought.

a^5, is a frame similar in form to an inverted T (see Fig. 10.) which is mounted on an axis turning in the bearings b^5, b^5, Fig. 2. c^5, Fig. 10 is a cam turning on the bearings b^5, b^5, Fig. 2. c^5, Fig. 10 is a cam turning on the stud d^5, affixed to the frame a^5, with the studs e^5, f^5, extending from one of its sides. The stud g^5, extending from the frame a^5 determines the quantity of motion of the cam in the direction toward it.

The spring h^5 attached to the frame a^5 always tends to move the cam toward the stud g^5. The bar i^5 is made flat and pointed at the upper extremity and has a vertical reciprocating motion in the guide j^5 j^5. The lever k^5 has a groove at its upper extremity in that edge which presents toward the wires l^5, l^5, l^5, and turns on the fulcrum m^5 attached to the frame a^5. n^5 represents the guide iron attached to the upper extremity of the frame a^5. o^5, Fig. 2 represents a spring inserted between the frame a^5 and the breast beam u. This spring presses the guide iron n^5 against the last of the series of wires l^5, l^5, l^5, Fig. 9, or that one which is to be acted on and thus keeps the frame a^5 in the same position relative to the said wire, that is the one to be acted on, although the position of the succeeding wires themselves as they successively approach this situation (being successively drawn out and placed under a new portion of the figuring warp) may at different times vary.

A frame formed similar to a lay turns on an axis in the bearings p^5, p^5, Fig. 2, affixed to the posts a, a—g^5 g^5 (see Fig. 9.) are the swords the tops of which rest against the frame v^5. x^5 is the cross rail to connect the swords together. An arm s^5, Fig. 4, is attached to this frame on which the cam t^5 acts to move it. Fig. 5, u^5, u^5, represents v^5 atached to the frame-work of the machine, on which the frame v^5, slides with a horizontal reciprocating motion. w^5 is a spiral spring attached at one extremity to the frame v^5 at the other to the breast beam u.

The bar x^5, Fig. 9, slides in the standard on the frame v^5 with a transverse reciprocating motion. y^5 represents a stop attached to the bar x^5, which determines the quantity of approach of the pliers toward the wires l^5, l^5, l^5. z^5 represents a piece of iron called the evener bolted to the bar x^5. To the bar x^5 the geared pliers are affixed as represented in Fig. 8. a^6, a^6 are standards screwed or otherwise affixed to the bar x^5 (one of which is removed in this figure) which support the axis of the blades of the pliers b^6 c^6.

Teeth are formed on these blades at the point of contact at their centers of motion which take into each other similar to the action of two cogged wheels, so that raising and depressing the long part of the blade b^6 opens and closes the pliers. d^6 represents a spring, which is attached to the bar x^5 and tends to raise the long part of the blade b^6 and open the pliers. The latch e^6 vibrates on the stud f^6 attached to the bar x^5 and is acted on by the spring g^6. The spiral spring h^6, Fig. 6, is affixed at one extremity to the bar x^5 at the other to the frame v^5. i^6, Fig. 9, represents an elbow playing on the stud j^6 with its upright arm resting against a stud u^6 extending from the bar x^5. The hook k^6 attached to the upright part of the elbow, acts on the stud e^5 and moves the cam c^5.

The wire l^6, connects the horizontal part of the elbow with the treadle m^6 the projection of which is acted on by the cam n^6, Fig. 4, affixed to the axle i^1. The spring o^6, Fig. 9, being attached at one extremity to the horizontal part of the elbow i^6, at the other to the frame v^5, serves to counteract the weights of the treadle m^6.

The latching wire p^6, Fig. 1, is hooked at the upper extremity and connected at its lower extremity to the treadle q^6, Fig. 4, which is acted on by the cam ball r^6 playing on the stud s^6, attached to the cam u^6. The standard t^6, Fig. 9, attached to the v, u^5 serves to release the latch e^6 from the blade of the pliers b^6.

Considering the position which the pliers or pincers assume after having deposited a wire under the figuring warp and returned again to a line with the wire next to be taken by them and parallel with the breast-beam, as the point of commencement, we shall describe the movements of the machinery specified under this operation.

The main axle i^1 turning, the notch of the cam n^6 comes around and suffers the spring o^6 to raise the treadle m^6, and horizontal part of the elbow i^6, which motion of the elbow carries its upright part away from the stud u^6 and allows the spring h^6, Fig. 6, to move the pliers toward the wires l^5 l^5 until the stop y^5 meets the standard of the frame v^5 and prevents its farther approach. As the upright part of the elbow is thus moved back the hook k^6, Fig. 9, acts on the stud e^5 and moves the cam c^5 toward the spring h^5, Fig.

10. This movement of the cam raises the bar i^5 and forces the point of it between the last of the series of wires l^5 l^5 or that one against which the guide iron n^5 rests and the one next in order from it, and separates one from the other and thus prevents more than one being acted upon at the same time.

When the bar i^5 has arrived at its greatest elevation the stud f^5 strikes against the lower arm of the lever k^5 and forces the grooved side of the reverse arm against the end of the wire l^5 operated as above described and forces it toward the pliers or forceps into the position seen in Fig. 9; at this instant the cam ball r^6, Fig. 4, acts on the treadle g^6 and depresses the latching wire p^6 which forces down the longer part of the blade b^6 closes the jaws of the pliers or forceps and pinches the wire tight between them.

As the longer part of the blade b^6 is thus depressed the shoulder of the latch c^6 locks onto its extremity and secures the grasp of the pliers on the wire after the action of the cam ball r^6 is released from the treadle m^6 depresses the horizontal part of the elbow i^6 and slides the bar x^5 in the standard of the frame v^5, which movement of the bar draws out the wire grasped by the pliers, from under the figure wrought over it and moves them into the position seen in Fig. 6. This motion of the elbow i^6 relieves the action of the hook k^6 from the stud e^5, Fig. 10, and allows the spring h^5 to move the cam c^5 back against the stud g^5, which movement of the said cam suffers the bar i^5 to fall and causes the stud f^5 to move the lever k^5 into the positions in which they are respectively seen in Figs. 6 and 10 and in which they are prepared to act on the succeeding wires. At the proper interval the cam t^5, Fig. 4, acts on the arms s^5 and slides the frame v^5, Fig. 6 on the v^5 u^5 u^5, and carries the pliers connected with the frame v^5 back toward the suttle boxes to the proper position to place the wire under the figuring warp, at the instant they arrive at their destination another notch in the cam n^6 suffers the spring o^6 to carry the upright part of the elbow i^6 away from the stud u^6 which movement of the said elbow allows the spring h^6, Fig. 6, to draw the pliers up and place the wire they grasp under the figuring warp; just as the pliers complete their motion in this direction the latch c^6 strikes against the standard f^6 which releases it from the blades b^6 and suffers the spring d^6 to open the jaws of the pliers and drop the wire. At this instant the cam n^6, Fig. 4, again acts on the treadle m^6 and carries the pliers back a short distance to prevent their coming in contact with the wires l^5, l^5, Fig. 9, as they return toward the breast bream; when this part of the cam has completed its action, the notch of the cam t^5, Fig. 4 suffers the spring w^5, Fig. 2, to draw the frame v^5, Fig. 9, back to a line with the next wire to be acted on, or in other words the position it assumed when we began to describe the operation. When the wires drop from the pliers they are not all in the same position—that is some are placed under the figuring warp farther than others, therefore to even them and prevent any failure in the operation of shifting them, which might occur from their irregularity, the evener z^5, as the pliers approach the warp to deposit the wire, strikes against the wire last deposited and drives it in as far as it is suffered to do, by the motion of the bar x^5; the wire deposited by this operation of the pliers, is driven in, in the same manner as the other—when the pliers come up to place in the next succeeding wire; all the wires being thus driven are left in an uniform position.

I do not deem it necessary to recapitulate the movements of this machine in their order of succession, as this will be apparent to every competent machinist, and the periods of the different parts of the process being nearly the same as in ordinary coach-lace looms. The pliers are armed with a wire while the weft is being inserted between the warp, and are prepared to move back simultaneously with the lay at the proper interval and place it under the figuring warp.

The rest v^6, Fig. 6, is attached to the breast beam and arranged over the lace, to prevent the lace back of the breast-beam from rising up and thus carrying the wires l^5, l^5, away from the pliers when the warp is raised to form the sheds. w^6 represents the standard of the temples (see Fig. 12) which is screwed to the inside of the breast-beam and under the cloth. x^6, x^6, are the guides of the temples which are fastened to the stand w^6 by the screw y^6 y^6. The cloth passes between the guides x^6 x^6 and is thus prevented from yielding when the wires l^5 l^5 are drawn out or pushed in.

The loom is put in motion and thrown out of gear as follows: The lever z^6, Fig. 1, turns on the stud a^7 and connects the clutch f^7 with the shifting bar b^7 which slides in the guides c^7, c^7, Fig. 1. The dotted lines at d^7, Fig. 7, represent a spring embedded in the bar b^7. The spiral spring e^7 is attached at one extremity to the bar b^7 at the other to the breast-beam u. The lever f^7 turns on the stud s^7, Fig. 6, attached to the breast-beam. h^7, Fig. 2, is a latch which vibrates on a stud attached to the post a, and has a handle at its upper extremity which the weaver grasps to stop the loom. The spring j^7, Fig. 7, being attached at one end to the latch h^7 at the other to the breast-beam always tends to bring the latch h^7 toward the lever f^7.

To put the loom in motion the weaver grasps the handle k^7 and moves the bar b^7 toward the right which movement of the bar moves the clutch f' Fig. 4 and connects the coggea wheel g' with the pulley a' at this instant the spring d^7 locks on the depending arm of the lever f^7 and secures the connection.

To stop the loom the weaver releases the clutch h^7 and suffers the spring c^7 Fig. 6 to move the bar b^7 and throw the loom out of gear. In the event of the weft being exhausted, on the bobbin in the shuttle, and also in case the shuttle does not arrive at its destination, it is desirable that the loom should be thrown out of gear, to prevent an imperfect place being made in the lace.—Stopping the loom when the shuttle stops in its passage to and fro is effected as follows;—The protecting rod l^7 Fig. 6 turns in the bearings on the v^5 u^5, u^5, and has the arms m^7 m^7, extending from it, the upper extremities of which rest against the shuttle binders y^3, y^3, Fig. 1.—n^7 represents a spring which acts on the arm m^7 and urges both the arms m^7, m^7, against their respective binders y^3, y^3,—o^7, Fig. 2, is the reverse arm extending from the rod l^7 and supports the bar p^7, which connects with the latch h^7 by means of an elbow and connecting rod which are hid in the drawing under the breast-beam.

When the shuttle enters the box properly it pushes out the binder y^3 and depresses the bar p^7 and causes it to escape the lower shell of the lay as it comes forward to beat up the weft:—but in the event the shuttle does not enter the box the spring n^7 raises the bar p^7 to meet the lower shell of the lay which as it comes forward strikes against the end of it and releases the latch h^7 and suffers the spring e^7 to throw the loom out of gear.

Stopping the loom when the weft is exhausted may be understood as follows; w^7 Fig. 11 represents the shuttle armed with the bobbin x^7 containing the weft.—y^7 is a spring affixed to the front side of the shuttle and is connected to the binder z^7 by the wire a^8 which has a screw with the nut b^8 to adjust the degree of resistance offered to the bobbin by the binder z^7.—c^8 is the guide wire which guides the weft off of the bobbin x^7.—d^8 is the stop wire sliding in holes through the guide wire c^8 and the stand e^8.—f^8 is a spiral spring encircled around the stop wire d^8 one end of which is attached to the wire d^8 by the pin s^8 inserted therein, the other abuts against the shuttle wood.—This spring yields to any gentle pressure on the end of the wire projecting beyond the guide wire c^8 and returns as far as suffered by the pin g^8 when that pressure is removed.—h^8 is a catch, which when the stop wire d^8 is pushed back even with the outside of the guide wire c^8 locks into a notch in the stop wire, and prevents its returning by the action of the spring f^8 until the said catch is again released.

The lever i^8 Fig. 6 turns on a stud attached to the breast-beam u, and extends along the side of the latch h^7 nearly to touch the shuttle guide.—When the stop wire d^8 is in the position seen in Fig. 11, the shuttle will pass to and fro without acting on the lever i^8 and no effect is produced on the stop motion.—But when the filling is nearly exhausted on the bobbin the spring j^8 recedes from the center as seen in the bobbin k^8 and as it comes around releases the catch h^8 and suffers the spring f^8 to throw forward the wire d^8, which as the shuttle enters the left hand shuttle box strikes against the extremity of the lever i^8 releases the latch h^7 and suffers the spring e^7 to throw the loom out of gear.

Contemplated variations in the arrangements of the parts to shift the wires over which the figure is wrought. The lever k^5 Fig. 10, which pushes the wires toward the pliers may be dispensed with, in which event the upright part of the frame a^5 must be removed to the reverse edge of the lace, and the bar i^5 pass between those ends of the wires which the pliers grasp to shift them, and the guide iron n^5 must also rest against the said ends, presented to the pliers or forceps.

The standards which support the reciprocating bar x^5, instead of being stationary with the frame v^5 may be affixed to an axle and vibrate in bearings, attached to the frame v^5.—The frame formed by the said axle and standards should be prevented from falling from the perpendicular toward the shuttle boxes, by a stop attached to the bearings in which the axle turns or to the frame v^5 which supports the said bearings. A spring is affixed to the frame v^5 which acts on the frame formed by the said axle and standards always tends to keep the said frame in a perpendicular position and against the aforesaid stop. A guide iron is affixed to the said frame, which guide iron when the pliers approach the wires to grasp them glide along the side of another guide attached to the frame a^5 and guides the pliers into the right position to take the last of the series of wires without acting on the one next in order to it.

Many parts of machinery have been described above without any intention of claiming them as a new invention, but merely for the purpose of leading to and more readily illustrating the design and operation of my improvements.

Having described my improvements in a loom to weave coach-lace, and shown by the foregoing description accompanying drawings, and model, the best mode of constructing and adapting the same that I am acquainted with; I desire to be understood that I do not intend to confine myself to that particular form, and arrangement and materials of the parts shown in the drawings and model by which I effect my improvements in looms to weave coach-lace; as different forms and arrangements of mechanism may be found capable of effecting the same object; but those which I claim as the peculiar features of my invention are separately and singly as follows;—

1. Dividing or separating the wires over which the figure is wrought, one from the other, by means of a pointed instrument passing between them.

2. Pushing the said figuring wires successively toward the pliers, forceps, or pincers by the means of pressure exerted on the reverse ends from those at which they are grasped by the pliers, forceps or pincers.

3. Withdrawing the said figuring wires from the figure wrought over them, and placing them under a new portion of the figuring warp by means of pliers, forceps or pincers.

4. A guide resting against the last of the series of wires to preserve the machinery which acts thereon in the same position relative to the said last of the series of wires.

5. Moving the machinery employed to shift the said figuring wires by means of eccentric wheels cams or wipers.

6. The stationary shuttle boxes, employed for purpose and in the manner set forth together with such variations of the parts thus claimed as may produce the same effect by means substantially the same.

In testimony of the above I have hereunto set my hand this twenty-seventh day of December, in the year eighteen hundred and thirty-six.

ERASTUS B. BIGELOW. [L. S.]

Witnesses:

David C. Murdock,

Charles W. Hartwell.

[First Printed 1914.]

Strom- und Gasvergleich!

Es ist ganz einfach, einen günstigen Strom- oder Gasanbieter zu finden:

stromgasbilliger.xilando.de

Tinte? Toner? Papier? Büro? Alles zum Tiefpreis:

tintetoner.xilando.de

Klar sehen!

Auf Versicherungsvergleich bestehen!

versicherungsvergleich.xilando.de

www.ingramcontent.com/pod-product-compliance
Lightning Source LLC
Chambersburg PA
CBHW080954170526
45158CB00010B/2804
9781076930903